SEMICONDUCTING COMPOUNDS

Papers presented at the
Schenectady Conference
June 1961

Published as a Supplement to the *Journal of
Applied Physics* in cooperation with the
American Institute of Physics

W. A. BENJAMIN, INC.

New York   1961

SEMICONDUCTING COMPOUNDS:

**Papers presented at the Schenectady
Conference, June 1961**

Manufactured in the United States of America
Library of Congress Catalog Card Number 61–18498

W. A. BENJAMIN, INC.
2465 Broadway, New York 25, New York

JOURNAL OF APPLIED PHYSICS

Supplement to Vol. 32, No. 10   OCTOBER, 1961

In This Issue

(continued on page iv)

The Journal of Applied Physics is published monthly at Prince and Lemon Streets, Lancaster, Pennsylvania.
Second-class postage paid at Lancaster, Pa.

CONTENTS—*Continued*

2–5 and 2–6 Compounds

2–6 Compounds

Committees

Sponsors

Air Force Office of Scientific Research

Office of Aerospace Research USAF

General Electric Research Laboratory

Executive Committee

W. W. Tyler, *Chairman*

H. Ehrenreich, *Secretary*

H. P. R. Frederikse, *Chairman,*
Program Committee

J. H. Crawford, Jr., *Chairman,*
Publications Committee

Advisory Committee

A. C. Beer	J. A. Krumhansl
J. L. Birman	B. Lax
R. G. Breckenridge	H. Levinste
H. Brooks	P. H. Miller, Jr.
E. Burstein	F. J. Morin
J. H. Crawford, Jr.	R. L. Petritz
H. G. Drickamer	D. C. Reynolds
H. Ehrenreich	A. Rose
H. Y. Fan	C. E. Ryan
H. P. R. Frederikse	W. W. Scanlon
G. R. Gunther-Mohr	R. E. Sellers, Jr.
R. R. Heikes	W. W. Tyler
E. O. Kane	R. F. Wallis

C. F. Yost

Program Committee

H. P. R. Frederikse	F. J. Morin
H. Ehrenreich	D. C. Reynolds
H. Y. Fan	W. W. Scanlon

B. Lax

Publications Committee

J. H. Crawford, Jr.	R. O. Carlson
Ethel McMillan	J. W. Cleland
M. Aven	R. R. Heikes
J. H. Becker	J. A. Krumhansl
A. C. Beer	F. Stern

Local Arrangements Committee

R. W. Bengtson	Moya Coan

A. S. Dahlhauser

SEMICONDUCTING COMPOUNDS

Journal
of
Applied Physics

Supplement to Volume 32, Number 10   October, 1961

Proceedings of the Conference on Semiconducting Compounds

Foreword

ON June 14–16, 1961, a conference on the physics of semiconducting compounds was held at the General Electric Research Laboratory in Schenectady, New York. This conference was sponsored jointly by the Air Force Office of Scientific Research, Office of Aerospace Research, USAF, and the General Electric Research Laboratory. This issue of the *Journal of Applied Physics* consists of papers presented at the conference.

During the past decade our understanding of semiconducting compounds has advanced rapidly. However, no conference devoted exclusively to papers and discussion on the physics of semiconducting compounds seems to have been held heretofore. The aim of the conference was to summarize both experimental and theoretical results in various classes of compounds, and if possible, to synthesize physical ideas common to these compounds. Emphasis was therefore placed on the 3–5 and 2–6 compounds with some discussion of the lead compounds and oxides. Several other compounds on which definitive work is available were also included. The conference was restricted to fundamental physical properties, such as band structure, transport, optical galvanomagnetic, and resonance phenomena. Within these categories of materials and subjects, the Conference Proceedings seem to constitute a fairly complete and impressive summary of our present knowledge and understanding of compound semiconductors.

A particular aim of this conference was to encourage strong interaction between workers in the various branches of semiconduction compound research. This was facilitated by arranging only six single sessions during the three days of the conference, which permitted everyone attending to hear all of the papers. Papers presented were of either 15, 20, or 30 minutes duration with approximately 5 minutes after each paper devoted to discussion.

Attendance at the conference was not restricted although prior registration was requested. In fact, an effort was made to bring together as many active workers in the field as possible and to have all of the laboratories which support research in compounds represented. Representation from groups without active programs but with strong interest in research trends in this field was also encouraged. The total attendance of the conference was 240. Although primarily a national conference, 25 foreign scientists were present to represent overseas laboratories active in semiconductor compound research. It may be of interest to note that the total attendance represented about 47 industrial laboratories, 20 universities, and 18 government laboratories. In addition, several U. S. Government offices involved with the funding of research were represented.

The organization of this conference originated with the formation of an Advisory Committee consisting of leading scientists from various fields of compound semiconductor research. This Advisory Committee served as a source of scientists for the Program and Publications Committees and in addition made valuable suggestions concerning possible papers, attendance, and all other aspects of the conference. They deserve much credit for the time and effort generously contributed, particularly in the formative stages of planning. The conference is particularly indebted to those members of the Program, Publications, and Local Arrangements Committee who made additional and specific contributions of time and effort. Committee Members are listed elsewhere in this issue.

In large measure the success of any conference depends on the quality of the papers presented. The authors represented in these Conference Proceedings are to be commended for the excellence of their contributions. In addition, the Chairmen should be complimented for providing expert guidance to the conference sessions.

The conference would not have been possible without the financial support of the AFOSR. The cooperation of members of the Directorate of Solid State Sciences, AFOSR, was indispensable in the planning of the conference. The expeditious publication of the Proceedings bespeaks the cooperation and effectiveness of the staff of the *Journal of Applied Physics*.

W. W. TYLER
H. EHRENREICH

3–5 Compounds: Band Structure and Transport Properties

W. W. Tyler, *Chairman*

Galvanomagnetic Properties of InSb

H. Weiss

Research Laboratory, Siemens-Schuckertwerke AG, Erlangen, West Germany

In discussing the galvanomagnetic effects of semiconductors with high electron mobility one has to distinguish four main groups of parameters which influence these effects: (1) For InSb with practically no magnetoresistive effect on a long rod, an increase in resistance by a factor 38 can be reached in 10 000 gauss with appropriate shape (field disk). (2) Layers periodically changing their electron concentration produce an anisotropy of magnetoresistance. For certain specimen orientations the Hall coefficient depends on the magnetic field and a planar Hall effect is observed. Near a step of concentration one measures apparent negative resistances which are caused not by a retrograde current, but merely by rotation of the current lines. (3) If there are more than one type of charge carriers, it is difficult to know the concentration and mobility characteristic of a special type. Because of the high mobility ratio in InSb it is possible to state a hole mobility of 620 cm²/v sec in 150 000 gauss for pure material at room temperature. (4) After elimination of the influences of the above-mentioned points 1 to 3 one cannot find in InSb any magnetoresistive effect of electrons in the conduction band up to 150 000 gauss. The Hall coefficient is magnetic field independent up to this value of the magnetic induction for n-type InSb.

T HE galvanomagnetic effects in InSb are of particular interest because of the high electron mobility of 76 000 cm²/v sec at room temperature. There are several reasons for both the magnetoresistive effect and the variation of the Hall coefficient with changing magnetic field. They may be divided into the following groups:

(1) Influence of the shape of the semiconductor,
(2) Inhomogeneous impurity distribution,
(3) Mixed conductivity, and
(4) Scattering mechanism, degeneracy, and band shape.

To understand the behavior of the electrons and holes one wants to know how they are influenced by a magnetic field, that is one looks for experimental data related to (4), the behavior of the charge carriers within a band. This behavior may be covered by one or more of the preceding influences to a great extent. In the following paper the above mentioned groups will be discussed one after the other until one gets some results concerning the "true" magnetoresistive and Hall effect within a band.

For the relation between electrical field strength $\mathbf{E}$ and current density $\mathbf{j}$ in an isotropic material with mobility μ and specific conductivity σ we have the following basic equation[1]:

$$\mathbf{E} = \frac{1}{\sigma}(\mathbf{j} + \mu[\mathbf{j} \times \mathbf{B}]), \qquad (1)$$

where B is the magnetic induction. According to Eq. (1) the Hall angle $\theta(\tan\theta = \mu B)$ is independent of the shape of the semiconductor for a right angle between $\mathbf{j}$ and $\mathbf{B}$. Let us regard the electrodes to understand the in-fluence of the shape of the semiconductor on magnetoresistivity. The electron mobility in metals is smaller by many orders of magnitude, so the metallic electrodes are equipotential lines. With no magnetic field the current lines are perpendicular to the metal-semiconductor boundary, in the magnetic field they are rotated by the Hall angle θ, but the direction of the electric field strength remains unchanged. Therefore no Hall voltage exists near the electrode within the semiconductor. It grows with increasing distance from the electrodes. The component of the current density $\mathbf{j}$ perpendicular to the electrodes in the magnetic field is reduced to $1/[1 + (\mu B)^2]$.

The highest magnetoresistive effect is shown by a circular disk, the so-called field disk. The resistance R_B of a field disk in a magnetic field B related to the resistance with no magnetic field R_0 is given by[1]

$$R_B/R_0 = (\rho_B/\rho_0)[1 + (\rho_0/\rho_B)(\mu_0 B)^2]. \qquad (2)$$

ρ_B/ρ_0 is the ratio of the specific resistance in field B to the specific resistance with no field, measured on a long rod. μ_0 is defined as the electron mobility for $B \rightarrow 0$. ρ_B/ρ_0 and μ_0 were both measured as functions of doping and temperature on single crystals made of InSb doped with Te. The specimens were cut perpendicular to the [111] direction, which was the direction of growth. ρ_B/ρ_0 decreases strongly and μ_0 decreases slowly with increasing electron concentration. R_B/R_0 calculated with Eq. (2) using experimental[2] values of ρ_B/ρ_0 and μ_0 is shown as a function of the specific conductivity for 6000 and 10 000 gauss (dashed line, Fig. 1). Compared with this curve are the highest measured values for InSb field disks. They were cut perpendicular to the growth direction. The highest experimental factor in

[1] H. Weiss and H. Welker, Z. Physik **138**, 322 (1954).

[2] H. Rupprecht, R. Weber, and H. Weiss, Z. Naturforsch. **15a**, 783 (1960); H. Welker, Intern. Halbleiterkonferenz, Prague, 1960 (to be published).

10 000 gauss amounts to 38, the calculated maximum is about 42.

Equation (2) shows that for a given mobility μ_0 one gets the highest value of R_B/R_0 for $\rho_B=\rho_0$. It is for this reason that the maximum of the curves in the figure does not lie at $\sigma=200$ $(\Omega\cdot\mathrm{cm})^{-1}$, but instead near $\sigma=300$ $(\Omega\cdot\mathrm{cm})^{-1}$, though the electron mobility is smaller than for intrinsic material. The maximum is shifted to higher σ values, if the magnetic field increases. So one gets a value of 200 for R_B/R_0 in 100 kgauss with intrinsic InSb and a value of 260 in 40 kgauss for $\sigma=800$ $(\Omega\cdot\mathrm{cm})^{-1}$, in good agreement between calculation and experiment.

The influence of the shape of the semiconductor on the galvanomagnetic effects can be reduced by long thin rods for the measurements. However, inhomogeneities are of importance too in a less apparent manner. Therefore, single crystals of Te-doped InSb provide strong anisotropy of the transverse magnetoresistance, also if no inhomogeneity can be detected by Hall measurements. The highest values of $\Delta\rho/\rho_0$ were observed if the specimens were cut parallel to the pulling direction (the [111] direction). The relative magnetoresistance for rods cut perpendicular to this direction showed values which were smaller by two orders of magnitude. For other directions the values lie between both limits, independent of the crystallographic orientation.[2]

The specimens cut parallel and perpendicular to the pulling direction differed not only by the values of transverse magnetoresistance but also by the dependence of the magnetoresistance upon the angle between the magnetic induction **B** and the current **j**. The Hall effect showed no anomalies.

The same typical behavior of the magnetoresistive effect was observed by H. Rupprecht with crystals pulled in the [113] and [100] directions. At the same time one essential feature was found: the anisotropy of $\Delta\rho/\rho_0$ in InSb is the smaller, the higher the pulling rate.[3] This confirms the suggestion that the greatly varying values of $\Delta\rho/\rho_0$ are caused by layers with changing concentrations of impurities.

Another proof for this idea is provided by the dependence of the anisotropy as a function of the electron concentration: The anisotropy is no more existent in the range of intrinsic conductivity as a result of the absence of impurities.

Formulas given by Herring[4] for a stratified medium allow us to compare our results with the model quantitatively. The calculations are valid for a material in which the electron concentration varies periodically with constant electron mobility. This assumption is allowed, because μ only changes slowly with electron concentration.[2]

At the border between two areas of different electron concentrations the Hall voltages must be identical on both sides to get $\mathrm{rot}\,\mathbf{E}=0$. This can only be reached by

[3] H. Rupprecht, Z. Naturforsch. **16a**, 395 (1961).
[4] C. Herring, J. Appl. Phys. **31**, 395 (1961).

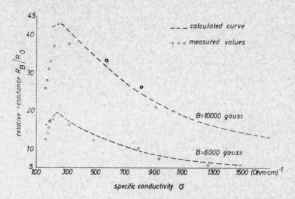

FIG. 1. Calculated and measured relative resistance R_B/R_0 of field disks as a function of the specific conductivity for 6000 and 10 000 gauss.

rotations of the current lines in both areas by equal but opposite angles in the magnet field as shown in Fig. 2(a). This means an increase in resistance caused by B. If there is no longitudinal and no transverse magnetoresistance in an homogeneous semiconductor one can deduce from Herring's formula the following resistance tensor for the stratified semiconductor:

$$|\rho|=\frac{1}{e\mu\langle n\rangle}\left[\begin{vmatrix} 1 & \mu B_z & -\mu B_y \\ -\mu B_z & 1 & \mu B_x \\ \mu B_y & -\mu B_x & 1 \end{vmatrix}+\begin{vmatrix} A & 0 & 0 \\ 0 & 0 & 0 \\ 0 & 0 & 0 \end{vmatrix}\right]. \quad (3)$$

$$A=\left(\langle n\rangle\left\langle\frac{1}{n}\right\rangle-1\right)\left[1+\frac{\mu^2(B^2-B_x^2)}{1+(\mu B_x)^2}\right].$$

The x axis stands perpendicular on the layers. $\langle n\rangle$ means the spacial average of the electron concentration n.

The new tensor is composed of the tensor of the isotropic material, corrected by the factor $n/\langle n\rangle$, and an additional tensor with only one term different from zero, which describes the strong anisotropy of the resistance in a magnetic field. For the current in the x direction we obtain

$$\Delta\rho/\rho_0=\left[1-\frac{1}{\langle n\rangle\left\langle\dfrac{1}{n}\right\rangle}\right]\frac{\mu^2(B_y^2+B_z^2)}{1+(\mu B_x)^2}$$

$$=\left[1-\frac{1}{\langle n\rangle\left\langle\dfrac{1}{n}\right\rangle}\right]\frac{(\mu B)^2\sin^2\varphi}{1+(\mu B)^2\cos^2\varphi}. \quad (4)$$

φ is the angle between current **j** and field **B**. As long as there is no longitudinal component of $\mathbf{B}(B_x=0)$, $\Delta\rho/\rho_0$ is proportional to B^2. A longitudinal component of the magnetic field reduces the rotation of the current lines for the same transverse component of **B**. At the same time the current has a component perpendicular to the plane of the drawing. For small fields with $(\mu B\ll1)$ one

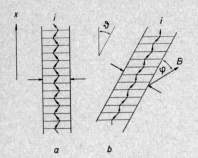

Fig. 2. (a) Current lines in magnetic field in a specimen with axis perpendicular to the layers. (b) Current with and without a magnetic field in a specimen at an angle ϑ between axis and normal on the layers. → ← Hall probes.

obtains: $\Delta\rho/\rho_0 \propto \sin^2\varphi$. In large fields one gets a different, almost singular dependence of $\Delta\rho/\rho_0$ upon the angle between current and magnetic field. Comparison of curves calculated after (4) with measured curves in Fig. 3 show that the model of layers with different electron concentrations is valid for InSb. For $\mu B = 5$ and a concentration difference of 20% in both layers of the same thickness, one gets $\Delta\rho/\rho_0 = 0.31$ from formula (4). That is the true order of magnitude, so the variations of concentration lie in the order of 20%.

From the tensor representation in (3) it is obvious that the Hall coefficient R_H is field independent for current directions parallel and perpendicular to the x axis. Under the same conditions there is no planar Hall effect which is directly connected to the longitudinal and transverse magnetoresistance in a homogeneous and isotropic material. In a specimen cut in the pulling direction with a measured transverse magnetoresistance of 26% the planar Hall effect only delivered a value of 4% for $\Delta\rho/\rho_0$.

The validity of the model of stratified structure is evident, if one cuts a specimen at an angle to the pulling direction and compares calculated and experimental data. From the resistance tensor in (3) $\Delta\rho/\rho_0$ and R_H can easily be calculated for any directions of field $\mathbf{B}$ and current $\mathbf{j}$. In rotating the coordinate system the first tensor on the right side in (3) retains its appearance with the transformed components of $\mathbf{B}$, but each term of the second tensor generally depends upon B^2.

We want to discuss a rod cut with an angle ϑ to the normal on the layers as shown in Fig. 2(b). Here already the current lines exhibit zig-zag behavior with no magnetic field. One obtains

$$\Delta\rho/\rho_0 = \gamma \frac{\mu^2(B^2 - B_x^2)}{1 + (\mu B_x)^2}, \tag{5}$$

where

$$\gamma = \frac{\cos^2\vartheta\left(\langle n\rangle\left\langle\frac{1}{n}\right\rangle - 1\right)}{1 + \cos^2\vartheta\left(\langle n\rangle\left\langle\frac{1}{n}\right\rangle - 1\right)}. \tag{6}$$

The maximum of $\Delta\rho/\rho_0$ is reached when the direction of $\mathbf{B}$ lies in the layers independent of ϑ and of the particular direction of $\mathbf{B}$ within the layers. For $\vartheta \neq 0$ there

is always a longitudinal magnetoresistive effect. $\Delta\rho/\rho_0$ vanishes for $\mathbf{B}$ perpendicular to the layers. For a difference of 20% in concentration formula (5) only differs from formula (4) by the factor $\cos^2\vartheta$ within 1% accuracy. This corresponds qualitatively to earlier experimental results.[2] If one measures the dependence of the relative magnetoresistance on the angle φ between current $\mathbf{j}$ and $\mathbf{B}$ with specimen b in Fig. 2, where $\mathbf{B}$ lies within the plane of the drawing, one obtains

$$\Delta\rho/\rho_0 = \gamma(\mu B)^2 \frac{\sin^2(\varphi+\vartheta)}{1 + (\mu B)^2 \cos^2(\varphi+\vartheta)}. \tag{7}$$

The shape of the curve is the same as for Eq. (4), but the maximum is shifted from the transverse position of $\mathbf{B}$ by the angle ϑ. Varying φ in such a manner that the magnetic induction for $\varphi = \pi/2$ stands perpendicular on the plane of the drawing, one gets the following relation:

$$\Delta\rho/\rho_0 = \gamma(\mu B)^2 \frac{\sin^2\varphi + \cos^2\varphi \sin^2\vartheta}{1 + (\mu B)^2 \cos^2\varphi \cos^2\vartheta}. \tag{8}$$

The maximum amplitude is identical with that of formula (7) but lies at $\varphi = \pi/2$ and the deviation of the curve $\Delta\rho/\rho_0$ versus φ from a $\sin^2$ curve is diminished by the factor $\cos^2\vartheta$ in the second term of the denominator and the additional term in the numerator. If $\mathbf{B}$ is rotated around the specimen in a plane which shows an angle $\psi \neq \pi/2$ against the plane of the drawing, the maximum of $\Delta\rho/\rho_0$ lies between $\pi/2$ and $\pi/2 - \vartheta$ and the shape of the curve between those of formulas (7) and (8). There is no value of φ for which $\Delta\rho/\rho_0$ vanishes completely. Figure 4 represents the dependence of $\Delta\rho/\rho_0$ upon φ for two single crystals with $R_H = 55$ cm³/amp sec cut from the same ingot with 20° and 45° against the pulling direction. The Hall coefficient is almost constant along the rods. The dependence of $\Delta\rho/\rho_0$

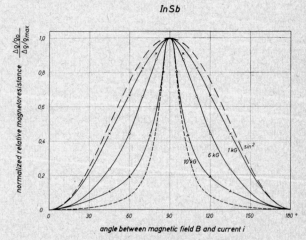

FIG. 3. Normalized relative magnetoresistance in dependence of the angle between current and magnetic induction $n = 5.4 \cdot 10^{16}$ cm⁻³. —— measured curves for 1, 6, and 10 kgauss; – – – calculated curve for $B \to 0$; - - - - calculated curve for 10 kgauss.

on the angle φ between current and magnetic field was measured in two planes perpendicular to each other for each specimen.

In agreement with calculations one gets the following result: The amplitudes of both maxima of one single rod are almost identical (10%). The half-width of the maximum positioned nearer to 90° is larger than that at a smaller φ. The minima do not lie at 0° and 180°, but between 150 and 170°. If φ_1 and φ_2 are the angles corresponding to the two maxima of one specimen, one obtains for the angle ϑ:

$$\cos\vartheta = 1/(1+\cot^2\varphi_1+\cot^2\varphi_2)^{\frac{1}{2}}. \quad (9)$$

The curves give $\vartheta = 43°$ for specimen I and $\vartheta = 60°$ for specimen II. From this we conclude that there is an angle of about 20° between the pulling direction and the normal on the layers. In good agreement one third specimen cut parallel to the pulling direction (not shown in figure) had both maxima at about 70°. This crystal was pulled without rotation.

According to formula (6) one calculates a ratio of the amplitudes of both specimens of 2.1, the measured ratio is 2. This good agreement between calculation and experiment points out that the layers must have a structure of high regularity within a crystal.

As mentioned above, the Hall coefficient generally depends on B. With Hall probes perpendicular to the plane of the figure the Hall coefficients are identical and independent of B for the specimens a and b in Fig. 2. If the probes are mounted in a plane which is rotated by an angle ψ against the plane of the figure, there exists the following relation between the Hall field strength E_H and the electrical field strength E_j parallel to the current with no magnetic field:

$$\frac{E_H}{E_j} = \mu B_\perp + \frac{\sin 2\vartheta}{2}\cos\psi\left(\langle n\rangle\left\langle\frac{1}{n}\right\rangle - 1\right)$$

$$\times\left[1+\frac{\mu^2(B^2-B_x^2)}{1+(\mu B_x)^2}\right]. \quad (10)$$

$B_\perp$ is the component of **B** perpendicular to the current and to the Hall probes. The second term contains the field dependence of R_H and the planar Hall effect. With no magnetic field one gets a field strength E_{HO} between the Hall probes:

$$\frac{E_{HO}}{E_j} = \frac{\sin 2\vartheta \cos\psi}{2}\left(\langle n\rangle\left\langle\frac{1}{n}\right\rangle - 1\right). \quad (11)$$

According to (10) the relative dependence of R_H is greatest for $\psi = 0$ and $\vartheta = \pi/4$:

$$\frac{\Delta R_H}{R_{HO}} = \frac{\mu B_\perp}{2}\left(\langle n\rangle\left\langle\frac{1}{n}\right\rangle - 1\right). \quad (12)$$

In this case R_H depends linearly on B. $\Delta R_H/R_{HO}$

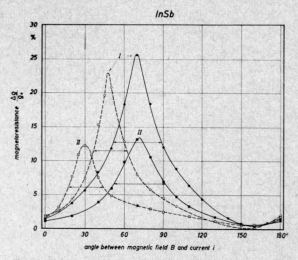

FIG. 4. Relative magnetoresistance of two specimens I and II as a function of the angle φ between current and magnetic induction. Maxima of comparable size belong to the same specimen. For curves – – – the specimens are rotated by 90° around their axis compared with curves ———; ←——→ half-width.

amounts to 0.03 in 10 000 gauss with $\mu = 50\,000$ cm²/v sec and 20% difference in doping. For other directions the field dependence of R_H is smaller. For $\vartheta = \pi/4$ and $\psi = 0$, the field strength E_{PL} of the planar Hall effect is also greatest:

$$\frac{E_{PL}}{E_j} = \frac{(\mu B)^2}{2}\frac{\sin^2[\varphi + (\pi/4)]}{1+(\mu B)^2[\varphi + (\pi/4)]}\left(\langle n\rangle\left\langle\frac{1}{n}\right\rangle - 1\right). \quad (13)$$

Because of the denominator dependence on B, E_{PL} is no longer proportional to $\sin^2(\varphi + \pi/4)$. E_{PL} nearly exhibits the angular dependence of φ with the period π in a small magnet field. This behavior is typical for homogeneous semiconductors crystallizing in the zinc-blende lattice with nonvanishing magnetoresistive effect.

The conditions near a single doping step are difficult to treat mathematically.[5] The current lines near the $n-n^+$ transition are rotated in the same manner as in the layer model. Therefore the Hall coefficient does not exhibit a step but increases continuously in trespassing the $n-n^+$ junction.

Schoenwald studied the potential in the vicinity of a step with a doping ratio of 1:6.8. Figure 5 shows the measured equipotential lines; **B** stands perpendicular on the drawing. Using the known mobility values, the current lines were designed. In the same figure the electrical field strength measured along the middle of the specimen with probes in the plane of **B** is given as a function of the position along the specimen. In the highly doped part it is smaller in field than with $B=0$, near the step it is negative. The data are identical for both negative and positive magnetic induction. The

[5] R. T. Bate, J. C. Bell, and A. C. Beer, J. Appl. Phys. **32**, 806 (1961).

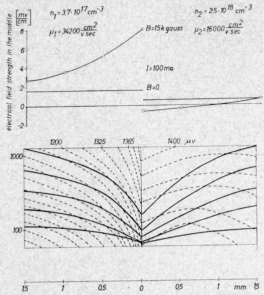

FIG. 5. Top: electrical field strength along the middle of a specimen with doping step for $B=0$ and $B=15\,000$ gauss. Bottom: – – – measured equipotential lines; ——— designed current lines.

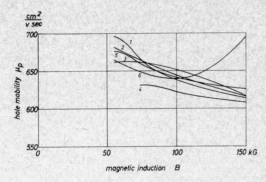

FIG. 6. Hole mobility μ_p as a function of the magnetic induction

No.	1	2	3	4	5	6	
Temperature	17	22	29	35	24	27	°C
R_H 77°K	+500	+500	+500	+500	+1100	−3000	amp sec/cm³

negative potential differences can easily be understood: Because of the high electron mobility the current lines are nearly equipotential lines and a small rotation is sufficient to produce a negative potential difference in the direction along the axis of the specimen. If one measures the changes in potential with probes in the plane of the drawing and averages for both positive and negative magnetic induction, one always obtains positive values for the variations in potential difference with magnetic field.

In the range of mixed conductivity both holes and electrons contribute to the transport of charge. For each type of charge carrier there exists an equation (1) which contains the parameters characterizing each type: conductivity: σ_n and σ_p, Hall coefficient: R_n and R_p. These four parameters still may depend upon the magnetic induction. If there is no interaction between holes and electrons, one can add the vectors of both current densities. Both equations are coupled by the common electrical field strength **E**. By this one obtains formulas for the dependence of the Hall coefficient on magnetic field in the mixed conductivity range, which agree with the results found by Hieronymus[6] in InSb:

(1) R_H is independent of B up to 150 000 gauss for intrinsic and heavily doped n-InSb.

[6] H. Hieronymus and H. Weiss, *Solid State Electronics* (1961) (to be published).

(2) R_H increases with increasing B for lightly doped n-InSb.

(3) R_H decreases with increasing B for p-doped InSb.

In discussing the experimental data one finds that R_n and R_p, the Hall coefficients for conduction and valance band, are both nearly independent of B up to 150 000 gauss. As for the electron mobility, one finds a value higher than 12 000 cm²/v sec in 150 000 gauss. Evaluation of the hole mobility as a function of B can also be done. Figure 6 represents the field dependence of the hole mobility μ_p for specimens with different doping. μ_p decreases from 710 cm²/v sec with no field to 620 cm²/v sec in 150 000 gauss.

The above discussed influences of shape, inhomogeneity, and mixed conductivity on the galvanomagnetic effects of InSb show good agreement between experimental and calculated data if one uses an electron mobility independent of B. In none of the experiments was it necessary to assume a magnetic field dependent electron mobility. Beyond that the magnetoresistive effect for electrons in InSb is smaller than 1% in 10 000 gauss for electron concentrations $>10^{16}$ cm⁻³. This was proven by measurements of the planar Hall effect[2] and experiments with crystals made with fast pulling rates. They were very homogeneous and nearly isotropic and had very small magnetoresistance. In contrast to the electrons the hole mobility is reduced by 14% in strong magnetic fields.

ACKNOWLEDGMENTS

The author wishes to thank H. Hieronymus, Dr. H. Rupprecht, and H. Schoenwald for valuable contributions to the present paper.

JOURNAL OF APPLIED PHYSICS   SUPPLEMENT TO VOL. 32, NO. 10   OCTOBER, 1961

Properties of Semi-Insulating GaAs

C. H. GOOCH, C. HILSUM, AND B. R. HOLEMAN

Services Electronics Research Laboratory, Baldock, Herts, England

Most samples of GaAs show properties similar to those of germanium and silicon, but it is possible to prepare GaAs with a resistivity at room temperature greater than 10^6 ohm-cm, and the electrical properties are then more like those of the wide band gap II–VI compounds, such as CdS. This form of material, known as semi-insulating GaAs, previously has not been studied thoroughly, partly because homogeneous samples were not available.

Measurements have now been made on semi-insulating GaAs, and results are reported for carrier concentration and mobility as a function of temperature. The interpretation of the results is sometimes complicated because even at room temperature the activation energy is about half of the intrinsic activation energy, and the carrier concentration can be close to the intrinsic concentration.

The dominant lattice scattering mechanism in GaAs is believed to be polar scattering, but even in the purest samples of semiconducting GaAs made thus far, impurity scattering is observed at room temperature. In a highly compensated material like semi-insulating GaAs, neither the Brooks-Herring nor the Conwell-Weisskopf theory of impurity scattering is likely to be valid. An initial study of carrier scattering has been made using measurements of transverse magnetoresistance and the field dependence of Hall coefficient. Some values for carrier lifetime are also reported.

I. INTRODUCTION

THE preparation of GaAs has now become a routine matter in many laboratories. It is possible to predict with fair accuracy the electrical properties of material made from pure gallium and arsenic, and most of the ingots produced are n type, with a carrier concentration between 10^{16} and 10^{17} cm^{-3}, and a resistivity below 1 ohm cm. But occasionally the same routine production yields an ingot with a resistivity greater than 10^6 ohm cm. We call this material "semi-insulating GaAs" (s.i. GaAs) to differentiate it from the more common "semi-conducting GaAs" (s.c. GaAs). Although it was first observed several years ago, few electrical measurements have been reported on it. In our laboratory a more complete investigation of s.i. GaAs is now being undertaken, and in this paper we give some of the first results. We have studied Hall coefficient and resistivity over the temperature range 300° to 900°K, the magnetic field dependence of resistivity and Hall coefficient, and carrier lifetime.

II. PREPARATION

There are several ways in which s.i. GaAs can be made. About one-third of horizontally grown ingots made from high-purity gallium and arsenic are found to be semi-insulating. Most ingots of s.c. GaAs which are exposed to an oxygen atmosphere at a temperature above 1300°K are converted to s.i. GaAs and often ingots prepared in alumina boats are semi-insulating. However the most reliable method of preparing the material is by floating-zone refining. The resistivity profile along a typical floating-zoned ingot after five zone passes is shown in Fig. 1. We see that there is no sudden transition from s.c. GaAs to s.i. GaAs, and that resistivities from 10^6 to 10^{-1} ohm cm can be found in the same ingot.

We should here differentiate between s.i. GaAs and the high-resistivity material made by diffusing copper into s.c. GaAs. Copper is an acceptor in GaAs, and the resistivity obtained after diffusion depends only on the compensation which the experimentalist can achieve. A complete range of resistivities from low-resistivity n type through high-resistivity n and p type to low-resistivity p type can be made in this way. If copper and similar metallic acceptors are carefully excluded from the preparation tube during doping with oxygen, or floating-zone refining, no particular care is needed to ensure the almost exact compensation of donors and acceptors that occurs in s.i. GaAs. There appears to be no tendency for the material to become p type—the compensation is automatic.

III. MEASUREMENTS OF HALL COEFFICIENT AND RESISTIVITY

A. Experimental Techniques

We have found that electrical measurements on s.i. GaAs can be strongly affected by the presence of conducting layers on the surfaces of the samples. The layers

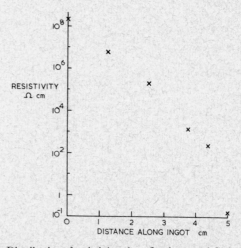

FIG. 1. Distribution of resistivity along floating-zoned GaAs ingot.

can be removed by heating the sample to 150°C in a good vacuum (better than 10^{-4} mm Hg), but they rapidly reform if the pressure in the system rises above 10^{-3} mm. All measurements described here were made with the sample in a vacuum, and a heating schedule was carried out before each series of observations. Since the samples were strongly photoconductive, the equipment was designed to exclude all light.

The samples were single crystals, about $6 \times 2 \times 1$ mm. Nonrectifying end contacts were applied either by alloying indium at a temperature of 450°C or by silver plating. The side contacts were tungsten probes, formed with a spark discharge. All voltage measurements were made with a vibrating plate electrometer, as the resistance between the probes and the sample was higher than 10^{10} ohm. Most samples taken from horizontally grown ingots were found to be inhomogeneous, and considerable selection was necessary to obtain specimens with a resistivity uniform to within 20%. Samples from floating-zoned ingots were usually sufficiently uniform.

For comparison two samples prepared by the floating-zone technique at Bell Telephone Laboratories were also studied. No significant difference was observed between those samples and samples prepared in this laboratory.

B. Results

For all samples Ohm's law was obeyed at fields up to 100 v cm^{-1}. At higher fields oscillating currents occurred and reproducible results could not be obtained. (These effects will be described more fully elsewhere by Dr. D. C. Northrop.) Measurements of Hall coefficient and resistivity were made over the temperature range 300–900°K. When samples were heated above 900°K irreversible changes in resistivity occurred.

Results for three typical samples are shown in Fig. 2. The activation energy of $R_H T^{\frac{3}{2}}$ is near 0.76 ev for most samples so far measured. All the samples were n type.

The temperature dependence of electron mobility, deduced from the data of Fig. 2, is given in Fig. 3. Sample 4 was taken from the same ingot as sample 3, but has a carrier concentration close to the intrinsic value at 300°K, and a rather smaller activation energy than the other samples. It appears that the mobility can be independent of temperature over a considerable range. Whelan and Wheatley[1] have reported a similar result for s.i. GaAs from 500° to 800°K.

IV. DISCUSSION

It is remarkable that we can obtain in so many samples carrier concentrations at 300°K within an order of magnitude of the intrinsic concentration. In Fig. 4 the calculated temperature dependence of intrinsic concentration is plotted as the solid line. It has been assumed in this calculation that the effective masses of electrons and holes are 0.072 m_0 and 0.5 m_0, respectively,

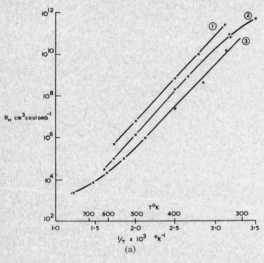

(a)

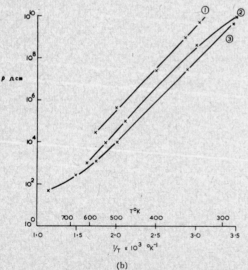

(b)

Fig. 2(a). Temperature dependence of Hall coefficient.
(b). Temperature dependence of resistivity.

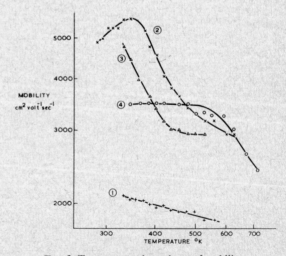

Fig. 3. Temperature dependence of mobility.

[1] J. M. Whelan and G. H. Wheatley, J. Phys. Chem. Solids 6, 169 (1958).

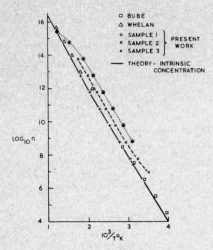

FIG. 4. Comparison of carrier concentrations in semi-insulating GaAs with calculated intrinsic concentration.

and that the energy gap is $1.52-4.3\times10^{-4}T$ ev. Results obtained for samples 1–3 are given in Fig. 4, together with data previously reported by Bube[2] and Whelan and Wheatley.[1] Since Bube quotes only the resistivity of his sample, the mobility has been taken as $3000\,\mathrm{cm^2/v\,sec}$.

It does not seem reasonable to conclude that these samples are uncompensated, particularly since s.i. GaAs can be made by oxygen doping. Further, the compensation cannot be exact in all samples, since some show a carrier concentration much greater than the intrinsic value. The large temperature dependence of carrier concentration indicates that the electrical properties of the material are determined by a deep impurity level. This may be either a donor or an acceptor, and there is as yet insufficient evidence for us to decide which. Allen[3] has suggested that GaAs becomes semi-insulating when the donor that is normally present is compensated by a deep acceptor. The center that acts as the acceptor can under some circumstances be electrically inactive, and the proportion that acts as an acceptor is determined by the donor concentration. As long as the total concentration of these centers is greater than the donor concentration the material becomes "automatically compensated" since each donor converts one center from an electrically inactive state to an acceptor. Allen considered only the case of material with a carrier concen-

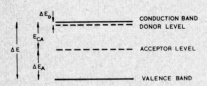

FIG. 5. Compensation of shallow donors and deep acceptors.
$$N_A \simeq N_D \cdot (= N) \quad N_A - N_D = \phi N$$
$$\alpha = N_C \exp[E_{CA}/kT]$$
$$n^2 + n \cdot \phi N - (n_i^2 + 2\alpha N) = 0.$$

[2] R. H. Bube, J. Appl. Phys. **31**, 315 (1960).
[3] J. W. Allen, Nature **187**, 403 (1960).

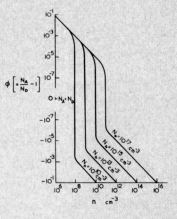

FIG. 6. The dependence of carrier concentration on the degree of compensation.

tration near intrinsic. We wish now to extend his calculations in order to explain the gradient of resistivity along a floating-zoned ingot (Fig. 1).

The model given by Allen is shown in Fig. 5. We can deduce the carrier concentration from the standard equations (see, for example, Spenke[4])

$$2nN_D{}^+ = N_c \exp(-\Delta E_D/kT)N_D{}^N, \qquad (1)$$

$$2pN_A{}^- = N_v \exp(-\Delta E_A/kT)N_A{}^N, \qquad (2)$$

$$N_D = N_D{}^+ + N_D{}^N, \qquad (3)$$

$$N_A = N_A{}^- + N_A{}^N, \qquad (4)$$

$$N_A{}^- + n = N_D{}^+ + p, \qquad (5)$$

$$np = n_i{}^2 = N_c N_v \exp(-\Delta E/kT). \qquad (6)$$

$N_D{}^+$ and $N_D{}^N$ are the concentrations of charged and neutral donors, respectively, and $N_A{}^-$ and $N_A{}^N$ are the corresponding terms for acceptors.

We assume that the donor activation energy is zero, that N_D and N_A are never less than $10^{11}\,\mathrm{cm^{-3}}$ nor greater than $10^{17}\,\mathrm{cm^{-3}}$, and that the acceptor level is deep enough for $\alpha [= N_c \exp(-E_{CA}/kT)]$ to be negligible compared with n. Then, if $N_A \approx N_D = N$ and $N_A - N_D = \phi N$,

$$n^2 + n\phi N - n_i{}^2 - 2\alpha N = 0. \qquad (7)$$

Our measurements give E_{CA} as 0.76 ev, so that $\alpha = 4\times10^4\,\mathrm{cm^{-3}}$. The degree of compensation ϕ can be either positive $(N_A > N_D)$ or negative $(N_D > N_A)$. The dependence of n on ϕ has been calculated from Eq. (7), and is illustrated in Fig. 6. In one region of the graph n is independent of N, and in another it is independent of ϕ. The variation of resistivity along the floating-zoned ingot can obviously be explained by an appropriate choice of ϕ and N, but there is not yet sufficient experimental evidence for a searching test of the model.

It is interesting to note that the measured value of E_{CA}, 0.76 ev, is just half of the optical energy gap extrapolated to 0°K. No reliable electrical determinations of the energy gap have been made on semiconducting

[4] E. Spenke, *Electronic Semiconductors* (McGraw-Hill Book Company, Inc., New York, 1958), p. 319.

GaAs since the purest material available is still extrinsic at 800°K and irreversible changes occur in GaAs above 900°K. There is also a complication in interpretation due to conduction in one of the upper minima of the conduction band.[5] Whelan and Wheatley have reported a value for ΔE extrapolated to 0°K of 1.58 ev but this was obtained on a semi-insulating sample. It is therefore not clear if they were measuring $\frac{1}{2}\Delta E$ or E_{CA}.

V. SCATTERING MECHANISMS

In most semiconductors a study of lattice scattering is complicated by carrier screening effects and carrier-carrier scattering. Investigations on s.i. GaAs will be free of these complications. On the other hand, it is doubtful if we can here apply the conventional treatments of impurity scattering; s.i. GaAs contains both donors and acceptors, and these are probably associated in pairs. We might as a first approximation consider each pair as a neutral impurity, and note that Erginsoy's theory of neutral impurity scattering[6] predicts a mobility that is independent of temperature. But a more sophisticated approach will certainly be necessary before we can attach much significance to this result. To obtain more evidence to enable us to identify the dominant scattering mechanism in s.i. GaAs we have made some measurements of the magnetic field dependence of Hall coefficient and resistivity.

VI. MAGNETORESISTANCE AND FIELD DEPENDENCE OF R_H

The variation of ρ and R_H with magnetic field H was studied at room temperature in fields up to 15 000 oe. The changes in R_H were less than 2% over this range. In contrast, the magnetoresistance effect was much larger than in semiconducting GaAs, as can be seen from Fig. 7. In all samples measured $\Delta\rho/\rho_0$ was proportional to H^2 but the constant of proportionality showed no obvious correlation with carrier concentration.

It would be interesting to see if this large magnetoresistance effect is predicted by theory. The theory of magnetoresistance effects in polar semiconductors has been given by Lewis and Sondheimer,[7] but their results, which are presented as the ratio of two infinite determinants, cannot be compared directly with the experimental values. We have made computations up to the fourth order of the determinants, but the convergence is still slow. At that stage the predicted magnetoresistance effect was quite large. An accurate computation is now being made.

The magnetoresistance effect in polar materials is strongly dependent on the ratio of the optical mode temperature θ to the lattice temperature T. We have observed that the temperature dependence of the magnetoresistance effect in s.i. GaAs is large. We might expect that the magnetoresistance effect would change if the mobility were strongly temperature dependent, but here the mobility is fairly constant (Fig. 8). Over the temperature range of our experiments the ratio θ/T changes from 1.0 to 1.3.

VII. CARRIER LIFETIME

Measurements of carrier lifetime in s.c. GaAs have shown[8] that there are large trapping effects in this material, and that the minority carrier lifetime can be several orders of magnitude less than the majority carrier lifetime. Trapping in s.i. GaAs has been investigated by Allen and Cherry,[9] who observed space-charge limited currents and estimated the trap density in their samples as only 10^{12} cm^{-3}. From measurements of photoconductive and photoelectromagnetic effects, we have confirmed that trapping is much smaller in s.i. GaAs than in s.c. GaAs. Typically, in the sample with the lowest carrier concentration both the hole and

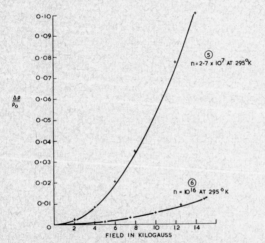

FIG. 7. Magnetoresistance in high and low resistivity gallium arsenide. Drawn curves are parabolic.

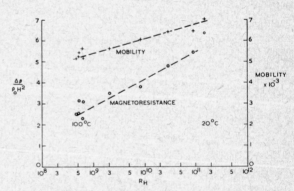

FIG. 8. Variation of magnetoresistance and mobility with temperature. Sample No. 7.

[5] L. W. Aukerman and R. K. Willardson, J. Appl. Phys. **31**, 939 (1960).
[6] C. Erginsoy, Phys. Rev. **79**, 1013 (1950).

[7] B. F. Lewis and E. H. Sondheimer, Proc. Roy. Soc. (London) **A227**, 241 (1955).
[8] C. Hilsum and B. R. Holeman, *Proceedings of the International Conference on Semiconductor Physics, Prague, 1960* (Publishing House of the Czechoslovak Academy of Sciences, Prague, 1961).
[9] J. W. Allen and R. Cherry, Nature **189**, 297 (1961).

electron lifetimes were rather greater than 10^{-7} sec. These results are reported more fully elsewhere.[10]

ACKNOWLEDGMENTS

We are grateful to Dr. F. A. Cunnell and W. Harding for preparing the gallium arsenide used in these experi-

ments, to Dr. D. J. Oliver for valuable discussions, and to Miss J. Devereux and Miss B. Taylor for assistance in the measurements. Samples made in the Bell Telephone Laboratories were kindly supplied by Dr. J. Whelan, and we would like to thank him for his cooperation. Permission to publish has been given by the Admiralty.

[10] B. R. Holeman and C. Hilsum, Conference on Photoconductivity, Cornell University, 1961.

JOURNAL OF APPLIED PHYSICS   SUPPLEMENT TO VOL. 32, NO. 10   OCTOBER, 1961

Study of Band Structure of Intermetallic Compounds by Pressure Experiments*†

A. SAGAR AND R. C. MILLER

Westinghouse Research Laboratories, Pittsburgh 35, Pennsylvania

The effect of hydrostatic pressure on the transport properties of n-type GaSb, InP, GaP, and p-type PbTe was measured to study their band structure. (1) The Seebeck coefficient, Hall coefficient, and resistance of three n-GaSb samples were measured as a function of hydrostatic pressure up to 17 000 atm between 200° and 400°K. The Seebeck coefficient α increased with pressure and approached a constant value at about 10 000 atm. The saturation value of α does not follow the simple $\frac{3}{2} \ln T$ relation for any of the samples; e.g., for a sample with $R_H(77°K) \approx 95$ coul^{-1} cm^3, the saturation value of α decreases with temperature. The contribution due to the phonon-drag effect has been considered as a possible explanation for this phenomenon. (2) The conductivity of p-PbTe increased almost exponentially with pressure both at 300° and 194°K; the Hall coefficient at 300°K decreased by about 5% at 8000 atm, while the conductivity increased by 55% at this pressure. (3) The resistance of n-InP samples increased with pressure; the pressure coefficient was found to be bigger for samples with higher impurity contents. (4) The resistance of an n-GaP sample decreased by about 3% at 10 000 atm.

INTRODUCTION

THE effect of hydrostatic pressure on the transport properties of many semiconductors has been studied by various investigators.[1] We have studied the transport properties of type n InP, GaP, GaSb, and p-type PbTe as a function of pressure. This work is still in progress and we present here the data we have at present.

EXPERIMENTAL

The equipment for measuring the resistivity and Hall coefficient vs pressure is described elsewhere.[2] For Seebeck coefficient vs pressure measurements, one end of the sample was soldered to a small heater and the other end to a copper block. The copper block was fixed to the plug of the pressure bomb. Chromel-Alumel thermocouples were soldered to the two ends of the sample. The thermocouple leads were brought out of the pressure bomb through the respective Chromel and Alumel "Bridgman-type" cones and were continued by the Chromel and Alumel wires to the cold junction. All

emf's were measured relative to Chromel and corrections[3] applied to obtain the absolute Seebeck coefficient. No corrections for the effect of pressure on the thermocouple characteristics were applied as this effect is small.[4] Measurements at normal pressure before and after a run using different sets of thermocouples, and with and without the pressure-transmitting fluid in the bomb, agreed to within 2%. The GaSb sample C was received from Dr. A. J. Strauss. The p-PbTe samples were prepared by Dr. J. McHugh by the Bridgman method.

n-InP

The magnetoresistance[5] and piezoresistance[6] measurements on n-type InP indicate that the conduction band minimum in this material lies in the center of the Brillouin zone. The magnetoresistance data of Glicksman[5] on highly doped samples indicate a significant (100) type anisotropy in the mobility tensor. The data by Edwards and Drickamer[7] on the absorption edge vs pressure suggest a (100) band at about 0.3 ev above the

---

* This work was supported in part by the U. S. Bureau of Ships.
† Work on GaP was done in collaboration with Dr. M. Rubenstein.
[1] R. W. Keyes, *Solid State Physics* edited by F. Seitz and D. Turnbull (Academic Press Inc., New York, 1960), Vol. 11, pp. 149–221.
[2] R. W. Keyes, Phys. Rev. **99**, 490 (1955); P. W. Bridgman, *Physics of High Pressures* (G. Bell and Sons, London, 1939).

[3] W. F. Roeser and H. T. Wensel, *Temperature*, edited by The American Institute of Physics (Reinhold Publishing Company, New York, 1941), p. 1308; J. Nyström, Arkiv Mat., Astron. Fysik, **34A**, No. 27, 1 (1947).
[4] F. P. Bundy, J. Appl. Phys. **32**, 483 (1961).
[5] M. Glicksman, J. Phys. Chem. Solids **8**, 511 (1959).
[6] A. Sagar, Phys. Rev. **117**, 101 (1960).
[7] A. L. Edwards and H. G. Drickamer, Phys. Rev. **122**, 1149 (1961).

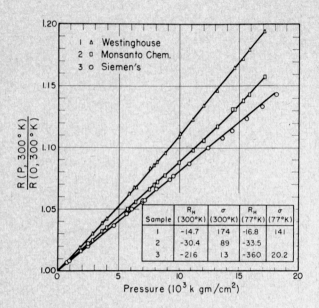

Sample	R_H (300°K)	σ (300°K)	R_H (77°K)	σ (77°K)
1	−14.7	174	−16.8	141
2	−30.4	89	−33.5	
3	−216	13	−360	20.2

FIG. 1. Resistance vs pressure for n-InP samples with different carrier concentrations at 300°K.

(000) band. We have measured the resistance of three n-InP samples of different carrier concentrations as a function of pressure. The results are shown in Fig. 1.

Small changes in the mobility of carriers with pressure, observed in many semiconductors[1] have been explained by Keyes[8] as primarily due to a change in the electron effective mass m^*. According to Keyes, the change of resistivity ρ due to pressure is given by

$$d(\log\rho)/dP = s \cdot d(\log Eg)/dP, \qquad (1)$$

where the mobility $\mu \approx m^{*-s}$. The observed pressure coefficient of resistance is larger for less pure samples and cannot be explained by adjusting the value of s in Eq. (1), because s decreases from $\frac{3}{2}$ for a pure sample with polar scattering to $\frac{1}{2}$ for a sample with impurity scattering. Furthermore, we cannot explain the non-linearity in the increase of resistance with pressure for the impure samples on the basis of the simple (000) conduction band model. However, we have neglected the interband effects[9] (e.g., scattering and the statistical distribution of carriers between the (000) and the (100) bands[7]) which would also contribute to an increase in the resistance of the sample with increasing pressure and would become quite important for impure samples in which statistical degeneracy sets in. Similar deviations in the effect of pressure on resistance of n-GaAs samples of different purity have been observed by Howard and Paul.[10]

[8] R. W. Keyes, *Semiconductor and Phosphors*, edited by M. Schön and H. Welker (Interscience Publishers, Inc., New York, 1958), pp. 236–246.
[9] M. I. Nathan, Tech. Report No. HP-1 (1958), Gordon McKay Laboratory, Harvard University.
[10] W. Howard and W. Paul (private communication); H. Ehrenreich, Phys. Rev. **120**, 1951 (1960).

n-GaP

The resistance of an n-type GaP single-crystal sample (R_H (300°K) = −17.7 coul^{-1} cm^3; $R_H\sigma$ = 84 cm^2 v^{-1} sec^{-1}) was measured as a function of pressure up to 10 000 atm at 300°K. The resistance decreased by 3% at 10 000 atm. The magnitude and the sign of this effect are similar to that found in n-type Si,[11] and suggest that the conduction band minima in GaP are along the [100] directions in k space. However, a part of the decrease in the resistance with pressure is perhaps due to the change in the scattering mechanism.

p-PbTe

The effect of pressure on conductivity of two p-type PbTe samples of similar carrier concentrations [R_H (300°K) = +2.5 coul^{-1} cm^3] was measured at 300° and 194°K. The results are shown on a semilog plot in Fig. 2. The pressure coefficient $d(\log \sigma)/dP$ is larger at higher temperature. Furthermore, the room temperature data on $\log \sigma$ vs P show a deviation from linearity. The Hall coefficient of one of these samples measured at 300°K decreased by about 5% at 8000 atm. The significance of these results is not yet fully understood. Recently Stiles et al.[12] have postulated a double valence band for this material. Their work indicates that in addition to the (111) valence band maxima there is another maximum at k=0; the energy separation between these bands is about 0.001 ev.

The observed pressure coefficient of conductivity in p-PbTe is due to (1) redistribution of carriers and (2) the change of the mobilities ($e\tau/m^*$) in these two bands with pressure. It is not possible at this stage to estimate these two effects separately. We feel that more data are needed on both n- and p-type PbTe before any consistent conclusions can be drawn about the band structure of this material.[13]

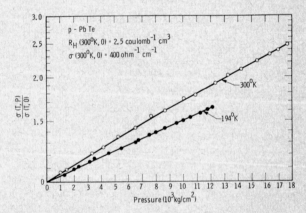

FIG. 2. Conductivity vs pressure for two p-PbTe samples with similar carrier concentrations at 300° and 194°K.

[11] C. S. Smith, Phys. Rev. **94**, 42 (1954).
[12] P. J. Stiles, E. Burnstein, and D. N. Langenberg, J. Appl. Phys. **32**, 2174 (1961).
[13] Recent work by Cuff et al. [K. F. Cuff, M. R. Ellet, and C. D. Kaglin, J. Appl. Phys. **32**, 2179 (1961)] on n-PbTe suggests

n-GaSb

Introduction

The conduction band of GaSb has been studied by various investigators[14–19] in the past few years. The accumulated data reveal that its conduction band consists of three sets of energy minima at different points in k space: (1) at the center of the Brillouin zone, (2) along [111] directions in k space and (3) along [100] directions in k space. These minima are situated at different energies as shown in Fig. 3. The properties of these bands will be denoted by subscript 0, 1, and 2, respectively. Although the existence of the (000) and the (111) minima has been established by direct experiments, the evidence for the (100) minima is indirect and is based on an analogy with the behavior of the (100) band in germanium. Zwerdling et al.[14] found from the magneto-absorption experiments that the lowest minimum of the conduction band is in the center of the Brillouin zone and obtained a value for the electron effective mass for this band $m_0{}^* = 0.047\ m_e$. The piezoresistance and Hall measurements by Sagar, et al.[15] revealed another set of energy minima along (111) directions in k space characterized by ellipsoidal constant-energy surfaces, situated above the (000) band in energy. The value of the energy separation between these two bands was estimated to be $\Delta E_{01} \approx 0.075$ ev. Keyes and Pollak[16] studied the effect of pressure on the three piezoresistance coefficients π_{44}, $(\pi_{11} - \pi_{12})$, and $(\pi_{11} + 2\pi_{12})$. They estimated the value of π_{44} for the (111) band for some highly doped samples and, considering the effect of statistical degeneracy on the value of π_{44}, obtained a value for the density of states mass for this band $m_1{}^*$ (D. S.)$= 0.40\ m_e$. No estimates for the anistropy of the mass tensor could be obtained from these experiments. The value for the room temperature mobility μ_1 for the (111) band for the samples measured was estimated to be about 300 cm^2 v^{-1} sec^{-1}; the mobility values for germanium samples with similar carrier concentrations are about 1500 cm^2 v^{-1} sec^{-1}. The reason for such low values for electron mobility in GaSb is not clear. The Hall mobilities at 77°K increase anomously with increasing carrier concentrations for Te-doped samples. This may at least partly be due to the fact that the low carrier concentration samples are very highly compensated. Strauss[17]

some striking similarities in the conduction and valence band structures of all the compounds in this group (i.e., PbTe, PbSe, and PbS). Our preliminary measurements on the effect of pressure on the conductivity on n-PbSe at 300° and 194°K give the following results: (1) $d(\ln\sigma)/dP = 3.6 \times 10^{-5}$ cm^2/kg and is independent of temperature. (2) No deviations from linearity within the experimental error ($< 0.5\%$) were observed in the data logσ vs pressure for n-PbSe.

[14] S. Zwerdling, B. Lax, K. J. Button, and L. M. Roth, J. Phys. Chem. Solids **9**, 320 (1959).
[15] A. Sagar, R. W. Keyes, and M. Pollak, Bull. Am. Phys. Soc. **5**, 63 (1960); A. Sagar, Phys. Rev. **117**, 93 (1960).
[16] R. W. Keyes and M. Pollak, Phys. Rev. **118**, 1001 (1960).
[17] A. J. Strauss, Phys. Rev. **121**, 1087 (1961).
[18] M. Cardona, J. Phys. Chem. Solids **17**, 336 (1961).
[19] W. Howard and W. Paul (private communication).

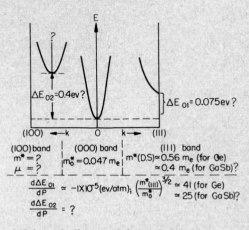

FIG. 3. Schematic diagram of the conduction band of GaSb, with estimated values for the parameters.

observed that this anomalous behavior disappears for Se-doped samples, though the starting material (p-GaSb) used was the same (i.e., had the same acceptor concentration) as that used for Te-doped samples. Thus the degree of compensation involved was the same for the Se- and Te-doped samples with similar carrier concentration. The reason for this difference in the behavior of Te-and Se-doped GaSb is not clear. Cardona[18] measured the reflectivity of a highly doped n-GaSb sample (doping agent not mentioned) at 300° and 90°K. Using Sagar's values for the parameters $m_1{}^*$, $b \equiv \mu_1/\mu_0$ and ΔE_{01}, he was able to get an excellent agreement between his calculated and experimental results at 300°K. He readjusted the value of ΔE_{01} to fit his data at 90°K. His results cannot be taken as a strong evidence for the correctness of the value of $m_1{}^*$ (D. S.)$= 0.56\ m_e$ used by Sagar, though they indicate a temperature dependence of ΔE_{01} and b. Howard and Paul[19] measured the resistance of n-GaSb samples up to 30 000 atm. They found that the resistance started to rise very steeply at about 20 000 atm, suggesting the existence of another band. Edwards and Drickamer[7] observed a red shift of the absorption edge at about 40 000 atm, suggesting the appearance of the (100) band. They estimate the value of $\Delta E_{02} \approx 0.4$ ev.

We have measured the Seebeck coefficient, Hall coefficient, and resistance of three Te-doped n-GaSb samples of different carrier concentrations as a function of pressure at 200°, 300°, and 400°K.

Results

The values of conductivity and Hall coefficient at 77° and 300°K for the samples used are given in Table I. The effect of pressure on the Seebeck coefficient α is shown in Fig. 4. The value of α increases with pressure and reaches a saturation value at about 12 000 atm. The saturation value of the Seebeck coefficient α_s vs temperature is plotted in Fig. 5. The data on α vs temperature at normal pressure is shown in Fig. 6. The

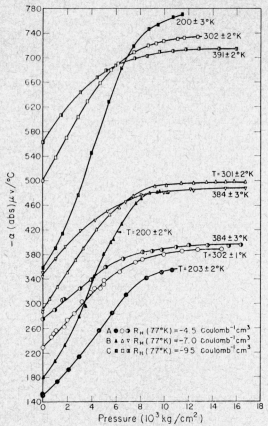

FIG. 4. The Seebeck coefficient vs pressure for three *n*-GaSb samples with different carrier concentrations at 390°, 300°, and 200°K.

effect of pressure on the resistance of these samples is shown in Fig. 7. The effect of pressure on the Hall coefficient of sample C at 300° and 194°K is shown in Fig. 8.

Discussion

Obviously it is quite futile to attempt to analyze the data on the transport properties of a three band system

TABLE I. Conductivity σ(ohm^{-1} cm^{-1}) and Hall coefficient R_H(cm^3 coul^{-1}) values for the *n*-GaSb samples at 300° and 77°K.

Sample	(300°K)		(77°K)	
	R_H	σ	R_H	σ
A	−5.1	550	−4.5	⋯
B	−10.4	250	−7.0	800
C	−163	15.9	−97.5	20.2
1-B	−64	38	−37.6	67

with most of the parameters poorly determined. The energy separation between these three bands change as a function of pressure. The value of ΔE_{01} decreases with pressure and is zero at about 8000 atms. For higher pressures the (000) band moves above the (111) band. The energy separations ΔE_{02} and ΔE_{12} also decrease with pressure, so that at high enough pressure the (100) band becomes important. (The rates of change of these ΔE's with pressure have not yet been precisely determined.) Thus at normal pressure, we are dealing with the (000) and the (111) bands. In the pressure range $\geqslant 10\,000$ atm, when almost all the carriers are transferred to the (111) band, we would be dealing with the properties of the (111) band only. In the pressure range $\geqslant 20\,000$ atm, we again have to deal with two bands, i.e., (111) and the (100). Thus, we can consider the saturation value of the Seebeck coefficient α_s in Fig. 5 as the value of α for the (111) band. We notice in Fig. 5 that α_s decreases with temperature for sample C and has a maximum for sample B. The data in Fig. 6 at normal pressure do not exhibit any such peculiarity.

A possible explanation, suggested by Pollak and Ure, for this temperature dependence of α_s for sample C could be the contribution of the phonon-drag effect. This contribution to α is given by[20] $\alpha_p = c^2 f\bar{\tau}/\mu T$, where a fraction f of the crystal momentum lost by the electrons is lost to the phonons; c is the sound velocity; $\bar{\tau}$ is the average relaxation time of the phonons which contribute to this effect; μ is the electron mobility, and T is the

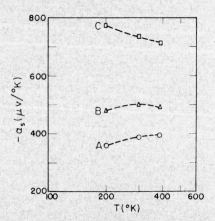

FIG. 5. The saturation value of the Seebeck coefficient vs temperature for the three *n*-GaSb samples. The saturation value of the Seebeck coefficient α_s is taken as the value of α at about 12 000 atm from Fig. 4.

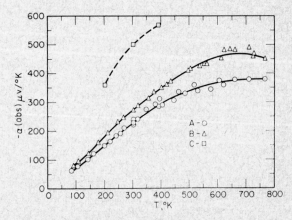

FIG. 6. The Seebeck coefficient vs temperature at normal pressure for the three *n*-GaSb samples.

[20] C. Herring, reference 8, pp. 184–235.

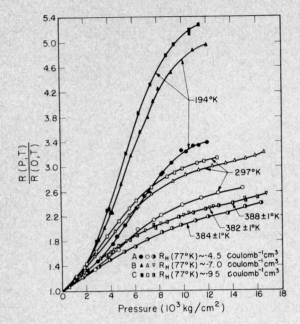

FIG. 7. Resistance vs pressure for the three n-GaSb samples at 194°, 300°, and 385°K.

temperature. It is difficult to determine α_p for the (111) band of GaSb due to lack of any estimates for $\bar{\tau}$ and f, and we will only examine the reasonableness of this explanation by comparing with the properties of n-type

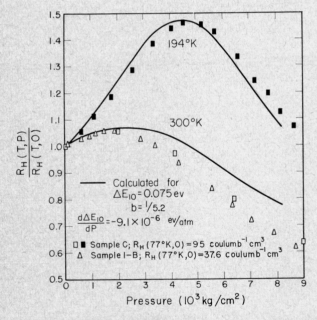

FIG. 8. Hall coefficient vs pressure for sample C at 194°C and 300°K. The solid curves are calculated using m_1*(D.S.)=0.56 m_e and assuming $\mu_H = \mu$ for both bands. The data on sample 1-B is taken from Fig. 5 of reference 15. Calculations shown in Fig. 5 of reference 15, were made assuming $\mu_{0H} = \mu_0$, but a value for $\mu_{1H}/\mu_1 = 0.76$ was used to take into account the mass anisotropy for the (111) band.[22] This correction has been omitted for the calculated curves in this figure. Classical statistics was used in the calculations.

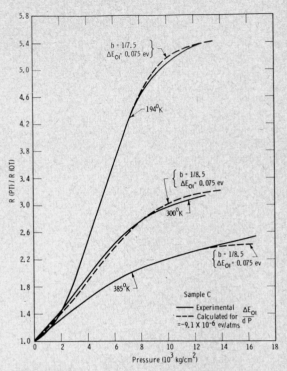

FIG. 9. Calculated and experimental curves for resistance vs pressure for sample C. Classical statistics was used in the calculations and a value of m_1*(D.S.)=0.56 m_e was assumed. The value of b was adjusted while ΔE_{01} was kept constant, to fit the data at different temperatures. Equally good agreement could also be achieved by adjusting the value of ΔE_{01} and keeping b constant with temperature.

germanium. The estimated value of the phonon-drag effect at 200°K in n-type germanium[21] with $n \simeq 10^{17}$ cm^{-3} is about 150 μv/°C. The value of μ (Ge)/$\mu^{(1)}$ (GaSb) $\gtrsim 5$. Keeping in mind that our GaSb sample C may be highly compensated, which would reduce considerably the value of $f^{(1)}$ (GaSb), we feel that a few hundred μv/°C for the value of $\alpha_p^{(1)}$ (GaSb) at 200°K is quite conceivable. The phonon-drag effect decreases with increasing temperature. Thus $(\alpha_s - \alpha_p^{(1)})$ would tend to follow the correct temperature dependence, i.e., increase with increasing temperature. The appearance of a maximum in α_s vs temperature for sample B is perhaps due to the "saturation effect" in the phonon-drag contribution which occurs in samples with high carrier concentrations.[20] The reason for the apparent absence of the phonon-drag contribution at normal pressure is not yet clear.

The calculations for the Hall coefficient and resistance vs pressure were made (as in reference 15) for sample C for which classical statistics could be used. The mobility ratio $b \equiv \mu_1/\mu_0$ was assumed to be constant with pressure. This involved neglecting (1) the change in m_0* and hence μ_0 with pressure and (2) the effect of interband scattering on μ_0. Incorporation of these effects in the calculations would have involved more unknown param-

[21] T. H. Geballe and G. W. Hull, Phys. Rev. 94, 1134 (1954).

eters and would not make the calculations any more fruitful at this stage. The calculations for the Hall coefficient vs pressure were made by assuming $\mu_H = \mu$ for both the bands.[22] The results of these calculations are shown in Figs. 8 and 9. The results in Fig. 9 indicate that if one of the parameters b or ΔE_{01} is assumed to be independent of temperature, then we have to assume that the other parameter increases with temperature. This was also noticed in Cardona's[18] results. It may be pointed out here that Sagar[15] had to assume just the opposite sign for the temperature dependence of these parameters to explain the observed temperature dependence of the piezoresistance coefficient π_{44}.

Comments

(1) We have failed to estimate the value of m_1^* (D. S.), from our Seebeck coefficient measurements due

to the appearance of the phonon-drag contribution to the Seebeck coefficient. (2) Different values of the parameters involved had to be used to fit the data on the Hall coefficient and resistance vs pressure. (3) Our experiments cannot determine the temperature dependence of b and ΔE_{01} separately.

ACKNOWLEDGMENTS

We wish to thank Dr. C. Herring for his very useful comments on the Seebeck coefficient data; and Dr. M. Pollak and Dr. R. Ure for many useful comments and discussions during the course of our work. We wish to thank Dr. A. J. Strauss and Dr. J. McHugh for providing some of the samples. We wish to express our appreciation to our machinist, Mr. J. K. Albrecht, for his excellent work on maintenance of the pressure machine. We also wish to thank Mr. E. Metz and Mrs. M. Harrison for their help in taking some of the data.

[22] C. Herring, Bell System Tech. J. **34**, 237 (1955).

JOURNAL OF APPLIED PHYSICS   SUPPLEMENT TO VOL. 32, NO. 10   OCTOBER, 1961

Band Structure Parameters Deduced from Tunneling Experiments

R. N. HALL AND J. H. RACETTE

General Electric Research Laboratory, Schenectady, New York

Measurements of the voltage and temperature dependence of tunneling in Ge and GaSb are presented which confirm the close proximity of the (000) and (111) conduction band edges in these materials. In the case of Ge, the energy separation of these edges is found to increase with increasing donor concentration.

Tunneling in the indirect semiconductor GaP shows no evidence for indirect (phonon-assisted) tunneling transitions. It is believed that tunneling in the junctions which we studied proceeds via deep-level impurities rather than between conduction and valence bands directly, thereby eliminating the requirement of wave number conservation.

Revised values for the zone-center longitudinal optical phonon energies as deduced from tunneling data in 3–5 and lead salt semiconductors are presented.

I. INTRODUCTION

ELECTRON tunneling in *p-n* junctions has been reported in a variety of different semiconductors.[1] Measurements of lightly doped junctions at low temperatures have revealed phenomena attributed to phonon excitation and to polaron annihilation.[2] The behavior of the tunneling current has also been used to measure the energy separation between two different band edges when tunneling involves transitions to more than one set of minima of the conduction band.[3]

The present work extends these observations with particular attention being devoted to a search for evidence indicating indirect tunneling in compound semiconductor junctions. While the data clearly confirm

the close proximity of the (000) and (111) conduction band edges in Ge and GaSb, similar behavior has not been found in the other 3–5 and lead salt semiconductors investigated. Particular attention has been devoted to the behavior of junctions constructed from GaP, which is believed to be an indirect semiconductor. The absence of structure indicating phonon-assisted indirect transitions or a reverse bias direct tunneling threshold is attributed to the limited solubility of donors in GaP which prevents the formation of a degenerate free-electron population in the conduction band.

II. (000) AND (111) CONDUCTION BAND EDGES IN Ge AND GaSb

Ge and GaSb are closely related semiconductors which exhibit interesting tunneling behavior due to the close proximity in energy of the (000) and (111) conduction band edges. In both semiconductors the (000) edge is characterized by a small effective mass, and

[1] R. N. Hall, *Proceedings of the International Conference on Semiconductor Physics, Prague, 1960* (Publishing House of the Czechoslovak Academy of Sciences, Prague, 1961), p. 193.

[2] R. N. Hall, J. H. Racette, and H. Ehrenreich, Phys. Rev. Letters **4**, 456 (1960).

[3] J. V. Morgan and E. O. Kane, Phys. Rev. Letters **3**, 466 (1959).

consequently the penetration factor for tunneling from these states to the light mass valence band, which is also located at the center of the Brillouin zone, is large. In Ge, conduction band states are found at somewhat lower energies at the zone edge in the (111) direction, whereas the situation is believed to be reversed in the case of GaSb. Tunneling between these (111) minima and the valence band involves an indirect transition (as contrasted with a direct transition in which no change in wave number occurs) such as a scattering process or the emission or absorption of a phonon to account for the change in wave number between the initial and final states. Other things being equal, the penetration factor for indirect tunneling is considerably smaller than that for direct tunneling.

Tunneling in Ge usually involves indirect transitions. Morgan and Kane showed,[3] however, that direct tunneling from the valence band to the (000) minimum can occur for reverse bias voltages V greater than $V_t = E_{000} - E_{111} - \zeta_n$, where $E_{000} - E_{111}$ is the energy difference between the two band edges and ζ_n is the Fermi energy, the location of the Fermi surface relative to the (111) edge. Because of the greater penetration factor for direct tunneling, a distinct increase in slope is observed in the reverse characteristic for voltages greater than this threshold. Morgan and Kane concluded from their observations that $E_{000} - E_{111}$ was distinctly less than the value, 0.154 ev, determined by Zwerdling et al.[4] The data which we report here, however, lead to values which are in good agreement with the 0.154-ev value.

Diodes were constructed by alloying Al, In, or In-Ga mixtures to wafers of Ge containing measured concentrations of P, As, or Sb. The direct tunneling threshold is clearly resolved in the case of Sb-doped junctions, but is difficult to determine from the current-

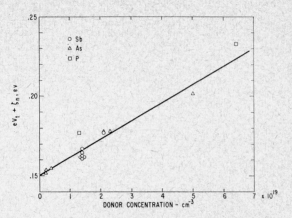

FIG. 2. Band edge separation for Ge as a function of donor concentration. The apparent concentration dependence may be partly spurious (see text).

voltage characteristic itself when the donor impurity is As or P. Nevertheless, it is clearly revealed by conductance-voltage plots[5] of the reverse characteristic, as shown in Fig. 1.

According to theory,[3] the conductance due to direct tunneling should increase linearly with voltage for $V - V_t \ll \bar{E}_\perp$ and become constant when $V - V_t \gg \bar{E}_\perp$, where $\bar{E}_\perp = \hbar F / 2\pi m_r^{\frac{1}{2}} E_{000}^{\frac{1}{2}}$. F is the "average" force on the electron produced by the junction field, m_r is the reduced mass ($0.02m$ for Ge), and E_{000} is the direct band gap. Estimated values for $\bar{E}_\perp$ range from 0.012 ev for a lightly doped junction ($N_d = 2 \times 10^{18}$ cm^{-3}) to 0.06 ev for a heavily doped junction ($N_d = 4 \times 10^{19}$ cm^{-3}). Our data clearly show the expected linearly increasing conductance beyond V_t. Values for $E_{000} - E_{111} = V_t + \zeta_n$ were calculated assuming a density-of-states mass of $0.55m$, and these are plotted as a function of donor concentration in Fig. 2. The data indicate an appreciable concentration dependence, which can be interpreted to mean that the (111) minima are displaced downward by an amount roughly equal to $0.5\zeta_n$. This concentration dependence may be somewhat exaggerated, however, since the junctions having the two highest donor concentrations were diffusion-broadened in order to reduce the conductance to values that could be measured with our equipment. For the lightly doped junctions, the ζ_n correction is small, and these data confirm the value deduced from IMO measurements.[4]

In n-type GaSb most of the electron population collects in the (000) minimum at low temperatures. At higher temperatures (room temperature and above for the impurity concentrations of concern here) most of these electrons shift to the higher (111) minima because of their much greater density of states. Because of the strong dependence of the tunneling probability upon effective mass, only those left in the (000) minimum can contribute appreciably to the tunneling current and the tunneling conductance shows a corresponding temperature dependence.

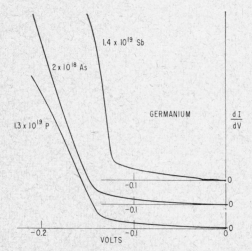

FIG. 1. Conductance as a function of reverse voltage of Ge junctions at 4.2°K.

[4] S. Zwerdling, B. Lax, L. M. Roth, and K. J. Button, Phys. Rev. 114, 80 (1959).

[5] J. J. Tiemann, Rev. Sci. Instr. 32, 1093 (1961).

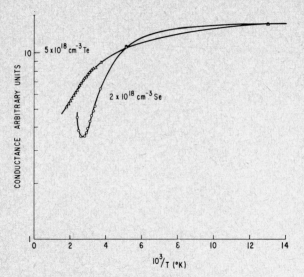

FIG. 3. Conductance at zero voltage vs temperature for GaSb junctions containing the indicated donor concentrations.

Figure 3 shows the temperature dependence of the conductance at zero voltage of two GaSb tunnel junctions formed by alloying Sn+1%Ge to n-type GaSb. The donor concentrations in the substrates were determined from Hall effect measurements, using Sagar's two-band model for the conduction band.[6] The change in conductance with temperature is clearly evident, and is generally in accord with the expected population shift. (The rise in conductance at high temperature exhibited by the lightly doped junction is due to a temperature dependent component of the excess current.) For lightly doped material such that Boltzmann statistics are valid, the junction conductance has the form

$$G_0/G = 1 + r \exp[-(E_{111} - E_{000})/kT],$$

where G is the junction conductance, G_0 is its value at $T=0$, and $r(r \gg 1)$ is the ratio of the densities of states for the two bands. The junctions shown in Fig. 3 are too heavily doped for the assumptions underlying this equation to be valid. Nevertheless, the more lightly doped junction has a maximum slope of 0.07 ev, which is only slightly less than Sagar's value of 0.08 ev for the band edge separation. The heavily doped junction, being more strongly degenerate, exhibits a smaller temperature dependence. These experiments provide further confirmation for the two-band model of n-type GaSb.

III. INVESTIGATION OF TUNNELING IN GaP

A wide variety of experimental evidence indicates that all of the smaller band gap 3-5 compounds are direct transition semiconductors, with the conduction and valence band edges located near the center of the Brillouin zone.[7] AlSb and GaP, however, appear to be indirect, with conduction band minima located in the

(100) directions as in the case of Si. Attention was therefore directed toward the construction of tunnel junctions from these two materials in the hope of observing structure indicative of phonon assisted indirect transitions analogous to those observed in Ge and Si[1] or a reverse bias threshold for direct tunneling as in Ge.[3]

In the case of AlSb we have been unable to prepare n-type crystals containing free electron concentrations greater than 3×10^{18} cm^{-3} and junctions which we prepared from them had tunneling currents that were much too small for these experiments. Higher electron and hole concentrations can be obtained in GaP, however, and p-n junctions were constructed which exhibit polaron and zone-center optical phonon structure similar to that of the direct transition 3-5 semiconductors.[1] Most of these junctions were prepared using Ga or Ag containing additions of S, Se, and Te as alloying materials to Zn-diffused GaP. Their electrical characteristics resembled those of a "backward diode," as illustrated in Fig. 4, being more highly conducting in the reverse than in the forward bias region. Junctions constructed using other alloys, including pure Ga, gave similar characteristics. A significant feature of these junctions is that the forward current rises rapidly beyond 0.2 v, whereas this rise should have occurred at a value approximately equal to the energy gap of the crystal.

The conductance minimum at the origin and the threshold for optical phonon excitation are evidence that the conductance of these junctions is largely due to electron tunneling through a thin space-charge layer. Furthermore, junction capacities of 2 μf/cm² and greater have been measured at zero bias for these junctions, indicating space-charge layer thicknesses of less than 50 A through which appreciable tunneling may reasonably be expected. However, none of the junctions exhibited structure in their electrical behavior which could be regarded as confirmation of the indirect nature of the band structure of GaP. We suggest that the tunneling current which is observed in these junctions involves transitions to deep-lying localized states or

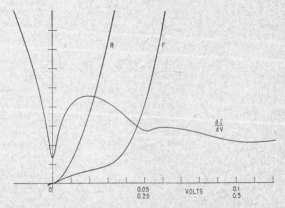

FIG. 4. Electrical characteristics of GaP junction at 4.2°K. F and R are forward and reverse current, respectively (same arbitrary scale), vs voltage (0.05 v/div). dI/dV is the conductance vs voltage (0.01 v/div).

[6] A. Sagar, Phys. Rev. 117, 93 (1960).
[7] H. Ehrenreich, J. Appl. Phys. 32, 2155 (1961).

impurity bands on the n-type side of the junction. Such states do not reflect the symmetry properties of the conduction band, and consequently the requirement of wave number conservation does not apply. The fact that large forward currents are always observed in these heavily doped junctions at only a few tenths of a volt bias indicates that such states are indeed present. We conclude that the absence of structure characteristic of indirect tunneling can only be regarded as evidence that the band edges occur at the same point in k space if the forward current rises to large values (compared with the tunnel current) at voltages that are reasonably close to the band gap of the semiconductor.

IV. TUNNELING IN LEAD SALT SEMICONDUCTORS

Rectifying junctions which exhibit tunneling have been prepared by alloying In, Ga, and Pb-In mixtures to p-type crystals of PbS, PbSe, and PbTe grown by sublimation. Rectification in these junctions appears normal and indicates that they are reasonably free of states within the energy gap. At room temperature they are more highly conducting in the forward than in the reverse direction, but as the temperature is decreased a forward drop develops due to the decreasing concentration of minority carriers, and eventually becomes comparable with the energy gap. At the same time, the reverse tunneling current increases in a manner consistent with the decreasing energy gap of the material. Tunneling is clearly evident at small forward voltages in all of these junctions, and in PbTe is sufficient to give rise to negative resistance regions with peak-to-valley ratios as high as 1.5 at 77°K.

Conductance-voltage recordings obtained at 4.2°K are reproduced in Fig. 5 for representative junctions made from each of these semiconductors. In each material, the polaron minimum at $V=0$ and the optical phonon threshold are observed. The slight overshoot near $V=0$ which appears in the curves for the PbSe and PbTe junctions is sometimes much more pronounced, but can be largely eliminated by electrolytic etching. For this reason, it is felt that it is a surface phenomenon rather than an intrinsic property of the junction. Approximately ten junctions of each material have been studied, and none of them has shown evidence for phonon-assisted indirect tunneling or for a reverse bias direct tunneling threshold. In the few cases where the latter phenomenon was thought to be observed in PbS, the effect was traced to a slightly rectifying ohmic contact, and was eliminated by extending the area of this contact by adding Ga with an ultrasonic tool. These experiments confirm, therefore, that these compounds are direct semiconductors in the sense that their conduction and valence band edges occur at the same points in k space.

The zone-center longitudinal optical phonon couples strongly to the tunneling electrons, and gives rise to an abrupt conductance increase from which its energy may

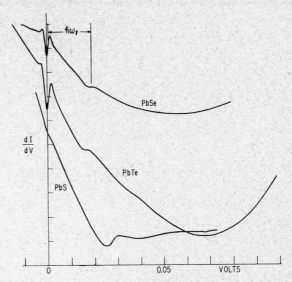

FIG. 5. Conductance vs voltage of PbS, PbSe, and PbTe junctions, measured at 4.2°K.

be determined. This energy, $\hbar\omega_l$, is taken to be the voltage difference between the points of greatest positive slope of the conductance characteristic, as indicated in Fig. 5. Phonon energies deduced from each group of diodes were 26.3 ± 0.4, 16.5 ± 0.6 and 13.6 ± 0.4 mev for PbS, PbSe, and PbTe, respectively.

V. PHONON ENERGIES IN 3–5 SEMICONDUCTORS

Values for the energy of the zone-center longitudinal optical phonon in a number of 3–5 semiconductors as deduced from tunneling data have been reported.[2] These values have been revised slightly on the basis of more careful and extensive measurements, and are summarized as follows:

Compound	InSb	InAs	InP	GaSb	GaAs	GaP
$\hbar\omega_l$ (mev)	23.8	30.1	45.0	29.2	34.9	51.0
	±0.4	±0.3	±1.5	±0.4	±0.5	±1.5

As in the previous section, these values correspond to the separation between the points of most positive slope of the conductance characteristic. This method of evaluation is to some extent arbitrary and may give rise to a small systematic error. These results are in reasonable agreement with recent optical data on InSb[8] and GaP[9] which give 24.4 ± 0.1 and 49.9 ± 1.0 mev, respectively. However, in the case of GaAs our value is appreciably smaller than that reported by Fray et al.[10] 36.4 mev.

ACKNOWLEDGMENT

The work reported in this paper was supported by the Electronics Research Directorate of the Air Force Cambridge Research Center, Air Research and Development Command, under contract.

[8] W. Engeler, H. Levinstein, and C. Stannard, Jr., Phys. Rev. Letters 7, 62 (1961).
[9] D. A. Kleinman and W. G. Spitzer, Phys. Rev. 118, 110 (1960).
[10] S. J. Fray, F. A. Johnson, J. E. Quarrington, and N. Williams, Proc. Phys. Soc. (London) 77, 215 (1961).

JOURNAL OF APPLIED PHYSICS   SUPPLEMENT TO VOL. 32, NO. 10   OCTOBER, 1961

Band Structure of the Intermetallic Semiconductors from Pressure Experiments*

WILLIAM PAUL

Division of Engineering and Applied Physics, Harvard University, Cambridge, Massachusetts

Three types of conduction band extrema in the (000), (100), and (111) directions in k space seem to determine many of the properties of the group 4 and group 3–5 semiconductors. Early experimental work on the pressure coefficients of the energy separations of these extrema from the valence band maximum energy, carried out on Ge (111), (000), (100), Si (100), and InSb (000), suggested that the three pressure coefficients might be independent of the specific element or compound in the group 4 and group 3–5 series. This work is discussed in detail, and the theoretical basis is briefly considered. All of the completed pressure measurements on these compounds are critically reviewed, and the correlation of unique pressure coefficients with specific band edges examined. It is demonstrated that pressure experiments can be planned to show up details of the band structure unavailable for study at atmospheric pressure. Particular attention is paid to GaP, and a new model for excess absorption occurring in n-type samples of this compound and in Si, GaAs, and AlSb is suggested. The application of similar techniques to PbS, PbSe, and PbTe is discussed, and results of electrical and optical measurements of energy gap and electron and hole mobilities presented.

I. INTRODUCTION

THE earliest experiments on the effect of pressure on the electrical resistivity of Ge[1a] demonstrated that the minimum energy gap between conduction and valence bands increased linearly at low pressures; later work[2] was interpreted to mean that at higher pressures this linear change decreased and possibly reversed sign, due to the increasing importance of some new set of states having a different pressure coefficient from those important at low pressures. In Si, pressure decreased the minimum energy gap.[3a] After theoretical and experimental work had identified the conduction and valence band "edges" in Ge and Si, investigation of Ge-Si alloys, interpreted by Herman, plausibly identified this new set of Ge states as a (100) conduction band set, possessing properties similar to the (100) set forming the conduction band extrema in Si.[4] Quantitative studies showed that the energy gap between these (100) extrema and the (000) valence band extremum decreased at about the same rate in Ge

and Si.[5] This observation, emphasized by the unexpected difference in sign of the effect for the (111) and (100) band edges[6] in Ge and Si, naturally led to some speculation regarding the uniqueness of the association of pressure coefficient and type of band edge in the group 4 semiconductors. Subsequent experiments that qualitatively supported this association broadened the extent of the speculation to include the group 3–5 and group 2–6 compounds and led to a cautious use of the correlation that does exist in the investigation of these materials.[7] It is the purpose of this paper to examine the sources of this speculation, the extent of its success, and its prospects for continued usefulness.

II. BASIS OF SPECULATIVE ASSOCIATION OF BAND EDGE PRESSURE COEFFICIENTS

The basis for any association of a unique pressure coefficient with the energy gap between the common valence band maximum and a particular type of conduction band minimum is primarily experimental. Since lattice dilatation would seem to be a more fundamental parameter than pressure, we shall on occasion quote experimental results in terms of either pressure or dilatation, or both. Theoretical arguments are naturally based on dilatation, while greater correlation exists among the pressure coefficients. The lattice compressibilities are different enough so that the one correlation does not imply the other.

* This work was supported by the Office of Naval Research.

[1] (a) P. W. Bridgman, Proc. Am. Acad. Arts Sci. **79**, 129 (1951); P. H. Miller and J. H. Taylor, Phys. Rev. **76**, 179 (1949); J. H. Taylor, Phys. Rev. **80**, 919 (1950); H. H. Hall, J. Bardeen, and G. L. Pearson, Phys. Rev. **84**, 129 (1951). (b) D. M. Warschauer, W. Paul, and H. Brooks, Phys. Rev. **98**, 1193 (1955); H. Y. Fan, M. L. Shepherd, and W. G. Spitzer, *Photoconductivity Conference at Atlantic City*, edited by R. G. Breckenridge, B. R. Russell, and E. E. Hahn (John Wiley & Sons, Inc., New York, 1956); W. Paul and D. M. Warschauer, J. Phys. Chem. Solids **5**, 89 (1958); A. Michels, J. van Eck, S. Machlup, and C. A. ten Seldam, J. Phys. Chem. Solids **10**, 12 (1959).

[2] W. Paul, Phys. Rev. **90**, 336 (1953); W. Paul and H. Brooks, Phys. Rev. **94**, 1128 (1954).

[3] (a) W. Paul and G. L. Pearson, Phys. Rev. **98**, 1755 (1955). (b) M. I. Nathan and W. Paul, Bull. Am. Phys. Soc. **2**, 134 (1957); W. Paul and D. M. Warschauer, J. Phys. Chem. Solids **5**, 102 (1958); H. Y. Fan, M. L. Shepherd, and W. G. Spitzer, *Photoconductivity Conference at Atlantic City*, edited by R. G. Breckenridge, B. R. Russell, and E. E. Hahn (John Wiley & Sons, Inc., New York, 1956); L. J. Neuringer, Phys. Rev. **113**, 1495 (1959); T. E. Slykhouse and H. G. Drickamer, J. Phys. Chem. Solids **7**, 210 (1958).

[4] F. Herman, M. Glicksman, and R. H. Parmenter, *Progress in Semiconductors*, edited by A. F. Gibson, P. Aigrain, and R. E. Burgess (John Wiley & Sons, Inc., New York, 1957), Vol. 2.

[5] M. I. Nathan, thesis, Harvard University (1958); Report HP1 (1958); M. I. Nathan, W. Paul, and H. Brooks, Phys. Rev. (to be published); see references 2 and 3.

[6] We shall frequently refer to a coefficient pertaining to the energy separation between the valence band and a band extremum X as the coefficient of the extremum X. However, this does not imply that we have any information regarding the absolute value or sign of the coefficient for the extremum X relative to the energy of an electron at infinity. This has particularly to be kept in mind when discussing "different" signs of effects for different band edges." In some cases, notably germanium, the actual deformation potential for a particular band edge can be found.

[7] A. L. Edwards, T. E. Slykhouse, and H. G. Drickamer, J. Phys. Chem. Solids **11**, 140 (1959); see reference 35.

TABLE I. Properties of group 4 and group 3–5 compounds. Compounds containing boron, nitrogen, thallium, and bismuth are not included. Conduction band minima are labelled "speculative" ("spec.") if the type is a systematic extrapolation or based on a pressure coefficient, and are unlabelled if the type is considered assured through measurement of cyclotron resonance, optical absorption, effective mass, etc. Temperatures of energy gaps are mixed; the room temperature gaps more easily allow comparisons where higher minima are present. Little attempt has been made to obtain the very latest values of the parameters in columns 1–3. On the other hand, columns 4–5 represent our best present assessment of the pressure coefficients. Numbers in parenthesis are references.

Compound	Lattice constant (A) (25°C)	Energy gap (ev)	Conduction band minima	$(dE_g/dP)_T$ (ev/kg cm^{-2})	$(dE_g/d \ln V)_T$ (ev)
C	3.567	5.3(300°K)	Δ_1(spec.)	$<10^{-6}$(31)	...
Si	5.43	1.21(0°K)	Δ_1	$+1.5\times10^{-6}$(3)	$+1.5$
		0.66(300°K)	L_1	5×10^{-6}(1)(2)	-3.8
Ge	5.66	0.803(300°K)	Γ_2'	12×10^{-6}(23)(25)(27)	-9
		0.85(300°K)	Δ_1	0 to -2×10^{-6}(19)(20)(21)	0 to $+1.5$
Sn	6.489	0.08(0°K)	L_1(spec.)	5×10^{-6}(31)	...
AlP	5.47	3.1(300°K)	Δ_1(spec.)	...	...
AlAs	5.66	2.16(300°K)	Δ_1(spec.)		...
AlSb	6.10	1.6(300°K)	Δ_1(spec.)	-1.6×10^{-6}(35)	...
GaP	5.47	2.2(300°K)	Δ_1(spec.)	$\begin{cases} -1.7\times10^{-6}(38) \\ -1.8\times10^{-6}(39) \end{cases}$	
		2.6(300°K)	Γ_1(spec.)	...	
GaAs	5.66	1.53(0°K)	Γ_1	$\begin{cases} 9.4\times10^{-6}(7) \\ 12\times10^{-6}(45) \end{cases}$	-7 / -9
		1.89(0°K)	Δ_1(spec.)	negative	...
GaSb	6.10	0.81(0°K)	Γ_1	$\begin{cases} 16\times10^{-6}(49) \\ 12\times10^{-6}(35) \end{cases}$	-9 / -6.75
		...	L_1	$\sim5\times10^{-6}$(47)	-2.8
		...	Δ_1(spec.)	negative (35)	...
InP	5.9	1.34(0°K)	Γ_1	4.6×10^{-6}(35)	-6.15
		...	Δ_1(spec.)	-10×10^{-6}(35)	$+7.45$
InAs	6.07	0.36(300°K)	Γ_1	$\begin{cases} 5.5\times10^{-6}(58) \\ 8.5\times10^{-6}(49) \\ 4.8\times10^{-6}(35) \end{cases}$	-3.3 / -5.1 / -2.9
InSb	6.49	0.2357(0°K)	Γ_1	$\begin{cases} 15.5\times10^{-6}(29) \\ 14.2\times10^{-6}(28) \end{cases}$	-6.7 / -6.1

Systematic Trends

Since the band structures of the 2–6 compounds are less well determined, we confine our attention, for the moment, to the group 4 and group 3–5 materials. They are listed in Table I, along with selected data about them. We note that the minimum energy gap decreases as the average atomic number increases. The minimum gap is greater in a compound than in its isoelectronic group 4 element. The valence band structure is similar in all of the compounds, and will not further concern us. On the other hand, the lowest identified states in the conduction bands are of three types[8]: (1) at the (000) position in the Brillouin zone (Γ_2' or Γ_1), (2) along the (100) directions (Δ_1), (3) along the (111) directions (L_1). There appears to be some systematic trend of the relative energies of these three minima with average atomic number. Thus the Δ_1 states are lowest in Si, are probably lowest in GaP, and perhaps also for diamond. The L_1 states are lowest in Ge, where the atomic number is higher, and they appear to be close to the extreme position in GaSb and in gray Sn. The Γ_1 minimum is lowest in InSb and tends to be low for compounds of high average atomic number. The variation is not entirely systematic, but if we knew

how to account correctly for electronic energy changes due to changes in ionicity (for want of a better term), it might well become so.

Alloy Studies

Where such examination is possible, alloys of group members show intermediate properties. However, strikingly nonlinear effects occur when the changeover in properties involves a change in conduction band extrema. An early example of this arose from the study of the optical energy gap in Si-Ge alloys.[9] From 0 to about 15% Si the lowest conduction band minimum is of the L_1 type, while from 15–100% Si, the Δ_1 states form the extrema. Measurements of magnetoresistance[9] have confirmed this interpretation, and have also shown that the mass ratio (and quite probably the masses) in the L_1 and Δ_1 minima in the alloys are very close to their values in the pure substances. Although changes in lattice constant accompany changes in alloy composition, this is not the prime cause of the change in energy gap; thus, for example, the energy gap changes by about 0.15 ev between 0 and 10% Si content, whereas the gap change that would result from the change in lattice constant by this composition is only 0.05 ev. For the (100) minima, a decrease in the lattice constant through alloying increases the appropriate

---

[8] The states Γ_2' and Γ_{25}' in the diamond lattice become Γ_1 and Γ_{15}, respectively, in the zinc blende. The Δ_1 minima may shift to the Brillouin zone edge point X_1, but for ease in writing, we shall refer to Δ_1 states only.

[9] See reference 4.

energy gap, whereas a decrease caused by pressure decreases it. Extrapolation from its variation in energy in Si-rich alloys is clearly useful in fixing the energy of the Δ_1 minima in pure Ge. A thorough study of the effects of alloying on all possible alloy systems of the group might establish a systematic behavior of the different minima with lattice constant and ionicity, and in fact, this seems a logical consequence of any systematic variation among the compounds themselves.[10] This systematic behavior, however, does not have to extend as far as a relative constancy of energy shift with dilatation for any one type of extremum, independent of substance studied, which is the question we are addressing ourselves to here.

Many of the properties of the Γ_2' or Γ_1, Δ_1, and L_1 extrema are similar in the different materials. The similarity in properties that depend only on the symmetry of the states is trivial. Less obvious are similarities in effective mass, yet the Δ_1 states seem to maintain the same mass ratio (and thus probably mass) between 100 and 15% Si in Si-Ge alloys, and analysis of pressure results in Ge involving higher minima, that are most probably of a Δ_1 type, requires an effective mass very similar to that of Si. The masses in all Γ_2' or Γ_1 states are small, and the matrix element between the light hole states and the Γ_2' or Γ_1 states appears to be constant for all members of the group,[11] so that, neglecting spin orbit interaction, the mass varies almost directly with the energy gap at $k=0$.

Bond Theory

Various authors have discussed the systematic variations of energy gap, carrier mobility, melting point, and hardness with lattice constant, average atomic number, and degree of ionicity.[12] The basis for these discussions has been the connection between bond strengths, bond lengths, and atomic constitution. Extrapolations from such systematic variations have been rather successful, especially in predicting new semiconductors, but the theory used does not give the details of band structure and band interaction, and fails to explain the changes in effective mass and carrier mobility associated with changes in the extrema of the conduction band. To choose one example, the violent effect of pressure on the conductivity of extrinsic GaAs,[13] which depends probably on a change of band extrema, would be inexplicable in this "bond" theory. However, it is not our purpose to examine critically the explanations offered for this systematic behavior, but simply to recognize that systematic variations do exist,

and that the explanation of some of them may be related to the pressure coefficient correlations that are the subject of this paper.

We can examine the predictions of a bond theory for the pressure coefficient of an energy gap. Decrease in lattice constant implies decreased bond length and so increased bond strength. Since the energy gap is the difference in energy between a bonding and an anti-bonding configuration, this implies an increased energy gap; the increase in energy gap in the Sn, Ge, Si, C sequence is consistent with this argument. In practice, a small decrease of the lattice constant produced by pressure may increase or decrease the energy gap, as is indicated by the opposite effects in Ge and Si. However, in both of these cases, the dielectric constant decreases with decreases in lattice constant,[14] from which we infer that there is a general tendency for the separation between conduction and valence bands to increase. Thus the predictions of the bond theory probably correlate with the behavior of the average separation of conduction and valence bands, but do not describe well the behavior (including pressure behavior) of the extrema of the bands, on which many of the critical properties depend.

Band Theory

Apart from some work of Parmenter on dilatational and alloying effects in Ge and Si,[15] the author knows of no attempt, from a band theory viewpoint, to estimate the perturbation of band structure caused by changes in lattice constant. It would appear, nevertheless, that satisfaction of the known experimental behavior of different band edges with dilatation is not an unfair test for any theory pretending to compute the band structure *ab initio*. It has been argued that the valence and conduction band electrons in our group of compounds move in an almost-free electron potential caused by the cancellation of the attractive periodic potential of the cores by a repulsive potential; this repulsive potential is introduced to simulate the result of orthogonalizing the valence electron wave function to the core wave function.[16] To the extent that this reduces the importance of the details of the charge structure of the core and yields a band structure only slightly perturbed from that of the empty lattice, we should expect similar band structures in all of the compounds. In terms of the absolute energies, the fluctuations in energy of the Γ_2', Γ_1, Δ_1, and L_1 minima are relatively small. The differences between the empty lattice energies and those in the actual lattice depend on the symmetry of the state considered. In general, one expects that states of s character will be raised in

[10] Solubility differences might thwart this study in specific cases.

[11] See references 43 and 53.

[12] See, for example, H. Welker and H. Weiss, *Advances in Solid State Physics*, edited by F. Seitz and D. Turnbull (Academic Press, Inc., New York, 1956), Vol. 3; O. Folberth and H. Welker, J. Phys. Chem. Solids **8**, 14 (1959).

[13] W. E. Howard and W. Paul (to be published); see reference 43.

[14] M. Cardona, W. Paul, and H. Brooks, *Solid State Physics in Electronics and Telecommunications*, edited by M. Désirant and J. L. Michiels (Academic Press, Inc., New York, 1960), Vol. 1, p. 206.

[15] R. H. Parmenter, Phys. Rev. **99**, 1759 (1955).

[16] J. C. Phillips and L. Kleinman, Phys. Rev. **116**, 287 (1959).

energy relative to those of p character, due to the action of the pseudorepulsive potential.

Presumably, the larger interatomic separation of the compounds of higher atomic number can be regarded as part of a scaling operation involving the mean radius of the host ion. It is not clear what the effect of a change in lattice constant on the energies of the different states will be, but it is very probable that it will be different for states of different symmetries and it is not implausible that it will be about the same for states of the same symmetry, especially if the constitutions of the cores are nearly the same. While this is encouraging, it is not apparent, even on a qualitative basis, what the relative changes will be. We might expect to obtain some clues by observing the changes in energy of the different states as we change from the empty lattice to the real one,[17] as this is qualitatively similar to the effect of pressure in increasing the ratio of the size of host ion to unit cell (we regard the host ion core as relatively incompressible). Then we find that in Ge, the behavior of the Γ_2', L_1, and Δ_1 state energies with respect to the Γ_{25}' state is not well reproduced; for example, the Δ_1 state energy increases rapidly over the Γ_{25}', opposite to the effect of decreasing the lattice constant.

Similarities in the dilatational (or pressure) coefficients for any particular set of minima may depend partly on their occupying similar positions in the overall band structure. This is assured by the nature of the experiments performed, which usually examine minima fairly close to the absolute maximum of the valence band. The minima might have a different pressure coefficient if that could be measured when the minima were far from being the lowest conduction band states. The latter type of measurement has not been carried out but, in certain specific cases, it seems that it is feasible (see Sec. IV).

These qualitative arguments are not, of course, the starting point of the present article, and are little more than an attempt to rationalize the experimental situation. Our conclusion is that there is no clear-cut reason why wide correlations of behavior under pressure of similar minima in the compounds should exist, even though experimentally it appears that the different types of minima can be identified by their pressure coefficient. We examine next the experimental situation.

Experimental Basis

Experiments carried out by Bridgman[1] established that the conductivity of n-type Ge decreased rapidly at pressures above $12\,000$ kg/cm^2 and went through a minimum near $50\,000$ kg/cm^2. Later work by Paul and Brooks[2] on pure and impure material showed that this effect was probably due to the growing importance of a second set of minima in the conduction band. The investigation of Ge-Si alloys,[4] interpreted by Herman,

suggested that it was very plausible that the new states needed were a Δ_1 set, which also formed the conduction band edge in Si. Smith's experiments[18] on the drift mobility of electrons at pressures up to $30\,000$ kg/cm^2 established that the observed conductivity decrease was a mobility effect, not a carrier density variation. The analysis of Paul and Brooks and a more complete study by Nathan[19] concluded, *inter alia*, that the pressure coefficient, relative to a fixed valence band maximum, of the Δ_1 states was between 0 and -2×10^{-6} ev/kg cm^{-2}. More recent work on magnetoconductance of n-type Ge at pressures up to $20\,000$ kg/cm^2 by Howard[20] has shown this pressure coefficient to be -1.5×10^{-6} ev/kg cm^{-2}. Optical absorption experiments of Slykhouse and Drickamer[21] to pressures of $100\,000$ kg/cm^2 have shown that the energy gap of Ge passes through a maximum with pressure near $50\,000$ kg/cm^2 and that at higher pressures the decrease[7] is at -1.2×10^{-6} ev/kg cm^{-2}.

To the author's knowledge, the symmetry of the higher states in Ge to which these pressure phenomena are attributed has never been directly established. The Harvard laboratory has been unable so far to reach pressures above $20\,000$ kg/cm^2 in nonmagnetic pressure vessels so that magnetoresistance measurements could be performed; probably the most feasible experiment would be a study of elastoresistance under pressure, following the methods described by R. W. Keyes and his collaborators.[22] However, this is very likely only a question of experimental tidiness as the Si-Ge alloy work makes it almost certain that the additional states are of the Δ_1 type. The energy of these states, found from extrapolation to 0% Si content, is roughly 0.22 ev. Nathan, Paul, and Brooks[19] find from the pressure data 0.18 ± 0.03 ev, Howard[20] 0.21 ± 0.03 ev, and Slykhouse and Drickamer[21] 0.2 ev.

The first measurements of the pressure coefficient of the $\Gamma_{25}'-\Delta_1$ gap in Si, by Paul and Pearson,[3a] gave a coefficient of -1.5×10^{-6} ev/kg cm^{-2}. Subsequent optical measurements by Paul and Warschauer[3b] gave -1.3×10^{-6} ev/kg cm^{-2}, and by Fan, Shepherd, and Spitzer[3b] $+5\times10^{-6}$ ev/kg cm^{-2}. Later Neuringer[23] found results in agreement with Paul and Warschauer, and Slykhouse and Drickamer,[21] in experiments to $140\,000$ kg/cm^2, determined a coefficient of -2×10^{-6} ev/kg cm^{-2}. Quite different measurements by Nathan and Paul,[3b] on the change in ionization energy of gold impurity in Si, determined the gap change at low

[17] F. Herman, Revs. Modern Phys. **30**, 102 (1958).

[18] A. C. Smith, thesis, Harvard University (1958); Report HP2 (1958); Bull. Am. Phys. Soc. **3**, 14 (1958).

[19] H. Brooks and W. Paul, Bull. Am. Phys. Soc. **1**, 48 (1956); Also see reference 5.

[20] W. E. Howard, thesis, Harvard University (1961); Report HP7 (1961).

[21] T. E. Slykhouse and H. G. Drickamer, J. Phys. Chem. Solids **7**, 210 (1958).

[22] R. W. Keyes, *Advances in Solid State Physics*, edited by F. Seitz and D. Turnbull (Academic Press, Inc., New York, 1960), Vol. 11.

[23] L. J. Neuringer, Phys. Rev. **113**, 1495 (1959).

pressures as -1.5×10^{-6} ev/kg cm^{-2}. It, therefore, seems to be fairly well established that the Si gap decreases with pressure, and we shall assume the rate to be -1.5×10^{-6} ev/kg cm^{-2}.

Thus we see that the behavior of the $\Gamma_{25}'-L_1$ gap and the $\Gamma_{25}'-\Delta_1$ gap in Ge is qualitatively different and that there is quite close quantitative agreement between the coefficient for the $\Gamma_{25}'-\Delta_1$ gap in Ge and the corresponding gap in Si.

The agreement persists if we convert the pressure coefficients into dilatational coefficients by computing $E_1=dE_g/d\ln V$, as we see by examining Table I.

Second Experimental Basis

The similar behavior of the Δ_1 states in Ge and Si under pressure form the first basis for our speculative scheme of pressure coefficients. The second is afforded by a comparison of the pressure coefficients of the Γ_2' minimum in Ge and the Γ_1 minimum in InSb. The pressure coefficient of the $\Gamma_{25}'-\Gamma_2'$ energy separation in Ge was first observed in results of Fan, Shepherd, and Spitzer.[1b] They found an increased pressure coefficient (over that appropriate for the $\Gamma_{25}'-L_1$ separation) for that part of the absorption edge attributed to direct transitions: Later, oscillatory magnetoabsorption experiments[24] showed that the experiments of Fan et al. were carried out at energies just below the edge. Paul and Warschauer[25] deduced a coefficient for the $\Gamma_{25}'-\Gamma_2'$ gap by assuming that the Γ_2' state was the only intermediate state in the indirect transition absorption into the L_1 extrema, and using the expressions of Bardeen, Blatt, and Hall[26] in analyzing the change of shape in their absorption edges with pressure. Neuringer[23] also reported a coefficient for this edge, also at energies slightly below that corresponding to the onset of direct transitions. Still later, Cardona and Paul[27] measured the direct transition absorption in very thin films well into the appropriate energy region. Their coefficient for the $\Gamma_{25}'-\Gamma_2'$ energy separation, measured at different absorption levels, was between 1.2 and 1.3×10^{-5} ev/kg cm^{-2} which is in substantial agreement with all of the earlier determinations despite their different shortcomings. Although the reasons for this agreement are not established, two can be considered: (1) that the absorption at energies just less than the direct gap is due to direct optical-phonon-aided transitions, and thus has the same pressure coefficient as the direct gap, (2) that the spectral resolution of the early experiments was low, so that at a nominal energy setting less than

the direct gap, some direct gap absorption spectrum was included.

Long[28] and Keyes[29] measured the pressure coefficient of the resistivity of intrinsic InSb and deduced therefrom the coefficient for the energy gap. Keyes' value, determined to the higher pressures, was 1.55×10^{-5} ev/kg cm^{-2}; Long's result was 1.42×10^{-5} ev/kg cm^{-2}.

Thus these two sets of results for the Γ_2' and Γ_1 states in Ge and InSb give coefficients two and a half times larger than that for the L_1 states in Ge, and an order of magnitude larger, and with the opposite sign, than the coefficient for the Δ_1 states. The agreement for the dilatational coefficients is however inferior (see Table I).

It therefore appears that the coefficients are grouped—so far—into three sets; the first, appropriate to the Γ_2' or Γ_1 state, of order approximately 2.5 in arbitrary units; the second, for the L_1 states,[1a,1b,2] of order 1; and the third of order -0.4. We shall not consider further the exactness of the agreement within any one set. On the one hand, the pressure coefficients have experimental errors almost sufficient to allow exact agreement. On the other, we do not need exact agreement for further progress, and so shall be content to keep a watching brief on the matter.

III. SPECULATIONS AND RESULTS

We now speculate that the pressure coefficients of the energies of the Γ_2' or Γ_1, L_1 and Δ_1 states with respect to the Γ_{25}' (or Γ_{15}) valence band maximum in all of the group 4 and group 3-5 semiconductors fall into three groups and that the coefficients are close to the values 12.5×10^{-6} ev/kg cm^{-2}, 5×10^{-6} ev/kg cm^{-2}, and -2×10^{-6} ev/kg cm^{-2}, i.e., in the ratio 2.5:1:-0.4. For ease of expression we shall refer to the $\Gamma_{25}'-\Gamma_2'$, $\Gamma_{25}'-L_1$, and $\Gamma_{25}'-\Delta_1$, energy gaps and pressure coefficients as Γ_2', L_1, and Δ_1 [or less precisely as (000), (111), and (100)] gaps and coefficients. In this section we shall briefly review the knowledge or speculation regarding the conduction band minima in the compounds, and the related pressure experiments.

Diamond

The conduction band extrema in diamond have not been clearly established. Systematic extrapolation from the Si band structure suggests either the Δ_1 conduction band state or some other energy extrema not encountered in the group 4 and group 3-5 compounds of larger atomic number. Optical absorption data of Clark[30] may be tentatively interpreted as indicating minima of Δ_1 type. Regarding the pressure coefficient, there is one rather inconclusive piece of evidence. Champion and Prior[31] report that they found no shift

[24] S. Zwerdling, B. Lax, K. J. Button, and L. M. Roth, J. Phys. Chem. Solids **9**, 320 (1959).

[25] W. Paul and D. M. Warschauer, J. Phys. Chem. Solids **5**, 89 (1958).

[26] J. Bardeen, F. J. Blatt, and L. H. Hall, *Photoconductivity Conference at Atlantic City*, edited by R. G. Breckenridge, B. R. Russell, and E. E. Hahn (John Wiley & Sons, Inc., New York, 1956).

[27] M. Cardona and W. Paul, J. Phys. Chem. Solids **17**, 138 (1960).

[28] D. Long, Phys. Rev. **99**, 388 (1955).

[29] R. W. Keyes, Phys. Rev. **99**, 490 (1955).

[30] C. D. Clark, J. Phys. Chem. Solids **8**, 481 (1955).

[31] F. C. Champion and J. R. Prior, Nature **182**, 1079 (1958).

in the absorption edge to pressures (nonhydrostatic) of about 10 000 kg/cm², although their spectrometer was capable of detecting a shift of 1 A, or a coefficient of about 2×10^{-7} ev/kg cm⁻². They also report a theoretical expectation of 10^{-8} ev/kg cm⁻² without quoting a source for their theory. Whatever the theory, and whatever the nature of the experimental stresses applied by Champion and Prior were, it would appear that the pressure coefficient is small. We note that this agrees with our systematic extrapolation to Δ_1-type minima.

Silicon

Only the Δ_1 states in the conduction band have been positively identified, and their pressure coefficient is one of the bases for our discussion.

Gray Tin

The band structure of gray tin is being actively investigated at atmospheric and elevated pressures. Magnetoresistance[32] and pressure[33] measurements both indicate complexity in the band structure, in which L_1 minima are probably involved. Paul has reported[34] a result of Groves on the variation of intrinsic resistivity with pressure to 2000 kg/cm² at $-40°C$; without mobility corrections, the energy gap is deduced to change at 5×10^{-6} ev/kg cm⁻², precisely the coefficient for the L_1 gap in Ge. Measurements at lower temperatures, however, give a temperature-dependent pressure coefficient of electron mobility which is possibly the result of a complex conduction and valence band structure.

It is the author's view that Groves' coefficient will prove to be that of an L_1 gap, and that it will therefore agree very well with the L_1 coefficient in Ge. The low electron mobility in n-type Sn and the magnetoresistance results support this contention. It is emphasized, however, that even if electrons in the L_1 states make the major contribution to the conductivity at $-40°C$, these are not necessarily the only low-lying conduction band states and they may not even be the lowest minima at this temperature or lower temperatures.

Aluminum Phosphide, Arsenide, Antimonide

Data pertinent to these relatively neglected substances are shown in Table I. For technical reasons, the conduction band extremum has not been investigated either by cyclotron resonance, magnetoresistance, or elastoconductance, all of which could potentially identify the lowest states. Systematic extrapolation from Si might suggest that the Δ_1 states are lowest in all three.

Optical absorption measurements under quasi-hydrostatic pressure conditions to 50 000 kg/cm² have recently been carried out by Edwards and Drickamer[35] on AlSb; they deduce a coefficient of -1.6×10^{-6} ev/kg cm⁻² from the shift of the frequency with pressure for a fixed (low) absorption coefficient. If the Δ_1 states are indeed lowest, this coefficient would correlate very well with our chosen one. No pressure experiments on AlP or AlAs have been reported.

Gallium Phosphide

The gallium compounds have provided a rich harvest for pressure studies. Data for GaP are listed in Table I. The lowest conduction band minimum has not been identified by any of the conventional methods although systematic extrapolation from Si suggests one of Δ_1 type. Spitzer et al.[36] have reported two optical absorption edges, the lower at 2.2 ev and apparently an indirect transition, the higher nearer 2.55 ev, and probably a direct transition. Extrapolation from absorption studies on GaAs-GaP alloys[37] suggests that the latter transition may involve a Γ_1 conduction band minimum.

Edwards, Slykhouse, and Drickamer[38] have measured the absorption spectrum to pressures of 50 000 kg/cm² and have deduced from the slope of curves of $h\nu$ versus P at constant absorption coefficient that (1) the lowest minima have a pressure coefficient of -1.7×10^{-6} ev/kg cm⁻² (2) there is probably a higher minimum which produces direct transition absorption, and whose energy gap has a positive pressure coefficient.

Thus, on the surface, these two studies qualitatively agree. However, the direct transitions in Spitzer's study supposedly occur above 2.55 ev and at absorption coefficients greater than 1000 cm⁻¹, while Edwards et al. estimate that the higher minimum is 0.1 ev above the lowest set, and that it contributes heavily to the absorption at atmospheric pressure for energies above 2.4 ev and absorption coefficients greater than 100 cm⁻¹. Severe discrepancies exist therefore in the interpretation of the edge. Our later discussion suggests that neither interpretation is well-established.

Zallen[39] has recently remeasured the GaP absorption edge over a wide energy range. At atmospheric pressure, his results agree within experimental error with those of Spitzer et al. At high pressures, he finds that the shape of the low-energy edge changes in such a way that graphs of $h\nu$ versus P at constant absorption might be interpreted to give either sign of pressure

[32] A. W. Ewald (private communication).
[33] S. H. Groves and W. Paul (unpublished measurements).
[34] Prague International Conference on Semiconductors (1960).

[35] A. L. Edwards and H. G. Drickamer, Phys. Rev. 122, 1149 (1961).
[36] W. G. Spitzer, M. Gershenzon, C. J. Frosch, and D. F. Gibbs, J. Phys. Chem. Solids 11, 339 (1959).
[37] H. Welker and H. Weiss, *Advances in Solid State Physics* edited by F. Seitz and D. Turnbull (Academic Press, Inc., New York, 1956), Vol. 3.
[38] A. L. Edwards, T. E. Slykhouse, and H. G. Drickamer, J. Phys. Chem. Solids 11, 140 (1959).
[39] R. Zallen and W. Paul (unpublished measurements).

coefficient. This is simply explained on the basis of a few reasonable assumptions. If we ignore phonon energy terms, and assume that only one intermediate state is involved (which may not be strictly accurate), we find that the absorption coefficient for parity-allowed indirect transitions is

$$\alpha^{\frac{1}{2}} = A(h\nu - E_g(P))/\Delta E(P), \qquad (1)$$

where α is the absorption coefficient, A a combination of constants, $E_g(P)$ the indirect gap, and $\Delta E(P)$ the energy separation of the lowest state from the intermediate state. Then, at fixed α, and close to $P=0$, $E_g(P)=E_g(0)$,

$$\left(\frac{\partial E_g}{\partial P}\right)_T = \left(\frac{\partial}{\partial P}(h\nu)\right)_\alpha - [h\nu - E_g(0)]\left(\frac{\partial \ln\Delta E}{\partial P}\right)_T. \qquad (2)$$

We can determine $(\partial(h\nu)/\partial P)_\alpha$ and $h\nu$ for various α, and then plot the first quantity against the second. The resultant line has slope $(\partial \ln\Delta E/\partial P)_T$ and intercept at $(\partial(h\nu)/\partial P)_\alpha = 0$ of

$$E_g - \left(\frac{\partial E_g}{\partial P}\right)_T \bigg/ \left(\frac{\partial \ln\Delta E}{\partial P}\right)_T.$$

From (2), it is clear that if $(\partial E_g/\partial P)_T$ is negative, and $(\partial \ln\Delta E/\partial P)_T$ positive, then the sign of $(\partial(h\nu)/\partial P)_\alpha$ can be either positive or negative. Carrying out the above procedure for measurements on a pure sample of GaP and assuming $E_g(0) = 2.2$ ev, Zallen finds

$$(\partial E_g/\partial P)_T = -1.8\times10^{-6} \text{ ev/kg cm}^{-2}$$

and

$$(1/\Delta E)(\partial\Delta E/\partial P) = 1.4\times10^{-5} \text{ kg}^{-1} \text{ cm}^2.$$

This pressure coefficient agrees very well with that of Edwards *et al.*, although derived by a different procedure, and is very close to that for the gap in Ge and Si. If the intermediate state is indeed a Γ_1 state, and we assign it the appropriate pressure coefficient, we shall require $\Delta E(0)\approx 1$ ev. This is quantitatively different from Spitzer's deduction; however, his experimental curves seem to us to be consistent with a second minimum at a somewhat higher energy, and our assumption of a single intermediate state of large pressure coefficient naturally leads to a high value of $\Delta E(0)$.

Spitzer *et al.* also report an absorption peak near 4μ, due to absorption in excess of the normal free carrier absorption. Similar excess absorption has been observed in Si,[40] GaAs,[41] and AlSb.[42] The explanation proposed by Spitzer *et al.* is that the absorption is due to transitions between the two conduction bands that give the

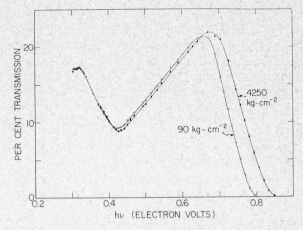

FIG. 1. Transmission curves at two pressures for a Ge-GaP sandwich, illustrating the small pressure coefficient of the transmission near 4μ. The shift of the germanium edge in this measurement is in agreement with that found in the work of Paul and Warschauer (reference 25). The magnitude of the Ge edge shift (not the energy gap change) is $\sim 9\times10^{-6}$ ev/kg cm^{-2}; if the transmission minimum in GaP were due to a $\Delta_1 \to \Gamma_1$ transition, its shift would be about 14×10^{-6} ev/kg cm^{-2} (Zallen and Paul, unpublished data).

two parts of the main absorption edge. This mechanism is favored over excitation or ionization of deep lying impurities.

To digress for a moment, at least partial resolution of the source of the absorption should be afforded by photoconductivity measurements. Photoionization should give a distinct effect, photoexcitation none, while free carrier interband transitions could yield slight photoconductance or photoresistance, depending on the mobilities in the two bands and the relaxation time between bands.

Measurement of the pressure coefficient of the two parts of the main absorption edge and of the excess absorption should unambiguously confirm whether interconduction band transitions cause the absorption. Preliminary measurements by Zallen[39] have not yet clearly established a coefficient for the higher minimum. Measurements of the transmission in the excess absorption region at pressures of 90 and 4250 kg/cm^2 are shown in Fig. 1. The shift of the "bump" with pressure is certainly at a rate smaller than 10^{-6} ev/kg cm^{-2}. If the minima involved were Δ_1 and Γ_1, and if our postulates regarding the Γ_1 state are correct, the rate might be expected to be 1.4×10^{-5} ev/kg cm^{-2}. It seems very unlikely, from Zallen's measurement, that $\Delta_1 \to \Gamma_1$ transitions cause the absorption. We shall return to this problem in Sec. IV.

Gallium Arsenide

GaAs has been intensively studied recently by Ehrenreich,[43] who has concluded that the lowest conduction band minimum is at the (000) position; presumably it is of the Γ_1 symmetry. Pressure measure-

[40] W. G. Spitzer and H. Y. Fan, Phys. Rev. **108**, 268 (1957).
[41] W. G. Spitzer and J. M. Whelan, Phys. Rev. **114**, 59 (1959).
[42] R. F. Blunt, H. P. R. Frederikse, J. H. Becker, and W. R. Hosler, Phys. Rev. **96**, 578 (1954). W. J. Turner and W. E. Reese, Phys. Rev. **117**, 1003 (1960).

[43] H. Ehrenreich, Phys. Rev. **120**, 1951 (1960).

ments by Howard[13] and Sagar,[44] Hall coefficient data, and data on GaAs-GaP alloys are analyzed to demonstrate the presence of a second set of minima, of the Δ_1 variety, roughly 0.35 ev above the Γ_1 set. Free carrier absorption supposedly involving interconduction-band transitions suggest an additional set of minima only 0.25 ev above the lowest set,[41] but as noted by Ehrenreich, the threshold for absorption into the Δ_1 minima may be reduced by 0.10 ev in highly doped samples, due to filling of the lowest states of the Γ_1 minimum.

Optical absorption measurements at high pressures by Edwards, Slykhouse, and Drickamer[7] have been interpreted to give a coefficient for the Γ_1 gap of 9.4×10^{-6} ev/kg cm^{-2}. Sagar[44] and Howard[13] have measured the extrinsic n-type conductivity as a function of pressure at room temperature. Howard's striking curve for this variation is shown in Fig. 2; the rapid resistivity increase, caused presumably by change-over of carriers from the Γ_1 minimum to a higher set, continues beyond 30 000 kg/cm^2. Ehrenreich assumed the coefficient of 9.4×10^{-6} ev/kg cm^{-2} for his analysis and deduced a coefficient of -1.4×10^{-6} ev/kg cm^{-2} for the higher set of minima.

This coefficient would be in fine agreement with the value we postulated for the Δ_1 minima. However, again there are disturbing features about the experimental situation. Unpublished measurements of Paul and Warschauer[45] gave a coefficient for the Γ_1 gap of $\sim 12 \times 10^{-6}$ ev/kg cm^{-2}. The difference between this value and that of Edwards et al. is significant for Ehrenreich's interpretation, and might radically alter his conclusions regarding identification of the higher

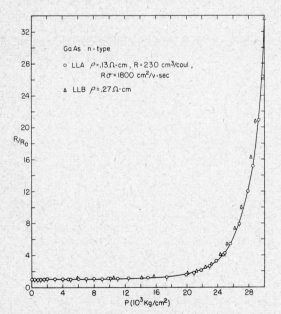

FIG. 2. Resistivity vs pressure for n-type GaAs, at room temperature (W. E. Howard and W. Paul, to be published).

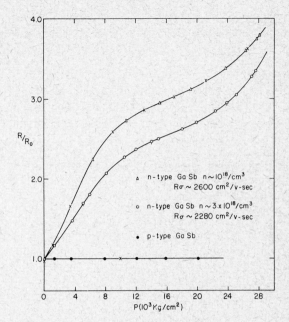

FIG. 3. Resistivity vs pressure for n-type GaSb, at room temperature (W. E. Howard and W. Paul, to be published).

minima.[43] Either coefficient, however, lies close to the value for the same minimum in Ge.

Furthermore, Edwards et al. determine a maximum gap near 60 000 kg/cm^2 whereas Ehrenreich's analysis predicts that the Γ_1 and Δ_1 minima are equal in energy near 33 000 kg/cm^2. Beyond the maximum gap, Edwards et al. determine that the energy gap decreases at a rate of -8.7×10^{-6} ev/kg cm^{-2}.

More recent quasi-hydrostatic measurements[46] to higher pressures seem to reach a plateau of resistivity a factor of 600 greater than that at atmospheric pressure.

To our mind, the situation in GaAs is not adequately explained. The coefficient for the Γ_1 minimum (whichever of the two determinations is nearer correct) agrees well enough with our postulates. However, in order to allow interpretation of the other data, this measurement should be repeated. We do not believe all of the higher energy minima have been accounted for, and suspect the positions of the Γ_{15} and the L_1 minima should also be considered.

Gallium Antimonide

Sagar has reported[47] measurements of resistivity, Hall effect, and piezoresistivity as a function of temperature and pressure. His results confirm previous deductions that the lowest conduction band minimum is in the (000) position, and presumably of Γ_1 symmetry. He finds that he can fit all his results reasonably well by assuming a second set of minima along the (111)

[44] A. Sagar, Westinghouse Research Rept. 6-40602-3-R1 (1959).
[45] W. Paul and D. M. Warschauer (unpublished measurements).

[46] W. E. Howard (unpublished measurements).
[47] A. Sagar, Phys. Rev. 117, 93 (1960); R. W. Keyes and M. Pollak, Phys. Rev. 118, 1001 (1960).

axes of the L_1 type and by postulating that the deformation potentials for the Γ_1 and L_1 states are similar to those for germanium. Sagar's model has been confirmed in part by Keyes and Pollak.[47]

Since Dr. Sagar's paper in this volume analyzes the present situation regarding GaSb in detail, we need not go into it here. Special attention should be drawn, however, to the unpublished measurement of Howard,[48] reproduced in Fig. 3, which shows that the resistivity increases sharply at pressures near 25 000 kg/cm². We suppose that a third set of minima is beginning to contribute; these states have not been positively identified, nor their pressure coefficient estimated from the electrical measurements.

Edwards and Drickamer[35] have reported measurements of the change in the optical absorption spectrum with pressure. For pressures up to 18 000 kg/cm², they find a shift of the frequency for fixed low absorption coefficient of 12×10^{-6} ev/kg cm^{-2}, which confirms their previous measurement,[38] and qualitatively agrees with earlier data of Taylor[49] on the optical absorption coefficient, which gave 15.7×10^{-6} ev/kg cm^{-2}. Thus the pressure coefficient for the Γ_1 minimum is in the range of values we postulated.

Edwards and Drickamer also report a change in slope of their curve of frequency versus pressure (at fixed absorption coefficient) which they attribute to a change in the extrema from the Γ_1 type to extrema in the (111) direction, approximately 0.09 ev higher in energy at atmospheric pressure, and with a pressure coefficient of 7.3×10^{-6} ev/kg cm^{-2}. At 45 000 kg/cm² the pressure coefficient passes through a maximum, and then decreases, indicating the existence of a third set of minima. These observations are generally consistent with the experiments of Sagar and of Howard. We shall not, however, use their coefficient quoted for the (111) minima as we feel sure that it is hard to obtain from the published data, and moreover, in cases where indirect optical transitions are involved, shape changes of the absorption edge tend to give misleading coefficients in $(h\nu)_\alpha$ versus P plots. The shape change we expect in this case will depend on $(E_c{}^{000}-E_c{}^{111})^{-2}$ where $E_c{}^{000}$ and $E_c{}^{111}$ are the energies of the (000) and (111) minima. This quantity is increasing rapidly with pressure above 18 000 kg/cm², which will tend to increase pressure coefficients estimated from iso-absorption curves. Similarly high values[50] (in our opinion) of the pressure coefficient of the L_1 states in Ge were obtained by Slykhouse and Drickamer,[21] Neuringer,[23] and Fan, Shepherd, and Spitzer,[3b] by neglecting the possibilities of shape changes. Paul and Warschauer[25] found that inclusion of such effects reduced apparent coefficients of 8×10^{-6} ev/kg cm^{-2}

by about half, bringing the gap coefficient close to that determined from intrinsic resistivity. It would appear that similar corrections would bring the coefficient for the (111) states in GaSb close to our postulated one for the L_1 type of extrema.

Although no coefficient is given for the third set of states, important above 45 000 kg/cm², we note that the coefficient is negative, which is the distinguishing feature of the Δ_1 type of extrema.

Indium Phosphide

Measurements of optical absorption[51] and magneto-resistance[52] show that the lowest states in this compound lie at (000), and the effective mass is low, consistent with Γ_1 states.[53] The pressure dependence of the optical energy gap has been measured recently by Edwards and Drickamer,[35] who find a coefficient of 4.6×10^{-6} ev/kg cm^{-2}. This coefficient is close to that we found for minima of the L_1 type in Ge and GaSb and is therefore *not* in agreement with our postulates.

The optical[51] and galvanomagnetic[52] measurements have also been interpreted to suggest the existence of higher minima. The interpretation of Edwards' measurements gives a maximum gap near 40 000 kg/cm², and a decrease at a rate[54] of -10×10^{-6} ev/kg cm^{-2} at high pressures. The latter decrease is far greater than we postulated for the Δ_1 states; it implies that the higher minima are 0.7 ev[54] higher in energy than the Γ_1 states at atmospheric pressure.

The pressure dependence of the n-type resistivity at room temperature has been measured by Sagar[55] and by Howard,[56] whose results qualitatively agree. Howard's measurements show that the resistivity of his purest sample increases linearly to 30 000 kg/cm² by 35%; this agrees roughly with the effects of changes in the elastic constants and in the effective mass of the Γ_1 minimum. The separation of the minima implied by the optical work is consistent with the fact that there is no evidence of any effect of higher minima on the electrical resistivity at high pressures, but is inconsistent with the indications of the presence of higher minima shown by the galvanomagnetic and optical experiments at atmospheric pressure.

Indium Arsenide

Experiments on infrared cyclotron resonance[53,57] indicate that the conduction band structure of this compound has an extremum at (000), probably of the Γ_1 type.[53] The pressure coefficient of the electron mobility and the energy gap were first measured, to

[48] W. E. Howard and W. Paul (unpublished measurements).

[49] J. H. Taylor, Bull. Am. Phys. Soc. 3, 121 (1958).

[50] It should be remarked that this statement may be debated by all of these workers. The work by Paul and Warschauer is the only one where the necessity for this sort of correction is asserted.

[51] R. Newman, Phys. Rev. 111, 1518 (1958).

[52] M. Glicksman, J. Phys. Chem. Solids 8, 511 (1959).

[53] H. Ehrenreich, J. Phys. Chem. Solids 12, 97 (1959).

[54] Our figures.

[55] A. Sagar (private communication). See this volume.

[56] W. E. Howard and W. Paul (unpublished measurements).

[57] B. Lax, Revs. Modern Phys. 30, 122 (1958).

2000 kg/cm², by Taylor,[58] who deduced a gap coefficient of 5.5×10^{-6} ev/kg cm⁻². Later optical measurements[49] by the same worker were interpreted to give an optical gap coefficient of 8.5×10^{-6} ev/kg cm². Recently published optical measurements by Edwards and Drickamer[35] to 50 000 kg/cm² are interpreted to give an optical gap coefficient of 4.8×10^{-6} ev/kg cm⁻² to 20 000 kg/cm² with a change-over to a coefficient of 3.2×10^{-6} ev/kg cm⁻² between 20 000 kg/cm² and 50 000 kg/cm². Taylor's variation[58] of electron mobility to 2000 kg/cm² at room temperature has been confirmed by DeMeis,[59] who has also extended the measurement to 30 000 kg/cm², without any spectacular change of the coefficient. His results are thus qualitatively consistent with those of Edwards and Drickamer. We note that all of the coefficients quoted are considerably lower than the one we have postulated for Γ_1 type minima.

Indium Antimonide

The pressure coefficients for InSb have been discussed already. These measurements have been extended to 30 000 kg/cm² in the Harvard laboratory, where it has been confirmed that there is no evidence for phenomena attributable to minima higher than the lowest (Γ_1) one at atmospheric pressure.

IV. SUMMARY AND PROSPECT

We must now examine the agreement between our speculative pressure (or dilatational) coefficients and experiment. In doing so we should weigh most heavily data on identified conduction band minima, by proved methods on good samples, and from several laboratories.

Minima of Δ_1 Type

From the last two columns of Table I, we see that the correlation of the pressure coefficients for Δ_1 minima is good. All of the pressure coefficients are negative, and, except for the single measurement at very high pressures on InP, in rather good agreement. The six measurements (on C, AlSb, GaP, GaAs, GaSb, InP) confirming our basis (Ge, Si) are all on minima labelled "speculative," but it seems hardly possible that all could be in error.

The discrepancy for the Δ_1 minima in InP found by Edwards and Drickamer is not, in our view, serious. The optical experiments to extremely high pressures of Drickamer and his collaborators are very informative, but need very careful interpretation. The technique is such that the fine details of the structure of absorption edges, and their variation at pressures below 20 000 kg/cm² may not be resolved.[35] Nevertheless, the experiments are superb in delineating the gross behavior of the extrema of the band structure, and giving guide lines for more detailed experiments. The interpretative

method usually used is to determine the change in frequency for fixed absorption coefficient. We emphasize, as we did under Sec. III in discussing GaP and GaSb, that this may lead to quite erroneous results in the case of indirect transition absorption. For InP, if the pertinent intermediate state for transitions into the Δ_1 state is the Γ_1 state, the shape change will be like that in GaP, leading to either sign of pressure coefficient for the Δ_1 states. The negative pressure coefficient, when measured, will be too small. That of Edwards and Drickamer is already an order of magnitude too large.

Changes of deformation potential and compressibility with volume have not been considered here, yet they clearly cannot be omitted. Such changes make it surprising that we get any correlation at all with Drickamer's coefficient at pressures above 50 000 kg/cm². If we agree to restrict our examination of correlations to pressures below 30 000 kg/cm², a reasonable (but still possibly high) limit for constancy of compressibilities and deformation potentials, the agreement among the Δ_1 coefficients is excellent.

Minima of L_1 Type

The correlation in Sn and GaSb is adequate. Further identification of the minima in gray Sn is required, and confirmation of the coefficient in GaSb.

Minima of Γ_2' or Γ_1 Type

In GaAs and GaSb, the correlation is adequate, while in InAs and InP it is certainly inadequate. However, on examining both the electrical and optical experiments,[35,49,58] it seems to us that at least two effects have not been considered, both of which alter the observed coefficient.

(1) The transitions in InAs and InP are presumably from Γ_{15} to Γ_1 and are allowed. Thus the absorption coefficient can be shown to be, under certain limiting conditions on the oscillator strengths,

$$\alpha = \frac{A m^{*\frac{1}{2}} E_g (\hbar\nu - E_g)^{\frac{1}{2}}}{n \hbar \nu},$$

where A contains constants only, and n is the refractive index. If we assume $E_g \approx \hbar\nu$, the band edge shape depends on $m^{*\frac{1}{2}}/n$, m will increase with pressure, n will decrease, and the slope of the edge will become steeper. As a result, if the energy gap is increasing, isoabsorption plots will give a spuriously low coefficient. (2) In semiconductors of low electron effective mass, n-type samples can be degenerate at room temperature. The increase of mass with pressure tends to reduce the degeneracy by lowering the Fermi level in the conduction band, and decreasing the photon energy required for the transition. This also reduces the apparent pressure coefficient of the energy gap. Crude estimates of the corrections necessary because of (1) and (2),

[58] J. H. Taylor, Phys. Rev. **100**, 1593 (1958).
[59] M. DeMeis (unpublished Harvard measurements).

which depend on the actual conditions of the experiment, indicate that they are considerable.

The samples used by Edwards and Drickamer are not too well characterized, so that we cannot estimate what corrections are required. However, for InAs, they report a *decrease* in the slope of the absorption curve by 50 000 kg/cm² by a factor of about 6. We do not see any explanation for this, but it seems quite apparent that isoabsorption plots cannot give coefficients of the energy gap in such circumstances.

Summary

In summary, we believe a large degree of correlation exists. It is not evident whether or not there is a systematic slow trend in the coefficients for any one minimum with starting energy gap or ionicity. The correlation of the dilatational coefficients is, if anything, poorer than that for the pressure coefficients. We have no explanation to offer for this.

Obviously, experiments to check and confirm the coefficients of Table I should be carried out. It would also be advisable if other energy separations could be measured under pressure. Chief among these are the "vertical" energy gaps supposedly investigated in the reflection spectra.[60] From such measurements we might expect further correlations, and also we might gain information on the rigidity of the whole valence band structure under dilatation.

We have said very little about the 2–6 or 1–7 compounds, although these have been investigated intensively by Drickamer and his co-workers, and by others. We believe it is too early to say, either from an experimental or theoretical standpoint, whether much correlation is to be expected. Space limitations forbid our considering the present evidence at any length. However, it is certainly valid to remark that we should, on the basis of our experience with the group 4 and group 3–5 compounds, look for internal correlations among the 2–6 compounds themselves, or indeed among any group of compounds that are similarly derived. Experience shows that different band edges very probably have different deformation potentials so that the pressure experiments have great value in sorting out effects in complex structures. The experiments on Ge, Sn, GaP, GaAs, GaSb, and InP demonstrate this fact, which is independent of the presence of correlated pressure coefficients, although correlation improves the technique. A few illustrative examples of investigations that are based on a knowledge of the pressure coefficients can be given.

Already published are experiments on hot electrons[61] and tunnel diodes.[62] Near infrared free carrier absorption involving interband transitions is a suitable example for our present purpose. Thus we have already demonstrated that the "extra" absorption in GaP[36] is probably not caused by a Δ_1 to Γ_1 transition. We intend to extend this examination to Si;[40] GaAs,[41] and AlSb.[42] Our investigation is based on the supposition[63] that the excess absorption observed by Spitzer and Fan in Si is caused by vertical $\Delta_1 \rightarrow \Delta_2'$ conduction band transitions. In the intermetallics, the degeneracy at X_1 splits into X_1 and X_3. The minima may or may not be on the zone boundary; the splitting has been calculated to be ~1.4 ev for BN[64] but may be smaller. Whether or not the minima are on the boundary, the transition from Δ_1 to Δ_1 (or X_1 to X_3) is allowed, whereas it is disallowed in Si. We speculate that the absorption in GaP and AlSb is caused by such transitions, and will give a small pressure coefficient, which can be tested. It is perhaps significant that no absorption of this type has been seen in Ge, GaSb, InP, and InAs; these materials do not have Δ_1 states lowest, and have no close minima vertically above the L_1 and Γ_1 states. A disturbing feature is the observation of excess absorption in GaAs. It is not inconceivable, however, that this could be caused by $\Gamma_1 \rightarrow \Gamma_{15}$ transitions. Again, we can suggest that the pressure measurement may be definitive.

As a second example of the use of pressure measurements in illuminating semiconductor properties, we wish to report our recent measurements on the series PbS, PbSe, and PbTe. We have been concerned in examining several aspects of their behavior: (a) any similarity in the conduction or valence band structures and scattering mechanisms, as shown by similarity in

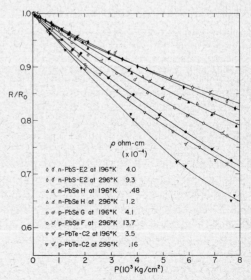

FIG. 4. Resistivity vs pressure, at 196°K and 296°K, of extrinsic samples of PbS, PbSe, and PbTe (L. Finegold, M. DeMeis, and W. Paul, unpublished data).

---

[60] H. R. Philipp and E. A. Taft, Phys. Rev. **113**, 1002 (1959); J. C. Phillips, J. Phys. Chem. Solids **12**, 208 (1960).

[61] S. H. Koenig, M. I. Nathan, W. Paul, and A. C. Smith, Phys. Rev. **118**, 1217 (1960).

[62] M. I. Nathan and W. Paul, Prague International Semiconductor Conference (1960).

[63] W. Paul (to be published).

[64] L. Kleinman and J. C. Phillips, Phys. Rev. **117**, 460 (1960).

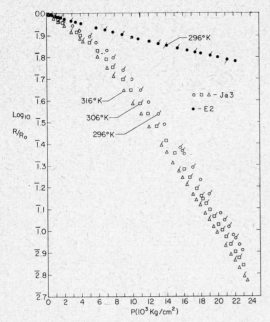

FIG. 5. Resistivity vs pressure for a near intrinsic ($J\alpha3$) and impure n-type sample ($E2$) of PbS, at 296°K (L. Finegold, M. De Meis, and W. Paul, unpublished data).

the pressure coefficients of the extrinsic conductivity; (b) any nonlinearity with pressure of the coefficients for the extrinsic mobility which may be evidence of complexity in the band structure; (c) any dependence of the pressure coefficients of the extrinsic mobility on temperature or impurity concentration, which gives us clues concerning the presence of different scattering mechanisms, particularly impurity scattering, and of multiple extrema having different pressure coefficients; (d) the pressure coefficient of the energy gaps, which allows us to separate the explicit and implicit (volume) contributions to the temperature coefficient of the gap; (e) the changes with pressure of the position and shape of the optical absorption edge, which gives similar information to (c) and (d). Part of this program has been completed, the rest continues.

Thus, we are unable to distinguish much qualitative difference in the pressure coefficients of the extrinsic conductivity of n-type PbS, n- and p-type PbSe and p-type PbTe shown in Fig. 4, although the magnitudes of the effects are different. We find nonlinearity in all of these compounds, but neither the magnitude nor the nonlinearity is sufficient in itself to establish a complex band structure. Thus, the size of the pressure coefficient of p-type PbTe at low pressures can be explained on the basis of a simple band structure, if we assume that the mobility is (roughly) quadratically dependent on the effective mass, and the effective mass changes are given by the pressure coefficient of the energy gap we find for PbS. The nonlinearity is similarly accounted for. This evidence is not conclusive, though, since we can think of combinations of circumstances that would obscure complexity in the band structure.

We are at present investigating these combinations (see below).

From Fig. 4 we see that there is no change in the pressure coefficient of extrinsic n-type PbS and n-type PbSe between 196° and 296°K. This is consistent with the absence of all scattering except lattice scattering, and suggests that, if the band structure is complex, the pressure coefficients of its parts must be similar. Otherwise, we should require that the additional extrema do not contribute much to the conductivity. A difference in the pressure coefficients of p-type PbSe and p-type PbTe at 196° and 296°K is found; for PbTe this is consistent with the model of a two-band valence band that has been suggested previously.[65] The result for PbSe suggests that its valence band structure is similar. We find measurable differences in the pressure coefficients for samples of n-type PbS of different impurity concentration at 196°K, but have done insufficient measurements to draw firm conclusions from these. We have determined the pressure coefficient of the energy gap of PbS from both measurements on photoconductivity and absorption spectrum under pressure, and on the intrinsic resistivity. The optical measurements carried out by Prakash[66] gave a coefficient of 8×10^{-6} ev/kg cm^{-2}, and some evidence (which has to be confirmed) of a change of shape of the absorption edge with pressure. Figure 5 shows a determination of the variation of the intrinsic resistivity with pressure at several temperatures.[67] The variation of the resistivity at low pressures is due primarily to mobility effects, but the rapid decrease of gap makes the sample more intrinsic at high pressures. If the curves for the different temperatures are corrected for the variation of the mobility of the electrons, determined at 23°C on an impure sample, and if (in lieu of such data for p-type PbS, our material being insufficiently impure) the correction for hole mobility is taken to be the same as that for the electrons, then a gap coefficient of 7×10^{-6} ev/kg cm^{-2} is obtained. The optical and electrical determinations are thus in fair agreement. One important consequence is that the deduced volume contribution to the temperature coefficient of the energy gap is only $+2.2\times10^{-4}$ ev/°K, assuming a volume thermal expansion coefficient of 6×10^{-5}/°K and a compressibility of 1.9×10^{-12} cm^2/dyne. This is to be compared with a total temperature coefficient of $+4\times10^{-4}$ ev/°K, and implies a positive explicit effect of temperature on the energy gap. This seems to us to be inconsistent with any of the present theories for this effect of the electron-phonon interaction.

ACKNOWLEDGMENTS

I am happy to acknowledge stimulating discussions on the subject of this paper with Professor Harvey

[65] J. R. Burke, Jr., B. B. Houston, Jr., and R. S. Allgaier, Bull. Am. Phys. Soc. 6, 136 (1961); R. S. Allgaier, this volume.
[65] V. Prakash and W. Paul (unpublished measurements).
[67] L. Finegold, M. DeMeis, and W. Paul (unpublished measurements).

Brooks, Dr. H. Ehrenreich, Dr. W. E. Howard, and Dr. G. Peterson.

The measurements on GaP were carried out by Mr. R. Zallen and on the lead salts by Dr. L. Finegold and Mr. M. DeMeis. All of us are grateful to Mr. J. Inglis and Mr. A. Manning for necessary machine work and to Mr. D. Macleod for fashioning the samples used in the optical and electrical investigations.

The samples of GaP measured in the new data reported were generously given us by the Monsanto Chemical Company and by Dr. W. G. Spitzer. For the PbS samples, we are indebted to Professor R. V. Jones of Aberdeen University and Dr. W. D. Lawson of the Radar Research Establishment; for the PbSe samples, to Dr. W. D. Lawson and Dr. A. C. Prior of R. R. E. and Dr. A. Strauss of Lincoln Laboratory, and for the p-type PbTe samples, to Dr. W. W. Scanlon of Naval Ordnance Laboratory.

JOURNAL OF APPLIED PHYSICS   SUPPLEMENT TO VOL. 32, NO. 10   OCTOBER, 1961

Energy Band Structure of Gallium Antimonide*

W. M. Becker, A. K. Ramdas, and H. Y. Fan

Purdue University, Lafayette, Indiana

Resistivity, Hall coefficient, and magnetoresistance were studied for n- and p-type GaSb. The infrared absorption edge was investigated using relatively pure p-type, degenerate n-type, and compensated samples. Infrared absorption of carriers and the effect of carriers on the reflectivity were studied. The magneto-resistance as a function of Hall coefficient for n-type samples at 4.2°K gave clear evidence for a second energy minimum lying above the edge of the conduction band; the energy separation is equal to the Fermi energy for a Hall coefficient of 5 cm³/coulomb. The shift of absorption edge in n-type samples showed that the conduction band has a single valley at the edge, with a density-of-state mass $m_{d1}=0.052\ m$. By combining the results on the edge shift, magnetoresistance, and Hall coefficient, it was possible to deduce: the density-of-states mass ratio $m_{d2}/m_{d1}=17.3$, the mobility ratio $\mu_2/\mu_1=0.06$, and the energy separation $\Delta=0.08$ ev between the two sets of valleys at 4.2°K. Anisotropy of magneto-resistance, observed at 300°K, showed that the higher valleys are situated along $\langle 111 \rangle$ directions. The infrared reflectivity of n-type samples can be used to deduce the anisotropy of the higher valleys; tentative estimates were obtained. Infrared reflectivity gave an estimate of 0.23 m for the effective mass of holes. The variation of Hall coefficient and transverse magnetoresistance with magnetic field and the infrared absorption spectrum of holes showed the presence of two types of holes. Appreciable anisotropy of magneto-resistance was observed in a p-type sample, indicating that the heavy hole band is not isotropic; this was confirmed by the infrared absorption spectrum of holes. The results on the absorp-tion edge in various samples seemed to indicate that the maximum of the valence band is not at $k=0$. However, it appears likely that transitions from impurity states near the valence band produced absorption beyond the threshold of direct transitions.

I. INTRODUCTION

INFORMATION on the band structure of GaSb has been obtained from various investigations. Roberts and Quarrington[1] found that the intrinsic infrared absorption edge extrapolated to 0.704 ev at 290°K and 0.798 ev at 4.2°K and had a temperature coefficient of -2.9×10^{-4} ev/°C in the range 100°–290°K. The shape of the absorption edge led the authors to suggest that either the minimum of the conduction band or the maximum of the valence band is not at $k=0$. Ramdas and Fan[2] attributed the absorption at high levels to direct transitions but found a temperature dependent absorption tail indicative of indirect transitions. They reported also effective mass values obtained from in-frared reflectivity measurements: $m_e=0.04\ m$ and $m_h=0.23\ m$. From studies of the resistivity and Hall coefficient in the intrinsic and extrinsic temperature ranges, Leifer and Dunlap[3] deduced $E_G(T=0)=0.80$ ev, $m_e=0.20\ m$ and $m_h=0.39\ m$. Zwerdling et al.[4] observed magneto-optical oscillations in the intrinsic infrared absorption which indicated that the absorption at high levels corresponded to direct transitions. By attributing the oscillations to Landau levels in the con-duction band, an electron effective mass $m_e=0.047\ m$ was obtained. Sagar[5] studied the temperature and pressure dependences of the Hall coefficient of n-type samples. The results were explained by postulating a second band with a minimum above the minimum of the conduction band. The second band was assumed to have minima along $\langle 111 \rangle$ directions by analogy with germanium, and piezoresistance effect was observed which supports the suggestion that the band has many valleys. Assuming the valleys to have the mass parame-ters as in germanium, Sagar estimated a density-of-states ratio of 40 and an energy separation of 0.074 ev at room temperature between the two conduction bands. The two-band model has since been used by other authors to interpret measurements on resistivity

* Work supported by Signal Corps contract.

[1] V. Roberts and J. E. Quarrington, J. Electronics 1, 152 (1955–56).

[2] A. K. Ramdas and H. Y. Fan, Bull. Am. Phys. Soc. 3, 121 (1958). The value of hole effective mass reported was in error and should have been $m_h=0.23\ m$. The experimental data used are shown in Fig. 8.

[3] H. N. Leifer and W. C. Dunlap, Jr., Phys. Rev. 95, 51 (1954).

[4] S. Zwerdling, B. Lax, K. Button, and L. M. Roth, J. Phys. Chem. Solids 9, 320 (1959).

[5] A. Sagar, Phys. Rev. 117, 93 (1960).

and Hall coefficient,[6] infrared reflectivity,[7] and pressure dependence of piezoresistance.[8] Taylor[9] observed that the infrared absorption edge shifted with pressure at a rate of 1.57×10^{-5} ev/atm up to 200 atm. Recently, Edwards and Drickamer[10] reported measurements extending to higher pressures. They found a rate of shift of 0.0120 ev/kilobar up to 18 kilobars which changed to 0.0073 ev/kilobar between 18 and 32 kilobars. Further more, the rate of shift leveled off and became negative in the range 32–50 kilobars. The results were explained by assuming that the conduction band has a similar structure as in germanium with a minimum at $k=0$, a set of $\langle 111 \rangle$ minima and a set of $\langle 100 \rangle$ minima lying at successively higher energies; with increasing pressure the $\langle 111 \rangle$ minima move away from the valence band at a slower rate than the $k=0$ minimum and the $\langle 100 \rangle$ minima move toward the valence band. Finally, Cardona[11] reported recently the observation of optical reflectivity peaks in the visible and ultraviolet regions which seem to be analogous to the peaks observed in germanium that had been attributed to $L_3 - L_1$ transitions.

The brief summary shows that a large amount of information has been obtained about the band structure of GaSb, especially regarding the conduction band. However, qualitative confirmation of the evolving band model and quantitative determination of the important parameters are still needed. Furthermore, little information is yet available about the valence band. The galvanomagnetic and infrared studies reported below are presented and discussed with emphasis on the band structure.

II. GALVANOMAGNETIC STUDIES

A. n-type GaSb

The results of magnetoresistance measurements on n-type samples give clear cut evidence for the existence of a second conduction band which is separated by a small energy from the lowest conduction band. Measurements were made on samples which had Hall coefficients ranging from -3 to -110 cm^3/coul. In these samples, the conduction electrons do not freeze out but become degenerate at sufficiently low temperatures. For the range of magnetic field used, the transverse magnetoresistance showed H^2 dependence as indicated by Fig. 1. Figure 2 shows the results plotted against the Hall coefficient R for the various samples. The magnetoresistance of the samples with $|R| > 5$ cm^3/coul decreased with decreasing temperature and became quite small at 4.2°K. This behavior is expected for carriers which become more degenerate with decreasing tem-

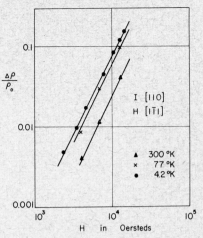

FIG. 1. Transverse magnetoresistance as a function of magnetic field in n-type samples. The 300°K and 77°K data were obtained on a sample having $R(300°K) = -4$ cm^3/coul and the 4.2°K data were given by a sample having $R(300°K) = -3.2$ cm^3/coul.

perature, in an energy band with spherical surfaces of constant energy in k space.[12] On the other hand, the samples with small Hall coefficients showed much larger magnetoresistance at 4.2°K than at room temperature. The 4.2°K curve shows a sharp rise with decreasing Hall coefficient. This is a clear indication that a second type of carrier comes in at sufficient carrier concentrations.[12] We estimate that the rise begins at $R \sim -5$ cm^3/coul, corresponding to an electron concentration of $n = 1.25 \times 10^{18}$ cm^{-3} and a Fermi level of

$$\zeta = (\hbar^2/2m^*)(3\pi^2 n)^{\frac{2}{3}} = 3.63 \times 10^{-15} n^{\frac{2}{3}}$$
$$\times (m/m^*) \text{ ev} = 4.21 \times 10^{-3} (m/m^*) \text{ ev.} \quad (1)$$

Taking $m^*/m = 0.047$,[4] we get $\zeta = 0.0895$ ev as the energy at which the second band lies above the minimum energy of the conduction band.

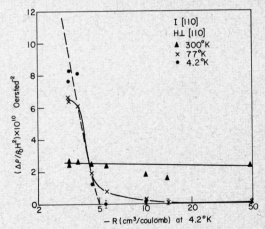

FIG. 2. Transverse magnetoresistance at three different temperatures plotted against Hall coefficient at 4.2°K, for n-type samples. The dotted curve is calculated for 4.2°K using (24) and the values of m_{d2}/m_{d1}, μ_2/μ_1, Δ given by (25) and (26).

[6] A. J. Strauss, Phys. Rev. **121**, 1087 (1961).
[7] M. Cardona, J. Phys. Chem. Solids **17**, 336 (1961).
[8] R. W. Keyes and M. Pollak, Phys. Rev. **118**, 1001 (1960).
[9] J. H. Taylor, Bull. Am. Phys. Soc. **3**, 121 (1958).
[10] A. L. Edwards and H. G. Drickamer, Phys. Rev. **122**, 1149 (1961).
[11] M. Cardona, Z. Physik. **161**, 99 (1961).

[12] A. H. Wilson, *The Theory of Metals* (University Press, Cambridge, England, 1953).

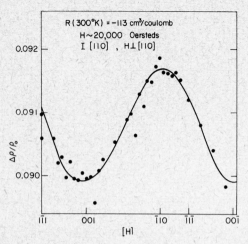

FIG. 3. Variation of transverse magnetoresistance with field orientation for n-type sample at 300°K.

Except for samples with small electron concentrations and at very low temperatures, we have to consider conduction in two bands, of which the higher band may be one of many valleys. In the weak field approximation which is justified by the H^2 dependence shown in Fig. 1, the magnetoresistance is characterized by the three parameters b, c, and d which can be determined from the measurements. These parameters are related to the components, $\sigma_{\alpha\beta\gamma\delta}$, of the magnetoconductivity tensor. The factors which determine the anisotropy of the transverse magnetoresistance and the magnitude of longitudinal magnetoresistance are given by[13]

$$-\sigma_0(b+c)=\sigma_{xxyy}+2\sigma_{xyxy},$$
$$-\sigma_0 d=\sigma_{xxxx}-(\sigma_{xxyy}+2\sigma_{xyxy}),\quad (2)$$

where x, y, z are the cubic axes of the crystal and σ_0 is the conductivity for $H=0$. The magnetoconductivity tensor, therefore $\sigma_0(b+c)$ and $\sigma_0 d$ of the individual valleys of various bands are additive. A valley with spherical surfaces of constant energy gives $(b+c)=d=0$ if the relaxation time is isotropic, which is usually a fair assumption. Thus the lower band should make little contribution to $(b+c)$ and d. Indeed, at low temperatures all the parameters of the lower band are negligible as shown by the smallness of the magnetoresistance effect of the samples with $|R|>5$ cm³/coul. We would then expect to find

$$b+c=(b_2+c_2)(\sigma_{02}/\sigma_0),\quad d=d_2(\sigma_{02}/\sigma_0).\quad (3)$$

Subscripts 1 and 2 will be used to indicate the lower and the higher bands, respectively. The parameter b is important for the transverse magnetoresistance. The following relation holds:

$$\sigma_{xxyy}=-\sigma_0 b-\sigma_0(\mu_H/c)^2,\quad (4)$$

where μ_H is the Hall mobility. Combining two such

[13] C. Herring and E. Vogt, Phys. Rev. **101**, 944 (1956).

equations, one for each band alone, we get

$$b=\frac{\sigma_{01}}{\sigma_0}b_1+\frac{\sigma_{02}}{\sigma_0}b_2+\frac{\sigma_{01}\sigma_{02}}{\sigma_0^2}(\mu_{H1}-\mu_{H2})^2\frac{1}{c^2},\quad (5)$$

where c is the velocity of light. Thus, even with two spherical bands having $b_1\sim 0$ and $b_2\sim 0$ at low temperature, there can be still a large b and a corresponding transverse magnetoresistance due to the last term. This is the cause for the sharp rise shown at 4.2°K in Fig. 2.

Figure 3 shows the variation of transverse magnetoresistance observed on an n-type sample at room temperature. A small longitudinal magnetoresistance, $(\Delta\rho/\rho H^2)_{110}{}^{110}$, was also observed. The results gave

$$(\Delta\rho/\rho H^2)_{110}{}^{110}=(b+c+\tfrac{1}{2}d)\sim 9.2\times 10^{-12}\text{ oe}^{-2};$$
$$b=2.21\times 10^{-10}\text{ oe}^{-2};\quad d=10.8\times 10^{-12}\text{ oe}^{-2}.$$
$$(6)$$

According to (3), these parameters are associated with the high band and they indicate that the high band consists of $\langle 111\rangle$ valleys. The effects were small in magnitude, the longitudinal magnetoresistance being about 1/25 of the transverse magnetoresistance. At 4.2°K, the ratio of the two is less than 1/30 and the longitudinal effect could not be detected at $H=7000$ oe. The fact that the ratio is very small does not necessarily mean that the higher band as well as the lower band has small anisotropy, since the parameter b can be much larger than b_1 and b_2 while $(b+c)$ and d are determined by (b_2+c_2) and d_2. The longitudinal magnetoresistance is given by

$$\left(\frac{\Delta\rho}{\rho H^2}\right)_{110}^{110}=\frac{\sigma_{02}}{\sigma_0}\left(\frac{\Delta\rho_2}{\rho_2 H^2}\right)_{110}^{110}=L_2\frac{\sigma_{02}}{\sigma_0}\left(\frac{\mu_{H2}}{c}\right)^2,\quad (7)$$

where L_2 depends on the anisotropy of the higher band. For $\langle 111\rangle$ valleys,

$$L_2=A\tfrac{2}{3}(2K+1)(K-1)^2/K(K+2)^2,\quad (8)$$

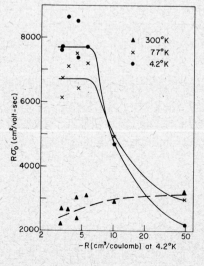

FIG. 4(a). Hall mobility at three different temperatures plotted against Hall coefficient at 4.2°K, for n-type samples.

where K is the anisotropy factor, $K = m_l \tau_l / m_t \tau_t$, and the factor A depends on the variation of relaxation time with energy and is of the order of unity. The Hall mobility of the crystal is given by

$$\mu_H = \frac{\sigma_{01}}{\sigma_0} \mu_{H1} + \frac{\sigma_{02}}{\sigma_0} \mu_{H2}. \quad (9)$$

Thus

$$L_2 = \left[\left(\frac{\Delta\rho}{\rho H^2} \right)_{110}^{110} \middle/ \left(\frac{\mu_H}{c} \right)^2 \right] \left(\frac{\mu_{H1}\mu_1 n_1}{\mu_{H2}\mu_2 n_2} + 1 \right)^2 \frac{\sigma_{02}}{\sigma_0}. \quad (10)$$

Equations (8) and (9) may be used to determine K from the data on the longitudinal magnetoresistance and Hall mobility, provided μ_1/μ_2 and n_1/n_2 are known. However, the right-hand side of (10) is very sensitive to the mobility ratio which has not been determined for the sample at room temperature. The carrier and mobility ratios at $4.2°\mathrm{K}$ have been evaluated and are given in (25) for the samples of low Hall coefficients. Anisotropy of magnetoresistance was not detected and an upper limit

$$\left(\frac{\Delta\rho}{\rho H^2} \right)_{110}^{110} \middle/ \left(\frac{\mu_H}{c} \right)^2 < 3 \times 10^{-3} \quad (11)$$

was deduced from consideration of the experimental accuracy. Substituting the values in (10), we get

$$L_2 < 4.8. \quad (12)$$

At $4.2°\mathrm{K}$, the carriers are highly degenerate and $A \sim 1$. The limit for L_2 for very large K is $\frac{4}{3}$. Thus the anisotropy of the higher band may be very large although magnetoresistance anisotropy was not detected.

The variation of the Hall coefficient with temperature is shown in Fig. 4(a) and 4(b). The data for $300°\mathrm{K}$ and

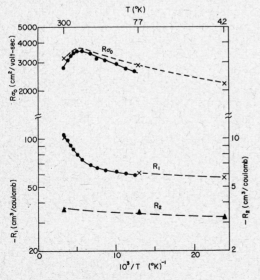

FIG. 4(b). Hall mobility and Hall coefficient as functions of temperature for n-type samples. Data for different samples are indicated by different symbols.

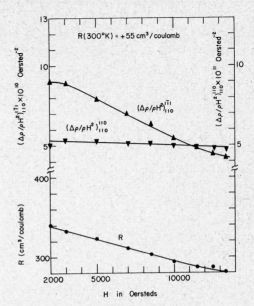

FIG. 5. Transverse and longitudinal magnetoresistances and Hall coefficient as functions of magnetic field for a p-type sample at $77°\mathrm{K}$.

$77°\mathrm{K}$ are in general agreement with previous measurements reported by Sagar[5] and Strauss.[6] As pointed out by these authors, the temperature dependence of R which is more pronounced in samples of smaller electron concentrations is qualitatively understandable on the basis of increasing share of carriers in the higher, low mobility band with increasing temperature. However, the increasing of $R\sigma_0$ with decreasing R, shown by the low temperature curves in Fig. 3, cannot be explained by the two-band conduction, since the $R\sigma_0$ varied in the range of $|R| > 5$ cm³/coul where all the electrons are in the lower band at $4.2°\mathrm{K}$. We suggest that the behavior is caused by impurity scattering which should be the dominant scattering mechanism. In fact, the drop of $R\sigma_0$ with decreasing temperature shown in Fig. 4(a) and Fig. 4(b) can only be attributed to the effect of impurity scattering. The n-type samples used in these experiments were doped with Te, starting with purest obtainable material which was p-type with $\sim 2 \times 10^{17}$ acceptors/cm³. Therefore we may expect in these degenerate samples a charge impurity concentration that exceeds the carrier concentration by $2N_A \sim 4 \times 10^{17}$ cm⁻³. The simple theory of impurity scattering predicts for degenerate carriers a mobility

$$\mu \propto n^{\frac{1}{3}}/N, \quad (13)$$

where N is the concentration of charged centers. For uncompensated samples $n = N$ the formula gives $\mu \propto n^{-\frac{2}{3}}$. Actually, the observed mobility of various degenerate semiconductors varies more slowly with the carrier concentration. Assuming $\mu \propto n^{-\frac{1}{3}}$ for the uncompensated case, we may expect for our samples

$$\mu \propto n^{-\frac{1}{3}}(n/n + 2N_A). \quad (14)$$

According to this expression, the mobility will increase with increasing n up to $n = 4N_A \sim 8 \times 10^{17}$ cm^{-3} which corresponds to $R \sim -7.8$ cm^3/coul. The explanation appears therefore to be reasonable. It should be pointed out that the data on Se-doped samples reported by Strauss[6] differed from the Te-doped samples in the variation of Hall mobility with R. The thorough understanding of scattering mechanisms in various kinds of samples requires further study.

B. p-type GaSb

Figure 5 shows the Hall coefficient and magnetoresistance as functions of the magnetic field for a p-type sample. The Hall coefficient and the transverse magnetoresistance decreased with increasing field. The weak field Hall mobility of the sample was 2700 cm^2/v-sec. The condition

$$(\omega_H \tau)^2 = (eH\tau/mc)^2 \sim (\mu H/c)^2 \sim (\mu_H H/c)^2 \ll 1 \quad (15)$$

is valid over the whole range of H. The decrease of the Hall coefficient and the transverse magnetoresistance is therefore a clear indication for the presence of two types of holes, similar to the case of germanium and silicon. The decrease of R and $\Delta\rho/\rho H^2$ occurs when the condition no longer holds for the small number of light holes with large τ/m. This effect was not observed at room temperature where the relaxation times of the carriers were much shorter, as can be judged from the measured Hall mobility of 690 cm^2/v-sec against 2700 cm^2/v-sec at 77°K. The fact that two types of holes are present at 77°K indicates that the two branches of the valence band must merge or come close together at the maximum.

It is interesting to note that a decrease in R or $\Delta\rho/\rho H^2$ has not been observed for n-type samples. This does not mean that the light hole mass is necessarily much smaller than the light electron mass. The impurities in n-type samples remain charged with decreasing temperature and the scattering effect prevents the relaxation times of the electrons from reaching sufficiently large values with decreasing temperature; as pointed out above, we expect at least a charged impurity concentration of $2N_A \sim 4 \times 10^{17}$ cm^{-3} even though some of the samples used had electron concentrations smaller by an order of magnitude. In the p-type sample, on the other hand, the holes freeze out with decreasing temperature and the charged impurity concentration at 77°K is small judging by the value of R.

The longitudinal magnetoresistance shown in Fig. 5 remained substantially constant, indicating that the effect was associated with the heavy holes, the effect being dependent on $(b+c)$ and d which are additive for various types of carriers according to Eq. (2). Figure 6 shows the variation of the transverse magnetoresistance with the field orientation. From these data and the longitudinal effect we get the following values:

$$(b+c) = 9.1 \times 10^{-12}, \quad d = -8.9 \times 10^{-12}, \quad b = 45.1 \times 10^{-12}.$$

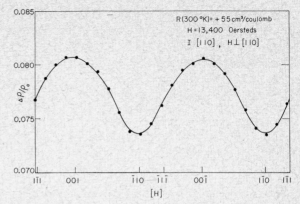

FIG. 6. Variation of transverse magnetoresistance with field orientation for a p-type sample at 77°K. Figures 5 and 6 give data for the same sample.

The relation $(b+c) \sim -d$ is consistent with a band with valleys along $\langle 100 \rangle$ directions.[14] On the other hand, it is also consistent with the behavior of the warped valence bands in germanium and silicon.[15]

III. INFRARED ABSORPTION

A. Effect of Carriers

1. Carrier Absorption in n-Type Sample

The long wavelength absorption in n-type gallium antimonide is shown in Fig. 7. Beyond ~ 5 μ the absorption increases smoothly as a function of wavelength. An absorption band can be seen in the range between 2 and 5 μ with a peak at 3.3 μ. This feature is very similar to the absorption band observed in n-type silicon[16] and in n-type gallium arsenide,[17] which has been attributed to transitions from the conduction band minimum to higher lying minima. According to this interpretation, the observed absorption band indicates the presence of energy band minima at ~ 0.25 ev above the minimum of the conduction band. This could be the $\langle 100 \rangle$ minima postulated by Edwards and Drickamer[10] to explain the pressure effect on the absorption edge. However, they estimate the postulated minima to be 0.4 ev above the minimum of the conduction band against the value of 0.25 ev indicated by the absorption band.

2. Carrier Effect on Reflectivity

Some time ago, W. G. Spitzer measured in this laboratory the reflectivity of n-type GaSb samples for the purpose of determining the carrier effective mass according to the method reported by Spitzer and Fan.[18] The measured reflectivity for one of the samples is

[14] M. Glicksman, *Progress in Semiconductors* (Heywood, London, 1958), Vol. 3, p. 3.
[15] J. G. Mavroides and B. Lax, Phys. Rev. **107**, 1530 (1957).
[16] W. Spitzer and H. Y. Fan, Phys. Rev. **108**, 268 (1957).
[17] W. G. Spitzer and J. M. Whelan, Phys. Rev. **114**, 59 (1959).
[18] W. G. Spitzer and H. Y. Fan, Phys. Rev. **106**, 882 (1957).

shown in Fig. 8. It is convenient to express the result of this type of experiment in terms of the effective mass for electric susceptibility, m_S which is defined by

$$-\chi = ne^2/m_S\omega^2, \quad (16)$$

where χ is the electric susceptibility which can be obtained from the optical measurements and n is the carrier concentration. Using the approximation $n = 1/Rec$, values of m_S/m 0.043, 0.039, 0.041 were obtained for three samples with Hall coefficients of -4.6, -3.4, -2.5 cm^3/coul, respectively; the accuracy of the determination did not justify a definite conclusion about the variation of m_S/m with Hall coefficient. The values of m_S/m are appreciably smaller than the effective mass value, $m_1 = 0.047\ m$, given by the magneto-optical measurement and the density-of-states mass value, $m_{d1} = 0.052\ m$, estimated below for the lowest conduction band. The discrepancy is understandable in the light of the two-band model. Considering two conduction bands we have

$$-\chi = n_1 e^2/m_{S1}\omega^2 + n_2 e^2/m_{S2}\omega^2. \quad (17)$$

The Hall coefficient is given by

$$R = R_1 \frac{1 + xyy_H}{(1+xy)^2} = \frac{1}{n_1 ec} \frac{\mu_{H1}}{\mu_1} \frac{1 + xyy_H}{(1+xy)^2}, \quad (18)$$

where

$$x = n_2/n_1, \quad y = \mu_2/\mu_1, \quad y_H = \mu_{H2}/\mu_{H1}.$$

Combining (17) and (18), we get

$$-\chi = \frac{1}{Rec} \frac{e^2}{m_{S1}\omega^2} \left[\frac{\mu_{H1}}{\mu_1} \frac{1 + xyy_H}{(1+xy)^2} \left(1 + x \frac{m_{S1}}{m_{S2}} \right) \right]. \quad (19)$$

The term in the brackets should account for the difference between the value of m_1 and the values of m_S cited above. The value of the term may be expected to be larger than unity, making the values of m_S appreciably smaller than m_1.

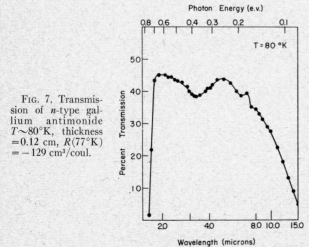

FIG. 7. Transmission of n-type gallium antimonide $T \sim 80°$K, thickness $= 0.12$ cm, $R(77°$K$) = -129$ cm^3/coul.

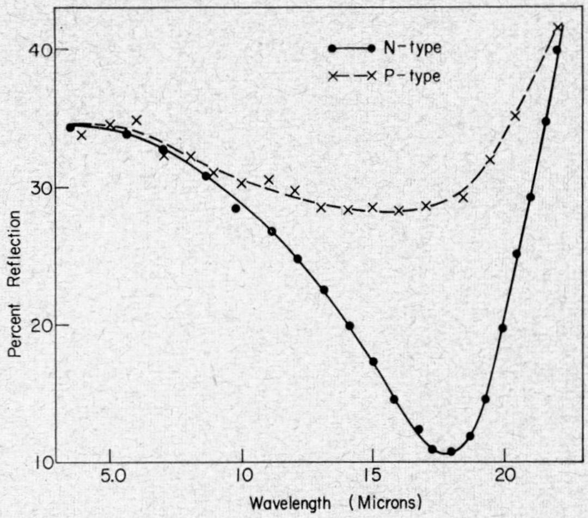

FIG. 8. Effect of free carriers on reflectivity of n-type and p-type gallium antimonide at 300°K: n-type sample, $R(300°$K$) = -3.4$ cm^3/coul, p-type sample $R(300°$K$) = +0.5$ cm^3/coul.

The value of μ_{H1}/μ_1 may be expected to be close to unity when there is sufficiently large carrier density for the Fermi level to be well inside the lower band. If the ratio of density-of-states masses, m_{d2}/m_{d1}, and the energy separation Δ are known for the two bands and if $y \approx y_H$ is known in addition, then x can be calculated from the Hall coefficient by using Eq. (18). With the additional knowledge of the m_1 value, the value of χ obtained from optical measurements provides an estimate of m_{S2}/m_{S1} which gives in turn $m_{d2}/m_{S2} = (m_{d2}/m_{d1})/(m_{S2}/m_{S1})$ in view of $m_{d1} = m_{S1}$. In case the higher band has many valleys of ellipsoids of revolution, the value of m_{S2}/m_{d2} is a measure of the ellipticity. Such an estimate depends on the reliable knowledge of the various parameters involved. Recently, Cardona[7] reported that his reflectivity curve measured at room temperature can be fitted by taking the values $\Delta = 0.08$ ev; ratio of density of states equal to 40, $y = \frac{1}{6}$, as estimated by Sagar, and by assuming for the second band $\langle 111 \rangle$ valleys with an ellipticity the same as in germanium. As shown below, we obtained from 4.2°K data on intrinsic absorption edge and galvanomagnetic effects a much higher value for the ratio of density of states. Values for y and Δ were also obtained, the value of y being also much smaller than the value $\frac{1}{6}$. Optical determination of χ at 4.2°K has yet to be made. The values of y and Δ at room temperature may be significantly different from the estimates obtained from 4.2°K data. Nevertheless, calculations were made using the values given by (23), (25), (26) in conjunction with the room temperature optical measurements. The calculated values of m_{d2}/m_{S2} are given in Table I for two values of m_1 (0.047 m and 0.052 m). Assuming that the second band has four $\langle 111 \rangle$ valleys each characterized by a longitudinal effective mass m_l and a transverse effective mass m_t we have

$$m_{S2}/m_{d2} = \delta^{\frac{2}{3}} r^{\frac{1}{3}} m_t / [3r/(1+2r)] m_t = 4^{\frac{2}{3}} (1+2r)/3r^{\frac{2}{3}}, \quad (20)$$

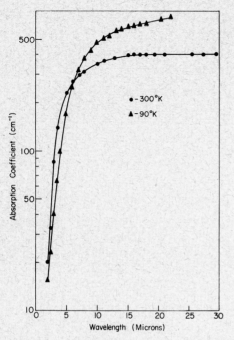

FIG. 9. Absorption spectrum of p-type gallium antimonide. $R(300°K) = +5.1$ cm³/coul, $R(77°K) = 11.0$ cm³/coul.

where $r = m_l/m_t$. The calculated values of r are also given in Table I. It should be emphasized that the calculated results given in the table should be regarded as no more than order of magnitude estimates.

3. Carrier Absorption in p-Type Sample

The long wavelength absorption in p-type gallium antimonide is shown in Fig. 9. The curve is similar to that observed in p-type samples of Ge,[19] InAs[20] and GaAs[21] indicating that the absorption is produced by interband transitions within the valence band. This interpretation finds confirmation from the galvano-magnetic measurement of p-type samples which showed the existence of two types of holes. The effective mass ratio of the two types of holes may be estimated from the analysis of the interband transitions if the energy

TABLE I. Calculation of m_{d2}/m_{S2} for the higher conduction band. Values $m_{d2}/m_{d1} = 17.3$, $\mu_2/\mu_1 = 0.06$, $\Delta = 0.08$ ev, are used. The values of $r = m_l/m_t$ are calculated assuming the conduction band has four $\langle 111 \rangle$ valleys of ellipsoids of revolution.

Experimental data (300°K)		For $m_1/m = 0.047$		For $m_1/m = 0.052$	
R(cm³/coul) $\quad$ m_S/m		m_{d2}/m_{S2} $\quad$ r		m_{d2}/m_{S2} $\quad$ r	
−4.6	0.043	2.84	4.1	3.41	6.75
−3.4	0.039	3.37	5.84	3.94	11.4
−2.5	0.041	3.13	4.58	3.67	8.85

[19] W. Kaiser, R. J. Collins, and H. Y. Fan, Phys. Rev. **91**, 1380 (1953); H. B. Briggs and R. C. Fletcher, Phys. Rev. **91**, 1342 (1953).
[20] F. Matossi and F. Stern, Phys. Rev. **111**, 472 (1958).
[21] R. Braunstein, J. Phys. Chem. Solids **8**, 280 (1959).

bands have spherical surfaces of constant energy. The absorption coefficient is given then by the expression

$$\alpha \propto \nu^{\frac{1}{2}} \left[\exp\left(\frac{m_L}{m_L - m_H} \frac{h\nu}{kT} \right) - \exp\left(\frac{m_H}{m_L - m_H} \frac{h\nu}{kT} \right) \right], \quad (21)$$

where m_H and m_L are the effective masses of the heavy holes and light holes, respectively. For $h\nu \gg kT$, the second term is negligible, and $ln(\alpha\nu^{-\frac{1}{2}})$ vs $h\nu$ plot should be a straight line, the slope of which gives m_L/m_H. If the expression is valid, the slope of such a plot should be proportional to $(1/kT)$. Although the data give approximately straight-line plots, the slope does not change much with temperature indicating that at least one of the hole bands is not spherical.

B. Intrinsic Absorption Edge

The intrinsic absorption edge in pure p-type samples is shown in Fig. 10. The steep rising part of the edge corresponds to

0.725 ev at 300°K, $\quad$ 0.80 ev at ~80°K,

0.81 ev at ~4.2°K.

These values are taken to be the threshold energies, $h\nu_d$, for direct transitions to the lowest conduction band. The value for liquid helium temperature agrees with that obtained by Zwerdling[4] et al. Figure 10 shows also the absorption edges at liquid helium temperature for two n-type samples of different carrier concentrations. The n-type samples show the shift of edge ex-

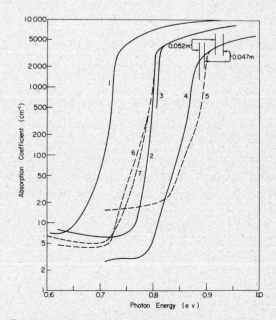

FIG. 10. Absorption edge in gallium antimonide. (1) p-type, T~300°K, $R(300°K) = 51$ cm³/coul; (2) p-type, T~80°K, $R(77°K) = 380$ cm³/coul; (3) p-type, T~4.2°K; (4) degenerate n-type, T~4.2°K, $R(4.2°K) = -5.55$ cm³/coul; (5) degenerate n-type, T~4.2°K, $R(4.2°K) = -3.19$ cm³/coul; (6) p-type, compensated, T~80°K; (7) p-type, compensated, T~80°K.

pected from the filling of the conduction band. The positions of the shifted edge as calculated for a single band model using the values 0.047 m and 0.052 m for the electron density-of-state mass m_{d1} are shown by the vertical lines. The sample of lower electron concentration, $R = -5.55$ cm³/coul, should have carriers only in the lower band according to Fig. 2. The data indicate that the edge position calculated for $m_d = 0.047$ m gives too large a shift whereas

$$m_{d1} = 0.052 \ m \qquad (22)$$

gives reasonable agreement. From the edge shift, 0.075 ev, of this sample, we get for the height of Fermi level in a sample of $R = -5$ cm³/coul

$$0.075(5.55/5)^{\frac{2}{3}} = 0.08 \ \text{ev} = \Delta, \qquad (23)$$

Δ being the energy difference between the minima of the second and the first bands.

The edge shift in the sample of higher electron concentration, $R = -3.19$ cm³/coul is considerably smaller than the estimates based on the single-band model, as is expected. The Fermi level given by the edge shift is 0.09 ev which provides an estimate of the electron concentration in the lower band: $n_1 = 1.48 \times 10^{18}$ cm⁻³. This information can be combined with Hall mobility and magnetoresistance data to determine the ratios of mobilities and of density-of-states masses for the two conduction bands. The parameter b for magnetoresistance is given by the expression (5) with $b_1 \sim 0$ for liquid helium temperature. We have seen that $(b+c)$ and d of the upper band are quite small compared to the third term of the right-hand side. Therefore, either the band is nearly isotropic or its parameters b, c, d must be small. In any case, we can neglect the term b_2 compared with the last term. It can be shown then

$$\frac{b}{(R\sigma_0)^2} = xy\frac{(1-y_H)^2}{(1+xy_Hy)^2}, \qquad (24)$$

where x and y are defined as in (18). We shall use the approximation $y_H = y$ which should not cause great error. Equations (18) and (24) can be used to calculate the values of x and y. Using the data $n_1 = 1.48 \times 10^{18}$ cm⁻³, $R = -3.19$ cm³/coul and $b/(R\sigma_0)^2 = 0.14$ we get

$$\mu_2/\mu_1 = 0.06, \quad n_2/n_1 = 2.67. \qquad (25)$$

Using the value of the Fermi level, $\zeta = 0.09$ ev, and the value $\Delta = 0.08$ ev, we get

$$m_{d2}/m_{d1} = (n_2/n_1)^{\frac{2}{3}}\zeta/(\zeta-\Delta) = 17.3. \qquad (26)$$

The shape of the absorption edge as shown in Fig. 10 suggests that the absorption begins with indirect transitions. After a steep drop, the absorption tails off extending to much longer wavelengths than is expected and the effect is more pronounced at room temperature than at the liquid nitrogen temperature. The behavior is similar to the case of phonon-assisted indirect transi-

tions in germanium and silicon with phonon-absorption transitions becoming reduced at lower temperatures. Furthermore, the absorption edge of the n-type samples is also difficult to reconcile with the assumption that the minimum of the conduction band and the maximum of the valence band coincide in k space. If the shift of the edge at high absorption level corresponds to the rise of the Fermi level in the conduction band, then the assumption predicts a very steep edge at 4.2°K with the absorption dropping as $\exp[h(\nu-\nu_\zeta)/kT]$ where $h\nu_\zeta$ corresponds to transitions at the Fermi level. The measured curves are much too sloping in comparison with this prediction. On the other hand, a sloping curve may be produced with the help of indirect transitions if the conduction band minimum and the valence band maximum are at different points in k space.

Measurements were made on p-type samples containing large and nearly equal concentrations of acceptor and donor impurities, in order to observe the impurity enhancement of indirect transitions. The impurity compensation kept down the carrier concentration and the background absorption due to carriers. The results are also shown in Fig. 10. It is seen that the absorptions in the two compensated samples appear to extend to about the same $h\nu$, ~ 0.72 ev, for both samples. This seems to be an evidence against the possibility that the effect was the result of the lowering of the conduction band minimum by the presence of impurities.

The data may be interpreted in the following way. The room temperature curve shows a clear change of slope at 0.627 ev. Subtracting the background carrier absorption α_c from the observed absorption α and plotting $(\alpha-\alpha_c)^{\frac{1}{2}}$ against $h\nu$, two straight line portions can be recognized which extrapolate to 0.627 ev and 0.690 ev, respectively. These two energy values appear to correspond to the onsets of indirect transitions with phonon absorption and phonon emission. Thus, we get

$$h\nu_i(R.T.) = 0.658 \ \text{ev}, \quad h\nu_p = 0.031 \ \text{ev},$$

where $h\nu_i$ is the energy gap and $h\nu_p$ is the phonon energy. We note that infrared measurements give 0.029 ev for the long wavelength, optical transverse mode.[22] The data for the compensated samples indicate

$$h\nu_i(L.N.) \sim 0.72 \ \text{ev}.$$

This gives $(h\nu_d - h\nu_i) = 0.08$ ev which is reasonably close to the estimate 0.067 ev for the same quantity at room temperature. Also, the threshold for phonon emission transitions should be $h\nu_i(L.N.) + h\nu_p = 0.75$ ev; curve 2, Fig. 10, indeed rises sharply near this energy. The rounding-off in the curve as it merges with the background absorption could be caused by impurity induced indirect transition. Furthermore, the n-type samples, curves 4 and 5, Fig. 10, having their thresholds

[22] G. S. Picus, E. Burstein, B. W. Henvis, and M. Hass, J. Phys. Chem. Solids 8, 282 (1959).

of direct transitions at 0.885 and 0.90 ev are expected to have thresholds at 0.805 and 0.82 ev for impurity-induced indirect transitions. Curves 4 and 5 in Fig. 10 show that the estimates are reasonably consistent with the experimental data. Thus, the suggested interpretation provides a satisfactory explanation for all the absorption edge observations. However, this interpretation requires that the valence band have off-center maxima, since we are certain that the lowest conduction band has its minimum at $k=0$. Furthermore, the presence of light holes at 78°K requires two degenerate bands at each maximum. It appears from the examination of the symmetry properties of the Brillouin zones[23] that such a band structure is unlikely.

An alternative interpretation may be suggested. The behaviors which cannot be understood on the basis of

direct transitions may be caused by excitations from impurity states in the range of ~0.08 ev from the valence band. We would have to assume that there are sufficient such states even in the purest p-type samples used. Transitions from these states produce the tail absorption seen at room temperature. At low temperatures, the states are depleted of electrons, resulting in a sharper absorption edge. In compensated samples, the states are occupied by electrons even at low temperature, giving a tail absorption. Finally, transitions from these states to the Fermi level in n-type degenerate samples begin at smaller photon energy than the direct transitions from the valence band, thus producing a sloping absorption edge. Measurements are being made with higher resolution. Preliminary results obtained indeed favor the second interpretation. Thus, we may accept tentatively that the maximum of the valence band is at $k=0$, having a warped heavy hole band which is degenerate with a light hole band.

---

[23] G. Dresselhaus, Phys. Rev. **100**, 580 (1955); R. H. Parmenter, Phys. Rev. **100**, 573 (1955).

JOURNAL OF APPLIED PHYSICS   SUPPLEMENT TO VOL. 32, NO. 10   OCTOBER, 1961

Lattice Absorption in Gallium Arsenide

W. COCHRAN
Cavendish Laboratory, Cambridge, England

AND

S. J. FRAY, F. A. JOHNSON, J. E. QUARRINGTON, AND N. WILLIAMS
Royal Radar Establishment, Great Malvern, England

A series of detailed measurements of the lattice absorption bands of gallium arsenide has been made over the wavelength range 10–40 μ and over the temperature range 20–292°K. These results can be interpreted in terms of multiple phonon interactions involving five characteristic phonon energies. These results, along with the known elastic constants, have enabled us to supply all the relevant data for a computation of the complete phonon spectrum using an extension of the shell model.

DURING the last nine months, measurements have been made of the lattice absorption spectrum of gallium arsenide[1] at the Royal Radar Establishment. The object of this work was to obtain as much information as possible about the vibrational spectrum of this material and is part of a general study of the lattice spectrum of 3–5 semiconductors. This is a continuation of similar investigations on silicon,[2] germanium,[3] and indium antimonide.[4]

Lattice absorption bands arise from the direct interaction of infrared photons and phonons in the crystal lattice. In the 3–5 semiconductors the strongest of these interactions is between a photon and a single long-wavelength optical phonon. This type of interaction is

responsible for the reststrahlen bands in these materials. However, the more important bands from our point of view are those that arise from the interaction of a photon with a pair of phonons. Two mechanisms are available for this type of coupling; one through anharmonic forces[5] and the other through second-order electric moments.[6] The anharmonic mechanism depends on the production of a single long wavelength optical phonon as an intermediate state, followed by its splitting into a pair of phonons. The second-order electric moment mechanism depends on the fact that a charge is induced on a particular atom when a neighboring atom is displaced from its equilibrium position and also that a dipole moment is produced when this atom is itself displaced.

In either case, energy and wave vector must be conserved between the initial photon and the two resulting phonons. The equations for the conservation of energy

---

[1] S. J. Fray, F. A. Johnson, J. E. Quarrington, and N. Williams (to be published).

[2] F. A. Johnson, Proc. Phys. Soc. (London) **73**, 265 (1959).

[3] S. J. Fray, F. A. Johnson, J. E. Quarrington, and N. Williams (to be published).

[4] S. J. Fray, F. A. Johnson, and R. H. Jones, Proc. Phys. Soc. (London) **76**, 939 (1960).

[5] D. A. Kleinman, Phys. Rev. **118**, 118 (1960).

[6] M. Lax and E. Burstein, Phys. Rev. **97**, 39 (1955).

TABLE I.

Energy dependence	Wave vector dependence	Temperature dependence
Single phonon interactions		
$E_1 = E^1$	$\mathbf{k}_1 = 0$	$(1+F_1) - F_1$
Two phonon interactions		
$E_1 + E_2 = E^1$	$\mathbf{k}_1 + \mathbf{k}_2 = 0$	$(1+F_1)(1+F_2) - F_1 F_2$
$E_1 - E_2 = E^1$	$\mathbf{k}_1 - \mathbf{k}_2 = 0$	$(1+F_1)F_2 - F_1(1+F_2)$
Three phonon interactions		
$E_1 + E_2 + E_3 = E^1$	$\mathbf{k}_1 + \mathbf{k}_2 + \mathbf{k}_3 = 0, \pm\mathbf{K}$	$(1+F_1)(1+F_2)(1+F_3) - F_1 F_2 F_3$
$E_1 + E_2 - E_3 = E^1$	$\mathbf{k}_1 + \mathbf{k}_2 - \mathbf{k}_3 = 0, \pm\mathbf{K}$	$(1+F_1)(1+F_2)F_3 - F_1 F_2(1+F_3)$
$E_1 - E_2 - E_3 = E^1$	$\mathbf{k}_1 - \mathbf{k}_2 - \mathbf{k}_3 = 0, \pm\mathbf{K}$	$(1+F_1)F_2 F_3 - F_1(1+F_2)(1+F_3)$

E^1 = photon energy; E_1, E_2 = phonon energies;
$\mathbf{k}_1$, $\mathbf{k}_2$ = phonon wave vectors; $\mathbf{K}$ = reciprocal lattice vector;
$F = 1/[\exp(E_1/kT) - 1]$.

and wave vector along with the expressions for the temperature dependence of the lattice bands are set out in Table I. The temperature dependence is given by Bose-Einstein statistics, and is shown as the difference between the forward process in which a photon is absorbed and the reverse process in which the photon is emitted.

Now let us consider a band due to the emission of a pair of phonons from branches A and B. For each discrete vibrational state of wave vector k, we can emit a phonon of energy $E(A,k)$ from branch A and a phonon of energy $E(B,-k)$ from branch B. Since $E(B,k)$ is equal to $E(B,-k)$ we can regard the energy associated with the state k, to be $E(A,k)+E(B,k)$ or $E(A+B,k)$. If S_1 is the number of states in the reduced zone for which the energy $E(A+B,k)$ lies in the range E_1 to $E_1+\delta E_1$, then for a sufficiently large crystal $S_1/\delta E_1$ should tend to a limit D_1 which we shall call *the density of combined states*, for the energy E_1 and the pair of branches A and B. We believe that the shape of the lattice bands is primarily governed by the density of combined states, unless the matrix elements for the interaction depend strongly on the wave vector.

Since the maxima in these combined density curves must be closely associated in energy with the maxima in the ordinary density of states curves for individual branches, we would expect the peaks in the lattice bands to have energies that are approximated to by simple combinations of a few characteristic phonon energies. This was shown to be the case, first for silicon,[2] and subsequently for germanium,[3] indium antimonide,[4]

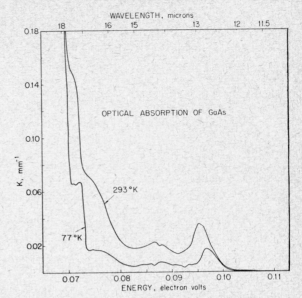

FIG. 2. Lattice absorption coefficient of high resistivity n-type GaAs vs wavelength from 10 to 18μ at 77° and 293°K.

gallium phosphide,[7] diamond,[8] and silicon carbide[9] and, as will be demonstrated in this paper, for gallium arsenide.[1]

We now wish to describe, first the measurements of the lattice absorption bands of gallium arsenide and their interpretation in terms of characteristic phonon energies, second, the attempts we have been making to measure the absorption through the reststrahlen band,

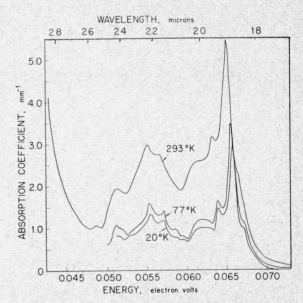

FIG. 3. Lattice absorption coefficient of high resistivity n-type GaAs vs wavelength from 18 to 28μ at 20°, 77°, and 293°K.

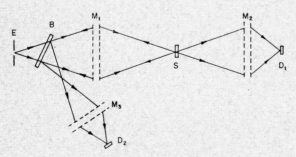

FIG. 1. Optical arrangement for simultaneous transmission and reflection measurements.

[7] D. A. Kleinman and W. G. Spitzer, Phys. Rev. **118**, 110 (1960).
[8] S. D. Smith and J. R. Hardy, University of Reading Report, Reading, England.
[9] L. Patrick and W. J. Choyke, Westinghouse Research Laboratory Paper 029-WOOD-P3.

TABLE II.

Position ev	Observed temperature dependence			Assignment	Calculated temperature dependence		
	20°	77°	292°		20°	77°	292°
0.0955	...	1.0	2.1	TO1+TO2+TO2 0.0324+0.0316+0.0316	...	1.0	2.6
0.0885	...	1.0	2.7	TO1+TO1+LA 0.0324+0.0324+0.0237	...	1.0	2.97
0.0860	...	1.0	3.0	TO2+TO2+LA 0.0316+0.0316+0.0228	...	1.0	3.05
0.0735	...	1.0	4.2	TO1+TO1+TA 0.0324+0.0324+0.0087	...	1.0	4.55
0.0716?				TO2+TO2+TA 0.0316+0.0316+0.0084			
0.0648	1.0	1.17	1.83	TO1+TO1 0.0324+0.0324	1.0	1.02	1.82
0.0631	1.0	1.14	1.88	TO2+TO2 0.0316+0.0316	1.0	1.02	1.85
0.0612	1.0	1.2	1.96	or ⎧ TO1+LO 0.0324+0.0288	1.0	1.03	1.85
				⎩ TO2+LO 0.0316+0.0296	1.0	1.03	1.84
0.058				LO+LO 0.029+0.029			
0.0565	1.0	1.16	1.98	TO1+LA 0.0324+0.0241	1.0	1.04	2.04
0.0548	1.0	1.14	2.01	TO2+LA 0.0316+0.0232	1.0	1.04	2.06
0.0510	1.0	1.2	2.20	LO+LA 0.0288+0.0222	1.0	1.05	2.25
0.048?				LA+LA 0.024+0.024			
0.0413				TO1+TA 0.0324+0.0089			
0.0398				TO2+TA 0.0316+0.0082			
0.038				LO+TA 0.029+0.009			

and finally, the extension of the shell model[10] to 3–5 compounds and its use to calculate curves of the density of combined states for gallium arsenide.

Before we describe the experimental results on gallium arsenide we should like to mention briefly the new technique we have developed for making simultaneous measurements of the transmission and reflection coefficients at normal incidence.[11] The optical arrangement is shown in Fig. 1. Light from the exit slit E is focused on the specimen S and the transmitted light is focused on the detector D_1. The light reflected from the specimen would normally refocus at E, but part of this is reflected from the beam splitter B and is focused on the second detector D_2. The specimen is mounted in a three position holder which is designed to allow first the specimen, then an aperture, and finally a silver mirror to be moved into the beam in turn. This enables us to calibrate the

system, and determine the reflection and transmission of the specimen. Corrections have to be applied for the reflection of the detector D_1 which causes the apparent transmission to be too large and the apparent reflection to be too small. These corrections can easily be found by using detector D_2 to measure the reflection of D_1. When the reflection coefficient r and the transmission coefficient t have been measured, the absorption coefficient is computed from the equation

$$[(1-r)^2/t]-t=2\sinh al. \qquad (1)$$

The measurements of the lattice absorption of gallium arsenide were made on several samples of high resistivity n-type material in the wavelength range $10–35\mu$ and over the temperature range 20–292°K. Sample thickness varied from 0.2 to 3 mm for these bands. The results are shown in Figs. 2, 3, and 4.

The bands shown in Fig. 2 we believe to be due to three phonon processes, while the strong bands in Figs. 3 and 4 we believe to be due to two phonon proc-

[10] F. A. Johnson and W. Cochran (to be published).
[11] S. J. Fray, F. A. Johnson, J. E. Quarrington, and N. Williams (to be published).

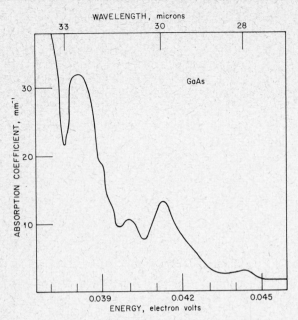

Fig. 4. Lattice absorption coefficient of high resistivity n-type GaAs vs wavelength from 28 to 35μ.

esses. The details of these bands along with their observed temperature dependence are set out in the first two columns of Table II. The second two columns give the probable assignment and the corresponding calculated temperature dependence. It will be seen that all the main peaks fit well into a simple scheme based on five characteristic phonon energies. These results, when compared with similar results for silicon and germanium, suggest that the vibrational spectrum of gallium arsenide is basically similar to that of germanium.

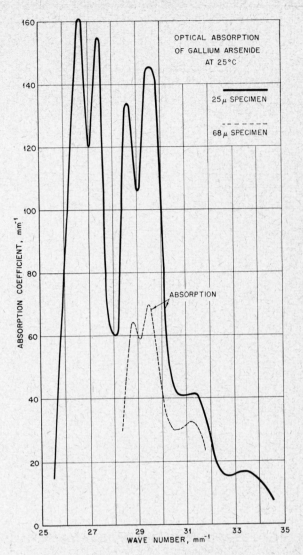

Fig. 6. Absorption coefficient vs wavelength range from 32 to 40μ (reststrahlen band).

Figures 5 and 6 show the preliminary results of our work on the reststrahlen band.[12] The first strong absorption peak comes where we should expect to find a single long wavelength LO phonon whose energy agrees well with that deduced from tunnel diode measurements.[13] The other two absorption peaks occur where we should expect to find the TO phonon. Perhaps the most surprising result was to find that over the range of specimen thickness 14–68 μ the transmission at each wavelength remained almost constant. We have obtained some much higher resolution data on these bands which indicates that each of the main peaks shown in Fig. 6 has a series of three to four fine structure peaks, but these measurements were very close to the limits

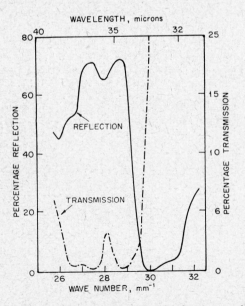

Fig. 5. Percentage reflection and transmission vs wavelength range from 32 to 40μ (reststrahlen band).

[12] S. J. Fray, F. A. Johnson, J. E. Quarrington, and N. Williams, Proc. Phys. Soc. (London) 77, 215 (1961).
[13] R. N. Hall, J. H. Racette, and H. Ehrenreich, Phys. Rev. Letters 4, 456 (1960).

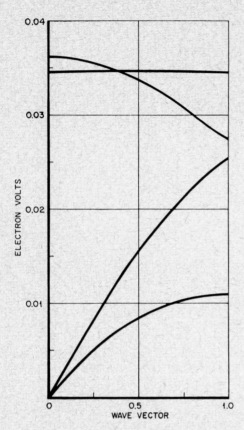

FIG. 7. Calculated curves of the energy vs wave vector (100 direction).

of sensitivity of our equipment, and consequently are not easily reproduced. We have no satisfactory explanation to offer for these results. We believe that the vibrational spectrum near $k=0$ of these thin samples becomes dependent on the size and shape of specimen. This work is continuing.

The last development we wish to discuss is the extension of the shell model[10] to the 3–5 semiconductors and its use to calculate the "density of combined states" for gallium arsenide. The shell model is based on the idea that the potential energy of a crystal must be a general quadratic function of both the displacement of the atom and their induced polarization, and that the correct

equations of motion are[14]

$$m\ddot{u} = -\partial V/\partial u$$
$$0 = \partial V/\partial P. \qquad (2)$$

In the case of germanium, Cochran[15] proposed a simple mechanical model which simulated both the normal short range bonding interaction forces, and the polarizations of the atoms. The polarization was simulated by assuming the atom could, for this purpose, be represented as a central massive core bound isotropically to a charged spherical shell of electrons and the polarization then arose from the relative displacement of core and shell. The potential energy of this type of model has the required quadratic form, and when used in conjunction with Eqs. (2) gave a very good fit to the known neutron data.[16] To extend this model to 3–5 semiconductors one has to allow for the different polarizibilities of the two atoms, and for the possibility that each atom may have a static ionic charge. The ionic charge is easily included by allowing the core and shell for each atom to have different charges and the difference in polarizibility is included by allowing the shell charges on the two atoms to differ and by allowing the isotropic bonding between core and shell to differ for the two atoms. The dispersion relations of this model depend on 10 independent parameters. Three of these are determined by the elastic constants which, incidentally, indicate an ionic charge of zero. A fourth is fixed by the refractive index, and two further constants are fixed by the reststrahlen band data. In view of the result that the ionic charge is zero we have assumed that the difference in polarizibility is adequately accounted for by allowing the two shells to have the same charge, but different bonding forces, to their respective cores. This eliminates one constant, and the remaining three were fixed from the general form of the lattice band results. We propose to use this model to calculate curves of the density of combined states which we can compare with the actual lattice bands. Unfortunately this calculation has not been completed in time for this conference, but the results in the 100 direction are shown in Fig. 7.

[14] V. S. Mashkevich, J. Exptl. Theoret. Phys. (USSR) 32, 866 (1957) [translation, Soviet Phys.—JETP 5, 707 (1957)].
[15] W. Cochran, Proc. Roy. Soc. (London) A253, 260 (1959).
[16] B. N. Brockhouse and P. K. Iyengar, Phys. Rev. 111, 747 (1958).

Galvanomagnetic Effects in III–V Compound Semiconductors*

A. C. BEER

Battelle Memorial Institute, Columbus, Ohio

The influence of various structural characteristics in the III-V compounds on galvanomagnetic properties is discussed. Evidence for the scattering of charge carriers by polar optical modes is reviewed, and the behavior of Hall and magnetoresistance coefficients is examined in regard to the conduction band structure. Unique characteristics, imparted by light masses in certain bands, include high mobilities and large magneto-effects associated either with transport in the band or with ionization energies of the impurity centers. The importance of avoiding inhomogeneities, either in specimen or in magnetic field, when measuring Hall coefficient or magnetoresistance in high-mobility materials is emphasized. Illustrations are given of the effects of nonuniformities in carrier concentration or in applied magnetic field on various galvanomagnetic phenomena.

THE influence on galvanomagnetic properties of charge-carrier scattering mechanisms and of band structure is exhibited in interesting ways in the different III-V compounds. In connection with the scattering, the fact that there exists two dissimilar atoms in the unit cell of the crystal, and that certain degrees of ionic bonding are present in the III-V semiconductors, renders it likely that polar optical modes might play a significant role in the interaction between the charge carriers and the lattice. That polar scattering may, in fact, predominate over acoustic phonon scattering at room temperature in such materials as InSb, InAs, InP, and GaAs is suggested by the theoretical calculations of mobility done by Ehrenreich.[1–4] Results on additional III-V compounds are available from calculations made by Hilsum.[5] Attempts to obtain further information on the scattering in many of these materials have caused measurements to be made of the transverse and longitudinal Nernst effects—the latter quantity being essentially the magnetothermoelectric power. Unfortunately, there is no unanimity in the interpretation of the results. For example, investigators such as Rodot[6] cited the behavior of the thermomagnetic effects at room temperature in indium antimonide as evidence for a predominance of scattering by polar optical modes, while Zhuse and Tsidil'kovskii[7] arrived at the conclusion that acoustic phonon scattering is the significant process. Nasledov and collaborators also favored the acoustic phonon process for scattering in indium arsenide[8] and

gallium arsenide.[9–11] The interpretation of thermomagnetic phenomena often is complicated by ambipolar effects, by mixed scattering mechanisms, and by the fact that for polar optical scattering the relaxation time approximation is applicable only under limiting conditions.[12]

Unique features in the band structure of many of the III-V materials are reflected in the galvanomagnetic properties. Of particular interest has been the importance of the nonparabolic nature of the dependence of energy on wave number for states away from the conduction band edge,[13] especially in the small bandgap materials such as indium antimonide and indium arsenide. In connection with the heavy-mass valence band, the possible importance of linear terms in k, caused by the lack of inversion symmetry in these compounds, has been discussed for the case of indium arsenide.[14] In GaSb and GaAs, it is known that a set of subsidiary conduction band minima exist at approximately 0.08 and 0.36 ev, respectively[4,15] at room temperature, above the principal minimum which has [000] symmetry. In a certain temperature region, therefore, electrons will be thermally activated from the principal conduction band to the subsidiary band, where the mobility is lower. As a result, a maximum is observed in the behavior of the Hall coefficient with temperature.

A number of interesting galvanomagnetic phenomena result from the low effective masses in certain bands and the resulting high mobilities found in such materials as indium antimonide, indium arsenide, and to a lesser extent, gallium arsenide. These low effective masses are also responsible for some unique characteristics associated with the donor levels. In indium antimonide, for example, the low conduction band effective mass together with a relatively large dielectric constant yields

* Supported in part by the Air Force Office of Scientific Research.

[1] H. Ehrenreich, J. Phys. Chem. Solids **2**, 131 (1957).
[2] H. Ehrenreich, J. Phys. Chem. Solids **9**, 129 (1959).
[3] H. Ehrenreich, J. Phys. Chem. Solids **12**, 97 (1955).
[4] H. Ehrenreich, Phys. Rev. **120**, 1951 (1960).
[5] C. Hilsum, Proc. Phys. Soc. (London) **76**, 414 (1960).
[6] M. Rodot, J. phys. radium **19**, 140 (1958); *Solid State Physics in Electronics and Telecommunications* (Proceedings of the Brussels Conference), edited by M. Désirant and J. Michiels (Academic Press, Inc., New York, 1960), Vol. 2, p. 680.
[7] V. P. Zhuze and I. M. Tsidil'kovskii, J. Tech. Phys. (U.S.S.R.) **28**, 2372 (1958) [translation: Soviet Phys.—Tech. Phys. **3**, 2177 (1959)].
[8] O. Emel'yanenko, N. Zotova, and D. Nasledov, Fiz. Tverdogo Tela **1**, 1868 (1959) [translation: Soviet Phys.—Solid State **1**, 1711 (1960)].

[9] O. Emel'yanenko and D. Nasledov, Fiz. Tverdogo Tela **1**, 985 (1959) [translation: Soviet Phys.—Solid State **1**, 902 (1959)].
[10] D. Nasledov, J. chim. phys. **57**, 479 (1960).
[11] O. Emel'yanenko, D. Nasledov, and R. Petrov, Fiz. Tverdogo Tela **2**, 2455 (1960) [translation: Soviet Phys.—Solid State **2**, 2188 (1961)].
[12] D. J. Howarth and E. H. Sondheimer, Proc. Roy. Soc. (London) **A219**, 53 (1953).
[13] E. O. Kane, J. Phys. Chem. Solids **1**, 249 (1957).
[14] Frank Stern, J. Phys. Chem. Solids **8**, 277 (1959).
[15] A. Sagar, Phys. Rev. **117**, 93 (1960).

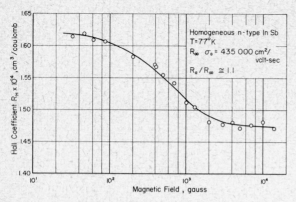

FIG. 1. Field dependence of Hall coefficient
in high-purity indium antimonide.

a value of the order of 7×10^{-4} ev for the ionization energy of a hydrogenic level.[16] This result is in agreement with the value of 6×10^{-4} ev reported by Nasledov for ultrapure specimens.[10] The conditions mentioned above, however, are responsible for a relatively large effective Bohr radius, i.e., for a substantial spreading of the electronic wave functions centered on different impurity atoms. Therefore, not many donor atoms can be accommodated in the crystal before one notices results of overlap of these wave functions. In the case of indium antimonide it appears from measurements by both Sladek and Putley that, for n-type impurity concentrations even as low as 10^{14} atoms cm^{-3}, the donor levels are effectively merged with the conduction band, and no carrier "freeze-out" occurs at low temperatures.[17,18] The effect of a magnetic field is to reduce the overlap of the donor wave functions. Because of the small effective mass in InSb, a significant carrier freeze-out can be achieved at moderate magnetic field strengths in material of donor concentrations in the 10^{14} cm^{-3} range.[17,19,20] In ultrapure specimens containing around 4×10^{13} extrinsic electrons per cm^{3}, Putley has measured "effective" ionization energies of around 2×10^{-4} ev in fields of 4000 gauss.[21] When cooled to 1.5°K, such a specimen exhibited sensitivity as a photoconductor for electromagnetic radiation having wavelengths of a few millimeters.

A high charge-carrier mobility in a semiconductor is useful not only because of the large augmentation of many galvanomagnetic phenomena, but also because the investigator can study the effects over the range from the weak to the strong magnetic field regions with the use of ordinary laboratory magnets. An example of

such measurements is illustrated in Fig. 1, which shows the Hall coefficient at 77°K in an InSb sample cut from a region in a single crystal which exhibited a high degree of homogeneity. The magnetic field in the region 400–14 000 gauss was monitored by an NMR gaussmeter. Below 90 gauss, the field was obtained by means of Helmholtz coils. It is apparent that in high-mobility materials galvanomagnetic measurements must be taken at very low field strengths to determine weak-field coefficients. We note in the figure that the weak-field plateau for the Hall coefficient is not achieved until the magnetic field is reduced to below 50 gauss. Data of the type shown are useful in that the strong-field Hall coefficient is directly related to the charge-carrier density—independent of the scattering mechanism. Thus, the measurements provide convenient means for determining the conductivity mobility. The weak-field Hall coefficient, on the other hand, is influenced by the charge-carrier scattering and the band structure. It, of course, yields the Hall mobility. Further information on scattering and band structure can be obtained from the magnetoresistance mobility and the directional magnetoresistance coefficients.

Magnetoresistance measurements in various III-V compounds have been carried out by a number of investigators. With no attempt to be complete, we shall cite in the references examples for indium antimonide,[22,23] indium arsenide,[24,25] indium phosphide,[26] gallium arsenide,[26,27] and gallium antimonide.[28] By making allowances for rather severe measurement problems, one may conclude with reasonable safety that the results are consistent with approximately spherical energy surfaces for the principal minima of the conduction bands.

The measurement problems alluded to above refer to the necessity for avoiding any perturbation of the Hall field in the specimen. In high-mobility materials, even a slight shorting of the Hall voltage can produce large magnetoresistive effects. Disturbances of the Hall field can occur through external contacts or by means of inhomogeneities in the measurement sample. The first problem, the so-called geometrical effects, has been discussed by a number of authors quite generally,[29] as well as in the special case for measurements on indium antimonide[23] and indium arsenide.[24] The question of inhomogeneities must be given special consideration in

[16] For a discussion of the hydrogenic model, see F. Stern and R. Talley, Phys. Rev. 100, 1638 (1955).

[17] R. J. Sladek, J. Phys. Chem. Solids 5, 157 (1958).

[18] E. H. Putley, Solid State Physics in Electronics and Telecommunications (Proceedings of the Brussels Conference), edited by M. Désirant and J. Michiels (Academic Press, Inc., New York, 1961), Vol. 2, p. 751.

[19] Y. Yafet, R. Keyes, and E. Adams, J. Phys. Chem. Solids 1, 137 (1956).

[20] R. Keyes and R. Sladek, J. Phys. Chem. Solids 1, 143 (1956).

[21] E. H. Putley, Proc. Phys. Soc. (London) 76, 802 (1960).

[22] G. L. Pearson and M. Tanenbaum, Phys. Rev. 90, 153 (1953).

[23] H. P. R. Frederikse and W. R. Hosler, Phys. Rev. 108, 1136, 1146 (1957).

[24] H. Weiss, Z. Naturforsch. 12a, 80 (1957).

[25] C. H. Champness and R. P. Chasmar, J. Electronics and Control 3, 494 (1957).

[26] M. Glicksman, J. Phys. Chem. Solids 8, 511 (1959).

[27] J. Edmond, R. Broom, and F. Cunnell, "Report of the meeting on semiconductors, Rugby, England, April, 1956" (The Physical Society, 1957), pp. 109–117.

[28] W. M. Becker and H. Y. Fan, Bull. Am. Phys. Soc. 6, 130 (1961). See also J. Appl. Phys. 32, 2094 (1961).

[29] J. R. Drabble and R. Wolfe, J. Electronics and Control 3, 259 (1957).

the case of the III-V compounds because of the possibility of anisotropic segregation of impurities. This phenomenon has been demonstrated by several investigators in the case of indium antimonide,[30,31] and may be expected to be important in a number of other compounds. That impurity concentration gradients can seriously affect galvanomagnetic voltages was pointed out sometime ago by Herring,[32] who has recently published a theoretical treatment of the effect of random inhomogeneities on electrical and galvanomagnetic measurements.[33] He also discusses the effect of macroscopic inclusions and the behavior of the laminar model. The problem of gross inhomogeneities was examined by Bate and Beer, and explicit solutions to the boundary value problem were obtained for the case where the carrier concentration varied exponentially along the specimen,[34] and where the variations in conductivity and Hall coefficient were approximated by step functions.[35]

When inhomogeneities exist, the current distribution in the sample is, in general, modified by the magnetic field. Thus, the galvanomagnetic voltages become dependent on probe position. In addition, the voltage changes due to the inhomogeneities may predominate over the effects which are characteristic of the material. This latter point is illustrated by the directional magnetoresistance measurements of Rupprecht et al. on tellurium-doped indium antimonide.[36] These investigators found that the important consideration was not the relation of the current to a general crystallographic direction, but rather the relation of the current to the specific direction [111] in which the crystal had been pulled. When these directions coincided, the transverse magnetoresistance was much larger. Further measurements by Rupprecht[37] on crystals pulled also in [100] and [113] directions revealed a similar anisotropy, namely, that the largest transverse magnetoresistance occurred when the current was in the direction of pull. The magnitude of the effect, which was largest for [111] directions of pull, became progressively less for the [100] and the [113] directions. Another interesting observation was the importance of the pulling speed in accenting the anisotropy. For the [111] pull, the $\Delta\rho/\rho_0$ dropped from a largest value of 0.7 for a speed of 0.25 mm/min, to a largest value of 0.02 for a pulling speed of 4.2 mm/min.

[30] J. B. Mullin and K. F. Hulme, J. Phys. Chem. Solids **17**, 1 (1960).
[31] W. P. Allred and R. T. Bate, J. Electrochem. Soc. **108**, 258 (1961).
[32] C. Herring (unpublished). See, for example, C. Herring, T. Geballe, and J. Kunzler, Bell System Tech. J. **38**, 657 (1959), p. 688.
[33] C. Herring, J. Appl. Phys. **31**, 1939 (1960).
[34] R. T. Bate and A. C. Beer, J. Appl. Phys. **32**, 800 (1961).
[35] R. T. Bate, J. C. Bell, and A. C. Beer, J. Appl. Phys. **32**, 806 (1961).
[36] H. Rupprecht, R. Weber, and H. Weiss, Z. Naturforsch. **15a**, 783 (1960).
[37] H. Rupprecht, Z. Naturforsch. (to be published).

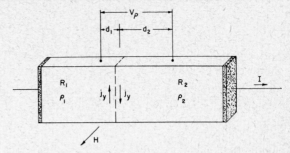

FIG. 2. Arrangement of resistivity probes in step-function model of inhomogeneous semiconductor. Length of specimen in the direction of current I is l, width in the y direction is w, and thickness in the H direction is b.

An inhomogeneous medium will usually produce an intermixing of Hall and magnetoresistance effects. This is illustrated by the expressions for the measured transverse magnetoresistance and Hall effects in the step function model when the current is normal to the plane separating regions of resistivities, ρ_1 and ρ_2, and respective Hall coefficients, R_1 and R_2[35]:

$$\left[\frac{\Delta\rho}{\rho_0}\right]_m = \left[\frac{\Delta\rho}{\rho_0}\right] + \frac{d_2R_2 - d_1R_1}{d_2\rho_2^0 - d_1\rho_1^0}\frac{(R_2 - R_1)H^2}{\rho_1 + \rho_2},$$

$$d_1, d_2 \ll w \ll l, \quad (1)$$

$$R_m = \frac{R_1\rho_2 + R_2\rho_1}{\rho_1 + \rho_2}, \qquad R \equiv R(H), \quad \rho \equiv \rho(H). \quad (2)$$

The arrangement of the resistivity probes is shown in Fig. 2; the Hall probes are located on the discontinuity at $y = \pm w/2$. In Eq. (1), the first term on the right is that due to the material alone, while the second term is that resulting from the inhomogeneity. We note that if R and ρ saturate, this term varies as H^2 and that it may be positive or negative. Thus, in strong magnetic fields this effect may predominate. In weak magnetic fields, it can also predominate if the material magnetoresistance coefficient $\Delta\rho/[\rho_0\mu_0^2H^2]$ is small, as it true of many n-type III-V materials. In the case of the Hall effect, the nonuniformities may introduce a significant magnetic field dependence. An example is shown by the data taken on a specimen in which a change in carrier density by a factor of 10 occurred in the current direction, as compared with the data measured on a uniform sample, shown in Fig. 3.

Thus, we see that measurement of magnetoresistance in strong magnetic fields is greatly complicated by the demands placed on the uniformity of the specimen. Actually, the situation is even more difficult when we are interested in extremely strong field strengths. For, as pointed out by Bate and Beer,[34] gradients in the magnetic field strength are in certain cases equivalent to gradients in the components of conductivity. This is evident from a comparison of the differential equations for the electric potential for the cases in question.

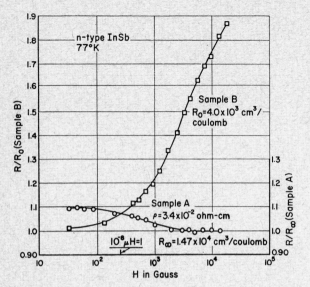

FIG. 3. Magnetic-field dependence of the normalized Hall coefficient of two high-purity n-type InSb samples. Carrier concentrations are: sample A: 4.25×10^{14} cm^{-3}; sample B: 5×10^{14} to 5×10^{15} cm^{-13} (after Bate *et al.*[35]).

(a) Nonuniformity in σ_t:

$$\sigma_t^{-1} d\sigma_t/dx \equiv K \neq 0, \quad d\beta/dx = 0,$$

$$\nabla^2 V + K \frac{\partial V}{\partial x} + K\beta \frac{\partial V}{\partial y} = 0.$$

(b) Nonuniformity in H:

$$d\beta/dx \neq 0, \quad d\sigma_t/dx = (\partial\sigma_t/\partial H)(dH/dx)$$

$$\nabla^2 V + \frac{1}{\sigma_t} \frac{\partial \sigma_t}{\partial H} \frac{dH}{dx} \frac{\partial V}{\partial x} + \left(\frac{\beta}{\sigma_t} \frac{\partial \sigma_t}{\partial H} + \frac{\partial \beta}{\partial H} \right) \frac{dH}{dx} \frac{\partial V}{\partial y} = 0,$$

where $\beta[\equiv RH/\rho_H]$ is the tangent of the Hall angle, σ_t are the diagonal components of the conductivity tensor, and $\pm\beta\sigma_t$ are the off-diagonal components. The forms of (b) at strong fields, $\beta > 1$, are as follows:

(b-1) $\quad \nabla^2 V + 2K \dfrac{\partial V}{\partial x} + K\beta(x) \dfrac{\partial V}{\partial y} = 0,$ if $\Delta\rho/\rho_0$ saturates,

(b-2) $\quad \nabla^2 V + K \dfrac{\partial V}{\partial x} + K\beta \dfrac{\partial V}{\partial y} = 0,$ if $\Delta\rho/\rho_0 \sim H$,

where $K \equiv -H^{-1}dH/dx$.

To illustrate the effect of a magnetic field gradient on the current distribution in the specimen, we consider case (b-1). The differential equation is then identical with that for the case of nonuniform conductivity and the expression given by Bate and Beer for the current density is[34]

$$j_x = \frac{I}{wb} \frac{\gamma/2}{\sinh\gamma/2} e^{-\gamma y/w}, \quad \gamma \equiv Kw\beta.$$

An important factor at strong fields is obviously β, the tangent of the Hall angle. Some recent measurements by Hieronymus and Weiss[38] on intrinsic InSb, which therefore possessed good charge-carrier homogeneity, suggest that values of β of 30 are readily possible. This result yields a value for γ of 0.75 for a specimen 1-cm wide in a field gradient such that $H^{-1}dH/dx$ is $2\frac{1}{2}\%$ per cm. The ratio of the current densities at the two edges of the sample is then $j_x(-w/2)/j_x(w/2) = 2.1$. Thus, we see that when large Hall angles are involved even a relatively small degree of nonuniformity can have a significant effect.

Another item of interest at strong magnetic fields in high-mobility bands is the possible appearance of quantum effects in the transport properties. For example, in n-type InSb at 77°K, Landau splitting energies, $\hbar\omega$, of around $3kT$ can be obtained at 20 kgauss. This has led numerous investigators[39-42] to measure magnetoresistance in the region where $\hbar\omega \gg kT$ (or $\hbar\omega \gg \zeta$, where ζ is the Fermi level in degenerate material) to obtain information on electron orbit quantization effects. The additional condition $\omega\tau \gg 1$—where τ is the carrier relaxation time, and the cyclotron frequency ω is given by eH/m^*c—is of course readily met in indium antimonide. Quantum effects can be responsible for a nonzero longitudinal magnetoresistance in isotropic solids, which does not saturate, and for failure of the transverse magnetoresistance to saturate.[43,44]

Most of the experimental results provide qualitatively a reasonably convincing argument for the observance of quantum effects in InSb. We know, of course, that any disturbance of the current distribution in the sample—whether by inhomogeneities in specimen or magnetic field, by contacting leads, or by surface conduction—will affect the saturation of the magnetoresistance and may produce significant results at large magnetic fields. Therefore, to secure data which can be used in quantitative studies of quantum effects imposes an extremely demanding task on the experimenter.[45]

We have indicated how inhomogeneities can dominate the galvanomagnetic behavior in high-mobility materials. That this may have been the state of affairs in the case of indium antimonide is suggested by the recent experiments of Weiss, who finds that a negligible magnetoresistance effect can be attributed to the elec-

[38] H. Hieronymus and H. Weiss, Solid State Electronics (to be published).

[39] J. C. Haslett and W. F. Love, J. Phys. Chem. Solids **8**, 518 (1959).

[40] R. T. Bate, R. K. Willardson, and A. C. Beer, J. Phys. Chem. Solids **9**, 119 (1959).

[41] Kh. I. Amirkhanov, R. I. Bashirov, and Yu. E. Zakiev, Doklady Akad. Nauk S.S.S.R. **132**, 793 (1960) [translation: Soviet Phys.—Doklady **5**, 556 (1960)].

[42] R. J. Sladek, J. Phys. Chem. Solids **16**, 1 (1960).

[43] P. N. Argyres and E. N. Adams, Phys. Rev. **104**, 900 (1956).

[44] E. N. Adams and T. D. Holstein, J. Phys. Chem. Solids **10**, 254 (1959).

[45] See, for example, discussion by Herring, Geballe, and Kunzler in Appendix A of reference 32.

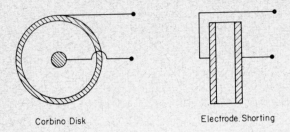

Corbino Disk   Electrode Shorting

FIG. 4. Mechanisms to produce shorting of the Hall voltage.

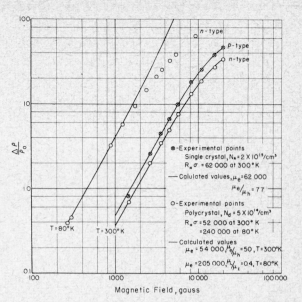

FIG. 5. Corbino magnetoresistance in indium antimonide (after Beer *et al.*[50]).

trons in the conduction band of indium antimonide.[46] Since the large magnitudes of the inhomogeneity effects result from perturbations of the Hall voltage, it is of interest to examine galvanomagnetic measurements under conditions such that the Hall field is absent. A common example is the Corbino disk[47] or the large width-to-length ratio specimen where the transverse field is shorted by the end electrodes[48] (see Fig. 4). The latter arrangement has been used by Goldberg to measure directly the magnetoconductivity in germanium.[49] Some unpublished data[50] on the Corbino magnetoresistance in InSb carried out at the Battelle Memorial Institute are shown in Fig. 5. It is readily shown that deviations from an H^2 dependence at strong fields result from a nonsaturation of the ordinary magnetoresistance. For the room temperature data on intrinsic InSb, this is attributed to the effect of the holes,[51] and the data shown are fitted using an appropriate electron-hole mobility ratio. In this case, spherical energy surfaces and acoustic phonon scattering were assumed. At 80°K, in the extrinsic range, the hole concentration is negligible and it appears likely that quantum effects may be responsible for the break away from an H^2 line. This point has been made by Frederikse and Hosler,[52] who found that the slope in a plot of their data at high magnetic fields approached a linear dependence in accordance with Argyres' predictions.[53]

Data at 77°K on a more pure sample of InSb are available from some recent measurements by Bate,[54] which also provided a check on the theoretical relation connecting the Corbino effect, transverse magnetoresistance, and conductivity mobility. The procedure

was to measure first the Corbino effect at 77°K as a function of magnetic field on a disk with soldered annular electrodes. Since the disk was 12.9 mm in diameter and 2.28 mm thick, with a hole 3.35 mm in diameter at the center, it was subsequently possible to cut out a sample in the shape of a rectangular parallelepiped, and then to measure directly the Hall effect, resistivity, and transverse magnetoresistance as functions of field at 77°K. A determination of the resistance of the contacts and connecting leads of the original Corbino disk was then made by subtracting from its measured resistance that resistance calculated, using its dimensions and the resistivity measured on the parallelepiped. The total resistance of the contacts and leads was about 35% of the disk resistance at zero field. The Corbino data were then corrected for contact effects by subtracting the contact resistance (assumed independent of magnetic field) from the resistance of the disk in the field. The resulting data, along with those of the transverse magnetoresistance, are shown in Fig. 6. When measurements are taken in the region where the Hall coefficient has saturated, it can be shown that a relation exists between the Corbino effect, the transverse magnetoresistance, and the conductivity mobility, as follows[50]:

$$\frac{\rho}{\rho_0}\left[\left(\frac{\rho}{\rho_0}\right)_c - \left(\frac{\rho}{\rho_0}\right)\right] \equiv F = \mu_0^2 H^2, \quad R_H \to R_\infty.$$

The quantity $(\mu_0 H)^2$ is shown as the "theoretical line" in Fig. 6, and values of F calculated from the transverse and Corbino magnetoresistance data are plotted for comparison. The agreement is seen to be excellent.

In the case of the Corbino disk, one can establish for a radial inhomogeneity in conductivity or in magnetic

[46] H. Weiss, J. Appl. Phys. **32**, 2064 (1961).

[47] For discussion of the Corbino effect, see E. P. Adams, Proc. Am. Phil. Soc. **54**, 45 (1915); H. Weiss and H. Welker, Z. Physik **138**, 322 (1954).

[48] O. Madelung, Naturwissenschaften **42**, 406 (1955).

[49] C. Goldberg, Phys. Rev. **109**, 331 (1958).

[50] A. C. Beer, T. C. Harman, R. K. Willardson, and H. L. Goering, "Research and development work on semiconducting materials of unusually high mobility," Office of Technical Services Rept. PB-121288 (unpublished).

[51] A. C. Beer, J. A. Armstrong, and I. N. Greenberg, Phys. Rev. **107**, 1506 (1957).

[52] H. P. R. Frederikse and W. R. Hosler, *Solid State Physics in Electronics and Telecommunications* (Proceedings of the Brussels Conference), edited by M. Désirant and J. Michiels (Academic Press, Inc., New York, 1960), Vol. 2, p. 651.

[53] P. N. Argyres, Phys. Rev. **109**, 1115 (1958); J. Phys. Chem. Solids **4**, 19 (1958).

[54] R. T. Bate (unpublished).

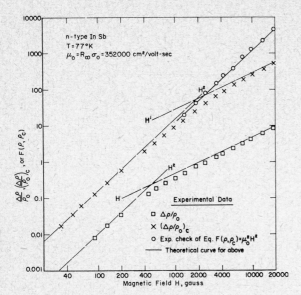

FIG. 6. Magnetic-field dependence of transverse magnetoresistance and Corbino magnetoresistance in InSb specimen. Data are also plotted to provide an experimental check of the theoretical relationship, valid when R_H saturates:

$$\frac{\rho}{\rho_0}\left[\left(\frac{\rho}{\rho_0}\right)_c - \left(\frac{\rho}{\rho_0}\right)\right] \equiv F(\rho,\rho_c) = (\mu_0 H)^2.$$

field that the potential does not contain the Hall angle, and therefore the Corbino magnetoresistance involves only $\sigma_t(r)$, where σ_t are the diagonal elements of the conductivity tensor, suitably averaged over the distance between the points where the potential is measured. There is no Hall field to produce the augmented effects at strong fields, and the potential assumes the simple form shown below.

$$V(r) - V(a) = -\frac{I}{2\pi b}\int_a^r \frac{dr'}{r'\sigma_t(r')},$$

where b is the thickness of the disk, and I is the current. The relationship is valid for any variation of σ_t with r—so long as the integral exists—and σ_t may depend explicitly on r or implicitly through $H(r)$.

A similar situation exists for a magnetoconductivity measurement when the impurity or magnetic field gradient is in the direction of the electric field. In this case, the expression for the potential is

$$V(x) - V(0) = -\frac{I}{wb}\int_0^x \frac{dx'}{\sigma_t(x')},$$

where wb is the cross-sectional area of the specimen normal to the x direction, and σ_t may depend on x either explicitly or implicitly through $H(x)$. On the other hand, if the gradients in the above example are in the y direction, one should measure an ordinary magnetoresistance effect with the current in the x direction to avoid distortions from the inhomogeneity effect.[34] In such a case, when β is independent of position, the expression for the potential can be put in the simple form

$$V(x,y) = -\left[I \bigg/ \left\{b\int_{-w/2}^{w/2} dy/\rho(y)\right\}\right][x+\beta y],$$

where the resistivity $\rho(y)$ is given by $\rho(y) \equiv \{\sigma_t(y)[1+\beta^2]\}^{-1}$. Hence, $V(x_1,y) - V(x_2,y)$ depends in this case only on an average of $\rho^{-1}(y)$.

Thus, we see that Corbino disk or magnetoconductivity measurements can be helpful in some cases in separating the contributions from nonuniformities in specimen or in magnetic field from the magnetoresistive behavior of the material proper. An experimental problem is to secure negligible resistance—or perhaps more important, uniform resistance—at the shorting contacts. Nevertheless, sensible results have been obtained in the case of indium antimonide, and it appears that the techniques may be useful in additional applications—especially where it is desirable to secure further support for data resulting from standard measurements.

ACKNOWLEDGMENTS

The author wishes to express his gratitude to several colleagues at the Battelle Memorial Institute, particularly I. Adawi, R. Bate, and J. Duga, for helpful discussions. He is also indebted to H. Rupprecht and H. Weiss for advance copies of articles in the process of publication.

3–5 Compounds: Band Structure, Electrical and Optical Properties

O. Madelung, *Chairman*

Magnetoreflection Experiments in Intermetallics

George B. Wright and Benjamin Lax

Lincoln Laboratory, Massachusetts Institute of Technology, Lexington 73, Massachusetts

Magnetoreflection experiments involving both intraband and interband transitions can provide valuable information about the electronic band structure of semiconductors. In the intraband experiments, performed near the plasma reflection edge, the application of a magnetic field splits the edge and results in the formation of two minima separated by the cyclotron frequency. It is thus possible to determine the cyclotron frequency directly, at room temperature, and for high carrier concentrations. When scattering losses are taken into account in the theory, it becomes possible to determine the carrier concentration, scattering time, and effective mass from the optical measurements alone. The theory of the effect is discussed for applied magnetic field transverse and parallel to the direction of propagation, and experimental results are presented for InSb, InAs, and HgSe. A consistent fit to Kane's theory for the variation of mass with concentration in InSb is obtained when previously published data have been corrected for reststrahl dispersion.

Interband magnetoreflection experiments can be useful in cases where high absorption coefficients or difficulty of preparing thin samples make transmission experiments unfavorable. This type of experiment yields information on energy band gaps, effective masses, and *g* factors. Experimental data for the direct transition in InSb are presented.

INTRODUCTION

INFRARED methods of determining the charge carrier effective mass in semiconductors have become increasingly popular in the last five years. Two methods involving no magnetic field depend upon the phenomena of free carrier infrared absorption, and free carrier plasma reflection.[1] Additional effects observed with the application of a magnetic field are cyclotron resonance,[2] Faraday rotation,[3] Voigt effect,[4] and magnetoplasma reflection.[5] All of these effects are different aspects of the same basic phenomenon, charge carrier intraband transitions, and may be described by one set of equations by adjusting the relative magnitudes of four parameters; the applied infrared frequency, the plasma frequency, the cyclotron frequency, and the charge carrier scattering frequency. Optical methods of determining effective masses involving *interband* transitions are interband magnetoabsorption,[6] interband Faraday rotation,[7] and interband magnetoreflection.[8] Papers dealing with each of the methods listed here were presented at this conference. This paper is con-

cerned chiefly with magnetoplasma reflection and interband magnetoreflection.

MAGNETOPLASMA REFLECTION

The free charge carriers in a semiconductor will have a complex index of refraction $(n-ik)$ in a magnetic field given by[5]

$$(n-ik)_{\pm}{}^2 = \kappa'\left[1 - \frac{\omega_p{}^2}{\omega(\omega \mp \omega_c - i\nu)}\right] \quad (1)$$

for propagation parallel to the static magnetic field. The $(\pm)$ sign refers to the sense of circular polarization relative to the magnetic field, and κ' is the dielectric constant of the crystal in the absence of free carriers. The effect on the index of refraction of the carriers is contained in the second term in the brackets, where $\omega_p{}^2 = Ne^2/m^*\epsilon_0\kappa'$, the plasma frequency squared, is proportional to carrier concentration N and effective mass m^*. The other two parameters are ν, the charge carrier scattering frequency, and $\omega_c = eB/m^*$, the cyclotron frequency. For the lossless case, both ν and k, the extinction coefficient, go to zero. Then the reflection coefficient is given by $R = (n-1)^2/(n+1)^2$. With no magnetic field, the second term reduces the index of refraction as the square of the wavelength, causing a decrease in the reflectivity. When $n=1$, at $\omega = [\kappa'/(\kappa'-1)]^{\frac{1}{2}}\omega_p$, the reflectivity is zero, and when $\omega = \omega_p$, the index of refraction is zero, and $R = 100\%$. For $\kappa' = 16$, as it may be in a semiconductor, these two frequencies are in the ratio 1.03, and the steep rise from zero to 100% reflectivity is the plasma reflection edge. If losses may be neglected, the value of n on the high-frequency side of the plasma minimum may be found

---

* Operated with support from the U. S. Army, Navy, and Air Force.

[1] W. G. Spitzer and H. Y. Fan, Phys. Rev. **106**, 882 (1957).

[2] E. Burstein, G. S. Picus, and H. A. Gebbie, Phys. Rev. **103**, 825 (1956); R. J. Keyes, S. Zwerdling, S. Foner, H. H. Kolm, and B. Lax, Phys. Rev. **104**, 1804 (1956).

[3] T. S. Moss, S. D. Smith, and K. W. Taylor, J. Phys. Chem. Solids **8**, 323 (1959).

[4] S. Teitler and E. D. Palik, Phys. Rev. Letters **5**, 546 (1960).

[5] B. Lax and G. B. Wright, Phys. Rev. Letters **4**, 16 (1960).

[6] For complete list of references see review by B. Lax and S. Zwerdling, Prog. in Semiconductors **5**, 223 (1960).

[7] B. Lax and Y. Nishina, Phys. Rev. Letters **6**, 464 (1961).

[8] R. N. Brown, J. G. Mavroides, M. S. Dresselhaus, and B. Lax, Phys. Rev. Letters **5**, 241, 243 (1960).

from the reflectivity as $n = (1 + \sqrt{R})/(1 - \sqrt{R})$, and a plot of n^2 versus λ^2 will yield κ' and m^*, if N, the carrier concentration, is known.[1] As the magnetic field is turned on, the charge carriers will rotate about the field axis, and the *effective frequency* of the right-hand circular polarization will differ from that of the left hand. The plasma edge shifts to higher frequencies for the polarization which rotates with the charge carrier, and to lower frequencies for the other polarization. The motion of the plasma minima is given by

$$\omega_{min\pm} \approx \omega_p \pm \tfrac{1}{2}\omega_c + \tfrac{1}{8}\omega_c^2/\omega_p. \qquad (2)$$

Thus, by following the motion of the minima with field, one obtains a direct measurement of the cyclotron frequency, independent of transport measurements. The two polarizations excite normal modes of the system in this configuration, so that the reflectivities are additive. This has the important experimental consequence that it is unnecessary to use polarized light to obtain a result which may be interpreted quantitatively. In Fig. 1 are shown preliminary data of F. L. Galeener,[9] taken on a sample of InSb at room temperature for propagation parallel to the magnetic field. The line shape is that expected from theoretical calculations when losses are included.

When the magnetic field is transverse to the direction of propagation, the expression for the index of refraction is considerably more complicated. For the lossless case it is given by[5]

$$n^2_\perp = \kappa'\left[1 - \frac{\omega_p^2}{\omega^2}\frac{(\omega^2 - \omega_p^2)}{(\omega^2 - \omega_p^2 - \omega_c^2)}\right], \qquad (3)$$

$$n^2_{\parallel}(B) = n^2(0),$$

where the subscripts refer to the polarization of the electric vector relative to the static magnetic field. In this case, one might resolve the electric vector into two counter-rotating vectors *in the plane of incidence*, phased so that no longitudinal component exists outside the sample. These two vectors would not excite normal modes, however, because of coupling at the sample interface. In fact, a polarization develops at the surface which is just the ac Hall effect. For $E \perp B$, two minima develop in the reflectivity, which move approximately as given by Eq. (2). More detailed investigation shows that, even for the lossless case, the motion of the minima is not linear in magnetic field at very low field strengths. The high-frequency minimum develops from the zero field minimum at $\omega = [\kappa'/(\kappa'-1)]^{\frac{1}{2}}\omega_p$, while the low-frequency minimum starts at $\omega = \omega_p$. For actual cases, it is more profitable to calculate the reflectivity numerically, including losses, than to attempt an analytical solution. Data taken on HgSe for the magnetic field transverse to the direction of propagation are shown in Fig. 2, where the solid lines represent the theoretically computed reflectivity, including losses. The carrier concentration, dielectric constant, effective mass, and scattering time have been determined by infrared magneto-reflection experiments alone. This work on HgSe, done in collaboration with A. J. Strauss and T. C. Harman and reported elsewhere,[10] has contributed to a demonstration that the conduction band of HgSe has a nonquadratic dependence on k similar to InSb. The effective mass and charge carrier scattering time, determined optically, are in excellent agreement with those found by Harman from transport measurements.

Reflection data for an *n*-type sample of InSb with $n = 1.8 \times 10^{18}$ cm^{-3} as determined optically and by transport measurements are shown in Fig. 3. It is of interest to compare the mass determined here with results published elsewhere.[1,11] Considering the interaction between the conduction and valence bands alone, E. O. Kane[12] found that the conduction band energy had a dependence on wave vector given by

$$E = \tfrac{1}{2}\hbar^2 k^2/m + \tfrac{1}{2}[E_g + (E_g^2 + 8k^2 P^2/3)^{\frac{1}{2}}], \qquad (4)$$

as long as the spin orbit splitting was much greater than E_g or kP. Here, P is the matrix component of

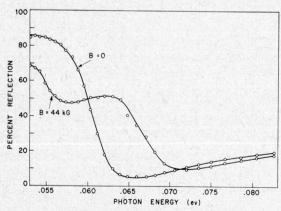

FIG. 1. Longitudinal magnetoplasma reflection in InSb. (After F. L. Galeener.)

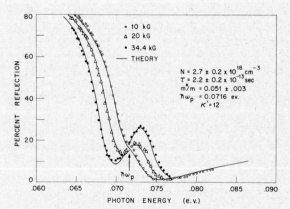

FIG. 2. Transverse magnetoplasma reflection in HgSe.

[9] F. L. Galeener (private communication).

[10] G. B. Wright, A. J. Strauss, and T. C. Harman, Bull. Am. Phys. Soc. **6**, 155 (1961).

[11] S. D. Smith, T. S. Moss, and K. W. Taylor, J. Phys. Chem. Solids **11**, 131 (1959).

[12] E. O. Kane, J. Phys. Chem. Solids **1**, 248 (1957).

momentum between conduction band and valence band Bloch functions at $k=0$. For sharp degeneracy, N and k are related by $N=k^3/3\pi^2$. Noting that we measure an effective mass defined as $m/m^*=m/\hbar^2 k \ (dE/dk)$, we obtain the relation between carrier concentration and effective mass given by

$$\left(\frac{m^*/m}{1-m^*/m}\right)^2 = 32.5\times10^{-32}E_g{}^2/P^4$$
$$+8.27\times10^{-30}N^{\frac{2}{3}}/P^2, \quad (5)$$

with E_g in ev and P in ev cm. Published data for InSb are shown plotted this way in Fig. 4. The straight line is drawn for $P=8.7\times10^{-8}$ ev cm, a value suggested by H. Ehrenreich for III-V compounds,[13] and supported by our measurements. A similar value was found for HgSe.[10]

In the determination of the effective mass from the zero field susceptibility, a possibility for error exists if dispersive mechanisms other than free carriers are present. For example, Moss et al. have shown that the reststrahl band in InSb contributes a susceptibility equal to $-4\lambda^2/(\lambda_0{}^2-\lambda^2)$, where $\lambda_0=54.6\ \mu$.[14] Then the expression for the index of refraction will have the form

$$n^2=\kappa'\left[1-\frac{\omega_p{}^2}{\omega^2}-\frac{4}{\kappa'}\frac{\lambda^2}{\lambda_0{}^2-\lambda^2}\right], \quad (6)$$

which for $\omega=2\omega_0$ will be very closely represented by

$$n^2=\kappa'\left[1-\frac{\omega_p{}^2+4\omega_0{}^2/\kappa'\omega_p{}^2}{\omega^2}\right]. \quad (7)$$

Then the mass found from the slope of the n^2 versus λ^2 curve will be too small by a factor $1+4\omega_0{}^2/\kappa'\omega_p{}^2=1.06$ for $\omega_p=2\omega_0$ and $\kappa'=16$. The results of Spitzer and Fan were converted by determining the plasma wavelength from the published data, and using the calculated value of κ' and the published carrier concentration to find the effective mass. Within the necessarily limited pre-

cision of this method, the resulting consistency with Kane's theory is quite good (Fig. 4).

The reflection curve of InAs with $n=5.3\times10^{18}$ cm^{-3} shown in Fig. 5 does not have theoretically predicted shape, apparently because of surface damage which could not be eliminated by polishing and etching. A shift of the plasma edge between front and back surface of a sample 1 mm thick indicated a carrier concentration change of 7% normal to the surface of the sample. The carrier concentration in the plane of the surface was checked by P. Maker using the position of the plasma edge as an indicator,[15] and found to be constant to within $\frac{1}{2}\%$. The effective mass for this sample deduced from the zero field data was $m^*/m=0.06$, $\kappa'=12.6$. The three basic magnetooptical experiments in the infrared involving free carrier susceptibility are cyclotron resonance, Voigt effect and Faraday rotation, and magnetoplasma reflection. Conditions for cyclotron resonance are most favorable when the cyclotron frequency may be brought to the operating frequency, and is simultaneously greater than the scattering frequency. At the same time, it is desirable that the plasma frequency of all carriers be far removed if troublesome corrections are to be avoided. For Faraday rotation and

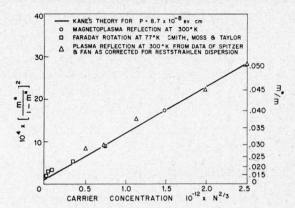

FIG. 4. Variation of effective mass with carrier concentration in InSb.

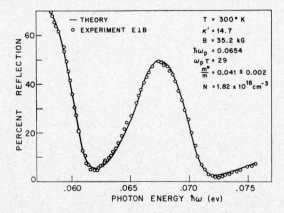

FIG. 3. Transverse magnetoplasma reflection in InSb.

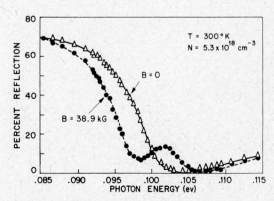

FIG. 5. Transverse magnetoplasma reflection in InAs.

[13] H. Ehrenreich, Phys. Rev. **120**, 1951 (1961).
[14] T. S. Moss, S. D. Smith, and T. D. H. Hawkins, Proc. Phys. Soc. (London) **B70**, 776 (1957).
[15] P. D. Maker, J. Baker, and D. F. Edwards, Bull. Am. Phys. Soc. **6**, 116 (1961).

Voigt effect, one wishes to be far removed from both the cyclotron frequency and the plasma frequency, and to have the scattering frequency much smaller than the operating frequency. The magnetoplasma reflection experiment is done near the plasma frequency for cyclotron frequencies much less than the plasma frequency, but greater than the scattering frequency. The experiments are thus seen to be complementary, and the choice of which one to perform will depend on material properties and instrumentation.

INTERBAND MAGNETOREFLECTION

A second type of magnetoreflection experiment involves interband transitions. When the magnetic field is applied, Landau levels form in the two bands, and cause peaks in the optical transition probability which have previously been observed in transmission.[6] For the transmission experiments, it has been necessary to prepare samples only a few microns thick, and to avoid a mounting of the sample which would cause strain on cooling to He temperature.[16] For higher energy transitions, such as the 111 direct transitions in Germanium and III-V compounds, the background absorption is so high that transmission experiments are not practical. To test the feasibility of interband magnetoreflection experiments, a spectrum was run on high purity InSb.[17,18] The experimentally obtained energy level versus magnetic field diagram obtained with a CaF_2 prism double pass monochromator is shown in Fig. 6 for a sample at liquid-nitrogen temperature. The data were obtained by setting the photon energy constant, and sweeping the magnetic field, so that the operating point in Fig. 6 is a horizontal line. At the points where the horizontal line intersects the Landau levels, characteristic deflections of the pen were obtained as shown in Fig. 7. The greatest peak-to-peak deflection is about 10% of the background level. These curves were taken with polarized light and the higher resolution obtained from a diffraction grating, and show fine structure not plotted in Fig. 6. From the zero field intercept of the lines, we deduce an energy gap of 0.228 ev at 80°K. The lines shown in Fig. 6 are attributed to transitions between the valence band and the $n=0$, 1, and 2, spin up and spin down conduction band states, and the splitting of the first two lines with field gives an electron g factor of 48. These transitions have been the subject of a much more detailed, high resolution study at helium temperature.[19]

In order to explain the line shape in reflection, it is necessary to obtain the relations for the dispersion as

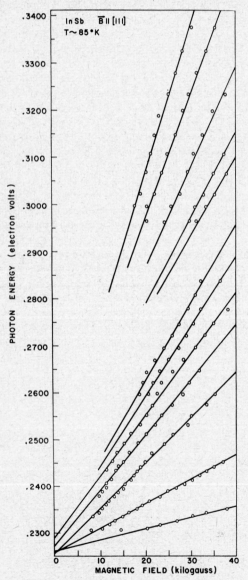

FIG. 6. Landau levels in InSb determined by transverse magnetoreflection.

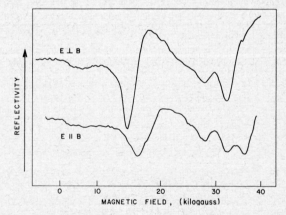

FIG. 7. Interband magnetoreflection in InSb.
$T\sim85°K$. $h\nu=0.2367$ ev.

[16] G. G. Macfarlane, T. P. McLean, J. E. Quarrington, and V. Roberts, Phys. Rev. Letters **2**, 252 (1959); D. F. Edwards and V. J. Lazazzera, Phys. Rev. **120**, 420 (1960).
[17] Purchased from Consolidated Mining and Smelting Company of Canada.
[18] G. B. Wright and B. Lax, Bull. Am. Phys. Soc. **6**, 18 (1960)·
[19] S. Zwerdling, W. H. Kleiner, and J. P. Theriault, J. Appl. Phys. **32**, 2118; S. Zwerdling and J. P. Theriault, Bull. Am. Phys. Soc. **6**, 147 (1961).

well as the absorption for interband transitions in the presence of a magnetic field. It can be shown[7,20] that when losses due to scattering are neglected, $(n-ik)^2$ is given by

$$(n-ik)^2 = \epsilon + \frac{A\omega_c}{2} \sum_n \left[\frac{-2}{\omega_n^{\frac{1}{2}}} + \frac{1}{(\omega_n+\omega)^{\frac{1}{2}}} + \frac{1}{(\omega_n-\omega)^{\frac{1}{2}}} \right], \quad (8)$$

where $\omega_n = (n+\frac{1}{2})\omega_c + \omega_g$ is the frequency corresponding to interband transitions in the presence of a magnetic field. Here ω_g is the energy gap frequency at zero field, and ω_c is the reduced cyclotron frequency of the two bands. To analyze a particular line, we shall make the assumption that a single transition is a small perturbation on the zero field case, and that if we singled out this line, we could integrate the effect of the rest of the terms. We could then rewrite Eq. (9) for $\omega > \omega_g$ and including losses as

$$n^2 - k^2 = \epsilon + A[2(\omega_g)^{\frac{1}{2}} - (\omega_g+\omega)^{\frac{1}{2}}] + \Delta, \quad (9)$$
$$2nk = A(\omega-\omega_g)^{\frac{1}{2}} + \delta,$$

where

$$\Delta = \mathrm{Re}\frac{A\omega_c}{2}\left(\frac{-2}{\omega_n^{\frac{1}{2}}} + \frac{1}{(\omega_n+\omega)^{\frac{1}{2}}} + \frac{1}{(\omega_n-\omega)^{\frac{1}{2}}} \right), \quad (10)$$

$$\delta = \mathrm{Im}\frac{A\omega_c}{2}\left(\frac{1}{(\omega_n-\omega)^{\frac{1}{2}}} \right).$$

We shall assume that we are far enough above the energy gap that

$$\omega - \omega_g \gg \nu.$$

In order to solve for the perturbed quantities in n and k over the zero field case, we can readily show that

$$\Delta = 2n_0\Delta n - 2k_0\Delta k,$$
$$\delta = 2k_0\Delta n + 2n_0\Delta k. \quad (11)$$

From this we obtain

$$\Delta n = \frac{1}{2}\frac{n_0\Delta + k_0\delta}{(n_0^2+k_0^2)},$$
$$\delta = \frac{1}{2}\frac{-k_0\Delta + n_0\delta}{n_0^2+k_0^2}. \quad (12)$$

From

$$R = 1 - \frac{4n}{(n+1)^2+k^2}, \quad (13)$$

---

[20] L. I. Korovin, Fiz. Tverdogo Tela 1, 1311 (1960).

we find the change in R for a change in n and k is given by

$$\Delta R = 4\frac{(n_0^2-k_0^2-1)\Delta n + 2nk\Delta k}{[(n_0+1)^2+k_0^2]^2}. \quad (14)$$

Substituting Eq. (12) in Eq. (14), we obtain the final result

$$\Delta R = 2\frac{(n_0^2-3k_0^2-1)n_0\Delta + (3n_0^2-k_0^2-1)k_0\delta}{(n_0^2+k_0^2)[(n_0+1)^2+k_0^2]^2}.$$

Noting that $n_0 > k_0$, we see that it is possible to account qualitatively for the sign of the change in reflection on this simple model. To perform a quantitative calculation of line shape, which should account for the decrease in amplitude with increase in magnetic quantum number, it is necessary to include the dependence of n_0 and k_0 on energy, as given by Eq. (9) if Δ and δ are neglected, and to include the scattering frequency ν, particularly if energies are near the absorption edge.

In conclusion, we have shown that infrared magneto-reflection measurements are useful in the determination of semiconductor energy band parameters. For transitions within the band, it has been possible to determine the transport properties by optical measurements alone, and to make measurements which would have been inaccessible by transmission techniques. The observation of interband magnetoreflection has demonstrated the possibility of measuring transitions such as the 111 direct transition in Ge and the III-V compounds which cannot be observed in transmission because of high background absorption. The use of bulk samples for reflection eliminates the problem of thin sample preparation, strain free mounting, and measurement of sample temperature encountered for transmission experiments.

ACKNOWLEDGMENTS

The authors would like to express their thanks to F. L. Galeener for permission to use his data on InSb, and to D. F. Edwards and P. D. Maker for collaboration in measurements on InAs. M. C. Plonko has been of major assistance in sample preparation and reflection measurements, and E. A. Curran has helped with the instrumentation. Our thanks also go to Dr. A. J. Strauss and T. C. Harman for help with the materials problem, and for many stimulating conversations over the period of the work.

Oscillatory Magnetoabsorption in InSb Under High Resolution

S. ZWERDLING, W. H. KLEINER, AND J. P. THERIAULT

Lincoln Laboratory,* Massachusetts Institute of Technology, Lexington 73, Massachusetts

Polarized oscillatory magnetoabsorption spectra of exciton formation transitions in InSb have been measured under high resolution at liquid helium temperature. An unstrained high-mobility sample 5 μ thick and magnetic fields up to 39.1 kgauss were used. Diffraction grating dispersion for the region $3.7 < \lambda < 6.0\ \mu$ made possible a resolution of 1–7×10^{-4} ev and the determination of absorption maxima to $\pm 5 \times 10^{-5}$ ev. Sixteen minima were resolved with $\mathbf{E} \perp \mathbf{B}$ for a field of only 5.0 kgauss. The initial interpretation of the detailed spectra obtained involved fitting theoretical spectra calculated from the theories of Kane, Luttinger, and Elliott-Loudon in order to account for band nonparabolicities, valence band degeneracy effects, and exciton binding energies, respectively. The results experimentally established the light-hole nonparabolicity, the valence band degeneracy effects for low magnetic quantum numbers, and the existence of high-field excitons in InSb.

POLARIZED high-resolution diffraction grating spectra have been obtained recently[1] in an oscillatory magnetoabsorption experiment on InSb conducted at liquid helium temperature, and the initial analysis and interpretation of the data have revealed new information about exciton formation and the electronic band structure of this semiconductor. The zero-field absorption edge occurred from 0.2346–0.2382 ev, and 16 clearly-defined absorption maxima were observed at photon energies greater than 0.2360 ev and a field of only 5.0 kgauss. The most intense absorptions persisted

even at 2.5 kgauss. The lowest absorption maximum shifted to higher energy with field at a slowly increasing rate. The shift of the second maximum to higher energy was greater but essentially linear, whereas the higher maxima shifted in the same direction at successively greater rates, but for almost every one, the rate decreased slowly with field.

Experiments of this kind on InSb, done several years ago, were conducted at room temperature with unpolarized radiation, using the much lower dispersion of a prism spectrometer.[2,3] No more than four broad absorption maxima were seen, and these were identified as direct interband transitions between $k=0$ valence and conduction bands. Recently, ten additional interband transitions were reported[4,5] as a result of a magnetoreflection experiment done at liquid nitrogen temperature.

The experiments to be described here have yielded extensive and detailed spectra obtained by measuring the transmission of an InSb sample using linearly polarized infrared radiation with photon energies larger than the forbidden gap and strong external magnetic fields. The sample studied was 5 μ thick and was prepared by grinding and polishing a section from an n-type single crystal ingot having 2×10^{14} impurities/cc and a mobility of 4.7×10^{5} cm^2/v sec at 77°K. The sample was attached at one end to a suitable copper support so as to avoid thermally-induced strains on cooling to liquid helium temperature. The optical cryostat containing the sample and its support was mounted on an electromagnet capable of producing fields up to 39.1 kgauss. The magnet was calibrated so as to produce particular field strengths with a precision of $\pm 0.1\%$. The double-pass grating monochromator used was equipped with an echelette grating blazed at 6 μ in the first order, and a transmission filter to eliminate higher grating orders. The resolution achieved ranged from 1–7×10^{-4} ev and

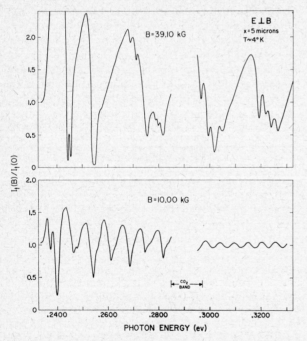

FIG. 1. High resolution spectra of oscillatory magnetoabsorption (OMA) in InSb at liquid helium temperature for fields of 10 and 39.1 kgauss. The polarization is $\mathbf{E} \perp \mathbf{B}$ and sample thickness is 5 μ. The spectra show that with increasing field, the absorptions become more intense, shift to higher energy, and develop doublet, quadruplet, and higher multiplet structure.

* Operated with support from the U. S. Army, Navy, and Air Force.

[1] S. Zwerdling and J. P. Theriault, Bull. Am. Phys. Soc. **6**, 147 (1961).

[2] S. Zwerdling, B. Lax, and L. M. Roth, Phys. Rev. **108**, 1402 (1957).

[3] E. Burstein, G. S. Picus, R. F. Wallis, and F. Blatt, Phys. Rev. **113**, 15 (1959).

[4] G. B. Wright and B. Lax, Bull. Am. Phys. Soc. **6**, 18 (1961).

[5] G. B. Wright and B. Lax, J. Appl. Phys. **32**, 2113 (1961), this issue.

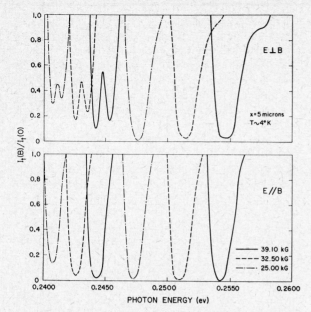

FIG. 2. Polarization anisotropy of the first two most intense absorptions in the OMA spectra of InSb. Note particularly the splitting of the lowest absorption for **E⊥B** and its absence for **E∥B**.

it was possible to locate the centers of intense minima to within 5×10^{-5} ev. Linearly polarized radiation from an AgCl sheet pile polarizer was used with propagation perpendicular to the magnetic field and with the electric vector oriented either parallel or perpendicular to the magnetic field. An appropriate wavelength range was scanned at constant field, the procedure being repeated for increments of 2.5 kgauss up to the maximum field available. The spectral range scanned was 0.2300–0.3325 ev. Typical spectra for fields of 10 and 39.1 kgauss with **E⊥B** are shown in Fig. 1. With increasing field, the absorptions increase in intensity, shift to higher energy, and develop multiplet structure. Since moderately intense absorption maxima persisted near the highest photon energy, the measurements have been extended recently to 0.4300 ev using a 3-μ first-order grating.

Pronounced polarization anisotropy was observed, the spectra for **E⊥B** containing considerably more structure than for **E∥B**. A striking example of this anisotropy is shown in Fig. 2 for the first two most intense absorption maxima for several field values. In the **E⊥B** spectra, a splitting of the first absorption maximum begins to appear just below 15 kgauss and increases with magnetic field, resulting in two well-resolved lines. On the other hand, the corresponding first absorption maximum in the **E∥B** spectrum shows no signs of a splitting. The second absorption maximum for **E⊥B** is centered somewhat higher in energy and also becomes broader with increasing field than for **E∥B**, indicating that it contains additional structure, unresolved probably due to dimensional broadening. The crystallographic anisotropy of the spectrum as a function of the orientation of the magnetic field relative

to the crystal axes is being studied and is expected to yield information about the nonspherical character of the band structure.

THEORETICAL BACKGROUND

Before presenting the results of our initial interpretation of these spectra, it will be useful to review briefly the theoretical basis of the interpretation. The InSb band structure at zero field according to Kane[6] is shown in Fig. 3. The diagram at the left shows the valence and conduction band energies versus k^2 for an "average" crystal direction, but omits the splitting of the double degeneracy. The diagram on the right is an enlargement of the region near $k^2 = 0$ and contains an inset table of Kane's band parameter values: the energy gap E_G, the spin-orbit splitting Δ, and the $k = 0$ band edge mass values for the light and heavy holes. The small gap is associated with important deviations from parabolicity for the electron and light hole bands, represented here by deviations from straight-line behavior. The effective masses and nonparabolicities of these two bands are similar and closely related according to Kane's theory. The region of interest for the interband transition approximation described below is indicated in the right-hand diagram as lying between the dash-dot lines.

For this initial interpretation of the spectra, all bands were assumed to be spherically symmetrical (warping in the valence band was neglected) and effects arising from the lack of inversion symmetry of the zinc-blende structure of InSb were omitted. Furthermore, effects of

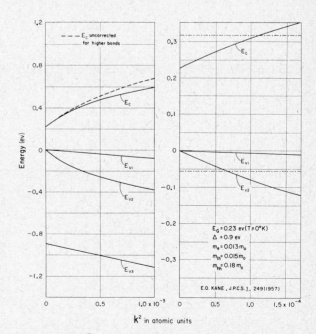

FIG. 3. The band structure of InSb at zero field according to Kane. The dash-dot lines in the figure at the right indicate the extent of the region investigated in the experiments described.

[6] E. O. Kane, J. Phys. Chem. Solids **1**, 249 (1957).

bands other than the conduction and valence bands were neglected except for the use of an empirical value for the heavy hole effective mass which differs from the free mass value of the simple Kane theory.

Consider next the energy level structure in the presence of an external magnetic field. Assume first that the bands are all parabolic and that spin and degeneracy effects may be neglected. The energy level structure will then consist of the familiar set of Landau levels in each of the three bands, as shown in the left half of Fig. 4. The spacings in the electron band and the light and heavy hole valence bands are characterized by the effective masses m_e, m_{lh}, and m_{hh}, respectively. Effects associated with the motion of electrons and holes parallel to the magnetic field are also neglected, since such effects are in any case modified significantly when Coulomb interaction is involved. If the spin is now introduced, the levels in these bands are all split as shown on the right side of Fig. 4, the splittings being given by the appropriate g-factors: g_e, g_{lh}, and g_{hh}. Actually, because of the valence band degeneracy at $k=0$, this model holds for the valence band only in the "classical limit" of high quantum numbers. Near the valence band edge, the positions and spacings of the magnetic levels deviate in a complicated way from those expected from Landau theory with g-factor splitting included, according to the theory of Luttinger and

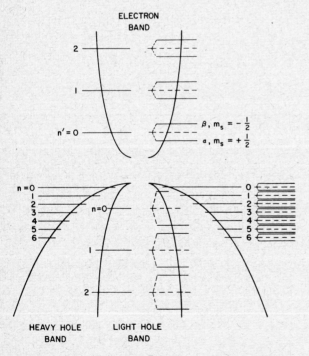

FIG. 4. Magnetic levels for a simplified model of the InSb band structure. Effects of nonparabolicity, valence band degeneracy and warping, and lack of inversion symmetry are omitted. Landau levels are shown in the left-half and their spin splitting in the right-half of the figure. Although this model is instructive, it applies only in the "classical limit" of high quantum numbers, and was not used for the present interpretation.

Kohn.[7,8] These deviations are the so-called "quantum effects" associated with the valence band degeneracy. The nonparabolicity in the band structure is manifested in the magnetic spectrum by the variation of the effective masses and g-factors with magnetic field and magnetic quantum number. The electron mass and light hole mass increase with field and quantum number, whereas their g-factors decrease.

So far, only the band structure of InSb has been considered, i.e., the energy levels of a one-electron problem. If the Coulomb interaction between the electron and hole is neglected, each transition observed would correspond to photon excitation of an electron from a valence band magnetic level to a conduction band magnetic level, i.e., an interband transition. However, the Coulomb attraction of the resulting electron-hole pair leads to a bound state structure, the exciton, and to a decrease in the transition energy by an amount equal to the binding energy of the exciton. All of the present experiments were conducted in the high-field region where the magnetic energy is large compared with the Coulomb binding energy. For InSb, the high-field region begins at 1.9 kgauss. A theory of direct transition exciton formation under these conditions has been developed by Elliott and Loudon[9] for $k=0$ conduction and valence bands which are spherical, nondegenerate, and parabolic. The "allowed" transitions which they discuss, those for which the conduction-valence band momentum matrix element does not vanish at the band edge, apply to the present experiments. Allowed transitions occur only to exciton levels with azimuthal quantum number $M=0$. Furthermore, the probability of forming the $M=0$ excitons associated with each interband transition is small except for the one of lowest energy. Thus, only one exciton formation transition should be observed for each interband transition expected in the absence of Coulomb binding. As the Landau quantum number increases, the average electron-hole separation of the exciton increases and the binding energy correspondingly decreases, so that, for sufficiently large quantum numbers, the binding energies become negligible.

INTERPRETATION

The initial interpretation of the present experimental results was done by fitting to the experimental spectra, theoretical spectra calculated by using a semi-empirical blending of Kane's theory which includes effects of nonparabolicity, Luttinger's theory[8] which treats valence band degeneracy effects, and the Elliott-Loudon exciton theory to account for the existence of exciton binding energies. The Elliott-Loudon theory was used in the absence of a more suitable one which would take account of the various energy band complications in

[7] J. M. Luttinger and W. Kohn, Phys. Rev. **97**, 869 (1955).

[8] J. M. Luttinger, Phys. Rev. **102**, 1030 (1956).

[9] R. J. Elliott and R. Loudon, J. Phys. Chem. Solids **15**, 196 (1960).

InSb, such as spin effects and the valence band degeneracy. The selection rules and transition probabilities[3,10,11] of the interband transition theory were used; effects of the Coulomb interaction on the integrated intensities of the transitions were omitted. Once suitable values of the parameters of the theory were found, i.e., values which gave a reasonable detailed fit for the energies of the transitions, these values were used to calculate relative intensities. A comparison between the latter results and the intensities of the absorptions in the experimental spectra showed good agreement, providing additional support for the theoretical interpretation.

The first step in evaluating the interband transition energies was to calculate the magnetic energy levels of the conduction and valence bands, including the effects of both nonparabolicity and the valence band degeneracy. At 10 kgauss, a satisfactory fit was found for the higher magnetic quantum number lines, where degeneracy effects are small, by assuming negligible exciton binding energy for these lines, but this could be done only if nonparabolicity was included for both the electron and the light hole. Consequently, the nonparabolicity of the light hole predicted theoretically is now established experimentally. The same satisfactory fit could not be maintained in the low quantum number region where degeneracy effects are large, if exciton binding energies were neglected. However, a reasonable fit was possible for both polarizations when exciton binding energies were included. In particular, the binding energy of the lowest light hole-electron transition exciton agreed very well with the value obtained from the calculation of Elliott and Loudon for the ground state binding energy of the lowest Landau level. For both polarizations, the observed spectral lines with low quantum numbers showed marked deviations from the regular spacing characteristic of the high quantum number region, giving direct experimental evidence for the existence of the expected valence band degeneracy effects.

By a reasonable fit we mean that almost all of the observed lines are well accounted for in position and relative intensity by calculated lines, although in many cases several theoretical lines are associated with an observed line. The number of theoretical lines associated with an observed line becomes smaller with decreasing magnetic quantum number. In the $\mathbf{E}\|\mathbf{B}$ spectrum, there is a one-to-one correspondence at low quantum numbers. Some of the predicted lines which are unresolved in the observed spectrum at 10 kgauss are clearly seen at higher fields. The consecutive strong transitions alternate in intensity in the $\mathbf{E}\perp\mathbf{B}$ theoretical spectrum, in excellent agreement with the experimental observations as shown in the lower section of Fig. 1. The theo-

retical analysis also showed that for large magnetic quantum numbers, the $\mathbf{E}\|\mathbf{B}$ spectrum was mainly determined by the heavy hole-electron transitions, whereas for $\mathbf{E}\perp\mathbf{B}$, the light hole-electron transitions predominate.

The same energy band parameters which gave a reasonable fit at 10 kgauss were used to calculate theoretical spectra at 22.5, 27.5, 32.5, and 39.1 kgauss and again gave a reasonable fit with the respective experimental spectra for both polarizations. The binding energy of the lowest light hole-electron transition exciton for each of these fields also agreed very well with the corresponding value obtained from the Elliott-Loudon theory. This excellent agreement provides strong experimental evidence for the existence of excitons in InSb in the high-field region.

An experimental binding energy is evaluated as the difference between the energy of the calculated interband transition line and that of the corresponding observed line (see Fig. 5). If the resulting value is to be significant, a unique identification of the observed line with an interband transition is necessary. This is a more stringent requirement than is needed if only a reasonable detailed fit between the observed and the calculated spectra is involved. As a consequence, it was possible to evaluate experimental binding energies for only a limited number of observed lines at various fields, all with conduction band Landau quantum numbers less than four.

Coupling between the spin coordinates and the position or momentum coordinates of the exciton is probably small. One would then expect that the exciton binding energies would not depend appreciably on the spin orientations of the electrons and holes. In order to investigate this dependence in the present data, a detailed exciton theory is not needed. A comparison of experimental binding energies at various fields for transitions differing only by a single electron or hole spin orientation indicated that the binding energies were indeed independent of the spin orientation to a good approximation. If this had not been so, it would have been less meaningful to apply the Elliott-Loudon theory, which neglects spin.

The Elliott-Loudon theory treats only those exciton transitions which have the same conduction and valence band Landau quantum number N, since only $\Delta N=0$ excitons occur in their simple model. It is not immediately clear how this theory should be applied to InSb, where transitions occur for both $n'-n=0$ and $n'-n=-2$ (quantum numbers in the notation of reference 10). In fact, the theory was found to apply well only to transitions with $\Delta\mathfrak{N}=0$, i.e., to those with the same magnetic serial number $\mathfrak{N}$ in the valence and conduction bands ($\mathfrak{N}=0,1,2,\ \ldots$). For electrons, $\mathfrak{N}=n'$; for light holes, $\mathfrak{N}=n$, whereas for heavy holes, $\mathfrak{N}=n-2$. To apply the theory, the identification $\mathfrak{N}=N$ was made. A comparison between the experimental binding energies and the corresponding calculated values for both light and

[10] L. M. Roth, B. Lax, and S. Zwerdling, Phys. Rev. **114**, 90 (1959).

[11] R. J. Elliott, T. P. McLean, and G. G. Macfarlane, Proc. Phys. Soc. (London) **72**, 553 (1958).

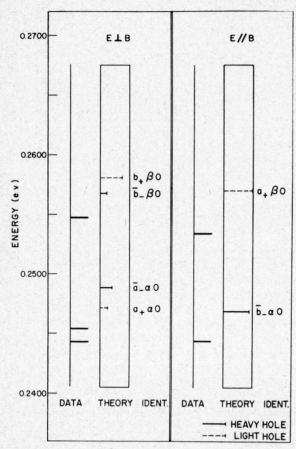

FIG. 5. Comparison of the first two principal absorptions in the experimental and theoretical spectra at 39.1 kgauss for both polarizations. The positions of the data lines indicate only the observed transition energies, whereas the positions of the theory lines indicate the calculated energies, their lengths being proportional to the interband transition probabilities. The transition identifications are shown in the notation of Roth.[10] In the $E \perp B$ diagram, the splitting of the observed line just below 0.2550 ev is unresolved probably due to dimensional broadening. The energy difference between corresponding theoretical and experimental lines is interpreted as exciton binding energy.

heavy hole excitons with $\mathfrak{N}=0$ showed excellent agreement. An approximate extension of the Elliott-Loudon theory was used for calculating the binding energies of excitons with $\mathfrak{N}>0$; in this extension, the binding energy of a higher level is given by the expression for the binding energy of the lowest level, but with B replaced by $B/(2N+1)$. For $\mathfrak{N}>0$, the result of a comparison using the latter calculation gave only fair agreement with experiment. The experimental binding energies tended to be smaller than the calculated values, although both showed the expected decrease with decreasing $\mathfrak{N}$. The agreement was poor for all $\Delta\mathfrak{N}\neq0$ excitons compared. Clearly, an exciton theory based on a model more appropriate for InSb is desirable.

A brief discussion of the energy band parameters used in this initial interpretation of the present experimental results is in order, but we wish to emphasize that the values should be regarded as tentative; attempts to

obtain a still better fit to the data are in progress. The value for the direct energy gap at liquid helium temperature was found to be $E_G=0.2357\pm0.0005$ ev. This value was obtained from the data by smoothly extrapolating the lowest exciton line to zero field and adding the zero-field exciton binding energy of 7×10^{-4} ev estimated from the effective mass hydrogenic model for the exciton in a dielectric medium. The extrapolated line extends from only 2.5 kgauss to zero field and has a very small slope, implying high reliability. The gap value is further confirmed by the coalesence point of the smooth extrapolation to zero field of the higher minima. The heavy hole effective mass $m_{hh}=0.18\,m_0$, as given by Kane, was used where required in the calculations. The value of the matrix element P as defined in Kane's theory was determined to be $P=1.53\times10^{-19}$ erg cm from the experimental heavy hole-electron transition energies in the high quantum number region of the $E\|B$ spectrum. This value of P led to the band edge effective mass values $m_e=0.0145\,m_0$ and $m_{lh}=0.0149\,m_0$. The band edge electron g-factor was obtained from the splitting of the first two lines of the $E\|B$ spectrum after taking into account the energy difference in the valence band levels involved; the g_e values found for fields greater than 17.5 kgauss were extrapolated linearly to zero field to yield the band edge value $g_e=-48$. From the parameter values above and Eq. (A-5) of reference 10 derived by Roth, the spin-orbit splitting of the valence band is calculated to be $\Delta=0.98$ ev. The parameter values cited, together with the assumptions that the Luttinger[8] parameter $q=0$ and the Dresselhaus, Kip, and Kittel[12] parameter $H_2=0$ for InSb, were used throughout the initial calculations reported in this paper.

In previous analyses of magnetoabsorption[1,3,10] and magnetoreflection[4,5] experiments, the first two observed lines were interpreted as being associated with a spin splitting of the lowest conduction band Landau level. The detailed interpretation made possible by the extensive structure in the present experimental spectra together with the theoretical analysis indicates that this is so, but also leads to the identification of the hole magnetic levels involved in the interband transitions associated with these lines. Formerly, the complications of the magnetic level structure of the valence band due to degeneracy were omitted. It was assumed that the transitions were from low-lying valence band levels whose shifts with magnetic field could be neglected; these were sometimes called heavy hole levels. The present interpretation shows that the transitions associated with the two lowest conduction band spin states are from different sets of magnetic levels, or "ladders,"[8] in the valence band. For example, Fig. 5 shows that for $E\|B$, the lower transition is from a heavy hole ladder (one which gives rise to heavy holes in the region of high

[12] G. Dresselhaus, A. F. Kip, and C. Kittel, Phys. Rev. **98**, 368 (1955).

quantum numbers), whereas the upper transition is from a light hole ladder, so that in general, the contribution of the valence band levels to the splitting is not necessarily small. Actually, for the parameters used here, the energy differences happen to be small for the valence band levels involved in these lowest transitions (although not for similar transitions to the pair of electron spin states associated with any of the higher electron quantum numbers). Therefore, the present experimental result for the electron band edge g-factor turns out to be nearly the same as that from previous estimates. The

present method gives, of course, a more reliable value for this quantity.

In summary, extensive and detailed experimental spectra described above have been obtained for InSb at liquid helium temperature, and an initial theoretical analysis and interpretation resulted in a reasonable fit between the experimental and the calculated spectra. It is essential, however, to include in the calculated spectra the nonparabolicity of both the electron and the light hole, effects of degeneracy on the valence band magnetic level structure, and exciton binding energies.

JOURNAL OF APPLIED PHYSICS   SUPPLEMENT TO VOL. 32, NO. 10   OCTOBER, 1961

Some Properties of p-n Junctions in GaP

H. G. GRIMMEISS AND A. RABENAU
Philips Zentrallaboratorium GmbH, Aachen, Germany

AND

H. KOELMANS
Philips Research Laboratories, Eindhoven, Netherlands

A new method in making single crystals of GaP and the preparation of diodes is described. The p-n luminescence and photoluminescence of undoped and Zn-doped GaP are investigated and the light output of the p-n luminescence as a function of temperature and excitation density is discussed. The spectral sensitivity of the p-n photovoltaic effect is recorded.

INTRODUCTION

WHEN a p-n junction is biased in the forward direction, the barrier potential, brought about by the gradient in carrier concentrations, is lowered. Consequently, more electrons and holes can diffuse through the p-n junctions and produce a strong rise in the concentration of minority carriers, a phenomenon well known under the name carrier-injection. The rise in the concentration of minority carriers gives way to an enhanced recombination of free carriers. This recombination can occur directly via a band-band transition or indirectly via a level within the forbidden gap. In p-n luminescence the recombination energy is given off as photons. It will be clear that the spectral distribution of the emitted light will depend on the position of the recombination level.

When looking for materials suitable for making p-n light sources, two requirements should be kept in mind. Firstly the material should (at least potentially) be a good phosphor, and secondly it should be a material which can be made into n- and p-type semiconductors of low resistivity, in order to keep ohmic losses low. Good conductivity requires shallow levels. These shallow levels, however, have bad trapping properties and are for this reason not very suitable for radiative recombination centers. Consequently, apart from the shallow "conductivity" levels, deep-lying "flourescence levels" should be introduced. They act as effective traps for minority carriers, thereby preventing nonradiative

recombinations via quenching levels which in actual phosphors are always present.

We have tried to find suitable materials among the III–V compounds because they can be made n type and p type and because several members show good fluorescence properties and thus probably fulfill the two requirements. We were able to show for instance that GaN shows photoluminescence and a weak electroluminescence when activated and coactivated with any of a large variety of elements.[1,2] The whole fluorescence picture closely resembles that of ZnS or CdS. The GaN samples, for instance, show photoconductivity, glow, and I. R. quenching. Unfortunately we did not succeed in making single crystals of GaN large enough to prepare p-n light sources. Better results were obtained with GaP[3,4] and AlP.[5] Both compounds show good p-n luminescence and the reverse effect a p-n photovoltaic response.

CRYSTAL GROWING AND DIODE PREPARATION

No suitable method of obtaining single crystals of GaP of sufficient size and purity, is described in the

[1] H. G. Grimmeiss and H. Koelmans, Z. Naturforsch. **14a**, 264 (1959).

[2] H. G. Grimmeiss, R. Groth, and J. Maak, Z. Naturforsch. **15a**, 799 (1960).

[3] H. G. Grimmeiss and H. Koelmans, Philips Research Repts. **15**, 290 (1960).

[4] H. G. Grimmeiss and H. Koelmans, Phys. Rev. (to be published).

[5] H. G. Grimmeiss, W. Kischio and A. Rabenau, J. Phys. Chem. Solids **16**, 302 (1960).

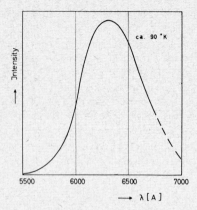

FIG. 1. Spectral distribution of the photoluminescence in undoped GaP at liquid air temperature (excitation with 3650 A).

literature. Growing them from the melt should be avoided because of the high phosphorus pressure at the melting point of GaP. Further, the temperature should be chosen as low as possible to reduce contamination by the container.

On the other hand, the sublimation rate of the pure material in the preferred temperature range ($<$1100°C) was too low. Vapor phase transport in presence of "carriers" is possible at comparatively low temperatures[6-8] but yields only whiskers or small crystals.

The method which we used and which was developed by I. Maak of our laboratory starts with polycrystalline material and a small amount of a carrier, e.g., iodine. The shape of the container, a quartz tube, as well as a controllable temperature gradient are chosen such that only one nucleus has the chance to grow and further nucleation during the process is prevented.

In this way at a maximum temperature of 1100°C single crystals up to 20–30 mm in length and 12–14 mm

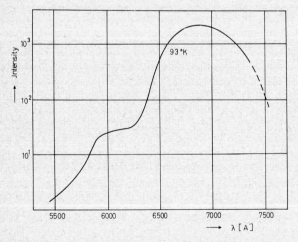

FIG. 2. Spectral distribution of the photoluminescence in GaP-Zn at liquid air temperature (excitation with 3650 A).

[6] G. R. Antell and D. Effer, J. Electrochem. Soc. **106**, 509 (1959).
[7] D. Effer and G. R. Antell, J. Electrochem. Soc. **107**, 25 (1960).
[8] M. Gershenzon and R. M. Mikulyak (private communication).

in diameter could be obtained at a growth rate of 12 mm/d. At low phosphorus pressures the crystals exhibit n-type conductivity, probably caused by P vacancies.

Until now diodes have been made exclusively with n-type disks, as a starting material. Zn was diffused into these disks at 800°C for half an hour. The desired shape of the diodes can be made by grinding and etching.

LUMINESCENT PROPERTIES

One of the first questions to settle was whether the luminescence in GaP could be described and interpreted in the way used in ZnS-type phosphors. We soon found out that the behavior of the p-n luminescence (just as the photoluminescence in GaN already quoted) closely corresponds to the luminescence behavior of the classical ZnS phosphors. As the III–V compounds are less ionic than ZnS, and consequently have a much smaller Franck-Condon shift, the analysis was in some way even simpler than in ZnS.

Our undoped GaP samples show only one activation energy both for n-type conductivity (0.07 ev) and for p-type conductivity (0.19 ev) in the temperature range between 90° and 500°K.[3] As could be expected, p-type conductivity can also be induced by the incorporation of Zn, for example, which gives acceptors at 0.4 ev from the valence band. These level positions are confirmed by optical measurements. Namely, in undoped n-type GaP we find with ultraviolet excitation an emission with a maximum at 6300 A or about 2.0 ev (Fig. 1), which corresponds well with the position of the Ga-vacancy level. Zn-doped GaP gives an emission at 7000 A or 1.8 ev (Fig. 2), again in close correspondence with the Zn-level position. In both cases the addition of optical and electrical measurements gives 2.2 ev which is approximately the band gap energy. It can be seen that the Franck-Condon shift is indeed small. Figure 3 shows that the emissions in p-n luminescence do not differ from the photoemission; both the p-n

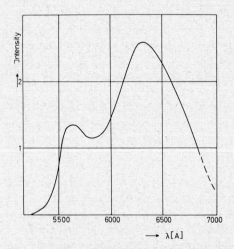

FIG. 3. Spectral distribution of the p-n luminescence in undoped GaP.

luminescence and the photoluminescence maxima are at 6200 A. An additional emission at 5650 A in the *p–n* luminescence spectrum can be seen. This additional maximum is also present in Fig. 4 which shows the results for Zn-doped GaP. Furthermore, the 7000-A emission known from the photoemission is also present. The reason for the appearance of an additional short-wave emission will be discussed later. It is thus clear that the spectral distribution of the emissions connected with the specific levels does not depend on the excitation mechanism. With this result in mind we tried to analyze our experimental results along the same lines as are usual in photoluminescence. We analyzed the recombination kinetics with the aid of the code system proposed by Klasens, with the assumption that the current through the *p–n* junction is proportional to the excitation density. Klasens[9] and Schön[10] showed that an analysis of the variation of luminescence intensity with excitation density and temperature can give valuable information about the recombination mechanism. In particular the use of the Klasens code number system enables one to sort out quickly from a vast number of potential possibilities that situation which applies to the experimental observation.

The analysis for GaP was simplified by the fact that the intensity varied superlinearly with the current. This enables one to consider only those cases in which superlinearity can occur. Figure 5 shows the behavior of a typical diode. The figure gives the light output as a function of current for different temperatures. It will be seen that the upper linear part and the superlinear part are practically independent of temperature. The lower linear part, on the other hand, shifts to higher values with decreasing temperature. At liquid-air temperature

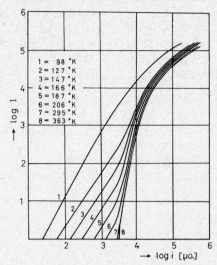

FIG. 5. The intensity of the yellow emission in GaP (6200 A) as a function of the current at different temperatures.

the superlinear part has practically vanished. Since both preparation and processing are rather complicated, the behavior of diodes, prepared in different ways, is not the same. This is illustrated by Fig. 6 which shows the

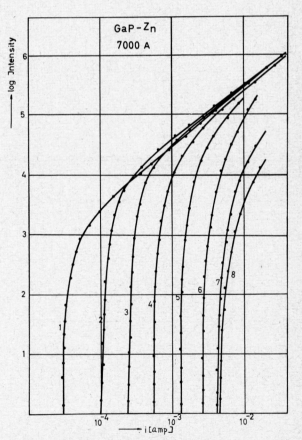

FIG. 6. The intensity of the red emission (7000 A) in GaP-Zn as a function of the current at different temperatures: (1) 98°K; (2) 113°K; (3) 127°K; (4) 143°K; (5) 166°K; (6) 203°K; (7) 253°K; (8) 293°K.

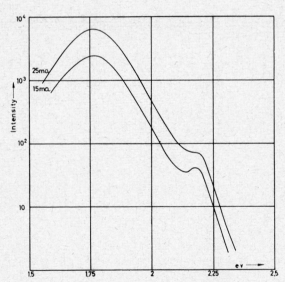

FIG. 4. Spectral distribution of the *p–n* luminescence in GaP-Zn at 20°C for two values of the current.

[9] H. A. Klasens, J. Phys. Chem. Solids **7**, 175 (1958).
[10] M. Schön, *Halbleiterprobleme IV* (Vieweg Verlag, Braunschweig, 1958), p. 283.

dependence for the 7000-A emission of a Zn-doped sample. The upper linear part is again approximately independent of temperature, but the superlinear part shows a rapid shift to higher current values with increasing temperature.

For this case we carried out a full analysis using a scheme with 3 levels—namely, acceptor due to Zn, shallow donors (0.07 ev), and a nonspecified deep lying killer level, which we have to assume because the efficiency is still rather low.

Several simplifications based on experimental results enabled us to sort out 3 sets of code numbers from the vast number which are possible in the Klasens codes system. The sets of equations connected with these sets of code numbers give a quantitative representation of the luminescence behavior of the diode. The close agreement between theoretical analysis and experiment shall be demonstrated for the shift of the superlinear part with temperature. The analysis reveals that superlinearity for both the 7000-A and the 5650-A emission should start at a current density $i_s = qn_0 k$, where q is the capture cross section of the killer center, n_0 is the equilibrium concentration of free electrons in the n-type part of the diode, and k is the concentration of killer centers. As a first approximation q and k may be considered to be independent of temperaure. Consequently the shift of the superlinear part should be connected by the change of n_0 with temperature. This means that a plot of i_s against T^{-1} should give a straight line with a slope

yielding the ionization energy of the shallow donor. Figure 7 shows that this is indeed the case. The activation energy is 0.07 ev, in close agreement with the value deduced from electrical measurements. It can be seen that at higher temperatures the line levels off. This effect is clearly due to electron exhaustion; it is also found in conductivity experiments.

As a further example we now discuss the temperature dependence of the emissions. Figure 8 shows a phenomenon well known in ZnS phosphors; namely, that with decreasing temperature the long wave emission goes through a maximum, with the short-wave emission growing at the expense of the long-wave emission at low temperature. Our analysis shows that a plot of the log of the intensity in the region of quenching $1/T$ should give a straight line with a slope related to the activation energy of the level causing the emission. For the green emission this was found to be true (0.07 ev). The activation energy deduced from the quenching of the red emission was somewhat too low.

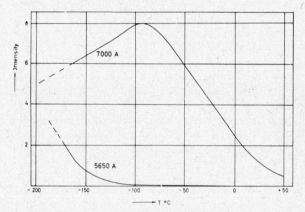

Fig. 8. Temperature dependence of the p-n luminescence in GaP-Zn.

In contrast to other authors,[8,11] we ascribed our green emission at 5650 A to recombination via a (shallow) level. This opinion is supported by the following arguments. An edge emission should (as a first approximation) start at energies greater than the band gap. In the actual case, however, the emission is definitely on the low-energy side. A band-band emission should shift (together with the band gap) to higher energies when the temperature is lowered. The maximum of our green emission, however, is practically stationary when the temperature is changed. Moreover, a band-band emission should not show much of a temperature quenching and especially it should not give the experimentally observed straight line with a small activation energy when plotted logarithmically against T^{-1}. Further, the intensity does not vary linearly with current as a band-band emission should, but again the variation is strongly superlinear.

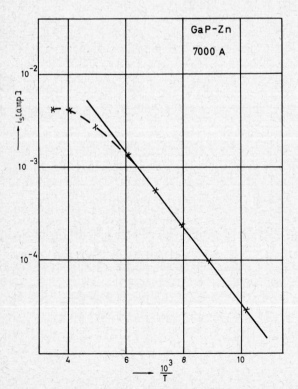

Fig. 7. Plot of logi_s against T^{-1}, where i_s is the value of the current in the middle of the superlinear range of the curves of Fig. 6.

[11] E. E. Loebner and E. W. Door, Phys. Rev. Letters 3, 23 (1959).

The reason why this emission is not found in photo-luminescence probably stems from the fact that with photoexcitation the excitation density is much smaller than with *p-n* injection excitation, with the result that luminescence is already quenched at liquid air temperature and for this reason cannot be detected.

The analysis of the recombination kinetics also gives information about capture cross section and level concentrations. For the actual determination of the capture cross section additional measurements of, for instance, decay times, are necessary. This type of measurements is now under way. A preliminary result is shown in Fig. 9 which shows the decay of the red Zn emission. At liquid air temperature the decay time is about 10 μsec. With increasing temperature the decay time decreases.

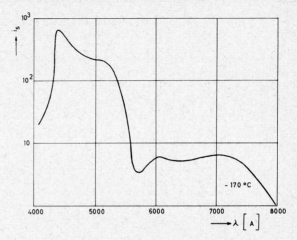

FIG. 10. Excitation spectra of the short circuit current i_s in GaP-Zn *p-n* junctions at liquid air temperature.

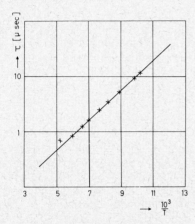

FIG. 9. Temperature dependence of the decay time of the red emission in GaP-Zn (7000 A).

This decrease is again related to an activation energy of 0.07 ev.

PHOTOVOLTAIC EFFECT

The photovoltaic response in GaP is rather interesting because the response in GaP is not restricted to the high-energy side of the band gap. Figure 10 shows that there is also an additional rather strong response on the long-wave side, which clearly shows two maxima, namely at 1.8 ev (7000 A) and 2.0 ev (6200 A). These maxima coincide with the luminescence emission maxima, and again the 1.8-ev maximum is clearly connected with the presence of Zn. Because we are dealing with reciprocal effects it is reasonable to connect the peaks in the long-wave photovoltaic response and the luminescence peaks to the same kind of levels. Since for the photovoltaic effect both mobile electrons and mobile

holes are essential, it must be concluded that the extrinsic response is due to a two-step excitation, with two optical steps or one optical and one thermal step.

The spectral photovoltaic response curve also shows a structure in the intrinsic region. Essentially there are two maxima; one at 2.4 ev and one at 2.8 ev. Similar results have been obtained by Spitzer and others,[12] who interpreted this as being due to direct and indirect band-band transitions. From infrared measurements these authors concluded the presence of an electronic absorption band which arises from transitions between two sets of conduction band minima with a separation of about 0.35 ev. Measurements of the fundamental absorption edge showed that the energy of separation between the [100] minima and the minimum of $k=0$ is indeed about 0.35 ev. The energy of separation between our maxima of 2.4 ev and 2.8 ev amounts to 0.4 ev, which value agrees well with the observations of Spitzer and co-workers.

Because of the larger band gap, the open circuit voltage in GaP may be expected to be higher than that obtained for Si. We measured an open circuit voltage up to 1.2 v at room temperature and up to 1.7 v at liquid air temperature. To our knowledge photovoltages of this magnitude have until now never been observed in single-junction cells. The short-circuit currents measured were up to 5 ma/cm². These values already indicate that GaP, even though it has a rather large band gap, might be an efficient solar cell.

12 W. G. Spitzer, M. Gershenzon, C. J. Frosch, and D. F. Gibbs, Phys. Chem. Solids 11, 339 (1959).

JOURNAL OF APPLIED PHYSICS   SUPPLEMENT TO VOL. 32, NO. 10   OCTOBER, 1961

Interband Faraday Rotation in III–V Compounds

BENJAMIN LAX

Lincoln Laboratory,* Massachusetts Institute of Technology, Lexington 73, Massachusetts, and National Magnet Laboratory,†
Massachusetts Institute of Technology, Cambridge 39, Massachusetts

AND

YUICHIRO NISHINA

National Magnet Laboratory,† Massachusetts Institute of Technology, Cambridge 39, Massachusetts

Experimental investigation of Faraday rotation in III-V compounds has exhibited a striking singularity at photon frequencies just below the energy gap. A quantum theoretical result associated with the direct transition has been developed to explain the phenomenon. The treatment has been extended to include forbidden transitions which are readily applicable to such materials as InAs, GaAs, and GaSb where interband transitions between the split-off valence bands have been observed. The treatment for observing Faraday rotation by reflection has also been considered and experimental results in InSb at optical frequencies will be presented. The calculations have also been performed for degenerate semiconductors at low temperature.

INTRODUCTION

FARADAY rotation has been very useful in studying the free carrier[1] and interband properties of electrons in III-V compounds. In extending these measurements to infrared photon energies close to the energy gap, apparent singularities in the rotation were discovered.[2] Such singularities were anticipated by Stephen and Lidiard[3] from the concept of a simple classical bound oscillator which indicated that the Faraday rotation contributed by bound valence electrons had a sign similar to that of the free carrier contribution. It is apparent from physical consideration that such a theory is an over-simplification and does not take into account the band properties of such semiconductors. In order to explain the experimental observations of the interband Faraday rotation in germanium, we have developed quantum theoretical analysis[4] which is equally applicable to the results found in III-V semiconductors. We shall also discuss briefly some preliminary observations made in InSb in the optical region by reflection techniques.

EXPERIMENTAL OBSERVATIONS

The initial experiments which were made in InSb demonstrated the basic phenomena of interband Faraday rotation quite clearly, as indicated in Fig. 1. This shows the measurements carried by transmission experiments through samples of InSb, several millimeters in thickness, with different carrier concentrations. At long wavelength, the rotation exhibits a behavior which is proportional to the square of the wavelength char-

acteristic of the rotation due to free carriers. At shorter wavelength, the rotation deviates from this pattern as one approaches the wavelength close to the energy gap. The rotation approaches zero and even goes negative quite rapidly. The existence of negative rotation in InSb has also been found by Smith, Moss, and Taylor.[1]

Negative rotation close to the direct energy gap in germanium was also found by Hartmann and Kleman,[5] who showed a decided singularity at these energies. Data very similar to InSb have also been observed by Cardona[6] in InAs as indicated in Fig. 2. He also studied GaAs, in which he observed that the interband contribution this time had a singularity in the Faraday rotation where the sign was identical to that of the free carrier and opposite to that found in InSb and InAs. This is shown in Fig. 3 where, at approximately 1μ, the Faraday rotation increases rather abruptly and shows a minimum at 1.5μ with a λ^2 behavior at longer wavelength.

THEORY

In the theoretical analysis, we can take two approaches to evaluate the interband Faraday rotation as a function of photon energy. The one initially developed[4] involves the application of Kramers-Kronig relation to evaluate the dispersive component of propagation constant, i.e., the phase constant β or index of refraction n, when the absorption coefficient α, in the presence of magnetic field, is known. This treatment requires certain implicit assumptions which are valid for the particular experiments under consideration. The tacit assumption is that the index of refraction n is essentially independent of frequency in the wavelength region of interest. Obviously this is not true, since in the region of the interband transition it must change. However, if the dispersive contribution of the interband transition is small compared to the dielectric constant, this assumption is appropriate. Actually it is

---

* This work was done in part at Lincoln Laboratory, operated with support from the U. S. Army, Navy, and Air Force.

† This work was done in part at the National Magnet Laboratory, operated with support by the U. S. Air Force through the Air Force Office of Scientific Research.

[1] S. D. Smith, T. S. Moss, and K. W. Taylor, J. Phys. Chem. Solids 11, 131 (1959).

[2] R. N. Brown and B. Lax, Bull. Am. Phys. Soc. 4, 133 (1959).

[3] M. J. Stephen and A. B. Lidiard, J. Phys. Chem. Solids 9, 43 (1959).

[4] B. Lax and Y. Nishina, Phys. Rev. Letters 6, 464 (1961).

[5] B. Hartmann and B. Kleman, Arkiv Fysik 18, 75 (1960).

[6] M. Cardona, Phys. Rev. 121, 752 (1961).

simpler to utilize the Kramers-Kronig method than the dispersion theory which follows; the latter, however, is more general.

The second method involves the application of dispersion theory in which the complex index of refraction is obtained from a single general integral as applied to a solid with energy bands. The origin of this theory has its physical basis in that the individual states in both conduction and valence band constitute a series of classical oscillators. Then the frequency for direct transition is the energy difference of the two companion levels with the same momentum vector. Then for the solid one merely sums, or actually integrates, over all the levels which constitute the two sets of bands. In the absence of magnetic field it can be shown that with the appropriate mathematical manipulation the interband contribution to the index of refraction is given by the following expression[7,8]:

$$(n-jk)^2 = \epsilon + \frac{2e^2}{m^2\hbar^4} \int \frac{|\mathbf{M}(p)|^2 \rho(p)dp}{\omega'(\omega'^2-\omega^2)}, \quad (1)$$

where $n-jk$ is the complex index of refraction, $\mathbf{M}$ is the momentum matrix for the interband transition, and $\hbar\omega' = \hbar\omega_g + p^2/2\mu$. μ is the reduced effective mass, p the momentum, $\hbar\omega_g$ the energy gap, ϵ the dielectric constant, m the free electron mass, and $\rho(p)$ the density of states.

The above result in this form has been used by Korovin[7] to derive the dispersion relations for the direct transitions for the semiconductors. The above integral is equivalent mathematically to that used by us in evaluating the dispersive component from Kramers-Kronig relation. We shall, however, rewrite Eq. (1) so that it can be applied to the case in which the magnetic field is included. In this case, we have to investigate the interband classical oscillator over the bands. And the integral corresponding to Eq. (1) is

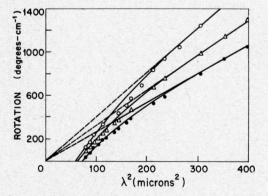

FIG. 1. Faraday rotation in n-type InSb at $H_0 = 15$ kgauss as a function of λ^2 showing the interband effect at lower wavelength. Electron concentrations: ○ 1.8×10^{16} cm^{-3}; ● 2.3×10^{16} cm^{-3}; △ 2.6×10^{16} cm^{-3}. (After Brown and Lax.[2])

[7] L. I. Korovin, Soviet Phys.–Solid State **1**, 1202 (1960).
[8] M. Suffczynski, Phys. Rev. **117**, 663 (1960).

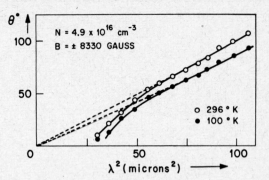

FIG. 2. Faraday rotation in n-type InAs as a function of λ^2. (After Cardona.[6])

found to be

$$(n-jk)^2 = \epsilon + \frac{4e^2\mu}{m^2\hbar^3} \sum_n \int \frac{|\mathbf{M}|^2\omega_c dp_z}{\omega_n(\omega_n^2-\omega^2)}, \quad (2)$$

where $\hbar\omega_n = \hbar\omega_g + (n+\frac{1}{2})\hbar\omega_c + p_z^2/2\mu$, $\omega_c = eH/\mu c$, the cyclotron frequency, and H the magnetic field is taken along the z direction.

If, for the present, we consider only parabolic bands, it can be shown that, for the allowed and direct transitions, the complex index of refraction for the circularly polarized waves is given by

$$n^{\pm} - jk^{\pm} = n_0 + \frac{cK\hbar\omega_c}{2\omega^2} \sum_n \left[-2(\omega_n^{\pm})^{-\frac{1}{2}} + (\omega_n^{\pm}+\omega)^{-\frac{1}{2}} + (\omega_n^{\pm}-\omega)^{-\frac{1}{2}} \right], \quad (3)$$

where $K = 2e^2(2\mu)^{3/2}(n_0m^2c)\hbar^{-7/2}|\mathbf{M}_d|^2$, $\omega_n^{\pm} = \omega_g + (n+\frac{1}{2}) \times \omega_c \pm \Delta\omega_d$; $\Delta\omega_d$ is the difference in ω_n depending on the sense of circular polarization, e is electron charge, m is free electron mass, and $\mathbf{M}_d$ is the momentum matrix element.

In these expressions the sums are defined over all frequencies above and below the energy gap. However, in order to explain the experimental results of Figs. 1–3, we need only consider the latter. It follows, therefore, that all quantities are real if losses are neglected and that the medium is purely dispersive with $k=0$. If we further assume that the interband component is small compared to the dielectric constant, and we follow the procedure outlined in a previous paper to evaluate the Faraday rotation,[4] $\theta = \frac{1}{2}(\omega/c)(n_+ - n_-) = \frac{1}{2}(\beta_+ - \beta_-)$. Then we find that we obtain the identical expression as before to give the following result:

$$\theta_d = K(\Delta\omega_d/\omega)[2\omega_g^{-\frac{1}{2}} - (\omega_g+\omega)^{-\frac{1}{2}} - (\omega_g-\omega)^{-\frac{1}{2}}]. \quad (4)$$

Here we assume that the rotation is linear in H, the applied magnetic field, and is related to H by a phenomenological parameter γ_d whose contribution comes from the $\Delta m_J = \pm1$. In a simple band this selection rule merely involves the differential change in orbital quantum number for the two senses of circular polarization. In an actual crystal with spin orbit coupling and complex valence band structure, it is difficult to estimate

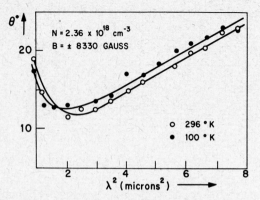

FIG. 3. Faraday rotation in n-type GaAs as a function of λ^2, showing positive interband rotation. (After Cardona.[6])

the magnitude and sign of this quantity without considering the details of the levels, i.e., Landau ladder and the sign of g factors. In a simple band, the sign should be that of the classical bound oscillator which tacitly neglects such complications and merely considers the change in orbital quantum number for the circularly polarized waves. In this simple case the Faraday rotation for the interband contribution should be of the same sign as that of free carriers. Furthermore, it is evident from Eq. (4) that there is a singularity at $\omega = \omega_g$. As the wavelength decreases and the photon energy approaches that of the energy gap, the rotation in InSb and InAs as well as in germanium becomes negative. This is opposite to the simple interpretation in terms of classical orbital contribution. GaAs, however, exhibits a sense of rotation contrary to others. The explanation suggested by Cardona[6] is that the g factor for GaAs is positive whereas in germanium and others it is negative. As pointed out by Lax[9] in considering the Faraday rotation in terms of an effective g factor, this is an oversimplification. The sign of the Faraday rotation has to be examined in terms of the individual transitions and in terms of the valence band parameters which are not yet determined.

In a manner similar to that applied to the direct transition, it is possible to obtain the result for the forbidden transition which is given by

$$\theta_f = C(\Delta\omega_f/\omega)[-2\omega_g^{\frac{1}{2}} + (\omega_g+\omega)^{\frac{1}{2}} + (\omega_g-\omega)^{\frac{1}{2}}], \quad (5)$$

where $C = \frac{1}{2}\hbar^2 e^2 (n_0 m^2 c)^{-1} \hbar^{-7/2} |\mathbf{M}_f|^2 (2\mu)^{\frac{3}{2}}$, $\mathbf{M}_f$ is the momentum matrix element similar to $\mathbf{M}_d$. The mathematical form of this expression is very similar to that obtained for the dispersion without magnetic field for the allowed direct transition (sign reversed) in germanium and the III-V compounds.[10,11]

If we assume that for a transmission experiment the absorption can be neglected for the energies well above the energy gap, then the Faraday rotation is Eq. (5)

[9] B. Lax, *Proceedings of the International Conference on Semiconductor Physics, Prague, 1960* (Publishing House of the Czechoslovak Academy of Sciences, Prague, 1961).
[10] H. R. Philipp and E. A. Taft, Phys. Rev. **113**, 1002 (1959).
[11] J. Tauc and A. Abraham, *Proceedings of the International Conference on Semiconductor Physics, Prague, 1960* (Publishing House of the Czechoslovak Academy of Sciences, Prague, 1961).

with the last term absent, since it becomes imaginary and responsible for the absorption term.

Consequently, to a good approximation, one can apply this particular analysis to describe the Faraday rotation in p-type material due to the interband transition between the valence bands split by the spin-orbit coupling. Such an experiment would be particularly interesting in highly-doped material which would be degenerate at low temperatures. We have extended this analysis to consider such a situation. The results are given for both allowed and forbidden transitions by the following equations:

$$\theta_d = K\left(\frac{\Delta\omega_d}{\omega}\right)\left\{2\omega_g^{-\frac{1}{2}}\left[1 - \frac{2}{\pi}\tan^{-1}\left(\frac{\omega_F}{\omega_g}\right)^{\frac{1}{2}}\right] - (\omega_g+\omega)^{-\frac{1}{2}}\right.$$
$$\times\left[1 - \frac{2}{\pi}\tan^{-1}\left(\frac{\omega_F}{\omega_g+\omega}\right)^{\frac{1}{2}}\right] + (\omega_g-\omega)^{-\frac{1}{2}}$$
$$\left.\times\left[1 - \frac{2}{\pi}\tan^{-1}\left(\frac{\omega_F}{\omega_g-\omega}\right)^{\frac{1}{2}}\right]\right\}, \quad (6)$$

where ω_F is the frequency in addition to ω_g for an electron in the conduction band to be excited to the Fermi energy level, and

$$\theta_f = C\frac{\Delta\omega_f}{\omega}[-2\omega_g^{\frac{1}{2}} + (\omega_g+\omega)^{\frac{1}{2}} + (\omega_g-\omega)^{\frac{1}{2}}]$$
$$+ C\frac{\Delta\omega_f}{\pi\omega}\left[2\omega_g^{-\frac{1}{2}}(\omega+\omega_g)\tan^{-1}\left(\frac{\omega_F}{\omega_g}\right)^{\frac{1}{2}}\right.$$
$$- (\omega_g+3\omega)(\omega_g+\omega)^{-\frac{1}{2}}\tan^{-1}\left(\frac{\omega_F}{\omega_g+\omega}\right)^{\frac{1}{2}}$$
$$\left. - (\omega_g-\omega)^{\frac{1}{2}}\tan^{-1}\left(\frac{\omega_F}{\omega_g-\omega}\right)^{\frac{1}{2}}\right]. \quad (7)$$

Despite the apparent singularity at ω_g, the energy gap, it can be shown analytically that this does not exist, but the singularity of Faraday rotation actually occurs at energy $\omega_g+\omega_F$. However, it should not be as sharp as in the nondegenerate samples. By using the original dispersion expressions of Eqs. (1) and (2), we can easily include losses due to scattering. In practice this is simple to do in the final result obtained so far. Whether the peak be in a form of a cusp as in the forbidden transition, or a sharp singularity as for the direct allowed case, the terms of interest are as follows:

$$\theta_d = K\left(\frac{\Delta\omega_d}{\omega}\right)\text{Real}\left[2\omega_g^{-\frac{1}{2}} - (\omega_g+\omega)^{-\frac{1}{2}}\right.$$
$$\left. - \left(\omega_g-\omega+\frac{j}{\tau}\right)^{-\frac{1}{2}}\right], \quad (8)$$

where τ is a relaxation time, and

$$\theta_f = C\left(\frac{\Delta\omega_f}{\omega}\right)\text{Real}\left[-2\omega_g^{\frac{1}{2}} + (\omega_g+\omega)^{\frac{1}{2}}\right.$$
$$\left. + \left(\omega_g-\omega+\frac{j}{\tau}\right)^{\frac{1}{2}}\right]. \quad (9)$$

To first approximation, these expressions apply both below and above the energy gap if we ignore the fine structures due to the interband transitions between the Landau levels. These expressions will give line shapes for the Faraday rotation more representative of the experimental results, particularly if a temperature dependence of τ is taken into account.

REFLECTION EXPERIMENTS

So far we have merely discussed the theory as applied to the transmission experiments. We have performed interband Faraday rotation experiments in InSb at optical wavelengths in the proximity of 2 ev corresponding to the transition between the $L_{3'} - L_1$ bands located at the edge of the Brillouin zone along the [111] direction. The experimental results are given in Fig. 4. The rotation of the plane of polarization upon reflection is indicated as a function of photon energy for two values of the static magnetic fields. The rotation shows two peaks corresponding to those associated with the transitions from the split $L_{3'}$ valence bands to the L_1 conduction band. The spin-orbit splitting deduced from these experiments give a value of approximately 0.4 ev.

The theoretical interpretation of this type of experiment can be obtained by considering the differential rotation of two circularly polarized waves which are reflected from the sample surface. The angle of rotation for each component of the two circularly polarized waves is given by

$$\tan\theta_\pm = 2k_\pm / (n_\pm{}^2 + k_\pm{}^2 - 1), \tag{10}$$

where $+$ and $-$ correspond to the right and left circular polarizations, respectively. The net rotation, which is small compared to unity, is given by

$$\theta = \frac{\theta_+ - \theta_-}{2} = \frac{(\Delta k)(1 - n^2 + k^2) + (\Delta n) \cdot 2nk}{(1 - n^2 - k^2)^2 + 4k^2}, \tag{11}$$

where $\Delta n = n_+ - n_-$ and $\Delta k = k_+ - k_-$. In this case Δn and Δk can be calculated for a direct transition from the dispersion relations in a manner analogous to that described by Eq. (3). This time, however, the Faraday rotation will have a contribution not only from the differential dispersive component, but from the differential absorption component as well. The important consequence of Eq. (11) is that in the reflection experiments there would be no rotation without absorption. The detailed analysis is rather involved. However, from the variation of Δn and Δk with wavelength, it is evident that a peak in Faraday rotation is associated with the corresponding reflection peak of the split bands in this material and qualitatively accounts for the shape of the curve obtained.

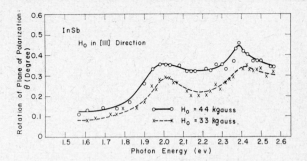

FIG. 4. Interband Faraday rotation in intrinsic InSb observed by reflection measurement.

DISCUSSION

The theory presented in this paper defines the nature of the interband Faraday rotation for the III-V compounds which resides in the differential dispersion of the two circularly polarized waves. At photon energies below the energy gap, this dispersion is the reactive component of the virtual transition between the valence and conduction bands. The feature of the problem which has not been explicitly evaluated is the phenomenological constant γ which represents the magnitude and sign of the rotation. The latter is proportional to the applied magnetic field. For our purposes, we need not evaluate γ theoretically, but can readily determine it from the experimental data in which the rotation has been studied as a function of magnetic field. As yet no detailed comparison has been made between theory and experiment in order to determine quantitatively the variation of rotation with photon energy. The experiments unfortunately have not been complete, in that they have not been properly designated to observe the rotation through the peak value and above the energy gap. Hence in order to explain the shape of the Faraday rotation vs. wavelength, it is necessary to include the loss as well. This we have indicated in Eqs. (8) and (9). Finally, the Faraday rotation which has been observed by reflection techniques can give information about the higher-energy bands of semiconductors and should be studied as a function of temperature to provide information about the bands and the scattering which determines the sharpness of the peaks. At low temperatures and high magnetic fields, it is hoped that fine structure and the anisotropy will be observed. This will permit the identification of the symmetry of the transitions involved. In principle, when individual transitions are studied the sign of the g factor should be determined from the Faraday rotation.

ACKNOWLEDGMENTS

We wish to thank Dr. Carl Stager for helpful suggestions in developing the experimental techniques and J. Halpern and R. Newcomb for their assistance with the experiment.

JOURNAL OF APPLIED PHYSICS   SUPPLEMENT TO VOL. 32, NO. 10   OCTOBER, 1961

Free Carrier Cyclotron Resonance, Faraday Rotation, and Voigt Double Refraction in Compound Semiconductors

E. D. PALIK, S. TEITLER, AND R. F. WALLIS

U. S. Naval Research Laboratory, Washington, D. C.

Measurements of cyclotron resonance absorption have been made in the far infrared spectral region from 25–150 μ on several III–V compounds at room and liquid-nitrogen temperatures using steady magnetic fields as high as 75 kgauss. For n-type InSb, InAs, InP, and GaAs, the data yield information concerning the conduction electron effective mass at the bottom of the band and its variation with magnetic field. Experiments have also been carried out on p-type InSb and corresponding information has been obtained for light holes. The dependence of the effective masses on both temperature and magnetic field can be satisfactorily interpreted in terms of Kane's theory for the band structure of these materials.

Measurements of Faraday rotation and Voigt double refraction have been made in the spectral region between 15 and 25 μ on a number of compound semiconductors at liquid-nitrogen temperatures. Either experiment gives the effective mass of the free carriers if their concentration is known. If both experiments can be performed, the results can be combined to give both the effective mass and carrier concentration directly without recourse to electrical measurements.

I. INTRODUCTION

THE magneto-optic studies have proven a powerful technique for determining the band structure of semiconductors. For many compound semiconductors, important information concerning effective masses of free carriers has been obtained through studies of cyclotron resonance absorption, Faraday rotation, and Voigt double refraction in the infrared spectral region.

II. CYCLOTRON RESONANCE

Infrared cyclotron resonance was first observed in InSb by Burstein, Picus, and Gebbie.[1] Cyclotron resonance in InSb and InAs was measured by Keyes[2] *et al.* Recently, Lax and Mavroides[3] have reviewed infrared cyclotron resonance work through 1960. At NRL the original work of Burstein and collaborators has been continued using the Bitter-type solenoidal magnets which provide steady fields as high as 100 kgauss. Infrared cyclotron resonance has been measured in the far infrared in several III-V compound semiconductors including n-type InSb,[4] InAs, and InP,[5] and GaAs.[6] Transmission measurements were made at room temperature and liquid-nitrogen temperature on thin sections glued to silicon or crystal quartz backings. In most cases, the experiments were performed using a reststrahlen monochromator, so that the wavelength was fixed and the magnetic field was swept. In some cases, the opposite was done using a prism monochromator. If sufficient energy was available, circularly polarized light was used to determine the sign of the

absorbers. To observe a distinct cyclotron resonance, it is necessary that

$$\omega_c \tau \gg 1, \qquad (1)$$

where $\omega_c = eH/m^*c$ is the cyclotron frequency and τ is the relaxation time. Here, e is the magnitude of the electron charge and H is the magnetic field. Unrationalized cgs units are used throughout this paper. In all cases, at liquid-nitrogen temperature $\omega_c \tau > 5$ and a distinct transmission minimum was observed. The effective mass m^* was determined by setting ω_c equal to the frequency of the radiation at the resonance condition. A plot of m^*/m versus H is shown in Fig. 1 for all four compounds. The results for InSb and InAs show a mass variation with field which is predictable from a knowledge of the effective mass at the bottom of the band m_0^*, band gap E_G, and spin orbit splitting Δ as discussed by Palik *et al.*[4] and Lax *et al.*[7] This mass variation is a consequence of the nonparabolic nature of the conduction band. For GaAs, InP, and InAs, the simple formula

$$m^* = m_0^*[1 + (\hbar\omega_c/E_G)(A-B)], \qquad (2)$$

obtained from fourth-order effective mass theory,[4] is reasonably valid and predicts a linear dependence of effective mass on magnetic field. In this equation,

$$A = 2[(1 + \tfrac{1}{2}x^2)/(1 + \tfrac{1}{2}x)](1-y)^2,$$
$$B = (1-y)(1-x)\{[(2 + \tfrac{3}{2}x + x^2)(1-y)/(2+x)^2] - \tfrac{2}{3}y\},$$
$$x = [1 + (\Delta/E_G)]^{-1},$$
$$y = m_0^*/m.$$

Equation (2) specifies the effective mass associated with the transitions from the lowest Landau level with spin up (assuming a negative g factor) to the next lowest level of the same spin. At room temperature, the corresponding transitions for spin down would give a small

[1] E. Burstein, G. S. Picus and H. A. Gebbie, Phys. Rev. **103**, 825 (1956).

[2] R. J. Keyes, S. Zwerdling, S. Foner, H. H. Kolm, and B. Lax, Phys. Rev. **104**, 1804 (1956).

[3] B. Lax and J. G. Mavroides, *Solid State Physics*, edited by F. Seitz and D. Turnbull (Academic Press, Inc., New York, 1960), Vol. 11.

[4] E. D. Palik, G. S. Picus, S. Teitler, and R. F. Wallis, Phys. Rev. **122**, 475 (1961).

[5] E. D. Palik and R. F. Wallis, Phys. Rev. **123**, 131 (1961).

[6] E. D. Palik, J. R. Stevenson, and R. F. Wallis, Phys. Rev. (to be published).

[7] B. Lax, J. G. Mavroides, H. J. Zeiger, and R. J. Keyes, Phys. Rev. **122**, 31 (1961).

TABLE I. Effective mass ratios obtained from cyclotron resonance at liquid-nitrogen temperature.

Material	Δ (ev)	E_G (ev)	m_0^*/m
InSb	0.9	0.225	0.0145
InAs	0.43	0.41	0.023
GaAs	0.33	1.53	0.071
InP	0.24	1.31	0.077

shift in the effective mass, but this shift is within the experimental errors for the samples studied.

In Fig. 1 the solid lines have been fitted through the experimental points for GaAs, InP, and InAs by using Eq. (2) and choosing m_0^* to give the best fit. To account for the mass variation in InSb, it is necessary to include higher order terms than contained in Eq. (2). This may be done using the Yafet theory[8] and the result is shown as the solid curve for InSb in Fig. 1. The low temperature effective mass ratios at the bottom of the band for InSb, InAs, InP, and GaAs are 0.0145±0.001, 0.023 ±0.002, 0.077±0.005, and 0.071±0.005, respectively. The values of Δ and E_G used in the analysis are tabulated in Table I together with the calculated values for the effective mass ratios. The above values for the effective mass ratios have been obtained for only one orientation of the crystallographic axes with respect to the magnetic field. However, the data are consistent with the theory of Kane[9] which predicts an isotropic mass at the bottom of the conduction band.

Measurements at room temperature gave transmission minima from which a room temperature mass could be inferred. For InAs, InP, and GaAs, there appeared to be little or no difference between the room temperature and low temperature masses. For InSb, the room temperature mass was about seven percent lighter than the low temperature mass. To a good approximation the conduction electron effective mass at the bottom of the band is given by the Kane formula[9]

$$\frac{m}{m_0^*} = 1 + \frac{2mP^2}{3\hbar^2}\left(\frac{2}{E_G} + \frac{1}{E_G + \Delta}\right), \quad (3)$$

where P is the momentum matrix element between valence and conduction band Bloch functions at $k=0$. The quantity P is probably temperature insensitive. Furthermore, the effective mass ratio is relatively insensitive to the value of Δ so that the temperature dependence of m/m_0^* arises primarily from that of the band gap E_G. It should be noted that, in InSb for example, lattice vibrations contribute about two-thirds of the observed temperature change in E_G and dilation only one-third.[10] Since lattice vibrations are ignored in the derivation of Eq. (3), one should include only the

[8] R. Bowers and Y. Yafet, Phys. Rev. 115, 1165 (1959).
[9] E. O. Kane, J. Phys. Chem. Solids 1, 249 (1956).
[10] T. S. Moss, Optical Properties of Semiconductors (Academic Press, Inc., New York, 1959).

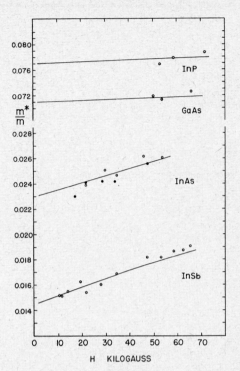

FIG. 1. Variation of low temperature effective mass ratio m^*/m with magnetic field H from cyclotron resonance data for several n-type III-V compound semiconductors.

dilational change in E_G in computing the temperature dependence of m/m_0^* to be consistent. Another contribution to the temperature dependence acting in the opposite direction arises from the thermal distribution of carriers over the states of the nonparabolic band.[11] The experimental data indicate that the dilational effect is more important than the thermal distribution effect in InSb; whereas in InAs, InP, and GaAs these effects roughly cancel.

Magnetoplasma effects can cause shifts in the observed position of the cyclotron absorption depending on the relative orientation of the sample, the magnetic field H, the Poynting vector S and the electric vector E of the radiation field. It is usually desirable to work in a frequency region such that $\omega \gg \omega_p$ and with the sample oriented to minimize magnetoplasma complications.[3]

The cyclotron resonance absorption observed at liquid-nitrogen temperature for InSb contained structure as shown in Fig. 2. It consisted of a main strong line, a weaker line to lower photon energy, and some absorption at still lower energy. At room temperature, the band was much broader containing a low energy tail not shown in Fig. 2. The two low temperature lines are due to the transitions from the two spin levels associated with Landau quantum number $l=0$ to the two spin levels associated with $l=1$. These transitions involve no change in the spin quantum number S or in the propagation constant k_z parallel to the magnetic field. The g factor for conduction electrons in InSb is

[11] R. F. Wallis, J. Phys. Chem. Solids 4, 101 (1958).

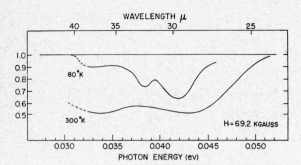

FIG. 2. Cyclotron resonance absorption in n-type InSb.

about -50 at the bottom of the band[12] and decreases in magnitude up into the band. This decreasing g factor causes the two transitions $(l=0,\ k_z,\ S=+\frac{1}{2}) \to (1,\ k_z,\ +\frac{1}{2})$ and $(0,\ k_z,\ -\frac{1}{2}) \to (1,\ k_z,\ -\frac{1}{2})$ to be separated. The magnitude of this separation can be calculated and is in agreement with the experimental results. The broad low energy absorption may be due to higher Landau level transitions. Measurements are now underway to observe this same effect in InAs.

Cyclotron resonance can be measured for holes as well as electrons. Far infrared cyclotron resonance of light holes has been observed in p-type InSb near liquid-nitrogen temperature in the spectral region 30–100 μ. Where possible, circularly polarized radiation was used to distinguish between hole and electron absorption. The light hole effective mass ratio was about 0.016, which is 10% heavier than the conduction electron mass and is in rough agreement with the estimate of Kane.[9]

III. FARADAY ROTATION AND VOIGT DOUBLE REFRACTION

For many materials it is difficult to satisfy experimentally the condition given by Eq. (1) for observing cyclotron resonance absorption. Very high magnetic fields, far infrared techniques and very thin samples may be required. It is frequently easier to utilize the dispersion caused by the cyclotron absorption to measure the Faraday and Voigt effects in a more easily accessible region in the infrared, generally between 5 and 25 μ. However, more information must be known about the sample such as carrier concentration, thickness, and refractive index. Under the condition $(\omega - \omega_c \tau) \gg 1$, the Faraday rotation for free electrons is given by the formula

$$\theta = \frac{\pi d}{\lambda}(n_- - n_+) = \frac{e^3 d N \lambda^2 H}{2\pi c^4 n m^{*2}},\quad (4)$$

where $n_\pm$ are the indices of refraction for $\pm$ circularly polarized light, N is the electron concentration, $\lambda = (2\pi c)/\omega$ is the wavelength, d is the sample thickness, and n is the index of refraction in absence of a magnetic

field. In this case plane polarized radiation is incident normally on a sample with the magnetic field parallel to the Poynting vector. Mitchell[13] and Stephen and Lidiard[14] suggested infrared free carrier Faraday rotation as a means of determining m^*. Smith et al.,[15] Moss and Walton,[16,17] Austin,[18,19] and Cardona[20] have used it to measure effective masses in InSb, InP, GaAs, InAs, and Bi_2Te_3. In all cases Eq. (4) applied.

Faraday rotation measurements have been made on an InAs sample containing 9.6×10^{16} electrons/cm^3 using low magnetic fields. The effective mass was found to be 6% heavier at room temperature than at liquid-nitrogen temperature. A similar result has been obtained by Cardona[20] who points out that such mass changes can be accounted for in terms of the thermal distribution of carriers over the nonparabolic band and the thermal dilation of the lattice. A comparison with the results given previously for cyclotron resonance indicates that the thermal distribution of carriers gives a more significant effect on the Faraday rotation values for the effective mass than on the cyclotron resonance values.

When the observing frequency ω is only slightly larger than ω_c, a correction term $\omega^2/(\omega^2 - \omega_c^2)$ should multiply the right side of Eq. (4). This term indicates that, for sufficiently large magnetic fields, the rotation θ does not vary linearly with field. Furthermore, the nonparabolic character of the energy bands leads to a mass variation with field. This mass variation causes a nonlinear dependence of θ on H which is in the opposite direction to that caused by the correction factor given above. In actual experiments, θ is found to vary linearly with H over a wider range than is indicated for Eq. (4) to hold, so that there is an apparent cancellation of the two nonlinear effects.

An extension of the equations of Stephen and Lidiard to the case $\omega_c \gg \omega$ yields an equation for rotation of the form

$$\theta = -2\pi e N d/nH,\quad (\omega_c - \omega)\tau \gg 1,\quad \omega^2 \gg \omega_p^2,\quad (5)$$

which is independent of wavelength and effective mass. For most materials, the conditions for Eq. (5) to be valid are not readily satisfied in the infrared region. However, it is possible to do the experiment for light mass materials such as InSb and InAs using magnetic fields up to 100 kgauss. Rotation at a fixed wavelength near 76 μ has been observed in a sample of InSb near liquid-nitrogen temperature. A number of magnetic fields were employed such that the cyclotron frequencies

[12] L. M. Roth and S. Zwerdling, Phys. Rev. 114, 90 (1959).

[13] E. W. J. Mitchell, Proc. Phys. Soc. (London) A68, 973 (1955).
[14] M. J. Stephen and A. B. Lidiard, J. Phys. Chem. Solids 9, 43 (1959).
[15] S. D. Smith T. S. Moss, and K. W. Taylor, J. Phys. Chem. Solids 11, 131 (1959).
[16] T. S. Moss and A. K. Walton, Physica 25, 1142 (1959).
[17] T. S. Moss and A. K. Walton, Proc. Phys. Soc. (London) 74, 131 (1959).
[18] I. G. Austin, J. Electronics and Control 8, 167 (1960).
[19] I. G. Austin, Proc. Phys. Soc. (London) 76, 169 (1960).
[20] M. Cardona, Phys. Rev. 121, 752 (1961).

were both considerably above and considerably below the observing frequency. For $\omega \gg \omega_c$, Eq. (4) is obeyed, as has been well established,[15] and for $\omega \ll \omega_c$, Eq. (5) is followed. Use of the two equations can yield m^* and N if n is known. For the InSb sample used, an effective mass ratio of 0.015 and a carrier concentration of 1.0×10^{15} carriers/cm³ were obtained.

Another technique for determining effective mass is the Voigt effect, which has been measured at NRL.[21,22] In this case, the magnetic field is in the sample plane and the radiation propagates through the sample in a direction normal to the magnetic field. In a convenient experimental arrangement, plane polarized light is incident upon the sample with the electric vector at an angle of 45° with respect to H. For $\omega \gg \omega_c, \omega_p$, the phase difference between the components with electric vectors parallel and perpendicular to the magnetic field is given by

$$\delta = \frac{2\pi d}{\lambda}(n_{11} - n_{\perp}) = \frac{e^4 N d H^2 \lambda^3}{4\pi^2 c^6 n m^{*3}}. \quad (6)$$

Here $n_{11} = n$ is the index of refraction for electric vector parallel to H and $n_{\perp}$ is the index of refraction for electric vector perpendicular to H. For ω somewhat closer to ω_c and/or ω_p, Eq. (6) will remain valid to a higher order of approximation if the right-hand side is multiplied by a factor $\omega^2/(\omega^2 - \omega_c^2 - \omega_p^2)$. The essential features of Eq. (6) to be emphasized are that δ depends on the inverse cube of the effective mass rather than the inverse square as in the Faraday effect and is an even function of the magnetic field. The resulting elliptically polarized light is measured with a quarter-wave plate

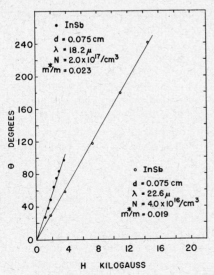

FIG. 3. Faraday rotation in two samples of n-type InSb at liquid-nitrogen temperature.

[21] S. Teitler and E. D. Palik, Phys. Rev. Letters **5**, 546 (1960).
[22] S. Teitler, E. D. Palik, and R. F. Wallis, Phys. Rev. **123**, 1631 (1961).

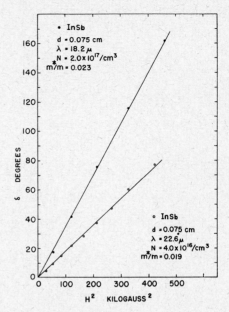

FIG. 4. Voigt double refraction in two samples of n-type InSb at liquid-nitrogen temperature.

and a polarizer in the same wavelength region as the Faraday effect. The combination of magnetic double refraction and properly oriented quarter-wave plate provides an effective rotation of linearly polarized radiation by an angle of $\delta/2$. Knowing N from Hall effect data, we have measured the Voigt effect in InSb, InAs, and GaAs for various carrier concentrations and determined effective masses consistent with values found by other workers. In general, the masses obtained from Voigt double refraction agree with the masses obtained from Faraday rotation.

An application in which the two effects are used to complement each other can be seen by dividing Eq. (6) by Eq. (4). Then

$$\delta/\theta = (eH\lambda)/(m^* c^2 2\pi) = \omega_c/\omega. \quad (7)$$

This equation permits the determination of m^* directly from laboratory measurements of θ and δ carried out at the same values of λ and H. For two samples of InSb, the results of Faraday and Voigt effect measurements are shown in Figs. 3 and 4. Here, θ versus H and δ versus H^2 are plotted. In each case, the dependence is linear. Using a form of Eq. (7) corrected for plasma effects, the liquid-nitrogen data have been analyzed using an iterative procedure[22] to yield the effective masses and carrier concentrations listed in Table II. Also shown in Table II are results for the two effects used independently when N was known for samples of InAs and GaAs. These results are consistent with those obtained by Smith *et al.*[15] and Cardona.[20]

The Faraday effect and Voigt effect are measured as a rotation of the plane of polarization and are of equivalent difficulty. As indicated by Eq. (7), δ is smaller than θ in the ratio ω_c/ω for the practical magnetic fields

TABLE II. Effective mass ratios obtained by Faraday and Voigt effects at liquid-nitrogen temperature.

Material	Carrier concentration	Faraday effect m^*/m	Voigt effect m^*/m	Combination of Voigt and Faraday effects
InSb	4.0×10^{16}			0.019
InSb	2.0×10^{17}			0.023
InAs	7×10^{15}	0.026		
InAs	1×10^{17}	0.030	0.031	
GaAs	4.3×10^{16}	0.076	0.071	

available. Consequently, for a new sample with N known, the Faraday effect would usually be measured first as it produces the largest rotation. However, if N is not known, the two experiments will yield both N and m^*.

ACKNOWLEDGMENTS

We wish to thank G. S. Picus and J. R. Stevenson for contributions to portions of work presented in this paper. We benefited from discussions with E. Burstein, F. Stern, and R. Toupin. Samples were kindly provided by the National Bureau of Standards, Naval Ordnance Laboratory, R.C.A. Research Laboratory, Services Electronics Research Laboratory, and Texas Instruments, Inc.

JOURNAL OF APPLIED PHYSICS   SUPPLEMENT TO VOL. 32, NO. 10   OCTOBER, 1961

Optical Absorption Edge in GaAs and Its Dependence on Electric Field

T. S. Moss

Royal Aircraft Establishment, Farnborough, Hants, England

Values of absorption constant covering the range 1 cm^{-1} to 10^4 cm^{-1} have been derived from transmission measurements made on single-crystal gallium arsenide. The absorption edge is very steep up to $\sim$4000 cm^{-1}, where there is a knee beyond which the absorption increases relatively slowly with photon energy. The energy bands have been calculated using Kane's theory. From these a theoretical absorption curve has been obtained which shows very good agreement with the experimental data.

Using semi-insulating material, it has been possible to measure the shift of the edge with applied electric field. The effect is small ($\sim$200-μ ev shift for 5000-v/cm field) but is in good agreement with theory.

INTRODUCTION

GALLIUM arsenide is one of the group III–V intermetallic semiconductors which were first studied by Welker[1] and his colleagues. It has many interesting properties, and is currently of much technological interest for making various solid-state devices, particularly transistors, parameteric diodes, and tunnel diodes. It shows promise of being the best solar battery material,[2] and for this application in particular it is necessary to have detailed information on the optical absorption up to high levels. Early measurements of absorption in the neighborhood of the edge have been published,[3] but they reached an absorption level of only 100 cm^{-1}. Some of this work has already been described in a recent publication.[4]

EXPERIMENTAL DETAILS

The material used for the study of the absorption edge was pure single-crystal GaAs containing $\sim$3$\times$10^{16} cm^{-3} excess electrons. Plane parallel samples were prepared by grinding with silicon carbide and polishing with diamond paste. Specimens from 1 cm down to 7-μ

thickness were used. The thickness of the thin specimens was found by measuring interference fringes in the 5 to 15-μ waveband.

For the measurements of the edge shift produced by an electric field, samples of very high resistance GaAs were obtained. This material contains about the same density of impurities as the above, but the free carrier density is only 10^7 cm^{-3} free electrons.[5] Specimens of this material were prepared in the same way except that they were etched after polishing, since this was found to give a considerable increase in specimen resistance. It was possible to prepare fairly thick specimens of resistance $>$10^{11} Ω, and to use applied voltages up to 10 kv. In order to avoid heating effects, the specimens were immersed in a liquid. Ligroin proved convenient for this purpose.

The radiation was provided by a tungsten lamp and a Leiss double-prism monochromator. The prisms were flint glass and a resolution of 3$\times$10^{-3} ev was used on the steepest part of the absorption curve. On the flatter part of the curve, at high K levels, the resolution was 6$\times$10^{-3} ev. Great care was taken to ensure spectral purity, and measurements could be made with insertion losses of up to 10^4:1 on the flatter part of the absorption

[1] H. Welker, Z. Naturforsch. **7a**, 744 (1952); **8a**, 248 (1953).
[2] T. S. Moss, *Solid State Electronics* **2**, 222 (1961).
[3] F. von Oswald and R. Schade, Z. Naturforsch. **9a**, 611 (1954).
[4] T. S. Moss and T. D. Hawkins, Infrared Phys. **1**, 111 (1961).

[5] J. W. Allen, Nature **187**, 403 (1960); W. R. Harding, C. Hilsum, M. E. Moncaster, D. C. Northrop, and O. Simpson, *ibid.*, 405 (1960).

curve. The detector system was an infrared photo-multiplier followed by an 800-cps amplifier.

ABSORPTION RESULTS

The measured values of absorption coefficient (K) are plotted in Fig. 1, from which it will be seen that the main part of the edge is very steep, K rising rapidly from 4 to 4000 cm^{-1}. Over this range the edge is exponential—as has been observed for many other materials[6]—with a slope ~ 100 ev^{-1}. The steepness of the edge indicates that no phonon-assisted transitions are involved, i.e., the absorption edge is due to the onset of vertical transitions.

CALCULATION OF ENERGY BANDS AND ABSORPTION COEFFICIENT

The E-k curves have been computed using Kane's theory.[7] They are given by

$$k^2 = [2m_c^* E(E-G)(E+\Delta)(G+2\Delta/3)] / [(E+2\Delta/3)G(G+\Delta)], \quad \text{a.u.[8]} \quad (1)$$

where G is the energy gap, Δ the spin-orbit splitting, m_c^* is the electron mass at the bottom of the conduction band, and E refers to the conduction band (E_c), the light-hole band (E_2), or the split-off valence band (E_3). The heavy-hole band (E_1) is assumed to be a simple, parabolic band corresponding to a mass $m_1^* = 0.68\ m_0$.[9]

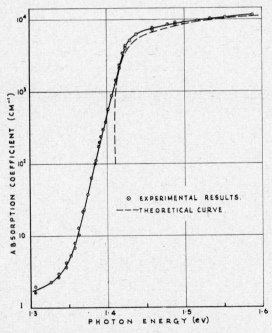

FIG. 1. Absorption in gallium arsenide.

[6] T. S. Moss, *Optical Properties of Semiconductors* (Butterworths Scientific Publications Ltd., London, and Academic Press Inc., New York, 1959, 1961), pp. 39, 86.

[7] E. O. Kane, J. Phys. Chem. Solids **1**, 249 (1957).

[8] In a.u., $m = e = h/2\pi = 1$.

[9] H. Ehrenreich, Phys. Rev. **120**, 1951 (1960).

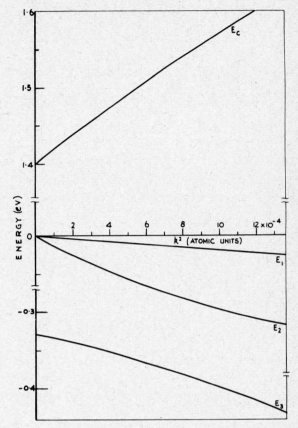

FIG. 2. E-k curves for gallium arsenide.

The curves have been computed using the parameters $Eg = 1.4$ ev, $\Delta = 0.33$ ev,[10] and $m_c^* = 0.072\ m_0$,[11] and are plotted in Fig. 2. The conduction band is slightly non-parabolic, the effective mass increasing with energy. Using the expression for effective mass which occurs in Faraday effect or dispersion experiments,[6] namely,

$$m^* = \hbar^2 (k\,dk/dE)_F, \quad (2)$$

where the subscript F means the value at the Fermi level, the electron mass should increase to $m^* = 0.082\ m_0$ at 0.1 ev above the bottom of the conduction band, i.e., for 3×10^{18} cm^{-3} carriers.

The masses of the light-hole and split-off valence bands are given by the slopes as $k \to 0$, namely

$$m_2^* = 0.085\ m_0, \quad m_3^* = 0.25\ m_0.$$

The absorption coefficient can be calculated directly from the $E-k$ curves.[7] Assuming direct, vertical transitions, we have

$$K = (4\pi^2/cnh\nu) \sum_j M_j^2 \rho_j, \quad \text{a.u.,} \quad (3)$$

where n is the refractive index and the summation is over the three valence bands $(j = 1, 2, 3)$.

[10] R. Braunstein, J. Phys. Chem. Solids **8**, 280 (1959).

[11] T. S. Moss and A. K. Walton, Proc. Phys. Soc. (London) **74**, 131 (1959).

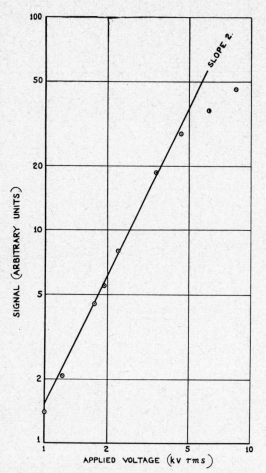

FIG. 3. Dependence of absorption edge on electric field.

The optical matrix element is

$$M_j{}^2=(2P^2/3)\{(A_cC_j+C_cA_j)^2+(A_cB_j-B_cA_j)^2\}, \quad (4)$$

where the bracketed term—which is always near unity —is computed for each k value from the coefficients

$$A=k\rho(E+2\Delta/3)/N,$$
$$B=2^{\frac{1}{2}}(E-G)\Delta/3N,$$
$$C=(E-G)(E+2\Delta/3)/N,$$

where N is a normalizing factor such that $A^2+B^2+C^2=1$, and $P^2=E(E+\Delta)/2(E+2\Delta/3)m_c{}^*$. For the parabolic E_1 band, $A_1=0$, $B_1=1$, and $C_1=0$. The density of states ρ_j is given in terms of the slopes of the $E-k$ curves by

$$\rho_j=k^2/2\pi^2(dE_c/dk-dE_j/dk). \quad (5)$$

The absorption coefficient plotted from Eq. (3) is shown in Fig. 1. The agreement with the experimental data is seen to be very good, particularly at short wavelengths where the calculated curve does not depend on the use of any arbitrary constants or adjustable parameters. In the neighborhood of the absorption edge, the fit has been improved by sliding the curve slightly sideways. This process gives a value for the energy gap at

292°K of $E_g=1.41$ ev, which is considered to be somewhat more accurate than the value of 1.4 ev assumed in the analysis.

The above absorption theory is essentially that for a perfect GaAs lattice, with no perturbations due to thermal vibrations or crystalline irregularities.[12] The presence of these could well explain the slope of the edge observed under the experimental conditions.

SHIFT OF EDGE WITH ELECTRIC FIELD

As the shift of the edge with electric field is quite small, it was essential to use very high fields; this necessitated using high resistance specimens in order to prevent heating. The best material obtained had a specific resistance $>10^8\Omega$ cm (when in complete darkness) and using this, it was possible to make specimens of resistance $>10^{11}\Omega$. As the temperature dependence of the edge in GaAs is fairly large ($dE/dT=500$ μev/°C), the measurements were always made with specimens in a liquid bath.

Measurements with dc electric fields and chopped radiation were inconclusive; therefore, a system using steady radiation and ac fields was developed.

It was assumed that the effect would be independent of the sign of the field (F), and would be proportional to F^2. The frequency used for the field, therefore, was made *half* that of the amplifying system, namely 400 cps. (Subsequent comparisons made with 800 cps and 400-cps fields confirmed this hypothesis, the effect with 800 cps being many times smaller than with 400 cps.)

The dependence of the observed signal at optimum wavelength on electric field is shown in Fig. 3, from which it will be seen that the points lie well on a line of slope 2 for rms voltages up to 4 kv. The reason for the tendency to saturate at the highest fields used is not completely understood; it is probably due in part to deterioration in the waveform from the amplifier used to supply the field when working near its maximum output.

The spectral dependence of the field-induced signal is shown in Fig. 4. The response peaks sharply at the wavelength where the zero-field transmission curve is steepest, having a width at half-amplitude which is $<kT$. As Fig. 4 shows, its shape and position are virtually identical with the derivative of the transmissions, thus proving that the signal observed is due to a pure shift of the absorption edge by the applied field. The absolute magnitude of this shift can be determined from the relative magnitudes of the field-induced signal and the differential transmission when they are measured under comparable conditions. In the experiments, the optical conditions were identical for the two measurements but it was convenient to use different waveforms, namely, sinusoidal for the applied voltage, and squarewave chopping for the zero-field transmission. The

[12] R. H. Paramenter, Phys. Rev. **97**, 587 (1955).

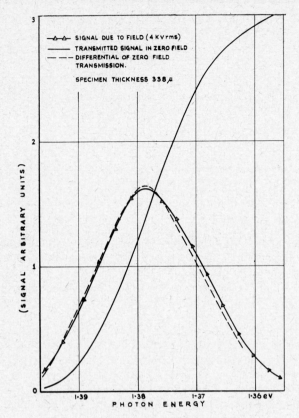

FIG. 4. Spectral dependence of change in transmission due to an electric field.

latter signal must, therefore, be reduced by $\pi/4:1$ to obtain the equivalent signal for sinusoidal chopping.

For 3 kv rms at 400 cps applied to a specimen 8.8 mm long, the observed shift was $\Delta E = 198 \pm 2$-μev. The peak-to-peak field at 800 cps in this case is

$$F = 3(2)^{\frac{1}{2}}/0.88 = 4.83 \text{ kv/cm.}$$

Thus,

$$\Delta E/F^2 = 8.5 \times 10^{-16} \text{ ev per (v/m)}^2. \qquad (6)$$

It has been shown theoretically by Franz[13] that at low absorption levels, for an exponential absorption edge defined by

$$A = A_0 \exp\alpha(\omega - \omega_0), \qquad (7)$$

[13] W. Franz, Z. Naturforsch. **13a**, 484 (1958).

the displacement is such that the edge becomes

$$A = A_0 \exp\alpha(\omega - \omega_0 + \alpha^2 e^2 F^2/12\hbar m^*); \qquad (8)$$

the shift is then

$$\Delta\omega = \alpha^2 e^2 F^2/12\hbar m^* \quad \text{rad/sec,}$$

or

$$\hbar\Delta\omega/eF^2 = \alpha^2 e/12 m^* \text{ ev per (v/m)}^2. \qquad (9)$$

The effective mass in this equation should presumably be the reduced mass of the electron-hole pair, i.e.,

$$1/m^* = 1/m_c + 1/m_1, \quad \text{or} \quad m^* = 0.065 m_0. \qquad (10)$$

From Fig. 1, $\alpha \equiv 98$ ev^{-1}, so that from Eq. (9) the expected shift is

$$\Delta E/F^2 = 9.3 \times 10^{-16} \text{ ev per (v/cm)}^2. \qquad (11)$$

This is less than 10% greater than the observed shift [Eq. (6)], so that the agreement between theory and experiment can be considered quite satisfactory. The experiment, therefore, may be a useful way of obtaining fairly accurate values of effective masses in high-resistance materials for which, at present, no other reasonable method is available.

CONCLUSIONS

Measurements have been made of the absorption coefficient of single-crystal GaAs in the neighborhood of the absorption edge. The results obtained agree well in absolute magnitude with values calculated from the $E-k$ curves which have been computed for this material.

A shift of the absorption edge with electric field has been observed. The good agreement found between the measured shift and theory indicates that this might be a useful method of measuring effective mass in insulators or near-insulators.

ACKNOWLEDGMENTS

Thanks are due to D. H. Roberts of the Plessey Company, C. Hilsum of S.E.R.L., and D. J. Dowling of Mining and Chemical Products, for providing the gallium arsenide used in this work, and to T. D. F. Hawkins for assistance with the measurements.

JOURNAL OF APPLIED PHYSICS   SUPPLEMENT TO VOL. 32, NO. 10   OCTOBER, 1961

Energy of Spectrum and Scattering of Current Carriers in Gallium Arsenide

D. N. Nasledov

Laboratory of Electron Semiconductors, Institute of Physics, Academy of Sciences, Leningrad, U.S.S.R.

Conductivity in impurity bands is observed in gallium arsenide at low temperatures. The impurity banding phenomena observed in *n*- and *p*-type crystals are markedly different due to the wide divergence in their effective masses. In the temperature interval where the crystal conductivity is determined by conduction in the impurity band, resistivity of *n*-type material decreases with increasing magnetic field.

The temperature dependences of carrier mobility indicate that at low temperature the impurity ions play the dominant role of the scattering centers. In order to investigate scattering processes more thoroughly, studies were made of the thermomagnetic Nernst-Ettingshausen effects which are very sensitive to the scattering mechanism. It was found that, in the case of nondegenerate specimens at low temperatures, scattering by impurity ions dominates, and at high temperatures by acoustical lattice vibrations; scattering by acoustical lattice vibrations becomes dominant even at low temperature in specimens of high degeneracy. Scattering was also investigated in indium arsenide and indium antimonide.

THIS report is a survey of a number of investigations on the electrical properties of gallium arsenide crystals carried out in the Laboratory of Electron Semiconductors of the Institute of Physics, Academy of Sciences, of the U.S.S.R. in Leningrad. We were aided in this work by O. V. Emelyanenko and T. S. Lagunova.

We shall present the results of the investigations of the energy band structure of these crystals and the thermomagnetic effects of Nernst-Ettingshausen, clarifying the scattering of current carriers. We will also present data on similar investigations of indium arsenide and indium antimonide as confirmation of certain conclusions reached.

The gallium arsenide crystals were obtained by slow cooling of an alloy, or by using the "Czochralski" method, and certain of them were alloyed.

The current carrier concentration at room temperature was from 2×10^{16} to 8×10^{19} cm^{-3}. The samples

manifested both an *n*- as well as a *p*-type conductivity.

The temperature interval at which the measurements were carried out was 1.5–900°K.

At all the temperatures except the highest, the material revealed an impurity conductivity. In the strongly alloyed *n*- and *p*-type samples, the Hall constant is independent of temperature down to 1.5°K. (Fig. 1: samples 3*p* and 8*n*). In pure samples, a maximum appears on the Hall constant R vs $1/T$ curves from 60° to 250°K. The smaller the electron or hole concentration in the sample, the lower the temperature at which the maximum is observed and the greater its relative magnitude. At uniform carrier concentrations, the relative maximum is greater for *p*-type samples than for *n*-type. In the purest of the *n*-type samples shown in Fig. 1 (sample 7*p*, $p = 1.5 \times 10^{17}$ cm^{-3}) the relative maximum is 7 and in an *n*-type sample (sample 9*n*, $n = 2.2 \times 10^{16}$ cm^{-3}) it is only 1.7.

The electron conductivity of pure crystals drops strongly when they are cooled to liquid helium temperatures. The conclusion can be reached from these results that there is impurity band conduction in the gallium arsenide crystals. Below 150–50°K, the conductivity in pure samples occurs in an impurity band and at higher temperatures in the conduction or valence band. The divergences in the Hall constant maxima for *p*- and *n*-type samples confirm the hypothesis that the formation and properties of the impurity band should depend on the effective mass of the current carriers. In gallium arsenide this dependence is manifested more strongly than, for example, in germanium and silicon, due to the wide difference of the electron and hole masses ($m_n^* \approx 0.05m$, $m_p^* \approx 0.5m$) if a hydrogen-like model of the impurity atoms is assumed. From the indicated differences of the effective masses, and equal degree of overlapping of the wave functions of the impurity atoms in *p*- and *n*-type samples should be observed at impurity concentrations differing by three orders of magnitude. By assuming that an approximately equal degree of overlapping leads to the same magnitude of the relative maxima of the R vs $1/T$ curves, it can be shown that these calculations are in

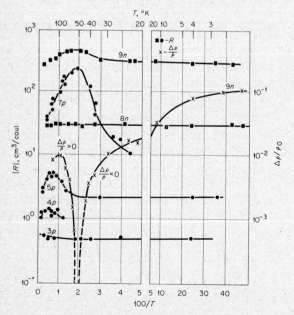

Fig. 1. Log Hall constant and log of magnetoresistivity vs reciprocal temperature for several *n*- and *p*-type GaAs specimens.

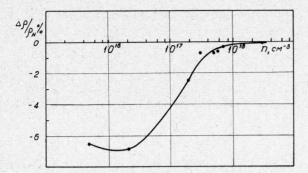

FIG. 2. Dependence of magnetoresistance on electron concentration in n-type GaAs at 4.2°K and in a field of 5300 oe.

qualitative agreement with the experimental results (Fig. 1, samples $9n$, $4p$).

The R vs $1/T$ curves exhibit the same temperature dependence as do the conductivity curves and the activation energy of the impurity levels in the investigated samples can be calculated (more exactly—the gap between the impurity bands and the conduction and/or valence bands). For acceptor impurities (Zn, Cd), this magnitude in samples with an impurity concentration of about 10^{17} cm^{-3} is about 0.02 ev and, for donors at $n = 2 \times 10^{16}$ cm^{-3}, several fold less. With an increase in the impurity content, the activation energy decreases and in strongly alloyed crystals of both the n- and p-type ($n \approx 10^{17}$ cm^{-3} and $p \approx 3 \times 10^{19}$ cm^{-3}, respectively) the impurity band completely overlaps the conduction or valence band.

It is of interest to note that in the temperature range below that of the Hall constant maximum ($T < 50$–80°K), the change in resistivity of all the n-type samples in a transverse magnetic field is negative, i.e., when the magnetic field is applied, the resistivity of the crystals decreases (Fig. 1). Although there is not yet any theoretical explanation for this phenomenon, the similarity of the inversion temperature for $\Delta\rho/\rho$ to that of the R maximum makes it possible to postulate that this phenomenon is associated with the conductivity in the impurity bands. The absolute effect decreases with an increase in the impurity concentration in the crystals (Fig. 2). A series of controlled experiments, in which samples with etched and polished surfaces, pressure or soldered contacts, and so on were used, indicated that the observed effect is not due to the surface effects, type of contact, or nonuniformities of the sample, but is apparently due to the volume properties of the uniform crystals.

In studying an n-type specimen, we encountered still another fact; the effective mass of electrons, as our measurements of the thermoelectromotive forces indicated, increases with the electron concentration (Fig. 3). The same increase in m^* is known to occur in InSb and InAs. Whether this is associated with the overlapping of the levels of the conduction and impurity bands observable from concentrations of the order of 10^{16}–10^{17}

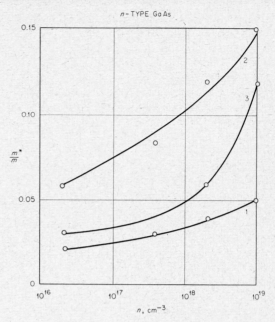

FIG. 3. The dependence of electron effective mass on electron concentration in n-type GaAs.

cm^{-3}, or to other factors, is not yet known. The effective mass of both electrons and holes in GaAs also increases with temperature; from 100° to 600°K the increase is by a factor of approximately 1.5. It can be assumed that the increase in effective masses is always due to an increase in the average current carrier energy occurring not only with an increase in temperature, but also when the current carrier concentration increases. In n-type GaAs samples, for example, due to the degeneracy of the electron gas, the position of the Fermi level shifts from the vicinity of $\zeta \approx 0$ in samples with a concentration of 10^{19} cm^{-3}. One of the possible explanations of the increase of effective masses could be given on the assumption that within the conduction band or the valence band of GaAs, there are additional bands with a higher density of states and a higher carrier effective mass, so that transitions into these levels might also result in an increase in m^*. Such a structure for the conduction band in particular was proposed by Ehrenreich. As Ehrenreich pointed out, however, such a mechanism should lead to an increase in the Hall constant with temperature; but in our experiments the increase in the Hall constant is observed only above 600°K and only in n-type crystals. This makes the hypothesis concerning the connection of the increase in m^* with the presence of additional bands within the conduction or valence bands of the crystal apparently inapplicable. The hypothesis of the strong nonparabolic nature of bands, which can occur according to Kahn because of interactions between the conduction and valence band levels, also seems implausible. If bands of such a nonparabolic nature actually exist, then because of the great width of the forbidden zone in GaAs, the extent should be very small—4–6 times less than in

D. N. NASLEDOV

InAs or InSb, while the experimental values of m^* given above are almost no different from those we obtained or those published in the literature for InAs or InSb.

In what follows, during the discussion of the scattering mechanism, we have used as a starting point the fact that the conduction and valence bands in GaAs have a very simple parabolic form. Although a more rigid examination of the transport phenomena should of course take into account the indicated characteristics of the effective masses of carriers in GaAs, this apparently only introduces corrections having no effect on the conclusions as a whole. An analysis of the kinetic effects in terms of a parabolic zone is considerably more simple.

As is known, the temperature dependence of the mobility of electrons and holes in gallium arsenide makes it possible to establish that the current carriers above room temperature are scattered by the thermal vibrations of the lattice (Fig. 4).

However, the mobility vs temperature curves do not make it possible to answer the question of what vibrations precisely dominate the scattering process—whether they are acoustical or optical. In scattering by the acoustical vibrations, the relation $\mu \propto T^{-\frac{3}{2}}$ must be obtained, whereas scattering by polar-optical modes leads to a range of T dependences, i.e., from $\mu \propto e^{\theta/T}$ to $\mu \propto T^{-\frac{1}{2}}$. Ehrenreich has attempted to answer this question by comparing the absolute magnitude of the experimental and theoretical mobilities and not the tempera-

ture dependence. Although he also has concluded that in GaAs and in a number of other group III–V compounds such as InSb and InP, scattering occurs predominantly by polar-optical vibrations of the lattice, his conclusion is not quite convincing since it is based on a calculation which uses a number of vaguely known parameters. Therefore, we decided to investigate the Nernst-Ettingshausen effect to elaborate the study of the scattering mechanism of the current carriers in GaAs.

The N-E electromotive forces appear in a sample along which a current of heat flows, if that sample is placed in a magnetic field. The transverse signal is proportional to $H \cdot \text{grad } T$ and the longitudinal signal to $H^2 \text{ grad } T$.

The coefficients determining these proportions are called transverse and longitudinal N-E constants. They

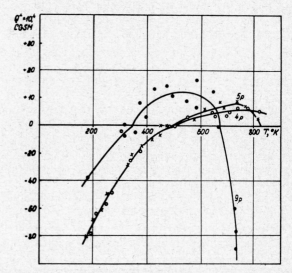

Fig. 5. The transverse Nernst-Ettingshausen coefficients for p-type GaAs and their dependence on temperature ($4p$–4.5×10^{18}, $5p$–1.0×10^{18}, $9p$–1.0×10^{17} holes/cm³).

only depend on mobility and on the scattering mechanism of the carriers. We shall not dwell on the physical analysis of the effects, which is very simple and straightforward, but shall give the formulas. Under the assumption that $\mu_x \propto l/v$, $l \propto v^2$ and defining the Hall mobility as $\mu_x = R\sigma$, we get

In the absence of degeneracy   In strong degeneracy

$$Q^{\text{I}} = \frac{k}{e} \frac{1-r}{2} \mu_x \qquad\qquad \times \frac{1}{\eta} \frac{\pi^2}{3}$$

$$Q^{\text{II}} = \frac{k}{e} \frac{1-r}{2} A_x \qquad\qquad \times \frac{1}{\eta} \frac{\pi^2}{3},$$

where $A \approx 1$.

Here, l is the mean free path, V the thermal velocity, r a certain number which depends on the scattering mechanism, and η is the reduced Fermi level given by

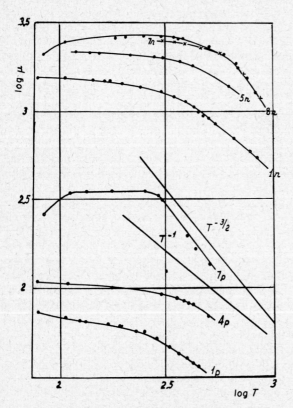

Fig. 4. The dependence of carrier mobility on temperature for n- and p-type GaAs.

$\eta = \zeta/kT$. For scattering by acoustical vibrations, $Q^{\perp}$ and $Q^{\parallel}$ are positive, whereas for optical mode scattering they are either zero or negative. The negative sign also appears for scattering by impurity ions.

In Fig. 5 are given the $Q^{\perp}$ vs T curves for three samples of p-type GaAs. The $R(T)$ and $\mu_x(T)$ curves for these samples were given earlier. The longitudinal effect in p-type GaAs, again because of the poor mobility of the holes, is small and could not be measured. Below room temperature, this effect was negative. The value of r calculated from the formulas given above is close to 4. This indicates that the holes are scattered at low temperatures primarily by the impurity ions. Above 300–400°K, the effect becomes positive and the magnitude of r, calculated at the maximum, is close to zero for all three samples ($r = 0.3$ to 0.2). This result, in our opinion, indicates that during the scattering of holes by thermal lattice vibrations, the acoustical branch is dominant. We would like to note that in the purest p-type sample ($9p$) the region of positive values of r begins at lower temperatures than in other samples, which should be expected, since scattering by impurity ions in it is weaker.

TABLE I. Calculations of the scattering parameter r as a function of temperature for three n-type GaAs samples.

	Samples		
T, °K	Sample $9n$	Sample $5n$	Sample 2^*n
100	1.8	4	2.5
200	1.7	4	2.5
300	1.6	3	0.8
500	1.3	1.3	0.3
700	1.2	0.6	0.1
800	1.3	0.2	0.0

A decrease in $Q^{\perp}$ and a transition to negative values at high temperatures is due to the onset of intrinsic conductivity. Investigations on n-type specimens lead to conclusions concerning the scattering mechanisms which are closely analogous to these. Figure 6 shows the $Q^{\perp}$ and $Q^{\parallel}$ vs T curves and also those of the temperature dependence of the current carrier concentration and of the mobility in an n-type sample, $5n$($n = 2.5 \times 10^{18}$ cm^{-3}).

All the measurements (the conductivity, the Hall effect, and the N-E effects) were carried out in a single apparatus. At lower temperatures the electron mobility is constant; the N-E effects are negative. The magnitude of r calculated by means of u_x and $Q^{\perp}$ was found to be equal to 4. This result indicated that the electrons are scattered by the impurity ions; the temperature independence of mobility is due to the degeneracy of the electron gas. With an increase in temperature the electron mobility begins to drop, and the N-E effects change to positive sign. This shows that scattering is by the acoustical branch of the thermal lattice vibrations.

Similar results were obtained for a large number of

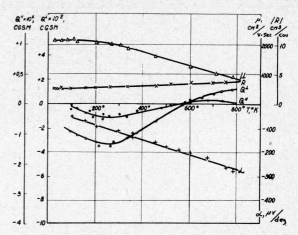

FIG. 6. The dependence of various transport coefficients on temperature for n-type GaAs (specimen $5n - n = 2.5 \times 10^{18}$ cm^{-3}).

n-type GaAs samples. The values of r for certain of them calculated from μ_x and $Q^{\perp}$ are given in Table I. (The calculations were made from the above-given formulas by taking into account the electron gas degeneracy.)

It appears to us that the results show that the scattering by the acoustical lattice vibrations play the dominant role in electron scattering in GaAs above 500–700°K, especially in degenerate samples. (In sample $9n$ the calculated values of r above 700–800°K are distorted by the appearance of intrinsic conductivity.) The results, however, do not exclude the possibility that at temperatures of the order of 300–500°K, optical lattice vibrations also participate in determining Q near the point where there is an inversion in sign.

It is of interest that in the purest sample, $9n$($n = 2.2 \times 10^{16}$ cm^{-3}), and the most impure sample, 2^*n ($n = 10^{19}$ cm^{-3}), the value of r at low temperatures is perceptibly less than 4. The same results were obtained for other n-type GaAs samples with similar carrier concentrations. It appears that this is due to the presence of additional scattering centers in the crystals examined, in addition to the ionized impurity atoms. In pure crystals, such centers may consist of oxygen or any other neutral atoms, and in strongly alloyed ones, lattice distortions caused by the introduction of impurity atoms are produced. In the last case the effect of screening of the impurity ions by electrons is very possible.

By comparing the values of r in the various samples, still another, at first unexpected phenomenon may be observed: the higher the electron concentration in the sample, the more strongly is scattering by thermal lattice vibrations manifested at lower temperatures. This is also especially apparent from the $Q^{\perp}$ vs T curves which make a transition to the positive values of $Q^{\perp}$ earlier, the more strongly alloyed is the sample (Fig. 7). This fact becomes understandable if it is remembered that the electron gas in $5n$ and 2^*n is degenerate, and the degeneracy is stronger the higher the electron concentration. In fact, the thermal electron velocities rapidly

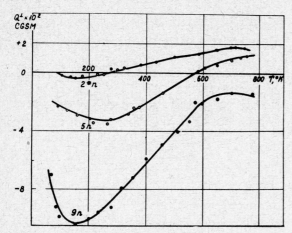

FIG. 7. Temperature dependence of transverse Nernst-Ettingshausen coefficients for n-type GaAs (specimens $9n-n$ $=2.3\times10^{16}$ cm^{-3}, $5n-n=2.5\times10^{18}$ cm^{-3}, $2*n-n=10^{19}$ cm^{-3}).

increase with the degree of degeneracy. Therefore, the role of lattice scattering, the effectiveness of which in the case of scattering by acoustical vibrations is proportional to the thermal carrier velocity, also increases. On the other hand, the role of Rutherford scattering by impurity ions either decreases or, if the increase in ion concentration is considered, remains almost constant. Therefore, in strongly degenerate samples, scattering by the lattice begins to have a noticeable effect at lower temperatures.

This interesting experimental result is confirmed by the measurement of $Q^{\parallel}$ vs T (Fig. 8) and also by the σ vs T curves which are presented for convenience along with the $Q^{\perp}$ vs T curves already cited (Fig. 9). We should note that both the $Q^{\perp}$ as well as the $Q^{\parallel}$ vs T curves clearly have extremum in the region of low temperatures; below 150–200°K the absolute values of $Q^{\perp}$ and $Q^{\parallel}$ begin to drop. This is in complete agreement with the theory of the effect: a decrease in $/Q/$ is caused by an increase in the degeneracy of the electron gas with a decrease in temperature.

The basic conclusions reached on the nature of the scattering of the carriers in GaAs are valid for other III–V compounds.

In Fig. 10 are shown curves of the temperature dependence of transverse and longitudinal N-E coeffi-

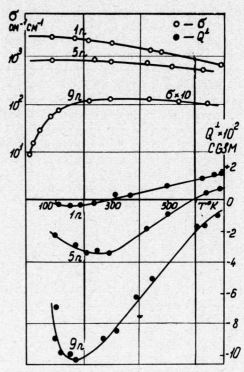

FIG. 9. Temperature dependence of conductivity and the transverse Nernst-Ettingshausen coefficient for n-type GaAs. ($1n-n=10^{19}$ cm^{-3}, $5n-n=2.5\times10^{18}$ cm^{-3}, $9n-n=2\times10^{16}$ cm^{-3}).

cients in n-type InAs samples with carrier concentrations of 1.6×10^{18} cm^{-3} (sample 2), 3×10^{18} cm^{-3} (sample 3), and 9×10^{19} cm^{-3} (sample 4). As in n-type GaAs, the effects are negative at low temperatures and positive at high temperatures. The higher the electron concentration in the samples, the more positive the N-E coefficients become at lower temperatures. As with GaAs, this is due to the degeneracy of the electron gas in n-type InAs, which is greater due to the smaller effective mass of the electrons.

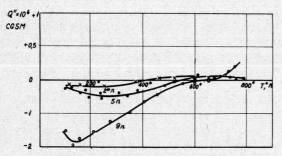

FIG. 8. Temperature dependence of longitudinal Nernst-Ettingshausen coefficients for n-type GaAs.

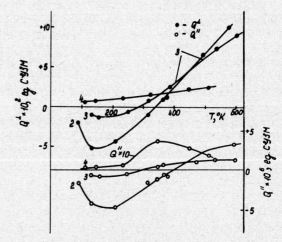

FIG. 10. The temperature dependence of transverse and longitudinal Nernst-Ettingshausen coefficients in n-type InAs. ($2-n=1.5\times10^{18}$ cm^{-3}, $3-n=3\times10^{18}$ cm^{-3}, $4-n=9\times10^{19}$ cm^{-3}).

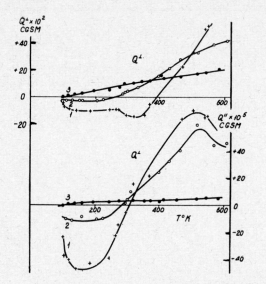

FIG. 11. Temperature dependence of the Nernst-Ettingshausen coefficients in n-type InSb. ($1 - n = 4 \times 10^{13}$ cm^{-3}, $2 - n = 3 \times 10^{17}$ cm^{-3}, $3 - n = 6 \times 10^{18}$ cm^{-3}).

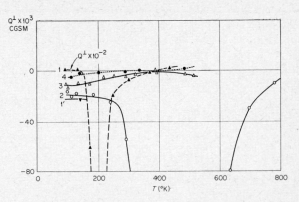

FIG. 12. The dependence of the transverse Nernst-Ettingshausen coefficient on temperature in InSb–AlSb alloys (1–InSb, 2–InSb·AlSb, 3–2.5 InSb·7.5 AlSb, 4–AlSb). The maximum values of the quantity $|Q_\perp|$ on the sections of those curves which lie off the graph are for specimen 1, 16 v per degree gauss, and for specimen 2, 0.3 v per degree gauss (cgs units).

The same results have been obtained recently using strongly alloyed n-type InSb samples (Fig. 11). The similarity of these curves to the previous ones is striking and becomes understandable if it is remembered that the N-E effects depend only on conductivity, on the current carrier scattering mechanism, and on their degree of degeneracy, while these characteristics in n-type GaAs, n-type InAs, and n-type InSb are very similar.

There are still difficulties in making a quantitative interpretation of the $Q^\perp$ vs T curves for n-type InAs and n-type InSb; the positive $Q^\perp$ values in the most impure samples are higher than would be expected according to simple theory, the formulas of which are cited above. It is possible that in strongly degenerate samples the screening effects and anisotropy play some kind of role. These questions will have to be investigated in more detail.

We shall present still another figure which demonstrates the N-E effects in p-type samples of InSb and AlSb and alloys based on these compounds (Fig. 12). The certain "complexities" of the curves, i.e., the presence of very pronounced minima, are only due to the onset of intrinsic conduction, which in p-type material has a stronger effect than in n-type crystals. With the exception of these regions of the curves, the results indicate the same thing—at lower temperatures, im-

purity ions play the predominant role in the scattering of carriers, and at high temperature acoustical lattice vibrations are the dominant scatterers.

The conclusions obtained using N-E effects are in partial contradiction with the results of Ehrenreich, if it is assumed that the electron scattering in GaAs and InSb occurs by the optical lattice vibrations. An analysis of the N-E effects, however, does not lead to any other conclusions except those reached. The positive sign of the effects observed in all the crystals at high temperatures indicates that the mean free path of the carriers *decreases* with an increase in their velocity; this is possible according to the existing theory only when scattering occurs by thermal acoustical lattice vibrations. The nonparabolic form of the band does not change this qualitative conclusion. The anisotropy of the bands, if it actually exists, according to the Tsidilkovskii calculations, apparently cannot change the sign of the effect. It is possible that the presence of certain types of current carriers of a single sign could result in a positive N-E effect without scattering by acoustical vibrations; but an investigation of the N-E effects themselves, as well as other investigations known to us, do not yet give any information on the presence of "light and heavy" electrons in the cited compounds. A further development of transport theory in III-V compounds must be made in order to reconcile the presence of conflicting data on the scattering mechanism of electrons in these materials.

JOURNAL OF APPLIED PHYSICS   SUPPLEMENT TO VOL. 32, NO. 10   OCTOBER, 1961

3-5 Compounds: Lead Salts

C. HERRING, *Chairman*

Electrical Properties of Lead Telluride

YASUO KANAI,* RIRO NII, AND NAOZO WATANABE

Electrical Communication Laboratory, Musashino-shi, Tokyo, Japan

Samples of *n*-type PbTe crystals with different electron concentrations ($n=2\times10^{17}\sim5\times10^{19}$/cc) were prepared by impurity doping or by heating in the vapor of lead. The electrical properties of these crystals were independent of the kind of impurities and depended only on the electron concentration. The electron mobilities of our samples were proportional to $n^{-\frac{1}{3}}$ at 77°K, and to $n^{-4/3}$ at 4.2°K. As the conduction electrons in our samples are degenerate at these temperatures, the experimental results mentioned above suggests that the conduction electrons in PbTe crystals are scattered mainly by acoustical mode at 77°K and by neutral imperfections at 4.2°K.

Since the electron mobility in PbTe becomes very large at 4.2°K (e.g., $1\sim4\times10^6$ cm²/v-sec), this material is quite suitable for the experimental studies of the quantum transport phenomena. To investigate such phenomena, the Hall and magnetoresistive effects of *n*-type PbTe crystals were measured at 4.2°K in a pulsed magnetic field up to 170 kgauss. In the specimens with relatively low electron concentrations (e.g., $n\simeq3\times10^{17}$/cc), the magnetoresistive effect had a single minimum at about 77 kgauss in a transverse magnetic field, and at about 55 kgauss in a longitudinal field. This minimum should correspond to the situation where the Landau level with $l=1$ coincides with the Fermi level in the PbTe crystals.

I. INTRODUCTION

THE physical properties of lead salts have been investigated over a long period of time, and at present the electrical, optical, thermal, and mechanical properties of lead salts are well established by many investigators[1] throughout the world. In spite of these advances, there remain some unsolved problems in the electrical properties of lead salts. One of the most interesting problems to be clarified hereafter is the scattering mechanism of conduction carriers in lead salts. The mobility of carriers in lead salts varies with temperature as $T^{-\frac{5}{2}}$ in the high-temperature region, where the thermal vibration of the crystal lattice will determine the value of mobility. In the low-temperature region, where the ionized impurity scattering should determine the value of mobility, experimental results show that this is not true, but carrier mobility is determined by the neutral imperfections. As a result of these peculiar characteristics, the mobility of carriers in lead salts becomes very large at low temperatures even in rather impure crystals. For example, we have observed such a large value as 4×10^6 cm²/v-sec as the electron mobility at 4.2°K in a lead-telluride crystal with electron concentration of about 6×10^{17}/cc.

The $T^{-\frac{5}{2}}$ temperature dependence of carrier mobility in PbTe was explained by Gershtein *et al.*[2] by assuming that the two-phonon process will predominate in the high-temperature region. Recently, similar temperature dependence of carrier mobility in PbSe was ex-

plained by Smirnov *et al.*[3] by assuming the change of effective mass with temperature in this material.

In this paper some experimental results concerning the electrical properties of *n*-type PbTe crystals will be reported. Our study consists of two parts. In the first part of the paper, the results of the measurements on electrical resistivity and Hall effect of *n*-type PbTe crystals with electron concentrations ranging from 2×10^{17} to 5×10^{19}/cc will be given for the purpose of obtaining some knowledge about the scattering mechanism at low temperatures. The second part of this paper concerns the so-called quantum transport phenomena. As mentioned above, the carrier mobility in PbTe becomes very large at liquid helium temperature. Hence, in a suitable magnetic field Landau levels are easily set up in the crystals, and, as in the case of the de Haas van Alphen effect, it is expected that oscillatory galvanomagnetic effects will be observed. In a stronger magnetic field, all of the conduction electrons will occupy the ground state ($l=0$) of Landau levels, and this situation is often called the quantum limit. To investigate the galvanomagnetic effects in the quantum limit, the electrical resistivity and Hall effects of *n*-type PbTe crystals with relatively low electron concentrations were measured at 4.2°K in a pulsed strong magnetic field (about 170 kgauss).

II. PREPARATION OF CRYSTALS AND MEASURING TECHNIQUES

Stoichiometric amounts of lead (Merck Company 99.99%) and tellurium (Nihon Kogyo Company, 99.999%) were weighted with the accuracy of 10^{-4}, and

---

* Present address: Sony Corporation, Research Laboratory, Hodogaya, Yokohama, Japan.

[1] For a review of physical properties of lead salts, see, W. W. Scanlon, Solid State Phys. **9**, 83 (1959).

[2] E. Z. Gershtein, T. S. Stavitskaia, and L. S. Stilbans, Soviet Phys.—Tech. Phys. **2**, 2302 (1957).

[3] I. A. Smirnov, B. Ya. Moizhes, and E. D. Nensbery, Soviet Phys.—Solid State **2**, 1793 (1960).

were put into a quartz crucible coated with carbon on its inner surface. After evacuation, the crucible was sealed off, and then heated at about 1000°C in an electric furnace for several hours. After the complete reaction of Pb and Te, the crucible was moved down slowly and the PbTe crystal was grown from the bottom of the crucible (Bridgman method). These PbTe crystals were usually p type and contained about $3\times10^{18}/cc$ holes. It was believed that the origin of these holes was caused by the deviation from the stoichiometry rather than by chemical impurities.[4] To produce n-type PbTe crystals, it was necessary to add some chemical impurities or excess Pb to the p-type crystals. Cu (an element in the 1st column of the periodic table), In (3rd), Ge (4th), Sb (5th), and Br (7th) were used as the chemical impurities. These elements were mixed with Pb and Te in the first stage of the preparation of crystals and melted together. To add the excess Pb to the p-type PbTe crystals, two different methods were adapted. In the first method, the same procedure was used as in the case of chemical impurities; in the second method, the p-type PbTe crystals were heated and converted to n-type crystals in the vapor of Pb.

From the resulting n-type PbTe crystals, small pieces of single crystals were cleaved out and shaped to the rectangular form, approximately $10\times3\times3$ mm^3. The surfaces of these samples consisted of the {100} planes of the crystals. Four probes, to measure the resistivity and Hall coefficient of the sample, and two current leads were all soldered with In.

The electrical resistivity and Hall effect were measured by conventional dc methods at both liquid nitrogen temperature (77°K) and liquid helium temperature (4.2°K). The samples were immersed directly in the refrigerant. The Hall effect measurement was usually performed with a magnetic field of 6.17 kgauss.

The measurement of resistivity and Hall effect in a strong magnetic field was performed only at 4.2°K and the samples were always immersed directly in the liquid helium. Here the samples were shaped approximately $7\times1\times1$ mm^3 and fixed in the center position of the coil which produced the magnetic field. The coil was made of copper wire 0.4 mm in diameter and had approximately the following dimensions: 8 mm i.d.,

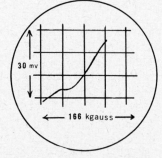

FIG. 1. Typical representation of oscilloscope trace showing resistivity change with magnetic field.

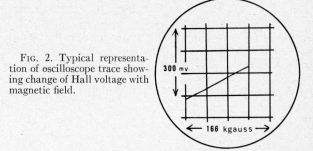

FIG. 2. Typical representation of oscilloscope trace showing change of Hall voltage with magnetic field.

22.4 mm o.d., and 20 mm length. The coil also was immersed directly in the liquid helium. A condenser bank with total capacity of 1100 μf was charged at about 1 kv and the stored charges then were discharged through a Sendaitron (a type of discharge tube which was developed at Tohoku University, Sendai, Japan). The value of the discharge current flowing through the coil was measured by a standard resistance and had a maximum value of about 400 amp. This current produced a pulsed magnetic field with its maximum strength at about 170 kgauss. The voltage drop between the resistivity or Hall probes was fed to the y axis on the screen of the storage tube. A voltage drop proportional to the current flowing through the coil was applied to the x axis. The typical representation of the resistivity or Hall voltage vs magnetic field curves is shown in Figs. 1 and 2. In Fig. 1 the resistivity change with magnetic field can be seen; Fig. 2 shows the Hall voltage vs magnetic field strength curve.

III. RESULTS AND DISCUSSION

A. Dependence of Hall Mobility upon Electron Density in PbTe at 77° and 4.2°K

Figure 3 shows the Hall mobility ($\mu=R/\rho$) vs electron density ($n=(Rec)^{-1}$, where R is the Hall coefficient)

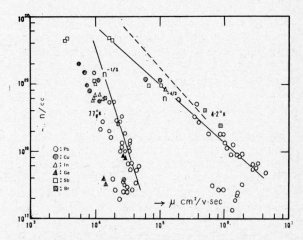

FIG. 3. Electron mobility, μ, vs electron density, n, curves for n-type PbTe crystals at 77° and 4.2°K. Elements added to the crystals as impurities are also shown.

[4] R. F. Brebrick and R. S. Allgaier, J. Chem. Phys. 32, 1826 (1960).

curves at 77° and 4.2°K for n-type PbTe crystals prepared by the method described in Sec. II. It can be seen that the mobilities have larger values for smaller electron density at the same temperature, and have larger values at 4.2°K than at 77°K. Using the values of effective masses of electrons in PbTe determined recently by Kuglin et al.[5] the degenerate temperature Td was calculated to be approximately 200°K for a sample with electron density of about 3×10^{17}/cc. Since the electron concentrations of our samples ranged between 2×10^{17} and 5×10^{19}/cc, our samples all were degenerate at 77°K and lower temperatures. The experimental results, which show the values of mobilities at 4.2°K to be larger than those at 77°K suggest that the scattering centers, which mainly determine the electron mobility at 77°K, are crystal imperfections of some type which decrease in number with decreasing temperature. Hence, the scattering centers at 77°K would not be the static imperfections such as chemical impurities or dislocations, but probably the thermal vibration of crystal lattice.

Assuming that the acoustical mode of lattice vibration is the main mechanism determining the electron mobility, the mobility μ will change its values as $\mu \propto \epsilon^{-\frac{1}{2}}$ with the electron energy, ϵ. In a degenerate semiconductor, the Fermi energy ϵ_F varies as $\epsilon_F \propto n^{\frac{2}{3}}$ with the electron density n, so the mobility μ will be expected to vary in our samples as $\mu \propto n^{-\frac{1}{3}}$. As can be seen in Fig. 3, this relation is in good agreement with the experimental result at 77°K. On the other hand, if the optical mode of lattice vibration is assumed to be the main factor of the scattering mechanism, then the mobility μ becomes independent of the electron energy ϵ below Debye temperature. As Debye temperature of PbTe is approximately 150°K,[6] the mobility should become independent of electron density, in contradiction to the experimental results at 77°K.

As shown in Fig. 3, the electron mobility μ in PbTe at 4.2°K varies as $\mu \propto n^{-\frac{1}{3}}$ with electron density n. In general, the collision time τ is expressed by: $\tau = 1/NvA$, where N is the number of scattering center, v is the electron velocity, and A is the cross section for scattering. As the electrons in our samples are degenerate at 4.2°K, the electron velocity v is proportional to $n^{\frac{1}{3}}$. Assuming that the ionized donor is the scattering center, we can put N equal to n. Then, the electron mobility can be expressed as:

$$\mu = e\tau/M_m = (\pi/3)^{\frac{1}{3}} \cdot (2e/h) \cdot (M_d/M_m) \cdot (1/A) \cdot n^{-\frac{1}{3}},$$

where M_m and M_d are the effective masses concerning the mobility and state density, respectively. Their values are calculated as $0.034M_0$ and $0.10M_0$ using the values of electron masses obtained by Kuglin et al.[5]

If we assume that the value of A is independent of electron energy, then the above expression is in good agreement with the experimental result at 4.2°K shown in Fig. 3. From the slope of the mobility vs electron density curve at 4.2°K, the value of A can be calculated as approximately 4×10^{-16} cm^2. Considering the fact that the value of A is of the order of the square of the atomic radius and is independent of electron energy, the ionized impurities in PbTe may be considered to be screened almost completely and, therefore, act as neutral impurities for the scattering of conduction carriers.

In the course of the experimental studies of the solid solutions between PbSe, PbTe, and SnTe or Bi$_2$Te$_3$, Bi$_2$S$_3$ and Sb$_2$Te$_3$, Joffe and Stil'bans[7] found that in crystals with positive ions replaced by other elements, the electron mobility decreased significantly but the hole mobility was influenced only slightly. An opposite regularity was found in crystals with negative ions replaced by other elements. From these experimental results, Shockley[8] suggested that in these compound semiconductors the wave functions of the conduction electrons would concentrate mainly around the positive ion site, and those of holes around the negative ion site. To confirm this idea, n-type PbTe crystals with various kinds of impurities such as Cu, In, Ge, Sb, and Br were prepared. In the PbTe crystals, it is supposed that the elements Ge and Sb replace the positive ion Pb, Br replaces the negative ion Te and Cu occupies the interstatial positions.[9] In opposition to the observations of Joffe and Stil'bans, as seen in Fig. 3, the differences in the kinds of impurity elements does not affect the electron mobility of our samples. This disagreement is probably due to the differences of the densities of foreign atoms between Joffe and Stil'bans' case and ours.

As seen in Fig. 3, the value of mobility at 77°K decreased more rapidly than $n^{-\frac{1}{3}}$ in the region of high-electron density, i.e., $n > 6 \times 10^{18}$/cc. If we assume that the decrease of electron mobility in this region is due to the increase of the impurity scattering, then, at 4.2°K the electron mobility in the same region of electron density should be much smaller than the experimental values. Since this is not the case, the origin of the rapid decrease of electron mobility at high-electron concentrations will be due to other mechanisms, probably the increase of the value of effective masses with the increase of electron energy. Assuming that the deviation from the $n^{-\frac{1}{3}}$ curve in the region of high-electron densities at 77°K is due to the increase of effective masses, a correction to the electron mobility vs electron density curve at 4.2°K can be made. The dotted curve in Fig. 3 represents the corrected curve

[5] C. D. Kuglin, M. R. Ellett, and K. F. Cuff, Phys. Rev. Letters 6, 177 (1961).
[6] D. H. Parkinson and J. E. Quarrington, Proc. Phys. Soc. (London) A67, 569 (1954).
[7] A. F. Joffé and L. S. Stil'bans, Repts. Progr. in Phys. 22, 194 (1959).
[8] W. Shockley, J. Phys. Chem. Solids 8, 14 (1959).
[9] J. Bloem and F. A. Kröger, Report of the Meeting on Semiconductors (The Institute of Physics and the Physical Society, London, 1956), p. 77.

for the data at 4.2°K and it appears that this curve becomes proportional to the inverse value of n, rather than $n^{-\frac{1}{3}}$.

The decrease of electron mobility at low-electron concentration (i.e., $n < 4 \times 10^{17}$/cc), which is seen in both curves (at 77° and 4.2°K) in Fig. 3, is probably due to the inhomogeneity of the crystals.

B. Galvanomagnetic Effects at High Magnetic Field

As shown in Fig. 3, the electron mobility in PbTe at 4.2°K reaches a value as large as 10^6 cm²/v-sec. Hence, in a moderate magnetic field the condition $\omega_c \tau > 1$ is fulfilled easily, so that the Landau levels are well defined, where $\omega_c = eH/M_c c$, ω_c is the cyclotron angular frequency, and M_c is the cyclotron mass. At low temperatures, where the conditions $\hbar\omega_c > kT$ and $\epsilon_F > \hbar\omega_c$ are fulfilled, it is expected that the oscillatory galvanomagnetic effects will be observed, as in the case of the de Haas van Alphen effects. For example, when a magnetic field is applied to the $\langle 100 \rangle$ direction of the crystal axis of PbTe, the value of the cyclotron mass becomes $M_c = 0.035 M_0$. If we assume that $\mu = 1 \times 10^6$ cm²/v-sec and $n = 6 \times 10^{17}$/cc, we can obtain for a magnetic field of 10 kgauss:

$$\omega_c \tau \equiv \mu H \times 10^{-8} = 100,$$
$$\hbar\omega_c \simeq k \times 40°K, \quad \text{and} \quad \epsilon_F \simeq k \times 300°K.$$

It can, therefore, be expected that the oscillatory galvanomagnetic effects will be observed in n-type PbTe at 4.2°K in a magnetic field of the order of 10 kgauss.

The electrical resistivity and Hall effect of n-type PbTe crystals were measured at 4.2°K in a pulsed-magnetic field which was produced by the method described in Sec. II. The electric current was applied in the [100] direction, and the magnetic field was in the [001] direction of the crystal axis. Typical results of the measurement are shown in Figs. 4 and 5. As seen in Fig. 4, the Hall voltage is proportional to the magnetic field strength, i.e., the Hall coefficient had a

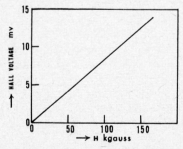

FIG. 4. Example of Hall voltage vs magnetic field strength curves of n-type PbTe crystals at 4.2°K. Electric current was applied in the [100] direction and the magnetic field was in the [001] direction of the crystal axis. Electrical resistivity of the sample $\rho = 5.62 \times 10^{-5}$ Ω-cm; Hall coefficient $R = -19.7$ cm³/coul at 4.2°K.

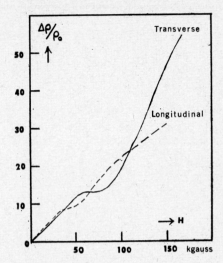

FIG. 5. Example of the transverse and longitudinal magnetoresistance vs magnetic field strength curves of n-type PbTe crystals at 4.2°K. The magnetic field was applied to the $\langle 100 \rangle$ direction of the crystal axis, and the sample was the same shown in Fig. 4.

constant value independent of the magnetic field. As seen in Fig. 5, the magnetoresistance in PbTe crystals ($n \simeq 3 \times 10^{17}$/cc) had a single minimum at approximately 77 kgauss in a transverse magnetic field. The amplitude of this minimum varied from sample to sample, but its position was about the same in many crystals with the same electron concentrations. Hence, it could be believed that the minimum of the electrical resistivity found at approximately 77 kgauss is the intrinsic behavior of n-type PbTe crystals.

It was calculated by many investigators[10] that the minimum of electrical resistivity in a crystal with well-defined Landau levels occurs approximately at the magnetic fields fulfilling the condition: $\epsilon_F = \hbar\omega_c(\frac{1}{2} + l)$, $l = 1, 2, 3, \cdots$. From the above condition, it can easily be seen that the position of the minimum is equidistant when plotted against $1/H$. Figure 6 shows transverse magnetoresistances for n-type PbTe crystals (Kuglin et al.[5]) and those of our investigation are plotted against $1/H$. It can be seen that the periodicity of the position of the resistivity minimum is quite good; it is also quite clear that the minimum at 77 kgauss corresponds to the situation $l = 1$. On the other hand, putting $l = 1$ in the above condition, the value of magnetic field corresponding to the resistivity minimum is calculated as approximately 35 kgauss. The reason for the discrepancy between this calculated value and the experimental data is not clear.

The resistivity change of the same sample in a longitudinal magnetic field is shown in Fig. 5. In this case, both the electric current and the magnetic field were applied in the [100] direction of the crystal axis. Here, one can see that a resistance minimum is also

[10] For a review of oscillatory galvanomagnetic effects, see, A. H. Kahn and H. P. R. Frederikse, Solid State Phys. **9**, 257 (1959).

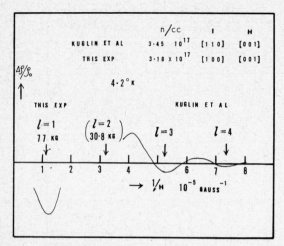

FIG. 6. The transverse magnetoresistance of n-type PbTe crystals at 4.2°K vs inverse value of magnetic field strength curve. This figure is drawn to show the periodicity of the position of the resistivity minimum, so the values of magnetoresistance are unimportant.

observed in the longitudinal case, but the position is at a lower magnetic field (i.e., at about 55 kgauss) than in the case of the transverse field.

As previously mentioned, in a magnetic field higher than 77 kgauss all conduction electrons are in the Landau level corresponding to $l=0$ (ground state). In such a case, it is said that the crystal is in a state of "quantum limit." The galvanomagnetic effect in the region of quantum limit has been studied theoretically by many investigators,[11] but their results are often inconsistent.

In general, the galvanomagnetic effect in the quantum limit depends upon the scattering mechanism and statistics of conduction carriers. Since our samples have the electron density $10^{17} \sim 10^{18}$/cc, the electrons are degenerate at 4.2°K even in a strong magnetic field. Some theoretical results concerning the resistivity change of degenerate semiconductors in the quantum limit are listed in Table I. Our experimental results show an approximately linear dependence of resistivity upon magnetic field strength in both the transverse and longitudinal cases, and a constant value of Hall coefficient in the quantum limit. These characteristics seem to be inconsistent with any of the theoretical results calculated on the various scattering mechanisms.

IV. CONCLUSIONS

The electrical resistivity and Hall effect of n-type PbTe crystals with different kinds and numbers of

[11] See, for example, E. N. Adams and T. D. Holstein, J. Phys. Chem. Solids **10**, 254 (1959).

TABLE I. The dependence of magnetoresistance on the magnetic-field strength of a degenerate semiconductor in the quantum limit.

Scattering mechanism	Transverse magnetoresistance	Longitudinal magnetoresistance
acoustical	$\propto H^{2a}$	$\propto H^{2a}$
ionized impurity	$\propto H^{\sim 0a}$	$\propto H^{-2a}$
point defect	$\propto H^5 \sim H^{1b}$	$\propto H^{2b}$

[a] See, P. N. Argyres, J. Phys. Chem. Solids **8**, 124 (1959).
[b] See, A. H. Kahn. Phys. Rev. **119**, 1189 (1960).

donors were measured at 77° and 4.2°K, and the relation between the Hall mobility and the electron density at both temperatures was determined. The Hall mobility was proportional to $n^{-\frac{1}{4}}$ at 77°K and proportional to $n^{-\frac{1}{3}}$ at 4.2°K. These results suggest that the conduction carriers are scattered mainly by the acoustic mode of lattice vibration at 77°K, and by some lattice defects with the cross section of the same order of atomic radius at 4.2°K.

Of the crystals with relatively low electron concentration, the resistivity and Hall effect were measured at 4.2°K in a pulsed strong magnetic field (170 kgauss max). When a transverse magnetic field was applied to the $\langle 100 \rangle$ direction of the crystal axis, a single minimum of electrical resistivity was found at about 77 kgauss. It is considered that this minimum corresponds with the situation where the Landau level, with $l=1$, coincides with the Fermi level. In a longitudinal magnetic field, a similar minimum of electrical resistance was also found at 55 kgauss. Except for the single minimum mentioned above, the increase of the electrical resistivity was approximately proportional to the magnetic field, and there was no anomaly in the Hall coefficient at the same magnetic field region. In a magnetic field stronger than 77 kgauss, the electrons in PbTe can be considered to be in the state of quantum limit. Our experimental results show that the increase of resistivity is approximately proportional to the magnetic field, and the Hall coefficient is independent of the magnetic field, even in the region of quantum limit.

ACKNOWLEDGMENTS

The authors wish to express their sincere thanks to Dr. T. Niimi of the Electrical Communication Laboratory, Tokyo, Japan, for his guidance and encouragement throughout the course of this work. One of the authors, Yasuo Kanai, is indebted to Dr. M. G. Hatoyama, Director of the Research Laboratory of the Sony Corporation, for his valuable criticism and important comments concerning the manuscript.

JOURNAL OF APPLIED PHYSICS   SUPPLEMENT TO VOL. 32, NO. 10   OCTOBER, 1961

Fundamental Reflectivity Spectrum of Semiconductors with Zinc-Blende Structure

Manuel Cardona

Laboratories RCA Ltd., Zurich, Switzerland

The fundamental reflectivity spectrum of several III–V and II–VI semiconductors is discussed and compared with the reflectivity spectrum of the group IV semiconductors. A general feature of these spectra is the presence of two peaks within the fundamental absorption region. The lower energy peak, which corresponds to a maximum in the refractive index, can be split into a doublet. This peak is probably due to direct transitions between the valence band extrema in the [111] direction at the edge of the Brillouin zone (L point) and the corresponding conduction band minima. The splitting of this peak corresponds to the spin-orbit splitting of the valence band extrema (L_3). The second peak corresponds to a maximum in the combined density of states for the transitions. The temperature dependence of these peaks is also discussed.

INTRODUCTION

THE fundamental reflectivity of germanium shows maxima at 2.1 and 4.4 ev.[1] The 2.1 ev max, which can be resolved into a doublet with components at 2.10 and 2.28 ev,[2,3] has been attributed to transitions between the L_3' valence band extrema and the L_1 conduction band minima, its doublet structure being due to the splitting of the L_3' valence band extrema by spin-orbit interaction. Similar structure has been found for the fundamental reflectivity of silicon, with maxima at 3.4 and 4.4 ev,[4] but these maxima do not show any doublet structure. Analogous reflectivity peaks have also been observed by Tauc and Abraham in the germanium-silicon alloys.[5] With increasing silicon concentration, the 2.1 ev germanium peak shifts linearly toward higher energies, but a break in the energy vs concentration curve occurs at 79 at. % of silicon. This break has been attributed to a crossing of the L_1 and L_3 conduction band minima, in agreement with the theoretical calculations which assign L_3 symmetry to the lowest [111] conduction band minimum in silicon.[6] The separation between the doublet components of the 2.1 ev reflectivity peak decreases with increasing silicon concentration and is no longer observable in pure silicon. This is understandable in view of the small spin-orbit splitting expected for this material (about 0.03 ev).

The higher energy (4.4 ev) reflectivity maximum in Ge and Si corresponds to a maximum in the absorption coefficient, probably produced by a maximum in the combined density of states for the transitions.[2] Hence, the energy at which this maximum occurs gives an indication of the width of these bands. Phillips[7] has suggested that this reflectivity maximum could be due to $X_4 \rightarrow X_1$ allowed transitions, in the [100] direction at the edge of the Brillouin zone. The double degeneracy of the X_1 state gives a very sharp maximum in the density of states at the energy corresponding to these transitions. Since the calculated energy distance between the X_4 and X_1 states is the same for germanium[8] as for silicon[6] (~ 4.2 ev), this would explain why these reflectivity peaks occur at the same energy for both semiconductors.

A similar structure in the fundamental reflectivity of the III–V compounds has been reported independently by Tauc and Abraham (GaAs, GaSb, InSb)[9] and by the author (GaSb, InP).[10,11] In all these III–V semiconductors two reflectivity peaks are found. The lower energy peak always shows a doublet structure which has an energy separation in agreement with the value predicted from the atomic spectra for the spin-orbit splitting at L_3.

In this paper we discuss the fundamental reflectivity spectrum of the III–V and II–VI compounds with zinc-blende structure and its temperature dependence. Measurements on GaP and AlSb show a reflectivity structure similar to that obtained for the compounds previously reported. A temperature coefficient approximately -4.5×10^{-4} ev $\times (°C)^{-1}$ for the reflectivity doublet is found for all compounds measured, except for GaP. This coefficient is, within the experimental error, the same as that found for germanium,[2] while the GaP coefficient agrees with the one found for the 3.4 ev reflectivity peak in silicon.[12] It is suggested that this might be due to a lowest [111] conduction band minimum with L_3 symmetry in both silicon and GaP.

In order to find out whether the II–VI semiconductors with zinc-blende structure have a similar reflectivity spectrum, ZnTe, CdTe, HgTe, HgSe, and ZnSe were measured. Tellurides and selenides were chosen for the measurements, since they should exhibit a larger spin-orbit splitting than sulfides. The reflectivity of these semiconductors also shows a peak resolvable into a doublet, with an energy separation which can be brought into agreement with the atomic spin-orbit splittings for the p-orbital states.

[1] H. R. Philipp and E. A. Taft, Phys. Rev. **113**, 1002 (1959).
[2] M. Cardona and H. S. Sommers, Jr., Phys. Rev. **122**, 1382 (1961).
[3] J. Tauc and E. Antoncik, Phys. Rev. Letters **5**, 253 (1960).
[4] H. R. Philipp and E. A. Taft, Phys. Rev. **120**, 37 (1960).
[5] J. Tauc and A. Abraham, J. Phys. Chem. Solids (to be published).
[6] L. Kleinman and J. C. Phillips, Phys. Rev. **118**, 1153 (1960).
[7] J. C. Phillips, J. Phys. Chem. Solids **12**, 208 (1960).

[8] F. Herman, Proceedings of the Prague Semiconductors Conference, Prague, Czechoslovakia, August-September, 1960, p. 20.
[9] J. Tauc and A. Abraham, Proceedings of the Prague Semiconductors Conference, Prague, Czechoslovakia, August-September, 1960, p. 375.
[10] M. Cardona, Z. Physik **161**, 99 (1961).
[11] M. Cardona, J. Appl. Phys. **32**, 958 (1961).
[12] M. Cardona (unpublished).

TABLE I. Optical gap (E_0), energies of the reflectivity peaks (E_1 and E_2), temperature coefficient of E_1, measured spin-orbit splittings of the valence band at the Γ point $[\Delta_0(\exp)]$ and at the L point $[\Delta_1(\exp)]$, and estimated values of the spin-orbit splittings at the L point $[\Delta_1(\mathrm{calc})]$ for several semiconductors with diamond and zinc-blende structure.

Material	$E_0(\mathrm{ev})$ 297°K	$E_1(\mathrm{ev})$ 297°K	$(dE_1/dT)\times10^4$ ev $\times$(°C)$^{-1}$	$E_2(\mathrm{ev})$ 297°K	$\Delta_0(\exp)$	$\Delta_1(\exp)$	$\Delta_1(\mathrm{calc})$
Ge[a]	0.67	2.10	$-(4.2\pm0.4)$	4.45	0.30	0.18	0.18
Si[b]	1.10	3.38	$-(2.7\pm0.6)$	4.4	...	...	0.03
InSb[e]	0.18	1.82	...	4.13	...	0.56	0.48
InAs[e]	0.33	2.52	...	4.72	0.43[d]	0.29	0.27
GaSb	0.70	2.02	$-(4.2\pm0.6)$	4.3	...	0.46	0.42
GaAs[e]	1.35	2.94	...	5.1	0.33[e]	0.26	0.22
InP	1.26	3.15	$-(4.2\pm0.6)$	5.0	...	0.14	0.13
AlSb	1.6	2.78	$-(5.0\pm1)$	...	0.75[e,f]	0.40	0.38
GaP	2.24	3.71	$-(2.5\pm0.6)$	...	...	...	0.07
HgTe	0.02	2.08	...	4.70	...	0.69	0.73
CdTe	1.5	3.29	...	...	...	0.55	0.64
ZnTe	2.1	3.57	$-(5.0\pm0.6)$	...	...	0.56	0.62
HgSe	0.6	2.82	$-4.3\pm0.6)$	...	...	0.31	0.41
ZnSe	2.67	4.7(?)	...	...	0.43	...	0.31

[a] See reference 2.
[b] See reference 12.
[e] See reference 9.
[d] F. Matoss and F. Stern, Phys. Rev. 111, 472 (1958).
[e] See reference 14.
[f] W. J. Turner and W. E. Reese, Phys. Rev. 117, 1003 (1960).

III–V COMPOUNDS

A. Measurements

Figure 1 shows the room temperature reflectivity spectra of GaSb, InP, and GaP. The measurement technique has been described elsewhere.[10] The GaSb sample of Fig. 1 was polished only, while the InP and GaP samples were polished and etched in a mixture of equal volumes of HNO_3 and HCl. Measurements on etched GaSb show a sharpening and a slight shift toward higher energies of the 2.0–2.5 ev doublet, while the value of the reflectivity at the 4.3-ev peak decreases. The 3.15-ev peak in InP does not show any structure at room temperature; however, a doublet structure with an 0.14 ev separation between the components has been seen at 90°K.[11] The reflectivity spectrum of GaP shows a maximum at 3.71 ev and indicates the existence of a second maximum beyond the short wavelength limit of our spectrometer. No structure has been observed in

the 3.71-ev max even at liquid nitrogen temperature. We estimate the error in the absolute reflectivity scale of Fig. 1 to be about 10%, due to misalignments in the optical system.

The reflectivity spectrum of AlSb samples was also measured. The samples were polished avoiding contact with water, since humidity alters the surface and covers it with a brown film which decreases the reflectivity. After polishing, the samples were dipped for a few seconds in the same etchant as that use for InP and GaP, rinsed in methyl alcohol, mounted inside a cryostat, and pumped down to about $10^{-3}\mu$. Figure 2 shows a section of the reflectivity spectrum (in arbitrary units) of an AlSb sample at 296° and 85°K which clearly exhibits a maximum with a doublet structure. The energy separation of the two doublet components is 0.4 ev. The low-energy component shifts with temperature at the rate $-(5\pm1)\times10^{-4}$ ev$\times$(°C)$^{-1}$. No temperature variation in the energy separation of both

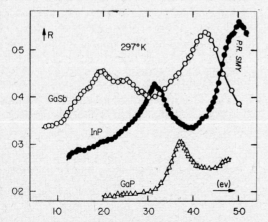

FIG. 1. Fundamental reflectivity spectra of GaSb, InP, and GaP at 297°K.

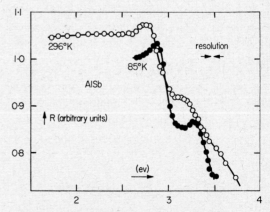

FIG. 2. Fundamental reflectivity spectrum of AlSb (in arbitrary units) at 296° and 85°K.

TABLE II. Atomic spin-orbit splittings for the p-one electron atomic wave functions (in ev).

Mg 0.01	Al 0.03	Si 0.06	P 0.10	S 0.16
Zn 0.11	Ga 0.21	Ge 0.35	As 0.51	Se 0.71
Cd 0.31	In 0.54	Sn 0.81	Sb 1.11	Te 1.46
Hg 1.15				

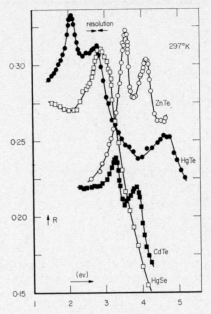

FIG. 3. Fundamental reflectivity spectra of ZnTe, CdTe, HgTe, and HgSe at 297°K.

doublet components can be detected. Unetched samples also show less pronounced maxima at the same energies. While the position of the maxima was reproducible, the shape of the curves changed depending on the degree to which the sample was etched and the length of time the sample stayed in contact with the atmosphere before being mounted in the vacuum cryostat.

The energies obtained for the low- and high-energy reflectivity maxima (E_1 and E_2, respectively), the temperature coefficient of E_1, and the separation between the doublet components of the low-energy reflectivity maximum [$\Delta_1(\exp)$] are listed in Table I together with the thermal energy gap E_0, and the spin-orbit splitting of the valence band at the center of the Brillouin zone [$\Delta_0(\exp)$] obtained from absorption measurements in p-type materials.

B. Discussion

Table II shows the atomic spin-orbit splitting in ev for the one electron p-orbital wave functions of several group II, III, IV, V, and VI elements, as found for the singly, doubly, triply, and quadruply ionized atoms, respectively.[13] The spin-orbit splittings in the valence band at the center of the Brillouin zone [$\Delta_0(\exp)$] observed in Ge, InAs, AlSb, and GaAs and listed in Table II have been explained by Braunstein.[14] He assumed that the atomic splittings have to be multiplied by a factor independent of the material in order to find the splitting in the solid, and that the valence band electrons at the center of the Brillouin zone in all III–V compounds spend 35% of their time around the cation and 65% around the anion because of the partially polar character of the binding. Since the spin-orbit splitting of the valence band at the Brillouin zone boundary in the [111] direction (L_3' for the diamond structure, L_3 for the lower symmetry zinc-blende structure) is roughly $\frac{2}{3}$ of the splitting at the Brillouin zone center, one can use the same procedure to estimate the L_3 splittings. The conversion factor from the atomic splittings to the splittings in the solids is 0.52, found from the measured 0.18-ev spin-orbit splitting at L_3' in german-

ium.[2] Table I lists the values of the energy separation between the doublet components of the low-energy reflectivity peak in the III–V compounds [$\Delta_1(\exp)$] and the estimated spin-orbit splitting at the L_3 valence band maximum [$\Delta_1(\text{calc})$]. The good agreement between $\Delta_1(\exp)$ and $\Delta_1(\text{calc})$ and the analogy with the germanium spectrum makes it reasonable to attribute this doublet to $L_3 \rightarrow L_1$ interband transitions. The 3.71-ev peak in GaP might be also due to $L_3 \rightarrow L_1$ transitions, the absence of structure being explained by the smallness of the estimated spin-orbit splitting in this material (0.07 ev) and the relative crystalline imperfection of our samples. However, it seems surprising that the temperature coefficient of this peak is only -2.5×10^{-4} ev $\times (°C)^{-1}$, while for all other III–V compounds measured, the coefficient is about -4.5×10^{-4} ev $\times (°C)^{-1}$. This difference could be explained if the transitions causing the reflectivity peak were other than the $L_3 \rightarrow L_1$. In silicon, where the 3.4-ev reflectivity peak is

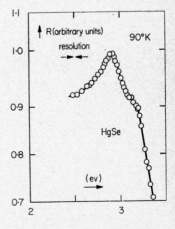

FIG. 4. Fundamental reflectivity spectrum of HgSe at 90°K.

[13] R. F. Bacher and S. Goudsmit, *Atomic Energy States* (McGraw-Hill Book Company, Inc., New York, 1932).

[14] R. Braunstein, J. Phys. Chem. Solids **8**, 280 (1959); (also, private communication).

probably caused by transitions between the L_3' valence band maximum and the L_3 conduction band minimum, the temperature coefficient of this peak is -2.7×10^{-4} ev $\times(°C)^{-1}$.[12] A similar situation may exist in GaP where the temperature coefficient of the 3.71-ev peak coincides with the one found for silicon. Measurements on GaAs-GaP alloys, which are at present being performed in our laboratory, will help clarify this question.

II–VI COMPOUNDS

A. Measurements

ZnTe, ZnSe, and HgSe single crystals were grown by the chemical transport reaction technique.[15] The ZnSe crystals, which can appear in two different modifications (cubic and hexagonal), were shown by x-ray techniques to have zinc-blende structure. The surfaces of growth of some of the crystals were good enough to permit reflectivity measurements without further treatment, although insufficient flatness introduced an error in the absolute value of the reflectivity. This error was estimated to be about 15%. Polycrystalline ingots of CdTe and HgTe were prepared by melting together stoichiometric proportions of the components. By breaking the ingots, shiny surfaces, with an optical quality sufficient to permit reflectivity measurements without polishing, were obtained. Figure 3 shows the room temperature reflectivity spectra for ZnTe, CdTe, HgTe, and HgSe. The spectrum of HgTe is very similar to the one observed for germanium; it exhibits two peaks, the lower energy one being a doublet. The higher energy peak has an intensity much smaller than that in germanium, but we believe this to be due to a less perfect surface. The irreproducibility of the peak intensity

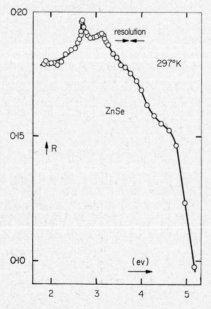

Fig. 5. Fundamental reflectivity spectrum of ZnSe at 297°K.

[15] R. Nitsche, J. Phys. Chem. Solids 17, 163 (1960).

for different samples confirms this hypothesis. The spectra of ZnTe and CdTe show only one doublet, but the existence of a second peak beyond the energy range of our spectrometer can be inferred from the reflectivity spectrum of ZnTe. A reflectivity peak with a less pronounced doublet structure was even observed in an evaporated layer of ZnTe. The room temperature reflectivity of HgSe shows a maximum with a slight structure. This structure becomes more pronounced at liquid nitrogen than at room temperature (see Fig. 4). No change in the energy separation of the doublet components with temperature was detected. The coefficients dE_1/dT are listed in Table I together with the room temperature energies of the low energy component of the doublet (E_1), the high energy reflectivity peak (E_2), the separation between the doublet components $[\Delta_1(exp)]$, and the optical absorption edge (E_0). The room temperature reflectivity spectrum of ZnSe is shown in Fig. 5. The energy of the lowest component coincides with the optical absorption edge and is therefore listed under E_0 in Table I. The doublet separation is listed under $\Delta_0(exp)$.

B. Discussion

Since we have assumed that the valence band electrons for the III–V compounds spend 65% of their time around the anion and 35% around the cation,[14] and in the semiconductors of the fourth group the same time is spent around each one of the atoms of the unit cell, it seems reasonable to extrapolate these figures and assume that in the II–VI compounds the valence band electrons spend 80% of their time around the anion and 20% around the cation. If we use the same conversion factor for the atomic splittings of Table I as for the III–V compounds (0.52), we find the L_3 spin-orbit splittings, which are listed under $\Delta_1(calc)$ in Table I. The reasonable agreement between the values of $\Delta_1(calc)$ and Δ_1 (exp) indicates that the reflectivity doublet in HgTe, CdTe, ZnTe, and HgSe is due to $L_3 \rightarrow L_1$ transitions between the valence and conduction bands. A lowest conduction band minimum at the L point with L_3 symmetry is excluded, since this would give a multiplet with 3 or 4 components for the reflectivity peak. This is confirmed by the values of the temperature coefficient of E_1 listed in Table I.

The energy of the first component of the reflectivity doublet in ZnSe shown in Fig. 5 coincides with the energy gap. It seems reasonable, therefore, to attribute this doublet to transitions between the valence and conduction bands at the center of the Brillouin zone. The energy separation between the doublet components (0.43 ev) is in good agreement with the value predicted from the atomic spectra $[\Delta_0(calc)\simeq(\frac{3}{2})\Delta_1(calc)=0.46$ ev]. The absence of a second peak attributable to interband transitions at the L point may be due to a poor surface condition. The change in the slope of the reflectivity curve at 4.7 ev could be due to these transitions.

This would be a reasonable extrapolation of the position of the peak in HgSe.

ACKNOWLEDGMENTS

The author wishes to acknowledge the skillful assistance of Heinrich Meier in preparing the samples and setting up the equipment. The GaP, InP, and GaSb samples were supplied by the RCA Laboratories, Princeton, New Jersey. The AlSb samples were obtained from the Federal Institute of Technology, Zurich, Switzerland. The samples of the II–VI compounds were prepared by H. U. Boelsterli in our laboratories. Thanks are also extended to Dr. J. Tauc for making available his results prior to publication.

JOURNAL OF APPLIED PHYSICS   SUPPLEMENT TO VOL. 32, NO. 10   OCTOBER, 1961

Band Structure and Transport Properties of Some 3–5 Compounds

H. EHRENREICH

General Electric Research Laboratory, Schenectady, New York

The experimental information relevant to the band structure of the compounds InSb, InAs, GaSb, GaAs, GaP, AlSb and some of their alloys is synthesized and interpreted in terms of a consistent theoretical picture which exploits the close relationship linking the band structures of the group 4 and 3–5 semiconductors. It is shown that the momentum matrix element determining the conduction band masses in those compounds whose edge is of symmetry type Γ_1, is nearly constant. Simple theoretical expressions, agreeing well with experiment, are derived for the corresponding effective masses and valence band spin-orbit splittings, following Kane's theory for InSb. The dominant scattering mechanisms determining the transport properties are reviewed briefly. Arguments are presented which show deformation potential scattering to be unimportant relative to polar optical mode scattering. A heuristic treatment indicates that the relaxation time approximation can be applied reasonably to polar scattering for temperatures $T > \hbar\omega_l/K$, where ω_l is the longitudinal optical frequency. Multiband transport effects are discussed with special reference to the Nernst effect in GaAs and the electron mobility in GaSb. An explanation of the high-temperature behavior of the Nernst coefficient in GaAs in terms of polar and intervalley scattering is proposed. The mobility in GaSb remains unexplained.

I. INTRODUCTION

THIS paper will attempt to review and synthesize experimental information relevant to the band structure of the 3–5 compounds InSb, InAs, GaSb, GaAs, GaP, AlSb, and some of their alloys and to provide or comment on interpretation of these experiments. It will also discuss transport properties of some of these materials as related to the dominant scattering mechanisms, the treatment of polar scattering in the relaxation time approxmation, and the role of simultaneous transport in several energetically close conduction bands. The other 3–5 compounds, chiefly those having wide band gaps, will not be included in this discussion because as yet there is not enough available information.

2. BAND STRUCTURE

Before making explicit statements concerning the band structure of the 3–5 compounds under consideration here, it may be useful to classify the various kinds of directly relevant experimental information that are available. We have done this in Table I by listing the various experiments and for each the material to which it has been applied. A brief statement concerning the conclusion to be drawn from the experiment follows.[1]

The reason for this somewhat unusual approach is that some classes of experiments have become sufficiently well categorized as to become associated with certain specific types of information. For example, the variation of the Hall coefficient in a magnetic field in p-type 3–5 compounds has been identified with degenerate germanium-like valence bands, as has the characteristic structure of the absorption peaks observed in free-hole infrared absorption. Similarly, the near constancy of the pressure dependence associated with the [000], [111], [100] band edges in the group 4 and 3–5 compounds, and particularly the negative coefficient associated with a [100] edge, has been used circumstantially to infer the location in k space of the lowest-lying conduction band edges in several compounds. Information concerning band gaps in other regions of the Brillouin zone is obtained by exploiting the analogy of the structure observed in the ultraviolet reflectivity spectrum of germanium, silicon, and the 3–5 compounds. Although the transitions at L may be identified with some certainty due to the characteristic spin-orbit splitting (Tau 61), the identification of the transitions at or near X is quite conjectural (Phi 60).

The guiding idea in this kind of interpretive think-

---

[1] In some cases the conclusions stated are those of the experimenters in the references, in others they are conclusions drawn by the present author or in other quoted theoretical work. It should be emphasized that the type of summary statement made here cannot be properly qualified. For a variety of reasons not all of the conclusions stated are equally reliable. The references given are intended to be representative and, in no sense, complete.

TABLE I. Experiments relevant to band structure.

Experiment	Material	References	Deductions
Fundamental optical absorption	InSb	Fan 56, Gob 60, Kan 57a	Direct transition
	InAs	Fol 54, Mat 58, Osw 59	Direct transition
	InP	New 58	Direct transition
	GaSb	Ram 58	Direct transition, indirect tail
	GaAs	Mos 61	Direct transition
	GaP	Spi 59b	Indirect edge and larger direct gap
	AlSb	Blu 54	Indirect edge
Fund. opt. absorption (pressure effects)	InAs	Tay 58, Edw 61	[000] cond. band edge
	InP	Edw 61	[000] cond. band edge (?); [100] minima nearby
	GaSb	Edw 61	[000] cond. band edge [111] and [100] minima nearby
	GaAs	Edw 59	[000] cond. band edge; [100] minima nearby
	GaP	Edw 59	[100] cond. band edge; [000] or [111] minima nearby
	AlSb	Edw 61	[100] cond. band edge
Fund. opt. absorp. (mag. field effects)	InSb	Bur 57b, 59, Zwe 57, 61 Rot 59	Direct gap, eff. mass values, band structure
	InAs	Zwe 57	Direct gap, eff. mass values, band structure
	GaSb	Zwe 59	Direct gap, eff. mass values, band structure
Fund. opt. absorp. and/or thermal gaps (alloys)	In(As,P)	Wei 56b, Ehr 59b	Same band in InAs and InP
	Ga(As,P)	Fol 55a, Ehr 60	Band edges differ in GaAs and GaP
	(Al,Ga)Sb	Bur 60	Band edges differ in Alsb and GaSb
	(Ga,In)Sb	Woo 59, 60	Same band edges in GaSb and InSb (?)
Ultraviolet reflectivity	InSb	Tau 61	Values for gaps $L_{3'} \rightarrow L_1$ and $X_4 \rightarrow X_1$ (?); spin orbit splitting of $L_{3'}$
	InAs	Tau 61	
	GaAs	Tau 61	
	GaSb	Tau 61, Car 61a	
Free carrier absorp. (electrons)	GaAs	Spi 59a	Positions of subsidiary cond. band minima relative to edge
	GaP	Spi 59b	
	AlSb	Tur 60	
Free carrier absorp. (holes)	InSb	Gob 60	Germanium-like valence bands; effective masses; spin-orbit splitting (in some cases)
	InAs	Mat 58	
	GaAs	Bra 59	
	GaSb	Ram 58	
	AlSb	Tur 59	
Free carrier reflect. and absorption	InSb	Spi 57	Effective mass
	InAs	Spi 57, Car 61b	
	InP	New 58	
	GaSb	Car 61c	Eff. masses; subsidiary cond. band.
	GaAs	Spi 59a, Ehr 60	Eff. mass
Faraday rotation	InSb	Mos 59a	Eff. mass; nonparabolic cond. band
	InAs	Aus 60, Car 61b	Eff. mass; nonparabolic cond. band
	InP	Mos 59b, Aus 60	Eff. mass
	GaAs	Car 61b	Eff. mass
Cyclotron resonance (microwave and infrared)	InSb	Dre 55a, Bur 56, Key 56	Eff. mass; energy surfaces
	InAs	Lax 60	Eff. mass; energy surfaces
Hall effect (electrons)	InP	Fol 55b	No multi-band conduction
	GaSb	Sag 60a	Subsidiary conduction band
	GaAs	Gra 58, Auk 60, Ehr 60	Subsidiary conduction band
Hall effect (holes)	InSb	Hro 55, Fre 57	Ge-like valence band
Magnetoresistence (electrons)	InSb	Pea 53, Kan 56, 57b	Spherical energy surfaces (cond. band) oscillations; eff. mass.
	InAs	Fre 58	Spherical energy surfaces (cond. band) oscillations; eff. mass.
	InP	Gli 59	Spherical energy surfaces (cond. band edge). Some [100] anisotropy
	GaSb	Bec 61	Spherical energy surfaces (cond. band edge). Subsidiary minima with small anisotropy
	GaAs	Gli 59	Spherical energy surfaces
Thermoelectric power	InSb	Bar 55	Electron eff. mass
		Wei 56a	Electron and hole eff. mass.
	InAs	Wei 56a	Electron and hole eff. mass.
	In(As,P)	Wei 59	Electron eff. mass increasing InAs $\rightarrow$ InP
	GaSb	Car 61d	Subsidiary minima
	GaAs	Edm 56, Ehr 60	Electron eff. mass
Piezoresistance	InSb	Key 55, Bur 57a, Pot 57 Tuz 58a	Spherical cond. band
		Pot 57, Key 60	Ge-like valence bands

TABLE I (*continued*)

Experiment	Material	References	Deductions
Piezoresistance	InAs	Tuz 58b	Ge-like valence bands
	InP	Sag 60b	Spherical cond. band
	GaSb	Sag 60a, Key 60	Spherical cond. band at edge [111] subsidiary minima
	GaAs	Sag 58	Spherical cond. band near edge
Resistivity vs. Pressure	InSb	Lon 55, Key 55, Ehr 57	Pressure dependent eff. mass
	InAs	Tay 55, Lon 56	Pressure dependent eff. mass
	InP	Sag 61	No evidence for subsidiary cond. bands
	GaSb	How 59, Sag 60a	Two sets of subsidiary conduction bands
	GaAs	How 59, Ehr 60	Subsidiary conduction bands

ing is to exploit to the fullest the close relationship, first pointed out by Herman (Her 55),[2] that appears to link the band structures of the group 4 and 3–5 compounds and to use information on well understood materials to deduce facts about other compounds.

At the time of writing, the only available detailed band calculations on the 3–5 compounds are for BN(Kle 60). Previously, Herman (Her 55) and Callaway (Cal 57) had mapped out general features of the band structure using a perturbation procedure based on the known structures of germanium and silicon. Further, Parmenter (Par 55) and Dresselhaus (Dre 55b) had discussed the symmetry properties of the bands in the zinc blende lattice using group theoretical techniques. With the help of such arguments and the experimental information to be discussed below, it is possible to indicate how the germanium band structure in the vicinity of the band edges must be modified to yield that for some of the 3–5 compounds. Figure 1 shows a sketch of the band structure of GaSb along the [100] and [111] directions neglecting spin-orbit effects. The gaps at Γ, L, and X, as well as their relative positions are believed to be to scale. The group theoretical symbols in parentheses correspond to those for germanium, as do the dotted lines. We note that the degeneracy at X, is removed and the intersection between the line Δ_1 and Δ_2, in germanium is no longer allowed in the zinc blende structure. It is accordingly likely that the related conduction band minimum in the 3–5 compounds occurs at X.

The experimental information to be discussed indicates that the conduction band structure for the compounds considered in this paper is similar to that shown in Fig. 1, the same conduction band minima at L, Γ, and X being present with Γ_1 lower than Γ_{15} but occurring at different relative energies. In BN, on the other hand, the conduction band curves corresponding to Γ_1 and Γ_{15} are interchanged (Kle 60).

Figure 2 schematically shows the relative positions of the lowest and subsidiary conduction band minima and those of the valence bands for the compounds under consideration, including spin-orbit splitting. Figure 2(c) shows the band structure at [000]. The effect of the linear k terms which remove the second-order degeneracy at $k=0$ is indicated schematically. These terms have almost negligible effect, although there is some evidence for their existence in InAs (Mat 58). Actual numerical values for many of the quantities defined in the figure are given in Table II, together with references.

We now turn to a discussion of the individual compounds.

InSb

This compound, remarkable because of its small electron effective mass and high mobility, was the first of the 3–5 group to be studied extensively. Cyclotron resonance experiments (Dre 55a) showed the conduction band to ge isotropic with a mass $m_n=0.013\,m$ and there-

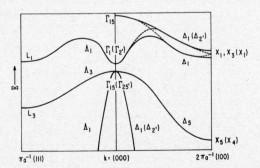

FIG. 1. Sketch of GaSb band structure obtained by modifying Ge band structure (indicated by dotted lines) and using numerical values of Table II. Symbols in parentheses refer to Ge symmetries; others to symmetries in zinc blende structure. Spin-orbit splitting has been neglected.

[2] See references in bibliography.

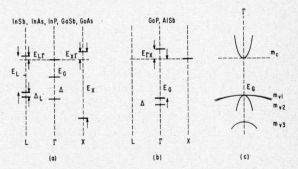

FIG. 2. (a), (b) Schematic energy positions of the valence and conduction band minima at L, Γ, and X. Δ and Δ_L denote, respectively, the spin-orbit splittings at Γ and L. Note that $E_{X\Gamma}$ is actually smaller than $E_{L\Gamma}$ for GaAs and probably InP. (c) Sketch of the band structure at [000].

TABLE II. Numerical values for the parameters defined in Fig. 2. Energies are in ev and masses in units of the electron mass. The asterisk denotes band gap values at or near helium temperatures. Other gaps represent room-temperature values. The identification of the numerical values for E_X with the gap at X is highly uncertain.

	InSb	InAs	InP	GaSb	GaAs	GaP	AlSb
E_G	0.2357*a	0.360b	1.29c	0.813*d	1.41e	2.2f	1.6*g
E_L	1.82h	2.53h		2.00h	2.94h		
$E_X(?)$	4.13h	4.77h		4.21h	5.08h		
$E_{L\Gamma}$				0.08–0.1i	>0.4		
$E_{X\Gamma}$			0.3(?)j	0.3–0.4j	0.36k		
$E_{\Gamma X}$						0.35l	0.3m
Δ	0.90n	0.43o	0.28n	0.86n	0.33p	0.12n	0.78q
Δ_L	0.56h	0.27h		0.48h	0.24h		
m_c	0.013r	0.021–0.027s	0.073t	0.047d	0.067–0.072k,s		
m_{v1}	0.25u	0.41o		0.23w	0.68k		∼0.4x
m_{v2}	0.015v	0.025o		0.06n	0.12k		
m_{v3}		0.083o			0.20k		

a Zwe 61.
b Zwe 57.
c New 58.
d Zwe 59.
e Mos 61.
f Spi 59.
g Blu 54.
h Tau 61.
i Sag 60a, Sag 61, Bec 61.

j Edw 61.
k Ehr 60.
l Spi 59.
m Tur 60, Fig. 3, present paper.
n Calculated values, present paper.
o Mat 58.
p Bra 59.
q Tur 59.
r Dre 55a.

s Car 61b.
t Mos 59b.
u Spi 57 and present paper.
v Kan 57a.
w Ram 58.
x Rei 58. Values for E_L, E_X, and Δ_L for InP, GaP, and AlSb are reported by Cardona (paper C1) at this conference.

fore likely to be located at $k=0$. Kane's theory (Kan 57a) based on the $k \cdot p$ approach gave considerable insight by providing explicit expressions for the conduction and valence band energies and wave functions vs k and succeeded in providing a quantitative explanation of the fundamental optical absorption (Fan 56, Gob 60). This indicated that both the conduction and valence band edges were located at $k=0$, and that furthermore the valence band edges are germaniumlike except for negligibly small linear k terms which removed the second-order degeneracy at $k=0$. These conclusions are corroborated by a variety of further experimental information. Magnetoresistance (Pea 53) and elastoresistance (Key 55, Pot 57) measurements support the idea of a spherical conduction band. The small conduction band effective mass is confirmed by thermoelectric power measurements and the so-called Burstein-shift (Bur 54) in the fundamental optical absorption. The subsequent oscillatory magnetoabsorption (Bur 57b, 59; Rot 59; Zwe 57, 61), infrared cyclotron resonance (Bur 56, Key 56), and free-carrier absorption experiments (Spi 57) have further elucidated the band structure of InSb in the immediate vicinity of the conduction and valence band edges.

The fact that the valence bands are nearly germaniumlike is further supported by the magnetic field dependence of the Hall effect (Hro 55, Fre 57) and piezoresistance effects (Pot 57, Tuz 58a, Key 60). The most detailed experimental exploration has been given in a paper of Gobeli and Fan (Gob 60) where optical absorption data in p-type material is used to obtain information concerning the warping of the valence bands. Reasonably accurate estimates of the valence band masses may be deduced from the infrared data of Spitzer and Fan (Spi 57), if the degenerate valence bands are taken into account in the analysis. Using Kane's theory (Kan 57a) to obtain a light mass of $0.015m$ one

finds that $m_{v1}=0.25m$. This is somewhat larger than the commonly quoted value of $0.18m$, and helps to reconcile the discrepancy between optical and thermal band gaps noted some time ago (Ehr 57). Faraday rotation (Mos 59a), magnetic susceptibility experiments (Bow 59), and detailed transport calculations (Ehr 57, 59a) also confirm the band structure just discussed. It has not been possible to discover evidence of subsidiary conduction band minimum in the [111] and [100] direction from either free-carrier optical experiments or pressure experiments. The reflectivity experiments in the ultraviolet region (Tau 61), however, do show a structure similar to that in Ge and the other 3–5 compounds. It is therefore likely that the qualitative features of the band structure shown in Fig. 1 for GaSb are also exemplified in InSb.

InAs

The band structure of InAs is closely related to InSb. Oscillatory galvanomagnetic measurements (Fre 58), cyclotron resonance experiments (Lax 60), infrared free carrier reflectibility (Spi 57, Car 61b), Faraday rotation (Aus 60, Car 61b), and magnetoabsorption measurements (Zwe 57), all indicate an electron mass in the range 0.015–$0.03m$. Effects associated with the nonparabolic form of the conduction band have been observed and analyzed by Cardona (Car 61b). The valence bands are germaniumlike. According to Matossi and Stern (Mat 58) the masses are $m_{v1}=0.41$, $m_{v2}=0.025$, and $m_{v3}=0.083$. These results, corresponding to conduction band mass $m_c=0.021$, were obtained from a detailed analysis of optical absorption.

There appears to be no direct experimental evidence for the existence of close-lying conduction band minima. Edwards and Drickamer (Edw 61) in optical experiments measuring the energy gap as a function of pressure do observe changes in slope above 20 000 atm, which might be interpreted as an intersection of the [000] and the [111] minima. If this were the case then at atmospheric pressure, the [111] minima would have to lie 0.1 ev above those at [000], and InAs would be expected to exhibit the same type of electrical behavior as GaSb. There is no evidence of such behavior. Reflectivity experiments in the ultraviolet (Tau 61) again show a structure very similar to that in InSb and germanium, and thereby indicate that the band structure in the [100] and the [111] directions is similar for all these compounds.

InP

The conduction and valence band edges appear to occur at $k=0$. This is strongly suggested by the small electron effective mass ($m_c=0.073$) obtained from the Faraday effect (Mos 59b), the fundamental optical absorption which appears to be direct (New 58) and the linear behavior with composition of the band gap in In(As,P) alloys (Wei 56b). Since InAs is known to have a band structure very similar to that of InSb, the linear variation of the band gap with composition suggests that InP is likewise similar. There is little direct evidence, either from free hole absorption or the magnetic field dependence of the Hall coefficient, bearing on the valence band structure, but from the preceding comments one certainly surmises it to be germaniumlike.

The evidence concerning the existence of subsidiary conduction band minima is conflicting. Glicksman (Gli 59) has reported magnetoresistance measurements which exhibited spherical symmetry for weakly doped n-type samples, but showed some [100] anisotropy for more heavily doped samples. This might be interpreted as evidence of an energetically close by set of [100] valleys. However, Sagar (Sag 60b) was not able to confirm this behavior either from elastoresistance measurements or from the observed behavior from the resistivity as a function of pressure. The measurements of optical band gap as a function of pressure (Edw 61) do show a reversal at 40 000 atm, which would place the [100] minima about 0.3 ev above those at [000]. This separation, however, is too large energetically to account for the magnetoresistance data.

GaSb

The band structure in the [100] and the [111] directions can be well determined from existing experimental information. A sketch (approximately to scale) is shown in Fig. 1. As already pointed out by Cardona (Car 61a), the band structure has a marked similarity to that of germanium, with the exception that the [000] minima are energetically lowest. The fact that this minimum is likely to occur at $k=0$ is confirmed by elastoresistance (Sag 60a) and magnetoresistance (Bec 61) measurements. The valence bands are probably germaniumlike (Ram 58). Free-hole absorption measurements have exhibited the transition between the heavy and light-

mass bands (Ram 58). The behavior of the fundamental optical absorption nevertheless leaves some doubt concerning the location of the valence band edge (Ram 58, Bec 61). However, if this edge were located at some other point in the Brillouin zone, then the extremum at [000] would be energetically close and the behavior of the resistivity as a function of pressure would reflect this fact. The data of Howard and Paul (How 59) show a negligible pressure effect for p-type material.

Sagar (Sag 60a), from elastoresistance measurements, concluded that subsidiary conduction band minima in the [111] directions lie about 0.08 ev above the [000] minima at room temperature. The existence of subsidiary minima has been confirmed by magnetoresistance (Bec 61), Hall effect (Sag 60a), and pressure (How 59, Sag 60a) measurements as well as experiments on the free carrier reflectivity (Car 61c) and thermoelectric power (Car 61d). Another set of subsidiary conduction band minima, apparently along [100] directions, is energetically removed from the edge by about 0.3–0.4 ev. There are three pieces of evidence for this: (1) A reversal of the shift of band gap with pressure is observed at 45 000 atm (Edw 61). (2) The band gap in (Al,Ga)Sb alloys (Bur 60) shows a break at about 60% composition as shown in Fig. 3. Further, the observed pressure shift in AlSb (Edw 61) indicates that the conduction band minima lie along [100] directions. Interpreting the break as due to the intersection [000] and [100] minima, one concludes that the [100] minima must lie energetically close to the conduction band edge of GaSb. (3) The behavior of the resistivity as a function of pressure (How 59) shows effects characteristic of a third set of close-lying conduction band minima.

Reflectivity experiments in the ultraviolet (Tau 61, Car 61a) again show the structure characteristic of germanium. It is thus possible to obtain the band gap at the point L and also (perhaps) near X. The information just discussed is sufficient to permit the determination of the gaps given in Fig. 1.

GaAs

The conduction and valence band edges both occur at $k=0$ and have the form sketched in Fig. 2(c). The [100] subsidiary conduction band minima lie 0.36 ev above the edge. The evidence for these facts, as well as the deduction of the numerical values given in Table II, has been discussed in detail in (Ehr 60). In this work data on the band gap in Ga(As,P) alloys were used to infer information concerning the symmetry and position of subsidiary conduction band minima in GaAs and GaP. More recent results of experimental work on these are shown in Fig. 3 together with the curve (Wel 56) and the extrapolations, indicated by the dotted lines, made in the original analysis. The conclusions, as originally made, are not changed appreciably by the new data.

Unfortunately, there is no direct information con-

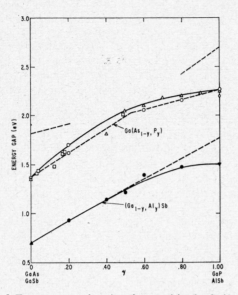

FIG. 3. Energy gap as a function of composition for $Ga(As_{1-y}P_y)$ and $(Ga_{1-y},Al_y)Sb$ alloys. Experimental data: ● Bur 60, □ Wei 60b, △ Fol 55a, ○ Wel 56, ◇ Loe 60, ▲ Lei 54. $Ga(As_{1-y}P_y)$: dashed curves refer to variation of gap as given in Wel 56, and linear extrapolations made in Ehr 60; solid curve indicates variation of gap as given in Wei 60. $(Ga_{1-y},Al_y)Sb$: dashed curve indicates linear extrapolation. Results of extrapolation procedure: In GaAs, [100] minima 0.4 ev above [000]; in GaP, [000] minimum 0.35 ev above [100]; in AlSb, [000] minimum 0.3 ev above [100].

cerning the [111] minima. The reflectivity data (Tau 61) in the ultraviolet range again shows the transitions at the point L and perhaps in the region around X and thus indicates that the band structure is qualitatively similar to that shown in Fig. 1 for GaSb.

GaP and AlSb

Relatively less detailed information exists concerning these materials. It is nevertheless possible tentatively to sketch out some of the features of the band structure in the immediate vicinity of the conduction and valence band edges. The valence band structure in AlSb is presumed to be nearly the same as that of germanium from observation of the split-off band in free hole absorption (Tur 59). There is no similar evidence for GaP, but the valence band structure is again assumed to be nearly germanium-like by analogy with the other 3–5 compounds. From the negative, siliconlike pressure coefficient associated with the optical band gap (Edw 59, 61), it may be inferred that the lowest conduction band minima lie along [100] directions in both materials, The extrapolations, shown in Fig. 3, of data on Ga(As,P) and (Ga,Al)Sb alloys, and the already established facts that the lowest conduction band minima in GaAs and GaSb are at [000], lead to the conclusion that the [000] minima lies about 0.4 ev above the [100] valleys in GaP and 0.3 ev above in AlSb. Free-carrier optical absorption experiments (Spi 59b, Tur 60) in n-type materials tend to confirm this deduc-

tion: transitions to subsidiary edges are observed at 0.35 ev in GaP and 0.29 ev in AlSb.

The effective masses in both materials are not yet well established.

BANDS AT $k=0$

The band structure at [000] and shown in Fig. 2(c) is of special importance in the 3–5 compounds, because, as already established, the conduction and valence band edges of InSb, InAs, InP, GaSb, and GaAs occur in this region of the Brillouin zone. Because of the closely related band structure of these compounds, Kane's theory (Kan 56, 59) for InSb can be immediately adapted to apply to all of them. From this theory it follows that the conduction band masses at the edge are given by

$$m/m_c = (E_P/E_G)\left[\tfrac{2}{3} + \tfrac{1}{3}E_G/(E_G+\Delta)\right]+1, \quad (1)$$

when E_G is the band gap, $E_P = (2m/\hbar)P^2$, P is the momentum matrix element $-i(\hbar/m)\langle S| P_z |Z\rangle$ coupling the s function corresponding to the conduction band with the p function corresponding to the light mass and splitoff valence bands, and Δ is the spin-orbit splitting of valence bands at [000]. Since m_c has been measured directly for a number of materials, and it is possible to estimate Δ reasonably for compounds in which it has not been directly observed, one can calculate E_P from the preceding equation for these materials.

The spin-orbit energy can be estimated with the help of a procedure suggested by Kane (Kan 57a). Since the spin-orbit splitting is determined overwhelmingly by the region inside the core, it is not affected very much by the solid. A direct comparison shows that for germanium $\Delta(\text{solid Ge}) = (29/20)\Delta(\text{atomic Ge})$. Since the 3–5 compounds have the same structure as Ge and very similar lattice constants, we assume the same scaling factor applies as well for the atomic constituents of the 3–5 compounds. We shall assume that the Szigeti (Szi 49) effective ionic charge $e^* = se$ determines the fraction of time s a valence electron spends on the group 3 constituent of the solid. Approximately, then

$$\Delta = (29/20)[s\Delta_{3,a} + (1-s)\Delta_{5,a}], \quad (2)$$

where $\Delta_{3,a}$, $\Delta_{5,a}$, respectively, are the atomic spin-orbit splittings of the group 3 and 5 constituents. The spin-orbit splittings calculated in this way are shown in Table II, except for InAs, GaAs, and AlSb, where the experimental values are given. The corresponding calculated values are 0.42, 0.31 and 0.61 ev, respectively, in good agreement with the experimental results. We may also compare Δ with the observed values of Δ_L given in Table II. It is seen that $\Delta_L \approx \tfrac{2}{3}\Delta$ as expected. This fact adds further credence to the deduction that the split peaks in the ultraviolet reflectivity experiments correspond to the $L_3 \rightarrow L_1$ transitions.

The values of E_P deduced from Eq. (1) and the experimentally observed values of m given in the lower portion of Fig. 4 are shown in the upper part of that figure. It is seen that E_P is constant within 20% with an average value $\bar{E}_P = 20$ ev. The calculated value for germanium, shown for comparison, is seen to be somewhat larger than typical values for the 3–5 compounds. The calculated triangular points, corresponding to effective mass values obtained from Eq. (1) using the computed Δ's and replacing E_P by $\bar{E}_P$, are seen to agree well with the experimental points. For GaP and AlSb the [000] conduction band masses are found to be, respectively, 0.12 and 0.09. The simplest formula,

$$m_c/m = E_G/\bar{E}_P, \qquad (3)$$

which is valid when $\Delta \ll E_G$ and $m_c \ll m$, also gives a good representation of the behavior of m_c for the various compounds and thus provides a method for making quick estimates. Equation (3) for the same $\bar{E}_P$ even appears to apply reasonably well to the 2–6 compounds.

The systematic trends exhibited by the compounds considered here provide further evidence that their band structure at [000] is indeed similar.

4. TRANSPORT PROPERTIES

This discussion will emphasize cases of n-type extrinsic conduction in materials characterized by a simple conduction band whose minimum lies at [000], since such measurements are more easily interpreted.[3] In addition, however, we shall consider the influence of subsidiary conduction band minima when relevant.

A. Scattering Mechanisms

The electron scattering mechanisms are conveniently divided into two categories, respectively involving the lattice and impurities. The lattice scattering mechanisms that must be considered are piezoelectric (Mei 53, Har 56), deformation potential (Bar 50), polar (How 53, Ehr 57), and nonpolar optical mode (Ehr 56) scattering. Of these, the first two involve acoustical phonons and the last two involve optical phonons. Piezoelectric and polar scattering, in particular, occur in the 3–5 compounds but not in germanium and silicon.

From detailed calculations of the mobility, polar scattering has been shown to be the dominant lattice scattering mechanism in InSb (Ehr 57, 59a), InAs (Ehr 59b, Cha 61), InP (Ehr 59b), and GaAs (Ehr 60, Hil 60) at temperatures above about 200°K. There is also some evidence that piezoelectric scattering dominates over deformation potential scattering in InSb from hot-electron experiments (Sla 60a) and measurements of the magnetoresistance in the quantum limit (Sla 60b). Deformation potential scattering can be shown to be too small, for example, by a factor of about 30 in InSb and a factor of about 10 in GaAs at room temperature to explain the magnitude of the mobility in pure samples.

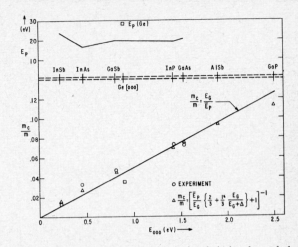

FIG. 4. Experimental ($\odot$) and calculated ($\triangle$) values of the [000] conduction band effective mass and calculated values of E_P as a function of E_{000}, the band gap at [000], for various 3–5 compounds and germanium. The solid line corresponds to Eq. (3) with $\bar{E}_P = 20$ ev. Note that the gaps shown for AlSb and GaP are larger than E_G for these compounds.

Whereas, however, all the parameters involved in the theoretical expression for the mobility due to polar scattering can and have been determined from independent experimental evidence without use of adjustable parameters, the estimates of the mobility due to deformation potential scattering are less certain because the deformation potential constant E_{1c} for the conduction band is not as well known. The quantity $E_{1c} - E_{1v} = -\kappa^{-1}(\partial E_G/\partial P)_T$, where κ is the compressibility and E_G is the band gap, can be easily determined experimentally. In order to estimate E_{1c}, it has in the past (Ehr 57, 60) appeared reasonable to suppose the valence edge to move rather more slowly than the [000] conduction band minima and to write approximately $E_{1c} \approx E_{1c} - E_{1v}$. To substantiate this statement we observe: (1) the quantities $E_{1c} - E_{1v}$ have almost the same magnitude for the [000] conduction band minimum of germanium as well as for the 3–5 compounds characterized by direct gaps. In InSb and GaAs they have been estimated to be (Ehr 57, 60) -7 ev, whereas in Ge and GaSb, $E_{1c} - E_{1v} = -9$ ev (Key 60). (2) In germanium, the motion of the [000] minimum with pressure can be determined absolutely because the deformation potentials for the [111] minima have been found from transport theory and measurements (Her 56, 59). The motion of the valence band edge with pressure deduced from these results in $\partial E_v/\partial P = -3.7 \times 10^{-6}$ ev/atm. (3) The pressure coefficients of the various band edges appearing in germanium, silicon, and the 3–5 compounds are very similar (Pau 61). It is therefore not unreasonable to expect the preceding pressure coefficient for the germanium valence band to apply to the valence bands of the 3–5 compounds as well. The deformation potential E_{1c} approximated by $E_{1c} - E_{1v}$ and calculated from the variation of E_G with pressure would therefore be too low by no more than 30–40%, but certainly not

---

[3] Some theoretical estimates of the mobility in p-type 3–5 compounds have been made by Hilsum (Hil 60) on the basis of polar scattering. In this connection see (Ehr 60), reference 25.

by the factor of 3 or so required to bring the calculated magnitude of the mobility due to deformation potential scattering in agreement with the experimental values (Kop 60).

The relative unimportance of this scattering mechanism and of nonpolar optical mode scattering in transport effects associated with the [000] conduction band minima of the 3–5 compounds has been previously discussed (Ehr 60, Sec. IV). However, deformation potential scattering would be expected to contribute significantly in bands having larger effective masses and to be dominant in intervalley scattering processes since polar scattering occurs most probably along forward directions.

The role of the impurity scattering mechanisms is understood rather less well than the mechanisms involving the lattice. Charged impurity scattering, as described by the Brooks-Herring formula (Bro 56), can be shown to account reasonably well for the low-temperature behavior of the highest mobility samples characteristic of a give carrier concentration in GaAs (Ehr 60), InAs (Cha 61), and InP (Ehr 59b). However, for samples having the same carrier concentration, these are large variations of the mobility from sample to sample, presumably due to the presence of some additional impurity scattering mechanism (Wei 60a).

The most frequently used method for elucidating the important scattering mechanisms in a given material has consisted of comparing calculated carrier mobilities with experimentally measured ones. The results of some of these calculations will be mentioned in the following section. There are a number of alternative methods that have recently been discussed in detail: (1) the Nernst coefficient (Bas 56, Her 58, Rod 59, Nas 60) may be directly related to the exponent r in the simple scattering law $\tau \sim E^r$, where τ is the relaxation time which depends on the energy E, in cases where such a law is valid. As will be seen below, some care must be taken in the interpretation of the data in terms of such a relationship. (2) The high magnetic field behavior of the magnetoresistance has been shown to depend sensitively on the scattering mechanisms (Ada 59). The measured results in InSb appear to be consistent with piezoelectric and/or polar scatterig (Ada 59, Sla 60a). (3) The behavior of the conductivity in strong electric fields may be related fairly directly to the scattering mechanisms (Str 58). Experiments on InSb (Pri 58, Kan 58, Sla 60b) are consistent with polar scattering as the principal mechanism causing energy loss above 78°K, and piezoelectric scattering at 4.2°K.

B. Transport Theory

The treatment of the electron transport properties involving polar scattering has involved difficulties because of the failure of the relaxation time approximation (to be denoted RTA) (How 53, Ehr 57). It has therefore been necessary to resort to variational techniques which have been extended to include correctly the combined effects of both polar and charged impurity scattering (Ehr 59a). These have given a quantitative account of the electron mobility in InSb (Ehr 59a), InAs (Ehr 59b, Cha 61), InP (Ehr 59b), and GaAs (Ehr 60, Hil 60). In the case of InSb it was important to include the effects associated with the nonparabolic conduction band, Fermi statistics, and electron-hole scattering at elevated temperatures. This approach has failed to explain the electron mobility in GaSb.

The inapplicability of the relaxation time approximation (RTA) to electron transport processes involving polar scattering is especially serious for the more complicated phenomena involving, for example, magnetic fields. Despite efforts in this direction, (Gar 59) no simply applicable variational techniques have yet been devised for treating such problems. Since it is known that a relaxation time exists for high-temperature $(T \gg \hbar \omega_l / K)$ polar scattering, and also can be defined for low-temperature scattering, it is of interest to inquire heuristically over what range of temperatures such approximations can actually be applied. Specifically, we ask whether, (1) relaxation time of the form $\tau = \tau_0 (E/\hbar \omega_l)^r$, can be fit to the available variational results for the mobility, thermoelectric power, and Hall coefficient if r is assumed to be an adjustable temperature dependent parameter, and (2) whether there exists a range of temperatures for which the function $r(T)$ is the same for all three transport properties. The quantity τ_0 here is regarded fixed by the requirement that at high and low temperatures the preceding expression for τ must reduce to the correct limiting solutions of the Boltzmann equation. In these limits, respectively, $r = \frac{1}{2}$ and 0 and

$$\tau_0 = [v_a M (h \omega_l)^{\frac{3}{2}} / 4\pi (2m_c)^{\frac{1}{2}} e^2 e^{*2} KT][(e^\vartheta - 1)/\vartheta], \quad (4)$$

where v_a is the volume of the unit cell, M is the reduced atomic mass, e^* is the effective ionic charge, ω_l is the longitudinal optical frequency, and $\vartheta = \hbar \omega_l / KT$. We consider the limit in which Boltzmann statistics are applicable and screening effects (Ehr 59c) can be neglected. For the calculations involving the mobility we use the Boltzmann limit of Eq. (15) of (Ehr 60) and determine r from the expression $G_1 e^{-z} = \frac{1}{2} \vartheta^{\frac{1}{2}-r} \Gamma(\frac{5}{2}+r)$ and the variational calculations for $G_1 e^{-z}$ (Ehr 59a). For the thermoelectric power we determine r by equating the simple expression for the transport term, $A = \frac{5}{2}+r$, with the variational solution (How 53). A similar procedure is followed for the Hall coefficient. Here $-necR_H = \Gamma(\frac{5}{2})\Gamma(\frac{5}{2}+2r)/\Gamma^2(\frac{5}{2}+r)$, and we determine r from comparison with the results of Lewis and Sondheimer (Lew 54).

The results are plotted in Fig. 5 as a function of ϑ. It is seen that for $\vartheta \leq 1$ all three methods of calculating r lead to approximately the same result and that in this range $0 < r < \frac{1}{2}$. For lower temperatures this agreement does not prevail because of the real failure of the RTA. In particular, it is impossible to find any solutions

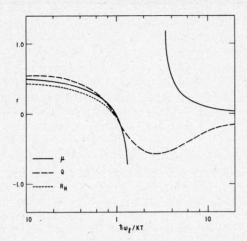

FIG. 5. r vs $\hbar\omega_l/KT$ obtained by equating variational solutions for the mobility μ, thermoelectric power Q, and Hall coefficient R_H with the corresponding expressions in the relaxation time approximation ($\tau \sim E^r$). Boltzmann statistics and simple parabolic bands are assumed.

r from the mobility calculations in the range $1.3 < \vartheta < 3.5$. The agreement at higher temperatures $T > \hbar\omega_l/K$, where polar scattering might be expected to be important or perhaps even dominant, provides some heuristic justification for using the RTA with the values of r determined here for more complicated transport phenomena as well. This useful fact will be employed in the discussion of the Nernst coefficient.

C. Multiband Transport

The close proximity of several conduction band minima in materials such as GaAs and GaSb leads to the possibility of simultaneous transport in several bands. Such effects are experimentally seen at elevated temperatures where it is possible to evaporate carriers from one minimum to another, or at elevated pressures if the energetic separation between different minima decreases with increasing pressure. Detailed analyses involving these effects have already been made in connection with the Hall effect (Gra 58, Auk 60, Ehr 60) and the pressure dependence of the electrical resistivity ρ (How 59, Ehr 60) in GaAs. The interpretation of the latter effects showed the mobility in the [100] valleys to be much smaller than that in the [000] minimum. Recent preliminary extension of this data to higher pressures (How 61) indicates the mobility in the [100] valleys may be 600 times smaller than that at [000], which is extremely surprising.[4] Similar types of analyses were performed by Sagar (Sag 60a) and Cardona (Car 61c) for GaSb. Here the mobility in the [000] minimum is about six times larger than that in the [111] valleys.

---

[4] In addition, the previous measurements of Howard and Paul (How 59) which were quoted in (Ehr 60) were repeated by Howard on purer samples. The new results agree more closely with the dashed curve in Fig. 4 of (Ehr 60) and thus increase the consistency between the pressure and Hall data (Auk 60). In the low-pressure range, too (Fig. 4 inset), the agreement between theory and experiment is markedly improved.

The pressure dependence of the resistivity (How 59) clearly indicates that above about 12 000 atm a third set of conduction bands, presumably the [100] valleys, begins to contribute. This effect increases the number of unknown parameters involved in the theory sufficiently that no unambiguous analysis along the lines of (Ehr 60) appears possible. Indications, however, are that the [100] valleys lie about 0.3 ev above the conduction band edge.

In the two-band case, band 1 here signifying the [000] minimum and band 2 the [100] valleys in GaAs or the [111] valleys in GaSb, the transport properties linear in the applied fields, namely the conductivity σ, Hall mobility μ_H, Hall coefficient R, thermoelectric power Q, and Nernst coefficient B are given by

$$\sigma = \sigma_1 + \sigma_2$$
$$\mu_H = \gamma_1 \mu_{H1} + \gamma_2 \mu_{H2}$$
$$R = \gamma_1^2 R_1 + \gamma_2^2 R_2$$
$$Q = \gamma_1 Q_1 + \gamma_2 Q_2$$
$$B = \gamma_1 B_1 + \gamma_2 B_2 - \gamma_1\gamma_2(Q_1 - Q_2)(\mu_{H1}/c - \mu_{H2}/c), \quad (5)$$

where $\gamma_i = \sigma_i/\sigma$ and the subscripts refer to the transport property to a particular band.

The single-band transport properties are seen to be weighted by powers of γ_i before being combined. The degree to which multiband effects are exhibited therefore depends on the conductivity in the second band, that is, on the carrier concentration and mobility relative to the first band. Even when γ_2 is very small, as in GaAs, the Hall coefficient will still show the effects of band 2 since the total number of carriers distributed between the conduction bands must remain constant in the extrinsic range.

In this case also the presence of the subsidiary minima may still be evident in the other transport properties because of intervalley scattering, which will affect electrons whose energy is such that they can be scattered by a phonon into the subsidiary minima, particularly when the density of states associated with band 2 is large relative to that in band 1. In GaAs intervalley scattering will not appreciably affect the simpler transport properties at atmospheric pressure because of the relatively large separation ($\Delta E = 0.36$ ev) between the [000] and [100] minima. At elevated pressures these effects are important because Δ is reduced. Furthermore, as will be seen, the Nernst effect is rather sensitive to intervalley scattering.

In GaSb these effects will be more important because of the close proximity of the [111] valleys [$\Delta E = 0.08$ ev] to the edge. For heavily doped material intervalley scattering may be important for most electrons contributing significantly to the transport process.

Nernst Effect in GaAs

In view of the approximate validity of the relaxation time approximation for polar scattering at temperatures above about 400°K, it is possible to comment on the reversal of sign observed in n-type GaAs at about 600°K by Emelyanenko and Nasledov (Eme 59, Nas 60) without resort to a variational treatment of the problem. This reversal has been interpreted as evidence of the importance of deformation potential scattering at higher temperatures, which is contradicted by the analysis of the scattering mechanisms just described. It is quite likely that this reversal can be accounted for by the presence of intervalley scattering into the [100] valleys. Since, as already indicated, the mobility ratio μ_1/μ_2 may be as large as 600, the effect of the second and third terms is the expression for B in Eq. (5) is unlikely to be important. A qualitative appreciation can be obtained from a calculation of the Nernst coefficient, assuming the validity of Boltzmann statistics and a relaxation time of the form $\tau \sim E^r e^{-E/\Delta E}$. The exponential factor serves to reduce τ for energies E comparable to the separation ΔE of the two minima. The result obtained is

$$B = (K/e)(\mu_H/c)[(\tfrac{5}{2}+2r)/(1+2KT/\Delta E) \\ -(\tfrac{5}{2}+r)/(1+KT/\Delta E)]. \quad (6)$$

For the n-type sample in reference (Eme 59) for which $n \approx 3 \times 10^{16}$ cm^{-3} and $\mu \approx 3000$ cm^2/v-sec at 300°K, we find $r \sim 0.5$ over the temperature range between 300° and 600°K and no evidence of a reversal of sign of r if we take $\Delta E = 0.36$ ev. This value is entirely compatible with polar and charged impurity scattering. Boltzmann statistics are not valid for the sample of (Eme 59) for which $n \approx 3 \times 10^{18}$ cm^{-3}. However, one of the principal effects of Fermi statistics will be a reduction of the effective gap. Assuming this to be given by $\Delta E - \zeta(T)$, where ζ is the Fermi level at temperature T, we find that the reversal in B should be shifted to lower temperatures by about the amount which is experimentally observed.

It is interesting to observe that a relaxation time having the same qualitative properties as that just described might also be used to represent the effects of nonparabolic bands if ΔE is replaced by the band gap E_G. The exponential factor would then represent the increased density of states which serves to scatter the electrons of higher energies more effectively.

Electron Mobility in GaSb

As already pointed out by Hilsum (Hil 60), the room temperature electron mobility, according to the elementary theory involving polar scattering, $(\mu \sim 44\ 000$ cm^{-1} v-sec) is more than an order of magnitude larger than the experimentally observed value $(\mu \sim 4000$ cm^{-1} v-sec). Multiband effects will reduce the calculated mobility appreciably even in lightly doped n-type samples at 300°K in which the Fermi level, if positive, is rather smaller than ΔE. From Eq. (5), we observe that μ_{H1} and not μ_H is to be compared with theory. For Sagar's sample 1B (Sag 60a), $(n = 1.65 \times 10^{17}$ cm$^{-3})$

μ_{H1} is about 20% larger than μ_H. The contribution due to intervalley scattering may be estimated roughly from a form of τ such as that used in Eq. (6). This produces a reduction by $(1 + KT/\Delta E)^{-\frac{5}{2}-r}$, or about a factor 2 at 300°K for $r = \frac{1}{2}$. Effects due to the nonparabolic nature of the conduction band may be obtained by calculating m^* for an electron in the nonparabolic band having energy KT, and finding the change in mobility. We find $m^*(300K) \approx 0.057$ and a further reduction of 30% in the mobility. The combined effect of these factors reduces the lattice mobility perhaps to 14 000 cm^2/v-sec. Charged impurity scattering for this sample (assuming no compensation) produces a room temperature mobility of 37 000 cm^{-1} v-sec. As in other 3–5 compounds with similar band structure deformation potential scattering is again too weak to produce an appreciable effect. Thus the total calculated mobility is still too large by a factor 3. It is accordingly likely that there are further, and as yet unexplored, scattering centers present in GaSb. If they can be found and eliminated, appreciable increases in the electron mobility may be expected.

BIBLIOGRAPHY

Ada 59　E. N. Adams and T. D. Holstein, J. Phys. Chem. Solids 10, 254 (1959).

Auk 60　L. W. Aukerman and R. K. Willardson, J. Appl. Phys. 31, 939 (1960).

Aus 60　I. G. Austin, J. Elect. and Control 8, 167 (1960).

Bar 50　J. Bardeen and W. Shockley, Phys. Rev. 80, 72 (1950).

Bar 55　R. Barrie and J. T. Edmond, J. Elect. 1, 161 (1955).

Bas 56　F. G. Bass and I. M. Tsidil'kovski, J. Exptl. Theoret. Phys. (U.S.S.R.) 31, 672 (1956) [Translation: Soviet Phys.—JETP 4, 565 (1957)].

Bec 61　W. M. Becker, A. K. Ramdas, and H. Y. Fan, J. Appl. Phys. 32, 2094 (1961), this issue.

Blu 54　R. F. Blunt, H. P. R. Frederikse, J. H. Becker, and W. R. Hosler, Phys. Rev. 96, 578 (1954).

Bow 59　R. Bowers and Y. Yafet, Phys. Rev. 115, 1165 (1959).

Bra 59　R. Braunstein, J. Phys. Chem. Solids 8, 280 (1959).

Bro 56　H. Brooks, Advances in Electronics, edited by L. Marton (Academic Press Inc., New York, 1956), Vol. 7.

Bur 54　E. Burstein, Phys. Rev. 93, 632 (1954).

Bur 56　E. Burstein, G. S. Picus, and H. A. Gebbie, Phys. Rev. 103, 825 (1956).

Bur 57a　F. P. Burns and A. A. Fleischer, Phys. Rev. 107, 1281 (1957).

Bur 57b　E. Burstein and G. S. Picus, Phys. Rev. 105, 1123 (1957).

Bur 59　E. Burstein, G. S. Picus, R. F. Wallis, and F. Blatt, Phys. Rev. 113, 15 (1959).

Bur 60　I. I. Burdigan, Soviet Phys.—Solid State Phys. 1, 1246 (1960).

Cal. 57　J. Callaway, J. Electronics 2, 330 (1957).

Car 61a　M. Cardona, Z. Physik 161, 99 (1961).

Car 61b　M. Cardona, Phys. Rev. 121, 752 (1961).

Car 61c　M. Cardona, J. Phys. Chem. Solids 17, 336 (1961).

Car 61d　R. O. Carlson, H. Ehrenreich, and S. Silverman (to be published).

Cha 61　R. P. Chasmar, J. Phys. Chem. Solids (to be published).

Dre 55a　G. Dresselhaus, A. F. Kip, C. Kittel, and G. Wagoner, Phys. Rev. 98, 556 (1955).

Dre 55b　G. Dresselhaus, Phys. Rev. 100, 580 (1955).

Edm 56　T. J. Edmond, R. F. Brown, and F. A. Cunnel, Rugby Semiconductor Conference (The Physical Society, London, 1956).

Edw 59　A. L. Edwards, T. E. Slykhouse, and H. G. Drickamer, J. Phys. Chem. Solids 11, 140 (1959).

Edw 61　A. L. Edwards and H. G. Drickamer, Phys. Rev. 122, 1149 (1961).

Edw 56   H. Ehrenreich and A. W. Overhauser, Phys. Rev. 104, 331 and 649 (1956).

Ehr 57   H. Ehrenreich, J. Phys. Chem. Solids 2, 131 (1957).

Ehr 59a   H. Ehrenreich, J. Phys. Chem. Solids 9, 129 (1959).

Ehr 59b   H. Ehrenreich, J. Phys. Chem. Solids 12, 97 (1959).

Ehr 59c   H. Ehrenreich, J. Phys. Chem. Solids 8, 130 (1959).

Ehr 60   H. Ehrenreich, Phys. Rev. 120, 1951 (1960).

Eme 59   O. V. Emelyanenko and D. N. Nasledov, Soviet Phys.—Solid State Phys. 1, 902 (1958).

Fan 56   H. Y. Fan and G. W. Gobeli, Bull. Am. Phys. Soc. 3, 111 (1956).

Fol 54   O. G. Folberth, O. Madelung, and H. Weiss, Z. Naturforsch. 9a, 954 (1954).

Fol 55a   O. G. Folberth, Z. Naturforsch. 10a, 502 (1955).

Fol 55b   O. G. Folberth and H. Weiss, Z. Naturforsch 10a, 615 (1955).

Fre 57   H. P. R. Frederikse and W. R. Hosler, Phys. Rev. 108, 1146 (1957).

Fre 58   H. P. R. Frederikse and W. R. Hosler, Phys. Rev. 110, 880 (1958).

Gar 59   F. Garcia-Moliner, Proc Roy. Soc. (London) A249, 73 (1959).

Gli 59   M. Glicksman, J. Phys. Chem. Solids 8, 511 (1959).

Gob 60   G. W. Gobeli and H. Y. Fan, Phys. Rev. 119, 613 (1960).

Gra 58   P. V. Gray and H. Ehrenreich, Bull. Am. Phys. Soc. 3, 255 (1958).

Har 56   W. A. Harrison, Phys. Rev. 101, 903 (1956).

Her 55   F. Herman, J. Electronics 1, 103 (1955).

Her 56   C. Herring and E. Vogt, Phys. Rev. 101, 944 (1956).

Her 58   C. Herring, T. H. Geballe, and J. E. Kunzler, Phys. Rev. 111, 36 (1958).

Her 59   C. Herring, T. H. Geballe, and J. E. Kunzler, Bell System Tech. J. 38, 657 (1959).

Hil 60   C. Hilsum, Proc. Phys. Soc. (London) 76, 414 (1960).

How 53   D. J. Howarth and E. H. Sondheimer, Proc. Roy. Soc. (London) A219, 53 (1953).

How 59   W. Howard and W. Paul (private communication).

How 61   W. Howard (private communication).

Hro 55   H. J. Hrostowski, F. J. Morin, T. H. Geballe, and G. H. Wheatley, Phys. Rev. 100, 1672 (1955).

Kan 57a   E. O. Kane, J. Phys. Chem. Solids 1, 245 (1957).

Kan 56, 57b   Y. Kanai and W. Sasaki, J. Phys. Soc. Japan 11, 1017 (1956); 12, 1169 (1957).

Kan 58   Y. Kanai, J. Phys. Soc. Japan 13, 967 (1958).

Key 55   R. W. Keyes, Phys. Rev. 99, 490 (1955).

Key 56   R. W. Keyes, S. Zwerdling. S. Foner, H. H. Kolm, and B. Lax, Phys. Rev. 104, 1804 (1956).

Key 60   R. W. Keyes and M. Pollak, Phys. Rev. 118, 1001 (1960).

Kle 60   L. Kleinman and J. C. Phillips, Phys. Rev. 117, 460 (1960).

Kop 60   Z. Kopec, Bull. acad. polon. sci. 8, 111 (1960).

Lax 59   B. Lax, L. M. Roth, and S. Zwerdling, J. Phys. Chem. Solids 8, 311 (1959).

Lax 60   B. Lax and J. L. Mavroides, Solid State Physics, edited by F. Seitz and D. Turnbull (Academic Press Inc., New York, 1960), Vol. 11.

Lei 54   H. N. Leifer and W. C. Dunlap, Phys. Rev. 95, 51 (1954).

Lew 54   B. F. Lewis and E. H. Sondheimer, Proc. Roy. Soc. (London) A227, 241 (1954).

Loe 60   E. E. Loebner and E. W. Poor, quoted in Wei 60.

Lon 55   D. Long, Phys. Rev. 99, 388 (1955).

Lon 56   D. Long, Phys. Rev. 101, 1256 (1956).

Mat 58   F. Matossi and F. Stern, Phys. Rev. 111, 472 (1958).

Mei 53   H. J. G. Meijer and D. Polder, Physica 19, 255 (1953).

Mos 59a   T. S. Moss, S. D. Smith, and K. W. Taylor, J. Phys. Chem. Solids 8, 323 (1959).

Mos 59b   T. S. Moss and A. K. Walton, Physica 25, 1142 (1959).

Mos 61   T. S. Moss, J. Appl. Physics 32, 2136 (1961), this issue.

Nas 60   D. N. Nasledov, J. chim. phys. 479 (1960).

New 58   R. Newman, Phys. Rev. 111, 1518 (1958).

Osw 59   F. Oswald, Z. Naturforsch. 14a, 374 (1959).

Par 55   R. H. Parmenter, Phys. Rev. 100, 573 (1955).

Pau 61   W. Paul, J. Appl. Phys. 32, 2082 (1961), this issue.

Pea 53   G. L. Pearson and M. Tannenbaum, Phys. Rev. 90, 153 (1953).

Phi 60   J. C. Phillips, J. Phys. Chem. Solids 12, 208 (1960).

Pot 57   R. F. Potter, Phys. Rev. 108, 652 (1957).

Pri 58   A. C. Prior, J. Elect. and Control 4, 165 (1958).

Ram 58   A. Ramdas and H. Y. Fan, Bull. Am. Phys. Soc. 3, 121 (1958).

Rei 58   F. J. Reid and R. K. Willardson, J. Elect. and Control 5, 54 (1958).

Rod 59   M. Rodot, J. Phys. Chem. Solids 8, 358 (1959).

Rot 59   L. M. Roth, B. Lax, and S. Zwerdling, Phys. Rev. 114, 90 (1959).

Sag 58   A. Sagar, Phys. Rev. 112, 1533 (1958).

Sag 60a   A. Sagar, Phys. Rev. 117, 93 (1960).

Sag 60b   A. Sagar, Phys. Rev. 117, 101 (1960).

Sag 61   A. Sagar and R. C. Miller, J. Appl. Phys. 32, 2073 (1961), this issue.

Sla 60a   R. J. Sladek, J. Phys. Chem. Solids 16, 1 (1960).

Sla 60b   R. J. Sladek, Phys. Rev. 120, 1589 (1960).

Spi 57   W. G. Spitzer and H. Y. Fan, Phys. Rev. 106, 882 (1957).

Spi 59a   W. G. Spitzer and J. M. Whelan, Phys. Rev. 114, 59 (1959).

Spi 59b   W. G. Spitzer, M. Gershenzon, C. J. Frosch, and D. F. Gibbs, J. Phys. Chem. Solids 11, 339 (1959).

Ste 57   F. Stern and R. M. Talley, Phys. Rev. 108, 158 (1957).

Str 58   R. Stratton, Proc. Roy. Soc. (London) A246, 406 (1958).

Szi 49   B. Szigeti, Trans. Faraday Soc. 45, 155 (1949).

Tau 61   J. Tauc and A. Abraham, Proceedings of the International Conference on Semiconductors, Prague, 1960, Czech. J. Phys. (to be published).

Tay 55   J. H. Taylor, Phys. Rev. 100, 1593 (1955).

Tay 58   J. Taylor, Bull. Am. Phys. Soc. 3, 121 (1958).

Tur 59   W. J. Turner and W. E. Reese, Bull. Am. Phys. Soc. 4, 408 (1959).

Tur 60   W. J. Turner and W. E. Reese, Phys. Rev. 117, 1003 (1960).

Tuz 58a   A. J. Tuzzolino, Phys. Rev. 109, 1980 (1958).

Tuz 58b   A. J. Tuzzolino, Phys. Rev. 112, 30 (1958).

Wei 56a   H. Weiss, Z. Naturforsch. 11a, 131 (1956).

Wei 56b   H. Weiss, Z. Naturforsch. 11a, 430 (1956).

Wei 59   H. Weiss, Ann. physik 4, 121 (1959).

Wei 60a   L. R. Weisberg and J. Blanc, Bull. Am. Phys. Soc. 5, 62 (1960).

Wei 60b   L. R. Weisberg, et al. ASTIA Rept. ERD-TN-60-759 (1960) (unpublished).

Wel 56   H. Welker and H. Weiss, Solid State Physics, edited by F. Seitz and D. Turnbull (Academic Press Inc., New York, 1956), Vol. 3.

Woo 59   J. C. Woolley, X. Evans, and C. M. Gillett, Proc. Phys. Soc. (London) 74, 244 (1959).

Woo 60   J. C. Wooley and C. M. Gillett, J. Phys. Chem. Solids 17, 34 (1960).

Zwe 57   S. Zwerdling, B. Lax, and L. M. Roth, Phys. Rev. 108, 1402 (1957).

Zwe 59   S. Zwerdling, B. Lax, K. Button, and L. M. Roth, J. Phys. Chem. Solids 9, 320 (1959).

Zwe 61   S. Zwerdling, W. H. Kleiner, and J. P. Theriault, J. Appl. Phys. 32, 2118 (1961), this issue.

NOTES ON CONFERENCE PAPERS

Several important pieces of experimental evidence reported at the Conference have direct bearing on the results presented in Tables I and II. The paper of Becker, Ramdas, and Fan, tentatively identified the fundamental optical absorption in GaSb as direct and attributed the additional structure as due to impurity levels. A 0.25-ev peak in the free carrier absorption was reported by these authors. This may be tentatively associated with a transition between the [000] and [100] valleys in GaSb. In addition, they observed structure in the magnetic field dependence of the hole Hall coefficient probably characteristic of Ge-like valence bands. There was also preliminary information from magnetoresistance experiments concerning subsidiary conduction band edges having [111] symmetry. Cardona reported ultraviolet reflectivity measurements on InP, GaP, and AlSb which exhibited structure similar to that in other group 4 and 3–5 compounds. His analysis of the spin-orbit splitting presented at the Conference is at variance with that reported in the present paper since he apparently assumed the same effective ionic charge for all 3–5 compounds considered. New determinations of the conduction band effective masses from cyclotron resonance, Faraday rotation, and the Voigt effect were reported by Palik, Teitler, and Wallis. These are in substantial agreement with the values listed in Table II. Data on the fundamental optical absorption in GaAs together with a detailed analysis following Kane's procedure for InSb were presented by Moss. These show quite convincingly that the band gap in GaAs is direct. Paul's paper discussed the systematics of the pressure coefficients of the various band edges.

Quite disturbing, in connection with the results of Tables I and II, was the failure to observe the expected pressure dependence of the free carrier peak in GaP which has been associated with transitions between the [100] and [000] minima. It is clear that the identification of this transition requires further consideration. The possibility, mentioned by Paul, that in GaAs other subsidiary conduction band edges are energetically close to the [100] minima cannot be discounted at the present time. However, the pressure and Hall effect measurements analyzed by the present author yield reasonably consistent results which might not be expected if other subsidiary minima were contributing whose behavior with pressure and temperaure might be thought to differ. The paper by Zwerdling and collaborators presented extremely accurate oscillatory magneto-optic results for InSb, some of which are already included in Table II.

JOURNAL OF APPLIED PHYSICS   SUPPLEMENT TO VOL. 32, NO. 10   OCTOBER, 1961

Optical Properties of Lead–Salt and III–V Semiconductors

FRANK STERN

United States Naval Ordnance Laboratory, White Oak, Silver Spring, Maryland

Dispersion relations for optical properties of solids, and a sum rule for the imaginary part of the dielectric constant, are reviewed and applications to semiconductors are described. Plasma frequencies, for which the real part of the dielectric constant vanishes, are associated with lattice vibrations, with free carriers, and with valance electrons. Gottlieb's infrared optical constants for LiF are in good agreement with the lattice vibration sum rule using unit effective charge. Dispersion relations for reflectivity have been used by Dixon to analyze optical constants of p-type PbTe in the infrared region where free-carrier effects dominate. He finds an effective mass at room temperature which rises from $0.1m$ to $0.3m$ as the carrier concentration is increased from 3×10^{18} to 10^{20} cm^{-3}. Morrison has measured the reflectivity of InAs, InSb, and GaAs and finds curves which agree well with those of Tauc and Abrahám. His analysis of these results using the dispersion relation predicts a plasma energy near 7 ev in these materials. Free carrier absorption in several III–V compounds and in PbS and PbTe is proportional to λ^p at long wavelengths, where p varies between 2, for InSb and AlSb, and 3, for InAs and GaAs. For n- and p-type PbS Riedl finds $p=2.4$. He observes additional structure near the intrinsic absorption edge of p-type PbTe, resembling that observed in n-type GaAs and GaP, which may be associated with a valence band about 0.1 ev below the band edge. The absorption coefficient near the absorption edge of all the lead compounds and of InAs is proportional to $e^{\hbar\omega/kT_{\rm eff}}$, with $80°\mathrm{K}<T_{\rm eff}$ $<120°\mathrm{K}$ for measurements taken at room temperature. Impurity effects on the energy gap are discussed in terms of a simple model.

I. INTRODUCTION

IN this paper some recent results concerning the optical properties of lead sulfide, lead selenide and lead telluride, and of the III–V semiconductors like indium arsenide and indium antimonide, will be discussed. Most of the general features of the optical properties of these materials have been rather extensively studied. The position of the absorption edge and its temperature dependence are well established for each of the materials. Reviews of the optical properties of both types of compounds are to be found in the books of Smith[1] and Moss.[2] The lead salts have been reviewed

by Scanlon,[3] and the III–V compounds by Welker and Weiss,[4] and by Hilsum and Rose-Innes.[5] Recent results for the pressure-dependence of the optical energy gap have been given by Edwards and Drickamer.[6]

Of the two groups of compounds, the lead salts have been studied much longer than the III–V compounds, and yet they are considerably less well understood.

---

[1] R. A. Smith, *Semiconductors* (Cambridge University Press, Cambridge, 1959).

[2] T. S. Moss, *Optical Properties of Semiconductors* (Butterworths Scientific Publications, London, 1959).

[3] W. W. Scanlon in *Solid State Physics*, edited by F. Seitz and D. Turnbull (Academic Press Inc., New York, 1959), Vol. 9, p. 83; J. Phys. Chem. Solids **8**, 423 (1959).

[4] H. Welker and H. Weiss in *Solid State Physics*, edited by F. Seitz and D. Turnbull (Academic Press Inc., New York, 1956), Vol. 3, p. 1.

[5] C. Hilsum and A. C. Rose-Innes, *Semiconducting III-V Compounds* (Pergamon Press, London, 1961).

[6] A. L. Edwards and H. G. Drickamer, Phys. Rev. **122**, 1149 (1961). See also W. Paul, J. Appl. Phys. **32**, 2082 (1961); A. Sagar and R. C. Miller, *ibid.* **32**, 2073 (1961).

Among the problems which remain are: (1) The band structure is not completely known, although magneto-resistance and elastoresistance measurements have given considerable information.[7] (2) The increase in energy gap on going from PbSe to PbTe has not yet been explained, although the indications are that it is PbTe which is out of order, rather than PbSe. (3) The mobility in the lead salts has not been quantitatively accounted for. (4) The static dielectric constant is unknown for the lead compounds because of the un-availability of samples of sufficiently low conductivity.

The chemistry of the lead salts is extremely interesting, and controls to a large extent the types of samples that may be prepared. The crystals can exist over a small range of compositions near the 50-50 composition. Because of the ionic nature of the compounds,[3] each excess metallic atom contributes a donor level and each excess nonmetallic atom contributes an acceptor level. Thus deviations from stoichiometry add carriers just as impurities do. In PbTe, for example, the maximum range of stability for undoped crystals occurs near 800°C, where the crystal can range from 1.5×10^{18} cm^{-3} n type to at least 8×10^{18} cm^{-3} p type.[8] Since the maxi-mum melting point of PbTe is on the p-type side, it is difficult to prepare good n-type crystals by pulling from the melt.[9] For PbS, the maximum melting point is on the n-type side,[10] while lead selenide more closely re-sembles PbTe.[11] At lower temperatures the range of stability of all the lead compounds narrows and approaches the 50-50 composition, but very long an-nealing times are required to reach equilibrium. Pre-cipitation of the excess constituent occurs as the samples are annealed, and has important effects on their properties.[12]

Because the deviations from stoichiometry which occur in the lead salts generally introduce donor and acceptor levels in concentrations of the order of 10^{17} cm^{-3} and higher, the impurity levels merge with the band edges,[13] and no freeze-out of carriers occurs at low temperatures. Thus these materials are degenerate and essentially metallic at low temperatures (but with high mobilities[14]) even in the absence of chemical impurities. Introduction of impurities should allow preparation of samples of lower carrier concentrations.

The chemistry of the III–V compounds, in com-parison with the chemistry of the lead salts, is much more favorable for the study of physical properties.

InSb can be prepared with carrier concentrations of the order of 10^{14} cm^{-3}. The other III–V's appear to have carrier concentrations of the order of 10^{16} cm^{-3} or more, unless they are compensated. It is not yet conclusively known whether this is the result of deviations from stoichiometry, as in the lead salts, or whether it is caused by foreign impurities that resist removal by zone refining and other purification techniques. The situation is complicated for the III–V's by the inappli-cability of the rather simple ionic model which appears satisfactory for the lead compounds.

In this paper we shall concentrate on several features of the optical properties of the lead salts and the III–V compounds on which recent work has been done. First we review the use of dispersion relations, and show how they have been applied to deduce the optical constants of the materials from spectral measurements of their reflectivity. Second, results for free-carrier absorption are presented. Finally, a brief discussion of the shape of the absorption edge, and of impurity effects on the edge, is given.

II. DISPERSION RELATIONS AND PLASMA FREQUENCIES

The dispersion relations connecting the real and imaginary parts of the dielectric constant $\epsilon = \epsilon_1 + i\epsilon_2$ are,[2,15] provided the wavelength is not too short:

$$\epsilon_1(\omega) = 1 + \frac{2}{\pi} \int_0^\infty \frac{\epsilon_2(\omega')\omega' d\omega'}{\omega'^2 - \omega^2}, \tag{1}$$

$$\epsilon_2(\omega) = \frac{2\omega}{\pi} \int_0^\infty \frac{\epsilon_1(\omega') d\omega'}{\omega^2 - \omega'^2}. \tag{2}$$

A brief history of dispersion relations (often called Kramers-Krönig relations) and a discussion of their connection with the physical requirement of causality, i.e., that no output can precede its input, have been given by Toll.[16] In this section we describe some general aspects of dispersion relations and plasma frequencies. Applications to compound semiconductors are described in Sec. III.

One application of the dispersion relations is in the derivation of a sum rule for an integral over the imagi-nary part of the dielectric constant. In particular, for frequencies ω far above the highest frequency at which $\epsilon_2 \neq 0$, one can write

$$\epsilon_1(\omega) = 1 - \left(\frac{2}{\pi\omega^2}\right) \int_0^\infty \epsilon_2(\omega')\omega' d\omega'. \tag{3}$$

This expression is not useful in practice, since the inte-gral must be taken to frequencies above those at which any absorption occurs, i.e., to x-ray frequencies for

[7] R. S. Allgaier, J. Appl. Phys. **32**, 2185 (1961).

[8] R. F. Brebrick and R. S. Allgaier, J. Chem. Phys. **32**, 1826 (1960). More recent data have been obtained by R. F. Brebrick and E. Gubner (unpublished).

[9] B. B. Houston, *Proceedings of the Fifth Navy Science Sym-posium, Annapolis, 1961* (to be published).

[10] J. Bloem and F. A. Kröger, Z. physik. Chem. **7**, 1 (1956).

[11] R. F. Brebrick and E. Gubner (unpublished); A. E. Goldberg and G. R. Mitchell, J. Chem. Phys. **22**, 220 (1954).

[12] W. W. Scanlon (unpublished).

[13] See Sec. V.

[14] R. S. Allgaier and W. W. Scanlon, Phys. Rev. **111**, 1029 (1958).

[15] H. Fröhlich, *Theory of Dielectrics* (Oxford University Press, Oxford, 1949).

[16] J. S. Toll, Phys. Rev. **104**, 1760 (1956).

most solids. There are, however, many cases in which different absorption mechanisms are well separated in frequency. It is then possible to find a frequency range which is high compared to the frequencies corresponding to one absorption mechanism, but low compared to the frequencies of all other absorption mechanisms. Then we can put

$$\epsilon_1(\omega) = \epsilon_\infty - \left(\frac{2}{\pi}\right)(\omega^2 - \omega_0^2)^{-1} \int_0^\infty \epsilon_2(\omega')\omega' d\omega', \quad (4)$$

where ϵ_2 here refers only to the contribution of a single mechanism, which is assumed to be peaked near ω_0, while ϵ_∞ gives the contribution of all other mechanisms to the real part of the dielectric constant, and is assumed to be independent of ω for the values being considered. The high-frequency dielectric constant for particles of charge q_i, mass m_i, and concentration N_i is found from a simple classical calculation to be approximately $\epsilon_\infty\{1 - [\omega_1^2/(\omega^2 - \omega_0^2)]\}$, where

$$\omega_1^2 = \sum (4\pi N_i q_i^2/m_i\epsilon_\infty). \quad (5)$$

We can write the sum rule in the form

$$(2/\pi)\int \epsilon_2(\omega')\omega' d\omega' = \sum_i (4\pi N_i q_i^2/m_i). \quad (6)$$

When $\omega^2 = \omega_p^2 = \omega_0^2 + \omega_1^2$, the real part of the dielectric constant vanishes. At this frequency electric fields can be present even when the electric displacement $\mathbf{D}$ vanishes. This leads to the possibility of collective oscillations of the medium. The frequency $(\omega_p/2\pi)$ is called the plasma frequency.

The probability that a charged particle will lose energy $\hbar\omega$ while traversing a medium of complex dielectric constant ϵ is proportional to[17]

$$\mathrm{Im}(-1/\epsilon) = \epsilon_2/(\epsilon_1^2 + \epsilon_2^2). \quad (7)$$

Thus if ϵ_2 is small when $\epsilon_1 = 0$, the probability of energy loss is large, and we say that the charged particle is strongly coupled to the plasma oscillations.

From the discussion leading to Eq. (4) it is clear that there can be several frequencies for which $\epsilon = 0$. In semiconductors there are at least three different regions that can be investigated:

1. The Ions. If the material is ionic or partially ionic, then the charge, mass, and concentration of each ion is to be inserted in Eqs. (5) and (6). The lattice plasma frequency will be in the infrared. For lattice vibrations the frequency ω_0 in Eq. (4) is the transverse optical frequency ω_t for small (but not too small) wave

vectors, also called the reststrahl frequency or the fundamental Raman frequency. The plasma frequency ω_p is the longitudinal optical frequency ω_l. The two frequencies are related by $(\omega_l/\omega_t)^2 = (\epsilon_0/\epsilon_\infty)$,[18] where ϵ_0 and ϵ_∞ are the dielectric constants below and above the dispersion region.

The charges q_i to be used in Eqs. (5) and (6) are effective charges, which will be different from $\pm e$ for most crystals.[19] The concept of effective charge must be used with care,[20] and the sum rule may need to be modified when realistic models for lattice vibrations are considered.[21]

2. The Free Carriers. The second category of plasma effects is that associated with free electrons and holes in a semiconductor. Since the concentration of charge carriers can be varied over many orders of magnitude in most semiconductors, the corresponding plasma frequencies can also be varied over a wide range, but will generally be in the infrared. For free carriers the value of ω_0 in Eq. (4) can be taken to be zero, and ω_1 in Eq. (5) equals the plasma frequency. Because of the ease with which the plasma frequency can be varied, and because ϵ_2 at the plasma frequency can be quite small in some cases, the free-carrier plasma provides a useful system for studying collective electron effects.

3. The Valence Electrons. The third major category of plasma effects involves the valence electrons of the solid, even though they may be bound. Since the number of valence electrons in a cubic centimeter of a metal or semiconductor is $\approx 10^{23}$, the valence-electron plasma frequency is considerably higher than that for the free carriers in a semiconductor, and generally lies in the ultraviolet. The optical constants in this region have been found for germanium,[22-23] and for silver.[24] The peak of the expression in Eq. (7), indicating the position of the plasma frequency, lies at 10 ev and at 3.75 ev in the two cases. These energies agree fairly well with those found from the energy losses of fast electrons traversing Ge and Ag foils.[25] There will in general be several different valence-electron plasma frequencies in a solid, as deeper shells of valence electrons contribute to the carrier concentration in Eq. (5).

[17] H. Fröhlich and H. Pelzer, Proc. Phys. Soc. (London) **A68**, 525 (1955); H. Fröhlich and R. L. Platzmann, Phys. Rev. **92**, 1152 (1953).

[18] R. H. Lyddane, R. G. Sachs, and E. Teller, Phys. Rev. **59**, 673 (1941).
[19] Note that the effective charge in (5) and (6) is bigger than the Szigeti effective charge [M. Born and K. Huang, *Dynamical Theory of Crystal Lattices* (Oxford University Press, Oxford, 1956), p. 112] by a factor $\frac{1}{3}(\epsilon_\infty + 2)$, which equals 1.33 for LiF.
[20] F. Matossi, Z. Naturforsch. **14a**, 791 (1959).
[21] A. D. B. Woods, W. Cochran, and B. N. Brockhouse, Phys. Rev. **119**, 980 (1960).
[22] H. R. Philipp and E. A. Taft, Phys. Rev. **113**, 1002 (1959); M. P. Rimmer and D. L. Dexter, J. Appl. Phys. **31**, 775 (1960).
[23] O. P. Rustgi, J. S. Nodvik, and G. L. Weissler, Phys. Rev. **122**, 1131 (1961). I am grateful for a preprint of this paper.
[24] E. A. Taft and H. R. Philipp, Phys. Rev. **121**, 1100 (1961).
[25] L. Marton, L. B. Leder, and H. Mendlowitz, in *Advances in Electronics and Electron Physics*, edited by L. Marton (Academic Press Inc., New York, 1955), Vol. 7, p. 183; C. J. Powell, Proc. Phys. Soc. (London) **76**, 593 (1960).

III. APPLICATION OF DISPERSION RELATIONS TO REFLECTIVITY

In the preceding section we described the dispersion relations connecting the real and imaginary parts of the dielectric constant. For analysis of optical measurements one wants a relation between the phase 2θ and the magnitude R of the complex reflectivity at normal incidence:

$$Re^{2i\theta} = [(n+ik-1)/(n+ik+1)]^2, \qquad (8)$$

where n and k are the real and imaginary parts of the index of refraction of the medium. Such dispersion relations have been derived in a fairly general context by Toll.[16] We can write his canonical phase shift in the form

$$\theta(\omega) = -\frac{\omega}{\pi} \int_0^\infty (\omega^2 - \omega'^2)^{-1} \ln R(\omega') d\omega', \qquad (9)$$

where we have dropped a number of terms given by Toll which do not apply in the present situation.[26] If the magnitude of the reflectivity is known over a wide enough range of frequencies, then the phase can be found from Eq. (9), and one can solve for $n(\omega)$ and $k(\omega)$ by using the real and imaginary parts of Eq. (8). This technique has been used previously to find the optical constants of a number of materials, including BaO,[27] Ge,[22–23] and Ag.[24] Its application to compound semiconductors is described below. We consider each of the three plasma regions of the previous section.

Lattice vibrations of LiF. Gottlieb[28] has measured the infrared reflectivity of LiF at several temperatures, and has found the optical constants in the region of the lattice vibration absorption, using the dispersion relation. LiF is particularly favorable for such measurements because the structure in the reflectivity occurs at relatively short wavelengths. In particular, the plasma frequency ω_p corresponds to a wavelength of 15.5 μ, and the absorption peak ω_0 corresponds to 32 μ.

The sum rule in Eq. (6) has been checked using the values of ϵ_2 given at increments of 3 cm^{-1} from 140 through 1400 cm^{-1} in Gottlieb's thesis.[28] We find that $(2/\pi)(2\pi c)^{-2} \int \epsilon_2 \omega d\omega$ equals 6.1, 6.0, 6.0, and 5.9×10^5 cm^{-2}, respectively, at 135, 210, 300, and 355°K. The corresponding calculated values from the right-hand side of Eq. (6) are 6.0, 6.0, 5.9, and 5.9×10^5 cm^{-2}, respectively, where we have taken the temperature dependence of the lattice constant into account,[29] and have used an effective charge of unity.[19]

It should be possible to carry out a similar analysis for the III–V compounds, whose reflection spectra in the infrared are known.[30–32] For GaP[32] and InSb[33] a detailed analysis of additional structure in the optical absorption spectrum has led to the determination of a number of points of the lattice vibration spectrum.

For the lead compounds, on the other hand, the situation is less satisfactory. A number of reflectivity measurements have been made[34] which show a rise in reflectivity near 50 μ, but no subsequent decline at longer wavelengths. Values of carrier concentrations for the samples have not been given, so it is not possible to estimate the extent to which free carriers influence the reflectivity.

Free carriers in p-PbTe. Measurements of reflectivity and absorption coefficient have long been used to obtain the free-carrier contribution to the dielectric constant of a semiconductor. Such measurements for the III–V compounds have given information about effective masses and their dependence on carrier concentration and temperature.[35–37] It is, however, possible to apply the dispersion relation of Eq. (9) to obtain the optical constants from the measured reflectivity alone.

The phase 2θ of the complex reflectivity, Eq. (8), of a sample containing free carriers can be compared with the phase $2\theta_0$ in a pure sample by using the relation

$$\theta(\omega) - \theta_0(\omega) = -\frac{\omega}{\pi} \int_0^\infty \frac{\ln R(\omega') - \ln R_0(\omega')}{\omega^2 - \omega'^2} d\omega', \qquad (10)$$

where R and R_0 are the reflectivities of the impure and the pure samples, respectively. Experimentally it is usually found that R becomes independent of carrier concentration as ω approaches $\omega_g = E_g/\hbar$, where E_g is the energy gap. For frequencies below ω_g there can be no electronic transitions in pure material, hence $\theta_0(\omega) = 0$ for $\omega < \omega_g$. Thus $\theta(\omega)$ can be found by evaluating the integral in (10) for the range $0 < \omega' < \omega_g$. Equation (10) is useful only if R_0 has become constant at the wavelengths for which free carrier effects become noticeable in the purest available sample. Fortunately, this condition is generally satisfied very closely.

This procedure has been applied to the reflectivity of p-type PbTe by Dixon,[38] who finds that the method

[26] Two of the additional terms given by Toll correspond to pathological behavior of the reflectivity at finite or infinite frequency, and can be excluded in our case.

[27] F. C. Jahoda, Phys. Rev. **107**, 1261 (1957).

[28] M. Gottlieb, J. Opt. Soc. Am. **50**, 343 (1960); Ph.D. thesis, University of Pennsylvania, 1959 (unpublished). I am grateful to Dr. Park Miller for sending me a copy of Dr. Gottlieb's thesis.

[29] B. Dayal, Proc. Indian Acad. Sci. **20A**, 138 (1944); J. Thewlis, Acta Cryst. **8**, 36 (1955).

[30] W. Spitzer and H. Y. Fan, Phys. Rev. **99**, 1893 (1955); H. Yoshinaga and R. A. Oetjen, Phys. Rev. **101**, 526 (1956), [InSb].

[31] G. Picus, E. Burstein, B. W. Henvis, and M. Hass, J. Phys. Chem. Solids **8**, 282 (1959) [InP, AlSb, GaAs, GaSb, InAs].

[32] D. A. Kleinman and W. G. Spitzer, Phys. Rev. **118**, 110 (1960), [GaP].

[33] S. J. Fray, F. A. Johnson, and R. H. Jones, Proc. Phys. Soc. (London) **76**, 939 (1960).

[34] J. Strong, Phys. Rev. **38**, 1818 (1931); W. M. Sinton and W. C. Davis, J. Opt. Soc. Am. **44**, 503 (1954); H. Yoshinaga, Phys. Rev. **100**, 753 (1955).

[35] W. G. Spitzer and H. Y. Fan, Phys. Rev. **106**, 882 (1957), [InSb, InAs].

[36] W. G. Spitzer and J. M. Whelan, Phys. Rev. **114**, 59 (1959), [GaAs].

[37] M. Cardona, Phys. Rev. **121**, 752 (1961), [InAs, GaAs]; J. Phys. Chem. Solids **17**, 336 (1961), [GaSb].

[38] J. R. Dixon, Bull. Am. Phys. Soc. **6**, 312 (1961).

gives useful results even for wavelengths beyond the reflectivity minimum. Previous workers[35-37] have restricted their attention only to the short-wavelength side of the reflectivity minimum, since the conventional method of analysis becomes considerably more difficult at longer wavelengths.

Tentative effective mass values for p-type PbTe found by Dixon at room temperature range from $0.1m$ to $0.3m$ as the carrier concentration is varied from 3×10^{18} cm^{-3} to 10^{20} cm^{-3}. Some uncertainty in these values arises because of the polycrystalline nature of the samples used at the high carrier concentrations. The rather large increase in effective mass with increasing carrier concentration tends to confirm Allgaier's conclusion[7] that there is a valence band whose edge lies about 0.1 ev below the top of the highest valence band, and suggests that the effective mass of the lower band is larger than that of the band edge.

The infrared reflectivity of n-type GaSb at room temperature and at 89°K has been measured by Cardona,[37] who has shown that the results are consistent with the existence of a second conduction band, separated from the [000] minimum by a gap of 0.08 ev at 300°K and 0.057 ev at 89°K.

A by-product of Dixon's work is the fact that the imaginary part of $(-1/\epsilon)$, given in Eq. (7), is quite small for his samples, indicating that plasma oscillations will be strongly damped in lead telluride at room temperature. It is easy to show for free carriers, using a classical model, that the peak value of $\mathrm{Im}(-1/\epsilon)$ is approximately equal to $\omega_p \tau / \epsilon_\infty$, where ω_p is the value of ω at which the real part of the dielectric constant vanishes, τ is the mean scattering time, and ϵ_∞ is the optical dielectric constant. Because of the high mobility in PbTe at low temperatures,[14] it should be possible to obtain a rather large value of $\mathrm{Im}(-1/\epsilon)$ at liquid nitrogen or liquid helium temperatures.

Valence electrons in III–V compounds. The reflectivity of several intermetallic semiconductors has been measured in the infrared, the visible, and the near ultraviolet portions of the spectrum by Tauc and Abrahám,[39] Cardona,[40] and Morrison.[41] Morrison extended his measurements to a photon energy of 6 ev, somewhat higher than the upper limit of the other workers. He used the dispersion relation, Eq. (9), and calculated the optical constants of InAs, InSb, and GaAs between 0 and 6 ev. The results were found to be rather sensitive to the extrapolation, and none of the extrapolations completely satisfied the condition, discussed in connection with Eq. (10), that the phase θ_0

for pure material should vanish for $\omega < \omega_g$. Morrison's results are of considerable interest, however, and suggest that the plasma energy in all three semiconductors is about 7 ev, and that the imaginary part of $-1/\epsilon$ reaches a peak well above 1. These values can be compared with those for Ge,[23] for which the peak in $\mathrm{Im}(-1/\epsilon)$ came at 10 ev and had a magnitude of 0.8. In order to establish the plasma energy in the III–V semiconductors more exactly, it will be necessary to measure their reflectivities further into the ultraviolet.

All the III–V compounds studied so far have two main peaks in the reflectivity, one near 2 ev and a larger one near 5 ev, and bear a strong qualitative resemblance to Ge.[22-23] In addition, the peak near 2 ev is found to be split.[42] Some of the principal features of the reflectivity are shown in Table I.

By analogy with Ge,[43] one can associate the reflectivity peak near 2 ev with transitions between the state L_3,[44] thought to be the highest valence band state at the endpoint of the $\langle 111 \rangle$ directions in the Brillouin zone, and the state L_1, the lowest conduction band state for the same wave vector. The splitting of this peak is associated with spin-orbit splitting of the state L_3. The peak near 5 ev may be associated with transitions between states at X, the endpoint of the $\langle 100 \rangle$ directions. In this case spin-orbit splitting may occur for the sphalerite structure,[44] but not for the diamond structure. It has not yet been observed.

Valence electrons in the lead compounds. Reflection measurements from the absorption edge to photon energies of about 3 ev, using light polarized both in and perpendicular to the plane of incidence,[45] have been made by Avery for PbS, PbSe, and PbTe.[46] The results have been summarized by Moss.[2] The imaginary part of the dielectric constant has a strong peak at about 2,

TABLE I. Principal features of the reflectivity of III–V compounds and Ge. All energies are in ev.

| | First peak | | Second peak | |
	Position	Splitting	Position	Reference
InSb	1.82	0.56	4.13	39,42
InSb	1.8	0.50	4.1	41
InAs	2.53	0.29	4.72	39,42
InAs	2.5	0.35	4.8	41
GaSb	2.00	0.48	4.22	39,42
GaSb	1.95	0.52	4.3	40
GaAs	2.94	0.26	~5.1	39,42
GaAs	3.0	0.20	5.1	41
InP	3.2	0.16	5.0	40
Ge	2.1	0.2	4.4	42

[39] J. Tauc and A. Abrahám, *Proceedings of the Conference on Semiconductor Physics, Prague, 1960* (Publishing House of the Czechoslovak Academy of Sciences, Prague, 1961), [InSb, InAs, GaAs, GaSb]. I am indebted to Dr. Tauc for an informative discussion of this experiment.

[40] M. Cardona, Z. Phys. **161**, 99 (1960), [GaSb]; J. Appl. Phys. **32**, 958 (1961), [InP].

[41] R. E. Morrison, M. S. thesis, University of Maryland, 1961; Phys. Rev. (to be published), [InSb, InAs, GaAs].

[42] J. Tauc and E. Antončík, Phys. Rev. Letters **5**, 253 (1960).

[43] J. C. Phillips, J. Phys. Chem. Solids **12**, 208 (1960).

[44] The properties of states at symmetry points in the Brillouin zone for sphalerite (zinc blende) crystals are given by R. H. Parmenter, Phys. Rev. **100**, 573 (1955), and by G. Dresselhaus, *ibid.* **100**, 580 (1955).

[45] D. G. Avery, Proc. Phys. Soc. (London) **B65**, 425 (1952).

[46] D. G. Avery, Proc. Phys. Soc. (London) **B64**, 1087 (1951); **B66**, 133 (1953); **B67**, 2 (1954).

TABLE II. Free-carrier absorption cross sections, α/N, for a number of compound semiconductors at room temperature. The values are for a wavelength of 9 μ, and vary with wavelength as λ^p. The values for PbTe (and probably for GaP) are affected by the presence of nearby interband absorption.

	Carrier concentration (10^{17} cm^{-3})	Cross section (10^{-17} cm^2)	p	Reference
n-InSb	1–3	2.3	2	52
n-InAs	0.3–8	3.6	3.0	53
n-GaAs	1–10	4.7	3.1	36
n-InP	0.4–4	4.5	2.6	54
n-GaP	10	31	1.7	55[a]
n-GaP	1	12	1.4	32[b]
n-AlSb	0.4–4	15	2	57
n-PbS	1	12	2.35	56
p-PbS	26	11	2.45	56
p-PbTe	9–36	10	1.1	56

[a] Dr. W. G. Spitzer (private communication) has pointed out that this sample had rather poor electrical properties. I am grateful to Dr. Spitzer for information about the GaP measurements.
[b] These values were estimated by the author of the present paper from Fig. 3 of reference 32, using a carrier concentration of 1.0×10^{17} cm^{-3}.

$2\frac{1}{2}$, and 3 ev in PbTe, PbSe, and PbS, respectively. The optical properties vary somewhat from sample to sample, n-type and p-type samples having somewhat different behavior. This is presumably caused by surface layers; n-type PbTe, for example, often develops a visible tarnish on exposure to air for an extended period.

IV. FREE-CARRIER ABSORPTION

The classical result for the absorption coefficient associated with free carriers can be written in the form

$$\alpha = 5.26 \times 10^{-17} (\lambda^2/n)[N\mu^{-1}(m^*/m)^{-2}] \text{ cm}^{-1}, \quad (11)$$

where λ is the wavelength of the light in vacuum, measured in microns, n is the index of refraction of the material at wavelength λ, N is the carrier concentration in cm^{-3}, m^* is the conductivity effective mass, m is the free electron mass, and μ is the mobility in cm^2/v-sec. If the relaxation time τ varies with energy, then Eq. (11) should also be multiplied by a factor $\tau_{av}\tau^{-1}{}_{av}$, which will be ≥ 1. If several energy surfaces contribute carriers, then the expression in brackets is to be summed over them. The classical expression gives quite good agreement with experiment for many materials, particularly at high temperatures.

Quantum-mechanical derivations of the absorption coefficient have been undertaken for a number of cases, the most elaborate being the study of free-carrier absorption in n-type Ge carried out by Meyer[47] and by Rosenberg and Lax.[48] Their results are in good agreement with experiment above room temperature, but in poorer agreement at lower temperatures, for which impurity scattering affects the mobility. When impurity scattering dominates, the quantum-mechanical treatments[47–49] show that the absorption coefficient

can vary as rapidly as the 3.5 power of the wavelength when $\hbar\omega \gg kT$. At long wavelengths the classical λ^2 dependence is found.

The III–V semiconductors and the lead compounds both have sufficiently strong polar character that polar scattering is the dominant scattering mechanism near room temperature.[50] Expressions for free carrier absorption in this case have been given by Visvanathan,[51] who finds that the absorption coefficient varies as the 2.5 power of the wavelength at short wavelengths.

A summary of some experimental results for free carrier absorption for a wavelength of 9 μ is given in Table II, using the experimental results of a number of authors.[32,36,52–57] The power p describing the wavelength dependence is found to have values between 2 and 3 in most cases.

From the classical expression, Eq. (11), we see that the free carrier absorption depends on the mobility. Thus a theory of free carrier absorption should be carried out in parallel with a theory of mobility, and should be able to fit both quantities over a range of conditions. The theory of mobility in the III–V compounds[50] gives satisfactory agreement with experiment, and could be extended to the lead compounds if the characteristics of the lattice modes were known. Extension of the theory to free carrier absorption when more than one scattering mechanism is present, and when degeneracy, screening, and complex band structure must be taken into account remains to be done.

Additional structure in the free-carrier absorption, has been observed for many semiconductors. The absorption in a number of p-type III–V compounds has been identified with transitions between various branches of the valence bands near the middle of the Brillouin zone, and has given information about spin-orbit interactions and valence band effective masses in these materials.[57,58] Matossi and Stern[58] concluded from the temperature dependence of the absorption in p-type InAs that there could be a small nonzero slope in the energy of the states degenerate at $k=0$. The unavailability of suffi-

[47] H. J. G. Meyer, Phys. Rev. **112**, 298 (1958); J. Phys. Chem. Solids **8**, 264 (1959).

[48] R. Rosenberg and M. Lax, Phys. Rev. **112**, 843 (1958).

[49] S. Visvanathan, Phys. Rev. **120**, 379 (1960).

[50] H. Ehrenreich, Phys. Rev. **120**, 1951 (1960) and references therein, and H. Ehrenreich, J. Appl. Phys. **32**, 2155 (1961).

[51] S. Visvanathan, Phys. Rev. **120**, 376 (1960).

[52] S. W. Kurnick and J. M. Powell, Phys. Rev. **116**, 597 (1959), [InSb].

[53] J. R. Dixon, *Proceedings of the Conference on Semiconductor Physics, Prague, 1960* (Publishing House of the Czechoslovak Academy of Sciences, Prague, 1961), and private communication; J. R. Dixon and D. P. Enright, Bull. Am. Phys. Soc. **3**, 255 (1958), [InAs].

[54] R. Newman, Phys. Rev. **111**, 1518 (1958), [InP].

[55] W. G. Spitzer, M. Gershenzon, C. J. Frosch, and D. F. Gibbs, J. Phys. Chem. Solids **11**, 339 (1959), [GaP].

[56] H. R. Riedl, Bull. Am. Phys. Soc. **6**, 312 (1961), [PbS, PbTe].

[57] W. J. Turner and W. E. Reese, Phys. Rev. **117**, 1003 (1960), [AlSb].

[58] F. Stern and R. M. Talley, Phys. Rev. **108**, 158 (1957); F. Matossi and F. Stern, *ibid.* **111**, 472 (1958), [InAs]; R. Braunstein, J. Phys. Chem. Solids **8**, 280 (1959), [GaAs]; A. K. Ramdas and H. Y. Fan, Bull. Am. Phys. Soc. **3**, 121 (1958), [GaSb]; G. W. Gobeli and H. Y. Fan, Phys. Rev. **119**, 613 (1960), [InSb].

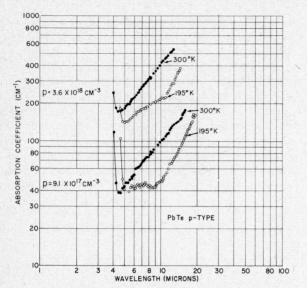

FIG. 1. Free-carrier absorption in p-type PbTe. These results were obtained by H. R. Riedl, reference 56.

ciently good p-type InAs has prevented further investigation of this interesting problem.

Recent measurements have exhibited another interband absorption mechanism, associated with non-vertical transitions between different branches of the conduction and valence bands. The results are summarized in Table III. There is also good evidence from electrical measurements[7] and from Dixon's effective mass determinations[38] described in Sec. III, that a second band edge is about 0.1 ev below the top of the valence band of PbTe. This is consistent with Riedl's results,[56] shown in Fig. 1 and in Table III, although other mechanisms cannot be entirely ruled out at present.

V. ABSORPTION EDGE

The exponential behavior found in the absorption edge of many ionic crystals,[59] often called the Urbach effect,

TABLE III. Properties of the additional free-carrier absorption seen at room temperature.

	Threshold energy (ev)	Peak cross section (10^{-17} cm²)	Postulated transition[a]	References
n-GaAs	0.25	0.5	$[000] \rightarrow \langle 100 \rangle$	36,50
n-GaSb	0.25		$[000] \rightarrow \langle 100 \rangle$	[b],50
n-GaP	0.31	12	$\langle 100 \rangle \rightarrow [000]$	55,50
n-AlSb	0.2	18	$\langle 100 \rangle \rightarrow [000]$	57,50
p-PbTe	0.08	4	$\langle 111 \rangle \rightarrow$?	56,7

[a] Dr. W. Paul, in the opening paper of this conference, proposed that the transition in GaP, and perhaps also in GaAs and AlSb, may be a vertical transition to a higher conduction band.
[b] Presented at this conference by W. M. Becker, A. K. Ramdas, and H. Y. Fan. These results were obtained at 80°K.

[59] W. Martienssen, J. Phys. Chem. Solids **8**, 294 (1959); D. L. Dexter, J. Phys. Chem. Solids **8**, 473 (1959); F. Urbach, Phys. Rev. **93**, 1324 (1953); F. Moser and F. Urbach, *ibid.* **102**, 1519 (1956).

is also found for the lead salts and for III–V semiconductors. The absorption coefficient is found to vary over one or more decades according to the relation

$$\alpha = A e^{\hbar \omega / k T_{\rm eff}}. \tag{12}$$

In KBr and KI the effective temperature is found to be $T_{\rm eff} = T/0.79$,[59] where T is the ambient temperature.

Experimental results at room temperature for a number of lead salt crystals and alloys, and for InAs, are shown in Fig. 2. The effective temperatures obtained for the curves in Fig. 2 range from 80° to 120°K. Low-temperature results at high resolution for InAs[60] indicate an effective temperature of 15°K or less for an ambient temperature of 78°K. Thus impurity effects cannot account for the room temperature results. The analogy with the results[59] for ionic crystals suggests that electron-hole interaction[61] should be considered in an attempt to account for the Urbach effect.

Shifts of the band edge caused by impurities are well known in semiconductors. The best-known such effect, first explained by Burstein,[62] is the increase in the optical energy gap produced by degeneracy in heavily doped samples. More recently, decreases in the optical energy gap have been found in compensated InAs,[63] and in p-type InSb,[52] InAs,[60] and other materials. A theory for the effect has been proposed by Eagles[64] on the basis of transitions from acceptor levels to the conduction band edge.

A simple theory of impurity effects in heavily doped semiconductors is possible by making use of a hydrogenic model, first applied to this problem by Stern and Talley.[65] In the earlier treatment[63,65] of impurity effects, we considered only the effects on one band

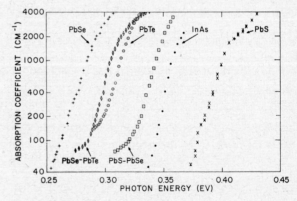

FIG. 2. Shape of the absorption edge in a number of compound semiconductors and two 50-50 alloys. The curve is a composite of results obtained by J. R. Dixon and J. M. Ellis, H. R. Riedl, and W. W. Scanlon.

[60] J. R. Dixon and J. M. Ellis (unpublished).
[61] See, for example, the article by T. P. McLean in *Progress in Semiconductors*, edited by A. F. Gibson (John Wiley & Sons, Inc., New York, 1960), Vol. 5, p. 53.
[62] E. Burstein, Phys. Rev. **93**, 632 (1954).
[63] F. Stern and J. R. Dixon, J. Appl. Phys. **30**, 268 (1959).
[64] D. M. Eagles, J. Phys. Chem. Solids **16**, 76 (1960).
[65] F. Stern and R. M. Talley, Phys. Rev. **100**, 1638 (1955).

edge—for example, the effect of donor impurities on the conduction band edge. A consistent treatment must take into account the effect of the impurities on both bands.[66] We attempt to give such a treatment for a typical material, one for which the conduction band effective mass is considerably smaller than the valence band effective mass.

At low donor concentrations, discrete impurity levels are present below each band edge.[67] As the donor concentration is increased, and becomes comparable to

$$N_{\min,c} = 3 \times 10^{23} (m_c/m\epsilon_0)^3 \text{ cm}^{-3}, \qquad (13a)$$

where m_c is the conduction band effective mass and ϵ_0 is the static dielectric constant, the impurity levels will form a band which merges into the conduction band and, in effect, lowers the band edge. Since the impurity distribution is random, the band edge will not be sharp, but will have a tail in which the density of states drops off rapidly. Lax and Phillips[68] have shown in the one-dimensional case that for high impurity concentrations it is possible for many purposes to consider the band edge to be simply shifted. This is their "optical model," which is equivalent to the earlier treatment of Stern and Talley.[65] In the three-dimensional case the effect of density fluctuations should be even less pronounced than in one dimension, since the lowering of the band edge is proportional to $N^{\frac{1}{3}}$ in three dimensions, but to N in one dimension.

When the donor concentration is increased further, it will finally exceed

$$N_{\min,v} = 3 \times 10^{23} (m_v/m\epsilon_0)^3 \text{ cm}^{-3}, \qquad (13b)$$

where m_v is the valence band effective mass. Further increases in donor concentration will cause both bands to shift together. Thus the reduction in the energy gap will reach a maximum as the donor concentration is increased.[69] This is consistent with the effects observed by Pankove in Ge.[70]

When acceptors are added, the critical concentration for the conduction band is exceeded first, and two effects will be present simultaneously. The first is a raising of the conduction band. The second is the introduction of discrete acceptor levels above the valence band. At first this will lead to an effective narrowing of the gap,[64] since the activation energy for acceptors is generally much larger than that for donors. When the acceptor concentration exceeds $N_{\min,v}$, both bands will again move together.

Finally, in perfectly compensated samples, our simple model suggests that the effects of donors and of acceptors on the conduction band approximately cancel. The donors have little effect on the valence band provided the carrier concentration is less than (13b), and the narrowing of the gap will be the result of the presence of discrete acceptor levels, as in p-type samples. This explanation differs from the one originally proposed.[63]

As a simple corollary of this discussion it follows for the lead compounds, if we take an effective mass of $0.2m$ and an estimated static dielectric constant of 70 in Eq. (13),[71] that there must be fewer than 7×10^{15} cm^{-3} donors or acceptors if the levels are to be separated from the band edge. If we consider the discussion of the chemistry of the lead compounds given in Sec. I, it is possible to understand why freeze-out of carriers has never been observed in these compounds at low temperatures.

VI. CONCLUSIONS

Measurements of the optical properties of the lead salts and the III–V semiconductors have contributed greatly to our knowledge of their electronic structure and lattice vibration spectrum. A powerful tool in analyzing optical measurements is the dispersion relation which allows optical constants to be determined from the values of reflectivity measured at normal incidence over a wide spectral range.

The plasma frequency associated with the free carriers in a semiconductor can vary over a rather wide spectral range, depending on the effective mass and the carrier concentration. This has been used to obtain information about effective masses, and can be used to study plasmas and their interaction with radiation and with charged particles.

ACKNOWLEDGMENTS

It is a pleasure to acknowledge my indebtedness to J. R. Dixon, J. M. Ellis, R. E. Morrison, H. R. Riedl, and W. W. Scanlon for making available unpublished results, for assistance in the preparation of this paper, and for many valuable discussions. I am grateful to R. S. Allgaier, R. F. Bis, R. F. Brebrick, D. P. Enright, E. Gubner, and B. B. Houston for making available most of the crystals used in this work, and for supplying information about their electrical and chemical properties.

[66] I am indebted to Dr. R. N. Hall for calling this omission in the earlier treatment to my attention.

[67] W. Kohn, in *Solid State Physics*, edited by F. Seitz and D. Turnbull (Academic Press Inc., New York, 1957), Vol. 5, p. 258.

[68] M. Lax and J. C. Phillips, Phys. Rev. **110**, 41 (1958).

[69] The Burstein effect, reference 62, must of course, still be taken into account.

[70] J. I. Pankove, Bull. Am. Phys. Soc. **6**, 303 (1961).

[71] E. Burstein and P. H. Egli, in *Advances in Electronics and Electron Physics*, edited by L. Marton (Academic Press Inc., New York, 1955), Vol. 7, p. 28.

JOURNAL OF APPLIED PHYSICS   SUPPLEMENT TO VOL. 32, NO. 10   OCTOBER, 1961

de Haas-van Alphen Effect in p-Type PbTe and n-Type PbS*

P. J. STILES,† E. BURSTEIN, AND D. N. LANGENBERG

University of Pennsylvania, Philadelphia, Pennsylvania

A study of the de Haas-van Alphen oscillations in the magnetic susceptibility has been carried out in p-type PbTe and n-type PbS. Measurements have been made on oriented single crystal samples with carrier concentrations of the order of $10^{18}/cm^3$ in pulsed magnetic fields up to 125 kgauss. The results for p-type PbTe indicate that the valence bands have a maximum at $\mathbf{k}=0$, and four equivalent maxima at the {111} Brillouin zone faces. The ⟨111⟩ ellipsoids have a longitudinal mass to transverse mass ratio of 6.4 ± 0.3. From the temperature dependence of the amplitude of the oscillations we obtain a value of $0.043\pm0.006\ m_0$ for the transverse effective mass. The data indicate that the (000) maximum and the (111) maxima lie within 0.002 ± 0.002 ev of each other. The data also indicate an effective broadening temperature of about $8°K$ which is attributed to inhomogeneities in the carrier concentrations in the samples investigated. Preliminary results on n-type PbS show a single isotropic Fermi surface cross section with a cyclotron mass of $0.14\pm0.04m_0$. The direct and indirect optical interband transitions are discussed in the light of these results.

I. INTRODUCTION

DETAILED studies of the Hall effect and magneto-resistance of the lead sulfide group of compounds have yielded valuable information about the energy band structure of these materials.[1] However, the transport properties involve combinations of effective mass and collision time parameters which cannot be unambiguously separated. In view of the high mobilities of the charge carriers at low temperatures, it was of interest to explore the possibility of carrying out various types of Landau level studies. As there is no appreciable freeze-out of carriers at liquid-helium temperatures, ordinary cyclotron resonance studies on bulk samples are not possible at microwave frequencies. However, the high mobilities of the carriers, even at relatively high carrier concentrations, do lend themselves to studies of Azbel-Kaner cyclotron resonance and de Hass-van Alphen (dHvA) and related oscillations. In fact, the available estimates of the effective masses and collision times indicate that the available materials behave, at low temperatures, very much like semimetals such as Bi.

We have observed well defined oscillations in the magnetic susceptibility in p-type PbTe and n-type PbS using pulsed magnetic fields.[2] (We have also observed comparable oscillations in preliminary work on Bi_2Te_3.) In the present paper we wish to report in detail the results of these observations. Oscillations in transport properties, e.g., magnetoresistance and Hall effect, have been reported previously in n-type PbTe[3] and other semiconductors. However, the measurements of susceptibility oscillations using pulsed fields have the advantage of greater simplicity because they do not require electrical contact and do not involve the current as an additional parameter.

Although our preliminary efforts to observe Azbel-Kaner cyclotron resonance in p-type PbTe were not successful,[2] we have recently observed resonances which we are investigating further.

In Sec. II, a brief description of the theory of the dHvA oscillations and their applications to semiconductors is given. Section III deals with the experimental technique and the method used for analyzing the data. The results for p-type PbTe and n-type PbS are discussed in Sec. IV and V, respectively. The last section contains our conclusions and some comments about the direct and indirect optical interband transitions.

II. THEORY

The dHvA effect has been reviewed in detail by several authors.[4,5] Only a brief description of the effect will be given here as an aid to understanding the experimental data.

When a magnetic field is applied to a system of charge carriers, the quasi-continuous distribution of states in momentum space collapses into a set of highly degenerate one-dimensional Landau sub-bands. It is found that, when the carrier concentration is high enough so that the system is degenerate, the magnetic susceptibility of the system has an oscillatory dependence on magnetic field. The period and amplitude of the susceptibility oscillations depend in a simple way on extremal cross-sectional areas and cyclotron effective masses at the Fermi surface. The effect thus provides a useful means of determining these properties.

In carrying out dHvA oscillation measurements with pulsed fields the signal is proportional to $dM/dt=$

* This work was supported in part by the U. S. Atomic Energy Commission and is a contribution from the Laboratory for Research on the Structure of Matter.

† Philco fellow.

[1] R. S. Allgaier and W. W. Scanlon, Phys. Rev. **111**, 1029 (1958); R. S. Allgaier, Phys. Rev. **112**, 828 (1958); R. S. Allgaier, Phys. Rev. **119**, 555 (1960); R. S. Allgaier, Czechoslov. J. Phys. (to be published); K. Shogenji and S. Uchiyama, J. Phys. Soc. Japan **12**, (1957); K. Shogenji, J. Phys. Soc. Japan **12**, (1957); K. Shogenji, J. Phys. Soc. Japan **14**, 1360 (1959).

[2] P. J. Stiles, E. Burstein, and D. N. Langenberg, Bull. Am. Phys. Soc. **6**, 115 (1961), and Phys. Rev. Letters **6**, No. 12 (1961).

[3] C. D. Kuglin, M. R. Ellett, and K. F. Cuff, Phys. Rev. Letters **6**, 177 (1961).

[4] D. Shoenberg in *Progress in Low Temperature Physics*, edited by J. C. Gorter (North-Holland Publishing Company, Amsterdam, 1957), Vol. 2, p. 226.

[5] R. G. Chambers, Can. J. Phys. **34**, 1395 (1956); A. B. Pippard, Repts. Progr. in Phys. **23**, 176 (1960).

$(dM/dH)(dH/dt)$. (dM/dH) is given by[6]

$$\frac{dM}{dH} = \frac{4VkT}{h^3}\left(\frac{ehH}{c}\right)^{\frac{3}{2}}\left(\frac{\partial^2 a}{\partial p_H{}^2}\right)_m^{-\frac{1}{2}}\left(\frac{2\pi cA}{ehH^2}\right)^2$$

$$\times \exp\left[-\frac{4\pi^3 ck}{eh}(T+T_B)m^*\right]$$

$$\times \cos\left(\frac{2\pi cA}{ehH}-2\pi\nu\mp\frac{\pi}{4}\right).$$

A is the extremal cross-sectional area of the Fermi surface perpendicular to the magnetic field and m^* is the cyclotron effective mass for the extremal cross section in question. T_B is an effective broadening temperature intended to take into account nonthermal broadening of the Landau levels, e.g., collision broadening. $(\partial^2 a/\partial P_H{}^2)_m$ is a dimensionless constant depending on the Fermi surface geometry, ν is a phase parameter, and V is the atomic volume. The other symbols have their usual connotations.

From the period of the oscillations in $1/H$, the extremal cross-sectional area of the Fermi surface perpendicular to the applied field can be determined. From the variation of the amplitude with magnetic field the product $m^*(T+T_B)$ can be determined, while m^* itself can be determined from the temperature dependence of the amplitude.

The necessary conditions for observing the dHvA effect in pulsed magnetic fields are contained in the requirement $E_F > k(T+T_B)$, where E_F is the Fermi energy. This imposes restrictions on the mobility, concentration range, and uniformity of carrier concentration throughout the sample. First, one must have a degenerate distribution of carriers. Second, the sum of the thermal energy kT and the energy uncertainty h/τ associated with the collision time should be small compared with the Fermi energy. Third, a practical consideration is that the variation of carrier concentration should be small enough so that the oscillations from each quasi-uniform portion of the crystal do not cancel out. Finally, eddy currents (which depend on the sample diameter and dH/dt) induced in the sample by the pulsed magnetic field should be small so as to avoid field inhomogeneities and sample heating.

III. EXPERIMENTAL METHOD

These experiments were done using the impulse method developed by Shoenberg.[4] In this method the sample is placed in a pickup coil and a large magnetic field is applied in a pulse a few milliseconds long. Oscillations in the susceptibility of the sample produce an oscillatory component in the voltage induced in the pickup coil. This oscillatory signal is amplified and displayed on an oscilloscope for analysis.

The samples were cylinders about 6 mm long and 3 mm in diameter. The penetration depth at the effective frequency of the applied magnetic field estimated from the sample conductivity is about ten times the radius of the sample so that eddy current shielding is small. The samples were cut with the cylindrical axes within 3° of a $\langle 100\rangle$, $\langle 110\rangle$, or $\langle 111\rangle$ direction. Since the lead salts cleave readily along $\{100\}$ planes, the crystals were aligned optically using light reflected from these planes.

The samples were inserted in a pickup coil wound on a nylon form. The samples were held in place in the nylon form with vacuum grease. Near the pickup coil was a bucking coil with enough turns so that the signal induced in it by the field pulse was slightly larger than the corresponding signal in the pickup coil. An adjustable fraction of the signal from the bucking coil was subtracted from the signal from the pickup coil, and the difference signal containing primarily the dHvA oscillations was fed to the vertical amplifier of a Tektronix 502 oscilloscope. A signal proportional to the magnetic field was taken from a 500-amp shunt connected in series with the magnetizing coil and was fed into the horizontal amplifier of the oscilloscope. The resulting X–Y plot was then photographed. The differential mode of this oscilloscope was used to permit addition of a dc voltage to the horizontal amplifier so that any section of the oscillations versus $B(=H)$ curve could be examined in greater detail.

In the latest measurements, a modified pickup and bucking coil system was constructed so that the sample and pickup coil could be rotated together in a predetermined plane by a simple pulley system. In this case, the main bucking coil was wound over the pickup coil and connected in series opposition with it in order to keep the torque on the rotating coil due to the field pulse at a minimum. Because this pickup coil was confined to a smaller volume and because some of the dHvA signal was lost in the bucking coil, the signal to noise was not as good as that obtained with the fixed-direction sample holder.

The determination of the extremal Fermi surface cross sections was carried out in the following way: From the photographs, the values of B at which the maxima and minima in dM/dH occurred were determined. Integers and half integers corresponding to successive maxima and minima were plotted versus the reciprocals of the fields at which they occurred. The slope of such a plot of the maxima and minima associated with a given cross section is proportional to the cross-sectional area.

For a given orientation of the magnetic field with respect to the crystal axes, the oscillations which appeared at the lowest magnetic fields corresponded to the smallest extremal cross section present. (For ellipsoidal energy surfaces, a small extremal cross section is associated with a small cyclotron mass, hence a relatively large dHvA amplitude.) Unambiguous data for

---

[6] W. J. Spry and P. M. Scherer, Phys. Rev. **120**, 826 (1960).

this cross section could be obtained from the field at which the signal emerged from the noise up to the field at which the larger extremal cross sections began to contribute signals with comparable amplitudes.

The unraveling of the experimental traces at the higher fields where several sets of oscillations were superimposed is a rather complicated task. If two sets of oscillations with comparable amplitude and not too different frequencies are superimposed, the resultant is an oscillation at the average of the two frequencies modulated at half the difference frequency (the beat frequency). If a long series of oscillations containing several beats can be observed, it is relatively easy to separate the two component oscillation series. However, in the present case the periods are large and often only a few oscillations corresponding to the same apparent period can be observed in the available field range. One is faced with a choice of either attempting to fit the observed traces in detail using two or three sets of periods, amplitudes, and phases as adjustable parameters, and taking into account the baseline curvature, or devising a more practical procedure. In the present work we have followed the second course. In dealing with the complex data obtained at higher fields, our procedure has been to choose groups of maxima and minima which are accurately periodic in $1/B$, i.e., give good straight lines when plotted versus $1/B$. The values quoted below for the cross sections were obtained from the slopes of these lines; the errors quoted are the standard deviations obtained by least square fitting straight lines to the points *chosen*. Where only one series of oscillations is present, e.g., at low fields, or where several series corresponding to quite different cross sections are superimposed, as at high fields with $\mathbf{B}\|$ $\langle 100 \rangle$, it is expected that the reliability of the quoted cross sections is indicated by the quoted errors. At fields and orientations where several not too different periods are present in the data, the derived cross sections will tend to be averages of the actual cross sections and their reliability should be judged in terms of the over-all model of the band structure suggested by the data.

The amplitude of a particular maximum is taken to be proportional to the vertical distance from the maximum to a straight line drawn through neighboring minima. Over the range of values of the magnetic field in which the amplitude could be measured, this approximation was calculated to be good to at least 2%.

The angle of rotation of the sample and pickup coil was determined by calibration of the rotation system at room temperature and then correcting for the contraction of the system. The rotation was checked further by observation of the movement of the pulley wire by a cathetometer as close to the lower pulley as possible. The two methods were quite consistent. When they differed, the values were averaged. The zero points of the angles θ and ϕ were approximately set before each experiment was begun and were finally determined by the symmetry of the results.

The temperature of the liquid-helium bath was determined by measuring the vapor pressure of the helium. We will discuss later some evidence indicating that the actual temperature of the carriers contributing to the dHvA effect may have been higher than the bath temperature in some of the experiments.

IV. RESULTS FOR p-TYPE PbTe

Our first results on p-type PbTe[2] were obtained on three samples with nominal hole concentrations of 3.0×10^{18} holes/cm³. We measured cross sections with the magnetic field in the $\langle 100 \rangle$, $\langle 110 \rangle$, and $\langle 111 \rangle$ directions. If, as previous workers[1] have reported, the valence band is composed of $\langle 111 \rangle$ ellipsoids of revolution, one should observe at most two different cross sections in each of the $\langle 110 \rangle$ and $\langle 111 \rangle$ directions and only one cross section in the $\langle 100 \rangle$ direction. In each direction we observed one additional large cross section. If the two smallest cross sections observed in the $\langle 110 \rangle$ direction are assumed to be associated with $\langle 111 \rangle$ ellipsoids, the ratio of these cross sections gives a value of the ellipsoid mass anisotropy ratio $K = m_l/m_t = 6.4$. A similar analysis of the data obtained in the $\langle 111 \rangle$ direction again yielded a value of $K = 6.4$, indicating that this is a plausible interpretation. The additional large cross sections measured in the three directions differed by about 25% among the measurements in the three directions.

We have arrived at a tentative conclusion about these large cross sections in the following way: If there are $\langle 111 \rangle$ ellipsoids with $K = 6.4$ located inside the Brillouin zone, about 4.9×10^{18} holes/cm³ in the ellipsoids would be required to account for the measured cross sections. If the band edge occurred at the $\{111\}$ Brillouin zone faces, only 2.4×10^{18} holes/cm³ would be required in the ellipsoids. Since the nominal hole concentration is 3.0×10^{18} holes/cm³ and some additional holes are required to explain the other measured cross sections, this strongly suggests four $\langle 111 \rangle$ ellipsoids rather than eight. A spherical section of the Fermi surface with a cross section equal to the average of the

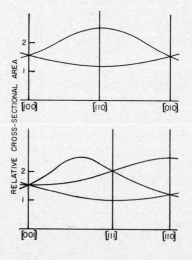

FIG. 1. The extremal cross-sectional areas of a set of $\langle 111 \rangle$ ellipsoids as a function of magnetic field orientation in the $\{001\}$ and $\{\bar{1}10\}$ planes, with $K = m_l/m_t = 6.4$.

larger measured cross sections would require 1.5×10^{18} holes/cm³. This suggests that these cross sections should be attributed to a single section of the Fermi surface, which by symmetry must be located at $\mathbf{k}=0$. This model consisting of four ellipsoids plus a section at $\mathbf{k}=0$ can thus account for all the measured cross sections with a total hole concentration of 3.9×10^{18} holes/cm³, in reasonable agreement with the nominal concentration.

In order to confirm that $\langle 111 \rangle$ ellipsoids were present, we performed experiments in which we examined the variation of the smallest cross section as the magnetic field was rotated in certain symmetry planes. Figure 1 shows the expected variation of cross-sectional area for $K=6.4$ in these symmetry planes. Figure 2 shows the variation of the smallest and largest cross sections as the magnetic field is rotated in a $\{110\}$ plane. Of the three

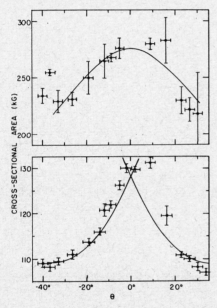

FIG. 2. The measured cross-sectional areas (in kG) for p-type PbTe which correspond to the largest and smallest ellipsoidal cross sections in a $\{110\}$ plane. The solid lines are the expected ellipsoidal cross sections for $K=6.4$ and $n=2.3 \times 10^{18}$ holes/cm³. θ is the angle between the direction of the magnetic field in the plane and a $\langle 110 \rangle$ direction in a $\{110\}$ plane.

possibilities, $\langle 111 \rangle$, $\langle 110 \rangle$, or $\langle 100 \rangle$ ellipsoids, only $\langle 111 \rangle$ ellipsoids have the smallest cross section in the $\langle 111 \rangle$ direction, as the data in this figure indicate. Thus the $\langle 111 \rangle$ ellipsoid model is confirmed. This was further checked by rotating the magnetic field in the $\{100\}$ plane. Results showing the proper symmetry are shown in Fig. 3.

The solid lines in Fig. 2 represent the largest and smallest ellipsoid cross sections expected for $n=2.3 \times 10^{18}$ holes/cm³ and $K=6.4$ The best fit was obtained with a value of $K=6.4 \pm 0.3$, in agreement with the value previously measured. The solid line in Fig. 3 is the expected value for the smallest cross section in a $\{100\}$ plane for $K=6.4$ and $n=3.9 \times 10^{18}$ holes/cm³. This is a rather poor fit. The dotted line represents the

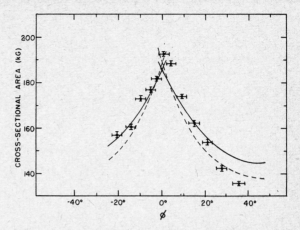

FIG. 3. The measured cross-sectional areas (in kG) for p-type PbTe which correspond to the smallest cross sections of the ellipsoids in a $\{100\}$ plane. The solid line is the expected cross-sectional area for $K=6.4$ and $n=3.9 \times 10^{18}$ holes/cm³. The dashed line is the expected cross-sectional area in a plane inclined from a $\{100\}$ plane $10°$ for $K=6.4$ and $n=4.0 \times 10^{18}$ holes/cm³. ϕ is the angle between the direction of the magnetic field in the plane and a $\langle 100 \rangle$ direction in the $\{100\}$ plane.

smallest cross section for a plane which is $10°$ off a $\{100\}$ plane, for $K=6.4$ and $n=4.0 \times 10^{18}$ holes/cm³. This indicates better agreement with the experimental results. The required misorientation is larger than expected but not unreasonable.

In order to estimate the difference in energy between the band edges at $\mathbf{k}=0$ and those at the $\{111\}$ Brillouin zone faces, we studied a sample with a lower carrier concentration. This sample was a $\langle 100 \rangle$ oriented sample with about 1×10^{18} holes/cm³. By comparing the ratio of the concentrations of holes in each band edge at these two total concentrations (assuming that the geometry of the Fermi surface was unchanged and that both bands were parabolic), we estimated that the band edge at $\mathbf{k}=0$ is higher in electron energy by 0.002 ± 0.002 ev. We also find from these calculations that the value of the effective mass of this band edge is $m^*=0.12 \pm 0.02 m_0$.

In some of our preliminary experiments we found a spurious dependence of mass on magnetic field. This we believe was due to heating of the lattice in a sample which was in poor thermal contact with the bath. In a range of magnetic field for which there appeared to be no heating, we obtained a value of $m_t=0.043 \pm 0.006\, m_0$. From the field dependence of the amplitude and this value of m_t, we determined that T_B is about $8°$K. This is probably due for the most part to variations in carrier concentration.

V. RESULTS FOR n-TYPE PbS

The data for n-type PbS were obtained on two cylindrically shaped crystals with the cylinder axes parallel to a $\langle 100 \rangle$ direction. They were cut from adjacent sections of the same boule with the cylinder axes perpendicular to the direction of growth of the boule. In each

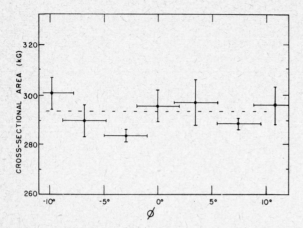

FIG. 4. The measured cross-sectional areas (in kG) for n-type PbS in a {100} plane. The dashed line is the mean. ϕ is the angle between the direction of the magnetic field in the plane and a $\langle 100 \rangle$ direction in the {100} plane.

case, the observed oscillations were associated with a single period.

The first sample yielded a value of 296 ± 3 kG for the cross section perpendicular to the $\langle 100 \rangle$ direction. A Hall measurement on this sample gave an electron concentration of $n = 3.0 \times 10^{18}$ holes/cm³. The second sample was then run in the rotating apparatus. The results are shown in Fig. 4. The mean cross section is 293 ± 5 kG. The effective mass determined from the temperature dependence of the amplitude is $m^* = 0.14 \pm 0.04$ m_0. The results incidate that the anisotropy in the observed section of the Fermi surface is small. It should be noted that Allgaier's[1] work on the magnetoresistance of n-type PbS suggests that there is no section of the Fermi surface which is highly anisotropic.

The number of electrons enclosed in a sphere whose cross section is equal to the mean measured cross section is 8.9×10^{17} electrons/cm³. A comparison with the measured concentration indicates that there is more than one section of the Fermi surface. Two interpretations are plausible: (a) The cross section observed is that of $\langle 111 \rangle$ ellipsoids with $K \sim 1$ and located at the Brillouin zone faces. This would require a total of 3.5×10^{18} electrons/cm³, which is reasonable. (b) There is a $\mathbf{k} = 0$ band edge corresponding to the cross section observed and also several nearly spherical ellipsoids not at $\mathbf{k} = 0$ whose cross sections were not observed.

VI. CONCLUSIONS

The results of the present investigation confirm the existence of the $\langle 111 \rangle$ ellipsoids in the valence band of PbTe previously proposed by other workers on the basis of transport studies. The results indicate that the ellipsoidal valence band maxima occur at the {111} Brillouin zone faces. Furthermore, they indicate the existence of a band edge at $\mathbf{k} = 0$ nearly degenerate in energy with the ellipsoidal band edges. The value obtained for the ellipsoid mass anisotropy, $K = 6.4$, is

about 50% larger than the value $K = 4.7$ derived by Allgaier[1] from magnetoresistance studies assuming an isotropic collision time. The difference between the two values may be due in part to the role played in magnetoresistance by the $\mathbf{k} = 0$ band maximum and in part to an anisotropic collision time.

The results for n-type PbS are still incomplete. However, the results do indicate that the Fermi surface in n-type PbS is composed of more than one section.

It is of interest to note that the existence of the $\mathbf{k} = 0$ valence band maximum nearly degenerate in energy with the (111) ellipsoidal band maxima in PbTe is consistent with some of the features of the room temperature optical data reported by Scanlon.[7] Scanlon has shown that the absorption edge of the PbS group of compounds involves both direct and indirect transitions. He derives values for the direct and indirect transition energies which differ by about 0.03 ev. When the phonon energy is taken into account, this yields a difference of about 0.01 ev between the direct and indirect energy gaps. This very small difference is within the experimental error and is consistent with the present model for the valence energy band structure of PbTe. Kuglin, Ellett, and Cuff[3] have reported ellipsoidal energy bands at the {111} Brillouin zone faces in n-type PbTe. Thus direct transitions are possible between $\langle 111 \rangle$ valence band and conduction band ellipsoids since they occur at the same values of $\mathbf{k}$, i.e., at the Brillouin zone {111} faces. Indirect transitions can take place either between the $\mathbf{k} = 0$ valence band edge and one of the (111) conduction band edges, which would involve $\mathbf{k} = 2\pi/a$ (111) phonons, or between one of the $\langle 111 \rangle$ valence band ellipsoids and a $\langle 111 \rangle$ conduction band ellipsoid lying at a different $\mathbf{k}$, which would involve $k = 2\pi/a$ (100) phonons. Indirect transitions can of course also take place, as suggested by Scanlon, between a (111) valence band edge and a (111) conduction band edge having the same $\mathbf{k}$ value which would involve $\mathbf{k} = 0$ optical phonons. The direct and indirect transitions involve essentially the same energy gaps. There will be, of course, small differences in the indirect transition energies due to the small differences in the energies of the phonons involved.

ACKNOWLEDGMENTS

The authors wish to express their appreciation to the research group at the Naval Ordnance Laboratory, in particular to R. S. Allgaier, W.W. Scanlon, and B. B. Houston, for many enlightening discussions and for supplying the crystals investigated. We also express our thanks to the group at Lockheed for the opportunity to exchange information and ideas. We thank B. N. Taylor for his work on the initial construction of the pulsed field facility. We also wish to express our appreciation to Dr. D. N. Stevens for his interest and encouragement in supporting this work.

[7] W. W. Scanlon, Solid State Phys. **9**, 83 (1959).

JOURNAL OF APPLIED PHYSICS   SUPPLEMENT TO VOL. 32, NO. 10   OCTOBER, 1961

Oscillatory Magnetoresistance in the Conduction Band of PbTe

K. F. Cuff, M. R. Ellett, and C. D. Kuglin

Lockheed Research Laboratory, Palo Alto, California

Oscillatory behavior of the transverse magnetoresistance has been used to investigate the conduction band structure of PbTe. From the oscillatory periods, it is established that the minima in the conduction band consist of four equivalent $\langle 111 \rangle$ prolate ellipsoids with a mass anisotropy of approximately 5.5 and located at the Brillouin zone edge. The decay of the oscillation amplitudes with temperature yields a transverse ellipsoid mass of $m_T^* = (0.030 \pm 0.005)m_0$. High field oscillations have been detected that point to the existence of a second band located at the center of the zone, having a mass of about 0.08 m_0, and lying within ± 0.002 ev of the ellipsoid minima. The phase of the oscillations suggests that spin splitting of the Landau levels may be quite large in PbTe.

I. INTRODUCTION

OSCILLATORY transport properties in metals and semiconductors are a direct result of the quantization of electron states in a magnetic field. In the presence of small magnetic fields, discrete electron states are not resolvable because of energy level broadening caused by the thermal motion of conduction electrons. However, at sufficiently high fields, the wave number **k** is no longer a good quantum number and the electrons occupy one-dimensional harmonic oscillator states due to their motion perpendicular to the magnetic field, while parallel to the field their behavior is unaffected by quantization. The electron states also have a spin degeneracy which is lifted by the magnetic field and under certain conditions the spin splitting of the oscillator or Landau levels is comparable to the level separation $\hbar\omega$. The energy, then, is the sum of these three contributions:

$$E = (n + \tfrac{1}{2})\hbar\omega + \frac{\hbar^2 k_z^2}{2m_z^*} \pm \tfrac{1}{2}\beta g_{\text{eff}} B \quad (n = 0, 1, 2, \cdots), \quad (1)$$

where ω is the cyclotron frequency, β is the Bohr magneton, and g_{eff} is the effective spectroscopic splitting factor.

The classical treatment of electrons becomes invalid and oscillatory phenomena should occur when the following conditions are satisfied:

$$\omega\tau \gg 1, \qquad (2a)$$

$$E_f > \hbar\omega, \qquad (2b)$$

$$\hbar\omega > kT, \qquad (2c)$$

where E_f is the Fermi energy and τ is the carrier relaxation time. If relation (2a) is to be satisfied for reasonable magnetic fields, the material used must have a very high mobility. Furthermore, it must remain degenerate down to low temperatures and the spacing of the Landau levels must be greater than their thermal broadening as given by (2c).

In a highly degenerate system, only a small fraction of the conduction electrons contribute directly to the transport processes, namely, those electrons for which $E \approx E_f$. Consequently, the electronic properties are quite sensitive to the density of states in the region near

the Fermi level. As the magnetic field in the system is increased, the level separation becomes greater and oscillations in the transport properties are produced as a result of the Landau levels passing through the Fermi surface.

Oscillatory susceptibility and transport properties have been extensively investigated for a number of metals and semi-metals, but few studies of semiconductors have been reported. Oscillatory magnetoresistance has been detected in n-type InSb[1–3] and both magnetoresistance[4,5] and Hall effect[5] oscillations have been observed in n-type InAs.

The low-temperature properties of the lead salt semiconductors (PbTe, PbSe, PbS) make them ideal substances for a study of the effects of electron quantization on the magnetic susceptibility and transport properties. Moreover, quantum effects in these materials should prove to be particularly interesting as they do not appear to have simple isotropic band structures. The lead salts are characterized by high mobilities at liquid-helium temperatures and furthermore, at low temperatures, there is no appreciable freeze-out of carriers so that a highly degenerate system can be achieved. These compounds also have fairly low effective masses so that the Landau level spacing can exceed thermal broadening for reasonable field strengths. In general, however, these crystals cannot be grown with a purity or homogeneity that will approach pure germanium or silicon. This carrier concentration inhomogeneity presents a serious drawback for oscillatory measurements as it effectively broadens the Fermi level and causes severe damping of the oscillations. Of the three lead salts, PbTe appears to be the most promising for the study of oscillatory transport properties as it has the highest mobility and probably the lowest effective mass. At 4°K, low doped crystals ($n \sim 10^{17}$/cm³) of n-type PbTe have mobilities as high as 3×10^6 cm²/v sec, while p-type samples of similar carrier concentration have mobilities as high as 4×10^5 cm²/v sec.

[1] H. P. R. Frederikse and W. R. Hosler, Phys. Rev. **108**, 1136 (1957).

[2] G. Busch, R. Kern, and B. Lüthi, Helv. Phys. Acta **30**, 471 (1957).

[3] Y. Kanai and W. Sasaki, J. Phys. Soc. Japan **11**, 1017 (1956).

[4] R. J. Sladek, Phys. Rev. **110**, 817 (1958).

[5] H. P. R. Frederikse and W. R. Hosler, Phys. Rev. **110**, 880 (1958).

Recently, oscillatory behavior has been observed in the susceptibility and in the transport properties of PbTe. Kanai et al.[6] have published results indicating oscillatory behavior in the transverse magnetoresistance in n-type PbTe and our preliminary work[7] in the same area has yielded information concerning the conduction band structure and electron effective mass. Stiles et al.[8] have investigated the valence band structure by de Haas-Van Alphen effect measurements, and distinct resistivity and Hall oscillations have been observed in p-type samples in this laboratory.[9] Comparison of the results for the conduction and valence bands indicates very similar structure.

II. SAMPLES

Single crystals of n-type lead telluride, approximately $1.5 \times 1.5 \times 8$ mm and each having its long dimension along either the $\langle 100 \rangle$ or $\langle 110 \rangle$ crystalline axes were used for the experiments. The samples were grown under a hard thermal gradient from a melt prepared with a 0.5% stoichiometric excess of lead. After being oriented by using the (100) cleavage faces as references and cut with a diamond saw, the samples were sealed in quartz tubing and annealed at 800°C for several hours. Following this anneal, the crystals were trimmed smooth with an air abrasive cutter and carefully inspected for imperfections. Preliminary measurements were then made, and the best specimens were selected and reannealed at lower temperatures in order to obtain greater homogeneity.

III. EXPERIMENTAL PROCEDURE

The crystals were mounted on a holder which allowed them to be rotated in a magnet capable of producing fields to 26 kgauss in a one-inch gap. The use of a dc preamplifier in conjunction with an X-Y recorder gave sufficient sensitivity to detect fractions of a microvolt. The amplified voltage taken across the potential probes was plotted on the Y axis of the recorder while the X axis was activated by a signal from a rotating coil gaussmeter.

Pairs of potential probes of 0.001-in.-diam Pt wire were attached about 1 mm apart to opposite sides of the specimen by means of a capacitor discharge. A number of current leads of 0.003-in. Pt wire were spot welded to each end of the crystal by the same technique.

Hall voltage (V_H) measurements were made using the "phantom probe" technique, i.e., a potential divider was connected across each set of potential probes and adjusted so that IR drops could be balanced out for both current directions. The data obtained from the

four combinations of current and field directions were averaged. The minimum value of the Hall voltage (when the directions of I and B are parallel) was used to check the alignment of the sample in the magnet gap.

The resistivity as a function of magnetic field $\rho(B)$ was monitored continuously on the X-Y recorder. Data were taken for all combinations of current and magnetic field directions on two sides of the sample and the results averaged. Above 1 kgauss, the magnetoresistance was nearly a linear function of the magnetic field with a small oscillatory component superimposed. This oscillatory part was obtained by biasing out the linear term with a signal from the rotating-coil gaussmeter while slowly sweeping the field and the resulting amplitudes were then replotted as a function of $1/B$. Since $\rho(B)$ was not quite proportional to B, the curves were often somewhat skewed; therefore, each curve was enveloped in order to obtain the appropriate amplitudes. The nodes were then determined from the intersection of the center line of the envelopes with the oscillatory curve, and the final curves (shown in Fig. 1 and 2) were plotted using the center line as the zero reference.

The samples were immersed in a liquid-helium bath at all times. The lowest temperatures were obtained by pumping directly on the helium and the temperature was determined from the helium vapor pressure. The oscillatory part of the magnetoresistance is a severely damped function of temperature and this, in conjunction with a carbon resistor placed next to the sample, was used to determine a safe maximum current for which there would be no appreciable heating of the sample.

IV. RESULTS AND INTERPRETATION

Measurements were made on three oriented single crystals having good homogeneity and high mobility. The orientation, carrier concentration, average zero field resistivity, and Hall mobility of each crystal is given

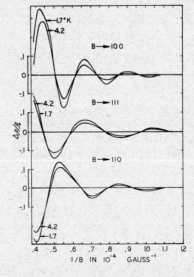

Fig. 1. Oscillatory component of the transverse magnetoresistance for sample 1 at 4.2 and 1.7°K for current along the [110] direction and for [100], [111], and [110] directions of the magnetic field. ρ_0 is the zero-field resistivity. This sample has an effective broadening temperature of 9°K.

[6] Y. Kanai, R. N. Nii, and W. Watanabe, J. Phys. Soc. Japan 15, 1717 (1960).
[7] C. D. Kuglin, M. R. Ellett, and K. F. Cuff, Phys. Rev. Letters 6, 177 (1961).
[8] P. J. Stiles, E. Burstein, and D. N. Langenberg, Bull. Am. Phys. Soc. 6, 115 (1956).
[9] K. F. Cuff, M. R. Ellett, and C. D. Kuglin (to be published).

in Table I. The carrier concentration n was obtained from the high field Hall coefficient $(R_\infty = 1/ne)$.

At helium temperatures and for the carrier concentrations used, lead telluride is a highly degenerate system. Therefore the Fermi energy, neglecting terms of order $(kT/E_f)^2$, can be written as

$$E_f = \frac{(\pi\hbar)^2}{2m_D^*}\left(\frac{3n}{\pi}\right)^{\frac{2}{3}}, \qquad (3)$$

where m_D^* is the density of states effective mass and n is the carrier concentration. Under the assumption that

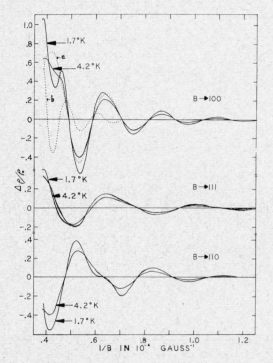

FIG. 2. Oscillatory component of the transverse magnetoresistance for sample 2 at 4.2 and 1.7°K for current along the [001] direction and for [100], [111], and [110] directions of the magnetic field. Curve a is the m_{100}^* oscillatory component extrapolated from the low field experimental curve. Curve b is the difference between the 1.7°K high field experimental curve and curve a. ρ_0 is the zero-field resistivity. The effective broadening temperature of sample 2 is only 6°K, and hence more structure is evident in the oscillations.

the band structure of PbTe can be approximated by a set of equivalent ellipsoids of revolution positioned in momentum space in a manner consistent with cubic symmetry, the density of states effective mass is defined as

$$m_D^* = (m_T^{*2}m_L^*)^{\frac{1}{3}}N_v^{\frac{2}{3}} = m_T^* N_v^{\frac{2}{3}} K^{\frac{1}{3}}, \qquad (4)$$

where m_L^* is the longitudinal mass in the direction along the axis of revolution and m_T^* is the mass perpendicular to the axis. The number of equivalent minima is given by N_v and $K = m_L^*/m_T^*$ is the mass anisotropy for a given ellipsoid.

Since the electronic properties of the system will de-

TABLE I. Sample orientation (long axis), high field Hall coefficient at 4.2°K, carrier concentration, zero field resistivity and mobility.

Sample	$i \to$	R_∞ (cm³/coul)	n (10^{17}/cm³)	ρ_0 ($10^{-6}\Omega$ cm)	$\left(\dfrac{\mu}{10^6 \text{cm}^2}\middle/ \text{v sec}\right)$
1	110	18.4	3.40	10.5	1.75
2	100	17.3	3.61	5.7	3.0
3	110	11.3	5.53	7.5	1.5

pend mainly upon the highest Landau level with respect to the Fermi surface, these phenomena will be quasi-periodic in the quantity $E_f/\hbar\omega$. Letting $E_f/\hbar\omega = \alpha/B$, the period in units of $1/B$ can then be expressed as α^{-1}. The cyclotron frequency ω is defined as eB/m_e^*, where m_e^* is the cyclotron effective mass. For an ellipsoid of revolution, the effective mass can be written as

$$m_e^* = m_T^*\left[\frac{K}{K\cos^2\theta + \sin^2\theta}\right]^{\frac{1}{2}}, \qquad (5)$$

where θ is the angle between the magnetic field and the major axis of the ellipsoid. Using (3), (4), and (5), α has the form

$$\alpha = \frac{\pi^2\hbar}{2e}\left(\frac{3n}{N_v\pi}\right)^{\frac{2}{3}}K^{1/6}[K\cos^2\theta + \sin^2\theta]^{-\frac{1}{2}}$$

$$= 31.4\left(\frac{n}{N_v}\right)^{\frac{2}{3}}f(K,\theta), \qquad (6)$$

for B in units of 10^4 gauss and n in units of 10^{18}/cm³.

Previous work[10,11] utilizing the anisotropy of weak field magnetoresistance in n-type PbTe has predicted the existence of a multivalley band structure having prolate energy ellipsoids $(K > 1)$ oriented along the $\langle 111 \rangle$ crystalline axes. Our data confirms these results and we find that there are four ellipsoids, thus locating the energy minima at the Brillouin zone edge.

Table II gives values of m_e^* and $f(K,\theta)$ for a $\langle 111 \rangle$ model with B along the principal crystalline axes. For

TABLE II. Expressions for m_e^* and $f(K)$ corresponding to a $\langle 111 \rangle$ model with magnetic field B along the principal crystalline axes.

B	Number of minima	m_e^*	$f(K)$
100	4 or 8	$m_T^*K^{\frac{1}{2}}[3/(K+2)]^{\frac{1}{2}}$	$K^{1/6}[3/(K+2)]^{\frac{1}{2}}$
110	2 or 4	$m_T^*K^{\frac{1}{2}}$	$K^{1/6}$
	2 or 4	$m_T^*K^{\frac{1}{2}}[3/(2k+1)]^{\frac{1}{2}}$	$K^{1/6}[3/(2K+1)]^{\frac{1}{2}}$
111	1 or 2	m_T^*	$K^{-\frac{1}{3}}$
	3 or 6	$m_T^*K^{\frac{1}{2}}[9/(K+8)]^{\frac{1}{2}}$	$K^{1/6}[9/(K+8)]^{\frac{1}{2}}$

[10] R. S. Allgaier, Czechoslov. J. Phys. (to be published).
[11] M. R. Ellett, K. F. Cuff, and C. D. Kuglin, Bull. Am. Phys. Soc. 6, 18 (1961).

TABLE III. Experimental α's obtained for magnetic field B along the principal crystalline axes, α_{100} calculated from the high field Hall coefficient using $K=5.5$, and the nonthermal damping factor β.

Sample	α_{100}	α_{110}	α_{111}	α_{100} calc		Nonthermal damping factor β in units of 10^4 gauss		
				4	8	$B\rightarrow100$	$B\rightarrow110$	$B\rightarrow111$
1	4.3	3.4	2.9	5.1	3.2	5.3	4.3	3.0
2	4.5	3.5	3.0	5.3	3.4	3.6	3.2	2.6
3	5.7	...	...	7.1	4.5	6.0	...	...

B along a $\langle 100 \rangle$ axis, a single period should be observed as the cross sections of all ellipsoids are equivalent. However, when the magnetic field is in either a $\langle 110 \rangle$ or a $\langle 111 \rangle$ direction, two cross sections exist. Depending upon whether the minima are at or inside the boundary of the first Brillouin zone, either four or eight ellipsoids are to be considered and the two possibilities are included in Table II. The oscillations become severely damped for increasing m_e^*, hence the low effective mass oscillations will predominate in the $\langle 110 \rangle$ and $\langle 111 \rangle$ cases for reasonably large K values. Using the periods corresponding to the predominant amplitudes, we then have for $K>1$: $\alpha_{100}>\alpha_{110}>\alpha_{111}$.

Table III summarizes the experimental α's and the α's calculated from (6) using the most probable anisotropy, i.e., $K=5.5$. Since the α's correctly satisfy the inequality of the preceding paragraph and a single period is observed in the [100] case, a $\langle 111 \rangle$ model is established. Further experimental evidence for the $\langle 111 \rangle$ model was obtained by rotating the sample with B in the (110) plane and noting the change in α. The minimum value of α occurred when the magnetic field coincided with a $\langle 111 \rangle$ crystalline direction as expected for this model. The ratios of the α's are very insensitive functions of K; however, for our maximum relative error in the α's a value of

$$K=5.5 \begin{array}{c} +2.5 \\ -1.5 \end{array}$$

is determined. A more accurate determination of K could be obtained through the direct observation of the high mass period for B in the [110] or [111] direction. However, we do not have sufficiently high field strengths available to clearly resolve these oscillations, and the best that can be said is that an analysis of the oscillatory curves implies a higher mass period consistent with the above determined anisotropy. From weak field magnetoresistance measurements, a parameter $K'=K\tau_T/\tau_L$ can be determined. At high temperatures where ionized impurity scattering can be neglected, the scattering should be fairly isotropic and $K'\approx K$. A value of K' of 4.0 ± 0.5 is obtained at 300°K, in fair agreement with the K determined from oscillatory measurements.

For all three samples, a discrepancy exists between the experimental period and the period as calculated from the carrier concentration. We find that

$\alpha_{N_v=8}<\alpha_{\exp}<\alpha_{N_v=4}$, that is, the number of carriers as determined from the high field Hall coefficient is not sufficient to account for an eight valley model while the number is too great for a four valley model. In order to agree with an eight valley model, a K close to unity would be required, and this is far outside our experimental error. However, considering the error in R_∞ and the α's, a K of 12 ± 3 would account for all the carriers in the four $\langle 111 \rangle$ minima. The upper limit for K determined from the ratio of the α's is close to this value; nevertheless, the most probable value of K indicates that a significant number of carriers are unaccounted for, thus indicating the presence of another minimum. The existence of this additional minimum is strongly suggested by the results of sample 2 where a second set of oscillations with an $\alpha_{100}=8.6\pm0.5$ is observed at 1.7°K. This second set of oscillations is clearly discernable at high fields in the [100] direction, but is obscured by the presence of higher mass ellipsoid oscillations in the [110] and [111] directions; see Fig. 2.

From the experimental α's, the number of carriers in the ellipsoids is found to be $2.8\pm0.2\times10^{17}$. If the number of carriers corresponding to the second set of oscillations is computed on the basis of a single, isotropic minimum, a value of $1.4\pm0.2\times10^{17}$ is obtained and thus the total carrier concentration as determined from the oscillatory periods is $4.2\pm0.4\times10^{17}$. The total number of carriers obtained from the high field Hall coefficient is $3.6\pm0.2\times10^{17}$. If the second set of oscillations which we observe were due to a set of equivalent minima N_v, the number of carriers would be increased by this factor and therefore the carriers accounted for by the oscillations would be far greater than the number determined from the Hall coefficient. Thus, the minimum must occur at $\mathbf{k}=0$ with some undetermined degree of warping.

Adams and Holstein[12] have investigated the conductivity for transverse magnetic fields in a system having spherical energy surfaces and for which $\omega\tau\gg1$. Using their expression for the oscillatory part of the conductivity in the case of lattice scattering, the periodic term in the transverse resistivity correct to second order in the scattering term can be written as

$$\frac{\Delta\rho_{0S}}{\rho_0}=\frac{5x}{\sqrt{2}}\left(\frac{\hbar\omega}{E_f}\right)^{\frac{1}{2}}\sum_{M=1}^{\infty}\frac{(-1)^M M^{\frac{1}{2}}e^{-2\pi M\gamma}}{\sinh(Mx)}$$

$$\times\cos\left(\frac{2\pi M E_f}{\hbar\omega}-\frac{\pi}{4}\right), \quad (7)$$

where ρ_0 is the zero field resistivity, $\gamma=(\omega\tau_c)^{-1}$, and $x=2\pi^2 kT/(\hbar\omega)$. In this formulation, τ_c is the collision broadening cut-off time and should be about the same order of magnitude as the zero field mean relaxation time. Adams and Holstein also include another oscil-

[12] E. N. Adams and T. O. Holstein, J. Phys. Chem. Solids 10, 254 (1959).

latory term in the conductivity which differs from (7) mainly by a $\pi/4$ phase shift in the argument of the cosine and an additional factor of $(\hbar\omega/E_f)^{\frac{1}{2}}$ in the amplitude. They state that this term should be dominant in the case of only a few oscillator levels and for oscillations greater than the smooth resistance, while the above expression is appropriate for small amplitude oscillations due to a number of quantum levels.

Although (7) was derived for isotropic energy surfaces, it has general features which are applicable in the case of ellipsoids. More specifically, the thermal damping term and the period are correct if $\hbar\omega$ is taken to be the correct cyclotron energy for the particular ellipsoidal cross section in question. Also, the amplitude of the oscillations for the ellipsoids should be similar to those predicted for spherical surfaces. Furthermore, Adams and Holstein have computed the relative amplitudes and phases of $\Delta\rho_{0S}/\rho_0$ for ionized impurity as well as lattice scattering and find essentially no difference in the result for these very dissimilar mechanisms. Thus, it is reasonable to assume that the lack of a detailed knowledge of the type of scattering in PbTe at helium temperatures should not introduce any serious error in the subsequent discussion of the observed amplitudes and phases.

Due to the high mobility of electrons in PbTe, resulting in $\omega\tau$'s of the order of several hundred for the range of magnetic fields used, collision broadening should only contribute a negligible amount of damping (Dingle temperature of about $0.1°K$) to the oscillatory behavior of the magnetoresistance. However, appreciable attenuation will result from both thermal and inhomogeneity broadening.

The largest damping effects encountered appear to be due to inhomogeneity broadening. If there are variations of the carrier concentration in the region between the potential probes, the Fermi surface will have a spatial variation and thus the oscillatory period will be a fluctuating function of position. In order to roughly account for this behavior, (7) must be averaged over a distribution function appropriate to the statistical variation in the carrier concentration. The exact form of this variation is not known but the experimental results indicate a simple exponential damping factor in $1/B$.

Taking the values of the effective mass from Table IV, which will be discussed later, (7) can be plotted as

$$\ln\left[\frac{\Delta\rho_{0S}}{\rho_0 B^{\frac{1}{2}}}\frac{\sinh x}{x}\right] \text{ vs } 1/B$$

in order to ascertain the form and magnitude of the nonthermal damping (see Table III for representative values). The slope was found to be approximately linear so that the corresponding term in the summation of (7) would be of the form $\exp(\beta M/B)$, where β is some constant and M the index of summation. If it is assumed that this damping is due totally to inhomogeneity broadening, we might expect that $\beta/B \propto \Delta E_f/\hbar\omega$, where ΔE_f

TABLE IV. Determination of the $\langle 100 \rangle$ and $\langle 111 \rangle$ effective masses from the temperature dependence of the oscillation amplitudes.

Sample	1		2		3	
	m_{100}^*	$1/B$	m_{100}^*	$1/B$	m_{100}^*	$1/B$
	0.040	0.45	0.035	0.54	0.052	0.42
	0.047	0.56	0.039	0.64	0.050	0.50
$B\to100$	0.044	0.67	0.036	0.76	0.053	0.60
	0.046	0.79	0.040	0.88	0.050	0.69
	0.043	0.91	0.036	0.98	0.048	0.77
	$\bar{m}_{100}^*=0.044$		$\bar{m}_{100}^*=0.037$		$\bar{m}_{100}^*=0.051$	
	m_{111}^*	$1/B$	m_{100}^*	$1/B$		
	0.036	0.51	0.033	0.52		
$B\to111$	0.032	0.69	0.031	0.68		
	0.031	0.87	0.025	0.86		
	0.026	1.05	0.027	1.03		
	$\bar{m}_{111}^*=0.031$		$\bar{m}_{111}^*=0.029$			

represents some effective broadening of the Fermi level due to inhomogeneities. The observed damping would then correspond to a 5–10% inhomogeneity. The changes in β with orientation are roughly consistent with the change in m_e^*. The β's correspond to a broadening temperature of 6–9°K for the samples measured.

When measurements were repeated on the crystals with potential probes in new locations, it was found that the damping factor often changed somewhat, yet the effective masses and periods as obtained using the previously described procedure were invariably the same within experimental error. This lends support to the view that inhomogeneity broadening is a significant damping factor. The thermal damping, caused by the inherent broadening of the electron energy levels due to the thermal motion of the electrons, has been used to determine the effective masses. This was done by taking the ratios of the oscillatory amplitudes for two different temperatures T and T' at a given magnetic field B. The effective mass is thus implicitly expressed as

$$\Delta\rho_{0S}(T)/\Delta\rho_{0S}(T') = x \sinh x'/x'\sinh x. \qquad (8)$$

Values for m_e^* were obtained by comparing amplitudes at 4.2° and 1.7°K taken half way between nodes and the results are given in Table IV. For the lowest carrier concentration samples, $m_{100}^* = (0.041\pm0.005)m_0$ and $m_{111}^* = m_T^* = (0.030\pm0.005)m_0$. The [100] effective mass for sample 3 is $(0.050\pm0.008)m_0$. Since sample 3 has a higher carrier concentration than 1 or 2, we might expect a small increase in the mass due to nonparabolic effects, although this value is less reliable than the others because the observed amplitudes are smaller.

The effective mass m_c^* of the $\langle 000 \rangle$ minimum could not be accurately determined by comparing amplitudes at 4.2 and 1.7°K as the oscillations were only a small perturbation on the 4.2°K curve. However, by using the [100] damping factors β_{100} and the known value for the ellipsoid [100] effective mass $(m_{100}^*)_e$ we have

$$(m_{100}^*)_c = (\beta_{100})_c/(\beta_{100})_e(m_{100}^*)_e \approx 0.08m_0,$$

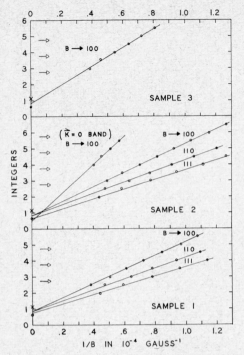

FIG. 3. The nodes of the oscillations are plotted at half-integer values vs $1/B$ for three magnetic field directions for samples 1, 2, and 3. The horizontal arrows indicate the positions of the resistivity maxima. The crosses on the vertical axes indicate the intercept expected for a phase of $-\pi/4$ (zero spin splitting of the Landau levels) while the circles give the intercept for a phase of $3\pi/4$ (spin splitting equal to the Landau level separation).

where we have used the fact that the β's are proportional to the cyclotron masses, and that $(\beta_{100})_c/(\beta_{100})_e = 2.0 \pm 0.3$. This value of $(m_{100}^*)_c$ is consistent with the large change in amplitude between 4.2 and 1.7°K.

There are several arguments which indicate that the $\langle 000 \rangle$ minimum lies close to the energy of the $\langle 111 \rangle$ minima. Using the ratios of α's and masses obtained for sample 2 with B in the [100] direction,

$$(E_f)_c/(E_f)_e = m_e^* \alpha_c / m_c^* \alpha_e = 0.95 \pm 0.2.$$

Hence, the Fermi level resides at approximately the same height with respect to the bottom of both bands. We can also form a ratio which gives the relative percentage of ellipsoid carriers for samples of different carrier concentration. Comparing the ratio (n_e/n_t) for sample 3 to the corresponding ratios for either samples 1 or 2,

$$(n_e/n_t)_3/(n_e/n_t)_{1,2} = 0.95 \pm 0.1,$$

where n_e is the number of carriers in the ellipsoids as determined from α_{100} and n_t is the total number of carriers given by R_∞. The percentage of carriers in the ellipsoids appears to remain practically independent of carrier concentration (and Fermi level), hence the ellipsoid and central minima must lie close to the same energy. Both of the previous arguments are consistent with the $\langle 000 \rangle$ minimum lying slightly above the $\langle 111 \rangle$ minima; however, because of experimental error we can

only conclude that the two bands lie within 0.002 ev of each other.

For the interpretation of the oscillatory periods and effective masses, it was assumed that our experimental curves are closely approximated by the first term in the summation of (7). The error accrued by neglecting higher terms is small, and although the experimental curves are somewhat skewed and sharpened at the higher field values, the curves become increasingly sinusoidal for lower magnetic fields. In Fig. 3, the nodes obtained from the oscillatory data are plotted at half-integer points versus $1/B$. The phase shifts cannot be obtained with great accuracy because the high field nodes are somewhat displaced due to higher harmonics and in addition the low field nodes cannot be obtained reliably because of weak amplitudes. However, to within experimental error it is found that the phase shift does not agree with the $-\pi/4$ value predicted by (7). The phases of the [100] and [110] periods are quite similar for all three crystals and are shifted about $120° \pm 30°$ from the phase given by Adams and Holstein. The [111] phases appear to be somewhat different and are shifted approximately $145° \pm 30°$. Since the $\langle 000 \rangle$ oscillations were obtained only roughly by graphical analysis, the phase cannot be given with any certainty.

It is possible to explain phase shifts of this nature if there is a large splitting of the Landau levels due to spin. Cohen and Blount[13] show that a term of the form $\cos(M\pi m_e^*/m_s^*)$ must be included in the summation of (7) in order to account for spin. Here, m_e^* is the orbital or cyclotron effective mass and m_s^* is a spin effective mass defined as: $m_s^*/m_0 = 2/g$. For small band gaps, large spin orbit coupling, and electrons residing in minima at points of low symmetry, Cohen and Blount have shown that $m_e^* \approx m_s^*$, i.e., the splitting of the spin and orbital levels is the same. Thus, under these conditions, the phase shift would be 180° from that predicted by (7). Since PbTe has a small band gap (0.2 ev at 4°K) and is almost certain to have a large (>1 ev) spin orbit interaction, it is quite possible that spin splitting is responsible for the phase shifts obtained from the ellipsoid oscillations.

The amplitudes of the transverse oscillations agree with the magnitudes predicted by Adams and Holstein to within a factor of two when the inhomogeneity damping is properly accounted for. Longitudinal oscillations were observed in samples 1 and 2 which agreed with the transverse behavior in regards to both period and phase, but are approximately $\frac{1}{8}$ the amplitude. The treatment of longitudinal conductivity by Argyres[14] predicts amplitudes a factor of five smaller than those given by (7).

It might be remarked that the behavior attributed to $\langle 000 \rangle$ oscillations cannot be adequately accounted for by the second harmonic in (7). The sum total effect

[13] M. H. Cohen and E. I. Blount, Phil. Mag. **5**, 115 (1960).
[14] P. Argyres, J. Phys. Chem. Solids **4**, 19 (1958).

of higher harmonics is to sharpen the oscillations and hence the higher harmonics are not individually distinguishable. Moreover, the amplitudes of the ⟨000⟩ oscillations are about a factor of three higher than is expected for a second harmonic. Although the assignment of the high field oscillations to a ⟨000⟩ minimum has been definitely suggested from several points of view, the evidence is, nevertheless, incomplete and hence the ⟨000⟩ minimum must necessarily be considered as a tentative conclusion.

ACKNOWLEDGMENT

We wish to thank V. J. King of this laboratory for the growth and preparation of the PbTe crystals.

JOURNAL OF APPLIED PHYSICS   SUPPLEMENT TO VOL. 32, NO. 10   OCTOBER, 1961

Valence Bands in Lead Telluride

R. S. ALLGAIER

U. S. Naval Ordnance Laboratory, White Oak, Silver Spring, Maryland

The magnetic field dependence of the Hall coefficient at 296° and 77°K, and the temperature dependence of the weak-field Hall coefficient and the resistivity between 296° and 77°K were studied in single-crystal samples of p-type PbTe having carrier concentrations ranging from 4.9×10^{17} to 1.7×10^{19} per cm³. The Hall data at 77°K are quantitatively consistent with magnetoresistance data which have previously established the presence of ⟨111⟩ ellipsoids in the valence band. They are not consistent with a low-temperature two-band model, proposed by Stiles from de Haas-van Alphen data at 4.2°K, unless the band edges lie at approximately the same energy (as Stiles found) and unless the carrier mobilities in the two bands are nearly alike. On the other hand, both the Hall and resistivity data above about 150°K do exhibit two-carrier effects suggesting the presence of a lower mobility band at an energy about 0.1 ev below those bands which are occupied at low temperatures.

INTRODUCTION

WE present in this paper measurements on several single-crystal samples of p-type PbTe having carrier concentrations from 4.85×10^{17} to 1.67×10^{19} per cm³. All carrier concentrations were calculated from the formula $p = 1/R_0 e$ using the value of the weak-field Hall coefficient R_0 measured at 77°K. We studied the Hall coefficient as a function of magnetic field intensity at 296° and 77°K, covering a range of $\mu_H H/C$ values from 0.001 to 1.5 (μ_H = Hall mobility in cm²/v-sec, H = magnetic field intensity in gauss, $C = 10^8$), and we studied the weak-field Hall coefficient and the resistivity as a function of temperature between 296° and 77°K. Our main purpose was to determine whether or not there is more than one occupied valence band in PbTe.

Studies of the cubically-symmetric lead-salt semiconductors PbS, PbSe, and PbTe have generally shown that all three compounds have similar properties.[1] It is quite well established that the carrier mobilities are very nearly proportional to $T^{-\frac{5}{2}}$ over a wide temperature range and that the extrinsic Hall coefficients are independent of temperature down to 4.2°K, even for carrier concentrations as low as 10^{16} per cm³.[1,2] It now appears, however, that the bands in PbTe differ from those in PbS and PbSe, and furthermore that more than one valence band may be contributing to the transport properties of p-type PbTe.

MAGNETORESISTANCE IN THE LEAD SALTS

Several papers have reported that in n- and p-type PbTe, the weak-field magnetoresistance coefficients b, c, and d obey the symmetry conditions ($b+c=0$, $d>0$) appropriate for a ⟨111⟩ ellipsoid-of-revolution multivalley model.[3–10] Our data on p-type PbTe with 3.5×10^{18} carriers per cm³, obtained on pulled single crystals grown by B. B. Houston, Jr. and R. F. Bis, precisely obeyed these conditions at 296° and 77°K.[7] By "precisely" we mean that whereas in the expression $b+c+xd=0$, x should equal 1, 0, or -1 for the ⟨100⟩, ⟨111⟩, and ⟨110⟩ multivalley models, respectively, our experimental values were $|x| \leq 0.03$. We recently found equally precise agreement with this model in a p-type sample with 5.0×10^{17} carriers per cm³. We have also found $|x| \leq 0.05$ at 296° and 77°K in an n-type crystal with 9.1×10^{17} carriers per cm³.[10]

These magnetoresistance results offer strong evidence of the presence of ⟨111⟩ ellipsoids in both the conduction and valence bands of PbTe. This type of energy surface

[1] For references through 1958, see W. W. Scanlon in *Solid State Physics*, edited by F. Seitz and D. Turnbull (Academic Press, Inc., New York, 1959), Vol. 9.

[2] R. H. Jones, Proc. Phys. Soc. (London) **76**, 783 (1960).

[3] K. Shogenji and S. Uchiyama, J. Phys. Soc. Japan **12**, 1164 (1957).

[4] R. S. Allgaier, Phys. Rev. **112**, 828 (1958).

[5] K. Shogenji, J. Phys. Soc. Japan **14**, 1360 (1959).

[6] H. N. Leifer, M. R. Ellett, K. F. Cuff, and R. S. Krogstad, Bull. Am. Phys. Soc. **4**, 362 (1959).

[7] R. S. Allgaier, Phys. Rev. **119**, 554 (1960).

[8] M. R. Ellett, K. F. Cuff, and C. D. Kuglin, Bull. Am. Phys. Soc. **6**, 18 (1961).

[9] C. D. Kuglin, M. R. Ellett, and K. F. Cuff, Phys. Rev. Letters **6**, 177 (1961).

[10] R. S. Allgaier, *Proceedings of the International Conference on Semiconductor Physics* (Publishing House of the Czechoslovak Academy of Sciences, Prague, 1961), p. 1037.

is the only one known which requires the conditions $b+c=0$ and $d>0$. However, the symmetry conditions for different kinds of energy surfaces are in a sense additive, and if carriers also occupied another set of ⟨111⟩ ellipsoids or a spherical energy surface (for which $b+c=0$, $d=0$), the observed symmetry condition would still be appropriate.

On the other hand, a smaller and more isotropic magnetoresistance which is almost the same in n- and p-type PbS and PbSe suggests the presence of approximately spherical energy surfaces in those bands.[7,10]

PREVIOUS EVIDENCE FOR A TWO-BAND MODEL AT ROOM TEMPERATURE

We have found that the Hall coefficient in a large number of samples of p-type PbTe is about 40% higher at 296°K than at 77°K,[10,11] whereas in n-type PbTe and in n- and p-type PbS and PbSe a 5–10% increase is commonly observed which can be ascribed to a change from degenerate to classical statistics.[12] This large increase in p-type PbTe samples suggested that, with increasing temperature, carriers were being transferred into a second valence band,[10] and upon re-examination of our previously published data on p-type PbTe,[12] we noted what might be interpreted as a droop in the conductivity versus temperature curves (relative to a $T^{-\frac{5}{2}}$ extrapolation) beginning at the same temperature

(~150°K) at which the Hall coefficient increase becomes noticeable. Thus it appeared that this second band had a lower mobility.

A LOW-TEMPERATURE MODEL FOR p-TYPE PbTe

Stiles, Burstein, and Langenberg have recently observed 30 or more de Haas-van Alphen oscillations in the magnetic susceptibility of p-type PbTe at 4.2°K,[13] using pulled crystals grown by Houston and Bis. They found that $\frac{2}{3}$ of the carriers are in four ellipsoids and $\frac{1}{3}$ are in an approximately spherical energy surface. From measurements on crystals with different carrier concentrations they concluded that the energy separation of the spherical and ellipsoidal band edges is quite small, perhaps only 0.002 ev.

EXPERIMENTAL

Our electrical measurements were made using conventional dc techniques. The samples used were obtained from pulled crystals with 3.5×10^{18} carriers per cm³ grown by Houston and Bis. Higher carrier concentrations were obtained from samples which had been prepared for a phase diagram study,[11] and in one case from a Na-doped pulled crystal. Lower carrier concentrations were achieved by a low-temperature vacuum anneal,

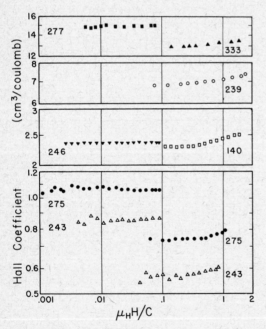

FIG. 1. Log-log plot of Hall coefficient vs mobility-magnetic field strength product $\mu_H H/C$ ($C=10^8$ when μ is in cm²/v-sec and H is in gauss) in p-type PbTe. Sample numbers identify data. Curves in the $\mu_H H/C$ range from 0.001 to 0.1 were obtained at 296°K, those from 0.1 to 1.5 at 77°K.

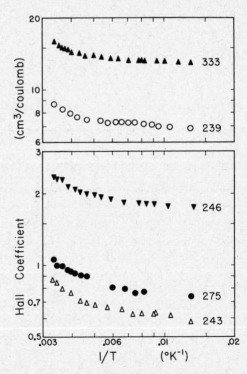

FIG. 2. Log-log plot of weak-field Hall coefficient vs reciprocal temperature in p-type PbTe from 296° to 77°K. Sample numbers identify data.

[11] R. F. Brebrick and R. S. Allgaier, J. Chem. Phys. **32**, 1826 (1960).

[12] R. S. Allgaier and W. W. Scanlon, Phys. Rev. **111**, 1029 (1958).

[13] P. J. Stiles, E. Burstein, and D. N. Langenberg, Bull. Am. Phys. Soc. **6**, 115 (1961). I am indebted to Mr. Stiles for communicating further details of this work as it progressed.

TABLE I. Characteristics of p-type PbTe Samples at 296° and 77°K.

Sample No.	p(cm^{-3}) $(1/R_{77°}e)$	R_0(cm³/coul) 296°K	R_0(cm³/coul) 77°K	$\left(\dfrac{R_0(296°K)}{R_0(77°K)}\right)$	$\mu_H(296°K)$ (cm²/v-sec)	$R_0(77°K)$ $\times \sigma(296°K)$ (cm²/v-sec)	$\mu_H(77°K)$ (cm²/v-sec)	$\left(\dfrac{R(\mu_H H/C=1)}{R_0}\right)$ 77°K
333	4.85×10^{17}	15.7	12.9	1.22	864	708	22 700[a]	1.03
277	4.96×10^{17}	15.1	12.6	1.20				
239	9.15×10^{17}	8.71	6.84	1.27	855	673	22 800	1.04
140	2.72×10^{18}	3.10	2.30	1.35	885	655	15 600	1.05
246	3.57×10^{18}	2.35	1.75	1.35	886	656	15 900	
275	8.46×10^{18}	1.05	0.740	1.42	861	606	9630	1.05
243	1.10×10^{19}	0.850	0.570	1.49	887	595	8040	1.07
334	1.67×10^{19}	0.593	0.374	1.59	871	549	6250	

[a] Estimated from other samples of like carrier concentration.

using a relationship between carrier concentration and annealing temperature determined by Scanlon.[14]

RESULTS

Pertinent characteristics of the samples investigated are given in Table I. The results are plotted in Figs. 1–3, all of which use logarithmic scales. The data are identified in the figures by sample number. Figure 1 shows the Hall coefficient versus $\mu_H H/C$ for seven samples of p-type PbTe. The points in the $\mu_H H/C$ range from 0.001 to 0.1 were taken at 296°K, those from 0.1 to about 1.5, at 77°K. Figure 2 shows the Hall coefficient versus $1/T$ from 296° to 77°K in five p-type samples, and Fig. 3 plots a "modified mobility" (the modification will be discussed below) versus $1/T$ from 296° to 77°K for five p-type and two n-type samples of PbTe.

DISCUSSION

A. Magnetic Field Dependence of the Hall Coefficient

Despite some scatter in the region of the smaller Hall voltages, it seems clear from Fig. 1 that the Hall coefficient does not depend on the magnetic field below about $\mu_H H/C = 0.1$ for all the carrier concentrations studied. Above this value, the Hall coefficient gradually rises, and at $\mu_H H/C = 1$ it is 3% (for 4.9×10^{17} carriers per cm³) to 7% (for 1.1×10^{19} carriers per cm³) higher than the weak-field value. The weak-field Hall coefficient for any cubically-symmetric ellipsoid-of-revolution model is[15]

$$R_0 = \frac{r}{ne}\frac{3K(K+2)}{(2K+1)^2} = \frac{r}{ne}f(K), \qquad (1)$$

where r is the well-known combination of scattering integrals, and $K = m_l/m_t$, the ratio of longitudinal to transverse mass. The strong-field limit is simply $R_\infty = 1/ne$. The anisotropy factor $f(K)$ in Eq. (1) is about 0.88 for $K = 4.2$ (the value derived from magneto-resistance data at 77°K[7]) while degenerate statistics

brings the factor r to within a few percent of unity for the samples we studied. The theory thus predicts an increase of about 10% in the Hall coefficient in going from weak to strong fields. Our data do not reach the strong-field limit, so we use instead an expression appropriate for the ⟨111⟩ model at any field strength when the statistics are completely degenerate and the current and field are parallel to cubic axes,[16]

$$\frac{R_H}{R_0} = \frac{1+[(2K+1)^2/3K(K+2)](\mu_H H/C)^2}{1+(\mu_H H/C)^2}. \qquad (2)$$

For $K = 4.2$ and $\mu_H H/C = 1$, $R_H/R_0 = 1.065$, which is in very good agreement with the ratio 1.07 observed in the sample with the highest carrier concentration for which $R(H)$ was measured, (see Table I), and which very closely approaches the approximation in Eq. (2) of completely degenerate statistics. As the carrier con-

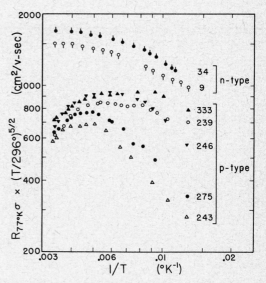

FIG. 3. Log-log plot of the product of the Hall coefficient at 77°K, the conductivity at temperature T, and the factor $(T/296°)^{5/2}$ vs reciprocal temperature for n- and p-type PbTe from 296° to 77°K. Data for n-type PbTe from reference 12. Sample numbers identify data.

[14] W. W. Scanlon (private communication).
[15] C. Herring, Bell System Tech. J. **34**, 237 (1955).
[16] L. Gold and L. M. Roth, Phys. Rev. **107**, 358 (1957).

centration is reduced and the statistics become less degenerate, the weak-field value will increase relative to R_∞ (which remains equal to $1/ne$) because the factor r in Eq. (1) increases, and hence, as actually observed, R_H/R_0 decreases.

We have dwelt at length on the agreement of the Hall data at 77°K with a $\langle 111 \rangle$ model in order to emphasize the point that if two bands were present and each were occupied by a substantial fraction of the total carriers (this situation is not likely to change much between 4.2° and 77°K), the Hall data would in general be quite different. The ratio of the weak- to strong-field field Hall coefficient for a simple two-carrier model with degenerate statistics is

$$\frac{R_0}{R_\infty} = \frac{f_a\mu_a{}^2 + f_b\mu_b{}^2}{(f_a\mu_a + f_b\mu_b)^2},\qquad(3)$$

where the f's and μ's are the fractions of the total carrier concentration and the mobilities in bands a and b. Unless $\mu_a \approx \mu_b$ or $f_a f_b \ll \frac{1}{4}$, this expression is considerably larger than unity, and we might reasonably expect to observe a decrease rather than an increase in the Hall coefficient in going from weak to strong fields. The best-known example of this behavior is in p-type Ge.

B. The Temperature Dependence of the Hall Coefficient and Resistivity

In the remaining discussion, we refer to all of the bands which are occupied at low temperatures simply as band 1, and to the band which becomes appreciably populated only above 77°K as band 2. Originally we had hoped to connect the two-carrier effects shown in Figs. 2 and 3 with the spherical band observed by Stiles at 4.2°K. However, as Table I shows, the ratio $R_{296°}/R_{77°}$ gradually increases from 1.22 to 1.59 as the carrier concentration increases. This behavior is characteristic of two bands which are separated by an energy so large that the fraction of the conductivity carried by band 2 at 77°K is very small (say <1%), even for the highest carrier concentration investigated. On the other hand, suppose that the energy difference were small enough that for some carrier concentration an appreciable fraction of the conductivity already originated in band 2 at 77°K. Then at 296°K so much more conductivity would take place in this band that the carrier distribution would be approaching, if not already past the point at which the two-carrier Hall coefficient takes on its maximum value (i.e., when the conductivities in bands 1 and 2 are equal). Further addition of carriers would increase the Hall coefficient (relative to the one-carrier value) at 77°K but not at 296°K, and the ratio $R_{296°K}/R_{77°K}$ would decrease. A pertinent example of this kind of Hall behavior may be seen in Fig. 1 of a paper by Sagar on GaSb.[17]

For small changes in the Hall coefficient and classical statistics it is approximately true that[18]

$$(R - R_1)/R_1 \propto e^{-\Delta E/kT},\qquad(4)$$

where R_1 is the Hall coefficient when all the carriers are still in band 1, and ΔE is the energy difference between the band edges. The stumbling block to using this formula in our case is the difficulty of determining R_1, since the low-temperature end of the Hall curves are not quite constant because of changing statistics. However, if we pick an R_1 which leads to a straight line on a semilog plot, we find that the slope of that line corresponds to a ΔE of about 0.1 ev. H. R. Riedl has recently observed some extra optical absorption on the low-energy side of the main absorption edge of p-type PbTe which first becomes evident at energies above about 0.1 ev.[19]

Figure 3 was designed to show the droop in the conductivity which we thought we detected in the old data referred to earlier.[12] We first computed the product $R_{77°}\sigma(T)$ which leads to the following result, assuming that all the carriers are in band 1 at 77°K:

$$\begin{aligned}R_{77°}\sigma(T) &= rf(K)[n_1 e\mu_1(T) + n_2 e\mu_2(T)]/n_0 e\\&= rf(K)[f_1\mu_1(T) + f_2\mu_2(T)],\end{aligned}\qquad(5)$$

where n_0 is the total carrier concentration. Thus, this quantity is proportional to the average conductivity mobility; if we used the Hall mobility, the increase in the Hall coefficient would have tended to mask the decrease in the conductivity. The ordinate of Fig. 3 plots the quantity $R_{77°}\sigma(T)$ times the factor $(T/296°)^{\frac{3}{2}}$ which therefore leads to a horizontal line when the mobility is proportional to $T^{-\frac{3}{2}}$. Figure 3 shows that for all the samples a decided droop occurs above about 150°K. For comparison we have also plotted data for two n-type samples of PbTe,[12] demonstrating that there is no corresponding effect in the conduction band. Figure 3 and Table I also reveal that the conductivity mobility at 296°K steadily decreases with increasing carrier concentration while the Hall mobility remains essentially constant. Both of these facts suggest that $\mu_2 < \mu_1$ (the two-carrier average Hall mobility formula discriminates against carriers of lower mobility). Consistent with this conclusion are Dixon's recent room-temperature optical reflection data on p-type PbTe which he analyzed to show that the effective mass increased by a factor of almost three as the carrier concentration increased from 3×10^{18} to 1.0×10^{20} per cm³.[20]

[17] A. Sagar, Phys. Rev. **117**, 93 (1960).

[18] L. W. Aukerman and R. K. Willardson, J. Appl. Phys. **31**, 939 (1960).
[19] H. R. Riedl, Bull. Am. Phys. Soc. **6**, 312 (1961).
[20] J. R. Dixon, Bull. Am. Phys. Soc. **6**, 312 (1961).

C. Magnetoresistance Anisotropy

We have begun a study of the magnetoresistance as a function of carrier concentration at 296° and 77°K. As yet we have seen little if any change in the anisotropy with changing carrier concentration at either temperature. At 77°K, this is consistent with the assumption that all the occupied band edges lie at nearly the same energy, as Stiles concluded from his results at 4.2°K. At room temperature there may be no large effects because there are not enough carriers in the second band or because this band consists of another set of ⟨111⟩ ellipsoids of similar anisotropy.

D. Hall Mobility at 77°K

Table I reveals a very strong dependence of the Hall mobility on carrier concentration at 77°K. This large effect cannot be a two-carrier effect since a large mobility ratio would be required, and this would disagree violently with the Hall data of Fig. 1. A careful analysis of the mobility-temperature data also shows that ionized impurity scattering cannot be the cause of more than a small part of the total decrease in the mobility at this temperature. We believe that the principal cause of the mobility variation is the effect of statistics on the lattice mobility; i.e., the average energy becomes a function of carrier concentration, and this affects the mobility through the energy dependence of the scattering time. Such an effect has been observed at room temperature in n-type PbTe samples with very high carrier concentrations.[21]

CONCLUSIONS

A low-temperature two-band model (⟨111⟩ ellipsoids plus a sphere) for p-type PbTe in which the mobilities of the two carriers are nearly equal and in which the two band edges occur at nearly the same energy is the simplest way of explaining the lack of any of the usual two-band effects in the magnetic field dependence and the carrier concentration dependence of the Hall data at 77°K. Above about 150°K, on the other hand, effects appear in the temperature dependence of both the Hall coefficient and the resistivity which are characteristic of a two-band model with an energy difference between the band edges of about 0.1 ev and with the lower mobility in the lower energy band.

ACKNOWLEDGMENTS

I am greatly indebted to Bland B. Houston, Jr. and Richard F. Bis, without whose crystal-growing efforts this work could not have been carried out. I also wish to record with thanks the many valuable discussions with Frank Stern, the advance information given me on their own work by Philip Stiles, Jack Dixon, and H. R. Riedl, and the much-needed experimental assistance of J. R. Burke, Jr.

[21] T. S. Stavitskaya and L. S. Stil'bans, Fiz. Tverdogo Tela 2, 2082 (1960) [translation, Soviet Phys.—Solid State 2, 1868 (1961)].

JOURNAL OF APPLIED PHYSICS   SUPPLEMENT TO VOL. 32, NO. 10   OCTOBER, 1961

Magnetotunneling in Lead Telluride

R. H. REDIKER AND A. R. CALAWA

Lincoln Laboratory,* Massachusetts Institute of Technology, Lexington 73, Massachusetts

Rotation of a large magnetic field (∼60 kgauss) in a plane perpendicular to the direction of junction current in PbTe tunnel diodes produces a periodic behavior of this current. Diodes in which the junction current flow is along the [100], [110], or [111] crystallographic axis have been investigated. The observed anisotropies are consistent with a crystal with cubic symmetry whose constant energy surfaces in k space are ellipsoids of revolution oriented along the ⟨111⟩ crystalline axes. The magnetotunneling results are interpreted in terms of the effective motion of each ellipsoidal valley as the magnetic field is increased, valleys oriented at different angles to the electric field contributing with different weights to the tunneling current. Quantitative comparison awaits theory. Heavier mass bands close in energy to these ellipsoids, although unimportant by themselves in tunneling, may be necessary to explain the apparent position of the Fermi level with respect to the band edges. The excess current has the same anisotropy as the tunneling current; however, the thermal current does not show this anisotropy and therefore must be of different origin. Hump current, which was observed in two diodes, disappeared in magnetic fields above 3 kgauss.

INTRODUCTION

ON the basis of the magnetotunneling effects in InSb,[1] it was predicted that similar large reductions in tunneling current with magnetic field would occur in tunnel diodes of other low-gap semiconductors.[1]

The marked decrease in tunneling current of PbTe tunnel diodes with application of magnetic fields up to 88 000 gauss has been observed.[2] In this paper the effects of magnetic fields on the tunnel current will be described for PbTe diodes fabricated so the current flow is along either the [100], [110], or [111] crystallographic axes and for magnetic fields both parallel

* Operated with support from the U. S. Army, Navy, and Air Force.

and perpendicular to the current flow. The theory of tunneling for the case of isotropic effective mass has been considered for both longitudinal[1,3–5] and transverse[3,4] magnetic fields. Several questionable approximations, however, have been perforce necessary in the solution for the transverse field.[6] For the presently assumed multivalley band structure of PbTe having prolate energy ellipsoids along the $\langle 111 \rangle$ axes,[7,8] it is impossible for the magnetic field to be longitudinal at the same time to the differently directed current components associated with each of the four ellipsoids. Thus there is no true longitudinal case. While no acceptable theory is presently available for quantitative comparison, the magnetotunneling experiments do yield information both about the tunneling process and about the band structure of PbTe.

DIODE FABRICATION

The diodes were fabricated using Bridgman-grown single-crystal Ag-doped PbTe which had approximately 3×10^{18} net acceptors per cm^3 and a Hall mobility of approximately 1000 cm^2 v^{-1} sec^{-1} at 77°K and 5000 cm^2 v^{-1} sec^{-1} at 4.2°K. The tunneling contact was made by alloying indium spheres into one face of an oriented wafer of this p-type material while at the same time ohmic contact was made to the opposite face by alloying a 0.005-in. thick thallium disk backed by a gold-clad tantalum tab. In order to obtain appreciable tunnel current and negative conductance at forward biases it was found necessary to utilize an alloy cycle of about 3-sec duration and a peak temperature of approximately 320°C. Longer times and/or higher temperatures resulted in little or no sensible tunnel current at forward biases.

The [100] oriented diode whose magnetotunneling is described below was fabricated elsewhere[9] by alloying to vapor grown p-type PbTe grown in an excess Te vapor.

DIODE CHARACTERISTICS

Typical I-V characteristics obtained at zero and higher magnetic fields are shown in Fig. 1. The largest peak-to-valley ratio obtained at zero magnetic field was 1.5 to 1 at 4.2°K. In order to explain the zero magnetic field I-V characteristic shown in Fig. 1 it appears necessary to postulate that there are other heavier mass

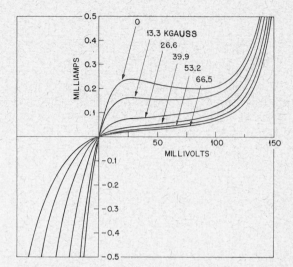

FIG. 1. Voltage-current characteristics at 4.2°K of a PbTe tunnel diode in different transverse magnetic fields. The junction current is parallel to the [110] crystallographic axis.

energy bands not too far in energy above the four $\langle 111 \rangle$ ellipsoidal minima assumed for both n- and p-type PbTe.[7,8] The net acceptor density in the base region is 3×10^{18} cm^{-3} and from results on other diodes we are led to believe that the net donor density in the n region is above 10^{19} cm^{-3}. Using a density of states effective mass of $0.1 m_0$ for the $\langle 111 \rangle$ ellipsoidal valleys the Fermi level would be 75 mv below the edge of the valence band and 170 mv above the edge of the conduction band. This degeneracy is inconsistent by a factor larger than 2 with the I-V characteristic of Fig. 1. The existence of other heavier-mass band edges would reduce the degeneracy of the material, while the tunneling current would still be produced mainly by the much lighter carriers in the $\langle 111 \rangle$ ellipsoidal valleys.

Capacitance measurements on several diodes yielded values for C/A of approximately 10 μf/cm^2. These values are significantly larger than the value of 1.6 μf/cm^2 calculated at zero bias using a net acceptor density in the base region of 3×10^{18} cm^{-3} and assuming a built-in junction potential of 0.2 v, a dielectric constant of 25[10] and an abrupt junction. A similar discrepancy in capacitance values has been reported for indium antimonide alloy diodes.[11] The zero-bias junction width calculated from the net acceptor density is 135 A and the average junction field is 1.5×10^5 v cm^{-1}. The junction width and average junction field (23 A and 9×10^5 v cm^{-1}) determined from the measured capacitance value yield values for tunnel current density which cannot be reconciled with the experimental tunnel current, indicating that simple diode theory cannot be used to determine junction width from capacitance measurements.

In order to investigate the effects of magneto-

[1] A. R. Calawa, R. H. Rediker, B. Lax, and A. L. McWhorter, Phys. Rev. Letters 5, 55 (1960).

[2] R. H. Rediker and A. R. Calawa, presented at the Symposium on Electron Tunneling in Solids, Philadelphia, Pennsylvania, January 30–31, 1961.

[3] R. R. Haering and E. N. Adams, J. Phys. Chem. Solids 19, 8 (1961).

[4] P. N. Argyres and B. Lax, J. Phys. Chem. Solids (to be published).

[5] P. N. Argyres, Bull. Am. Phys. Soc. 6, 345 (1961).

[6] P. N. Argyres (private communication).

[7] C. D. Kuglin, M. R. Ellett, and K. F. Cuff, Phys. Rev. Letters 6, 177 (1961).

[8] R. S. Allgaier, Phys. Rev. 112, 828 (1958).

[9] At General Electric Research Laboratories, Schenectady, New York.

[10] T. C. Harman (private communication).

[11] C. A. Lee and G. Kaminsky, J. Appl. Phys. 31, 1717 (1960).

resistance in all PbTe tunnel diodes fabricated in our laboratory, a "dummy" was fabricated in which the tunneling contact was replaced by an ohmic contact. This "dummy" had zero magnetic field resistance of less than 0.5 ohms compared to over 20-ohm resistance in the tunneling region at zero magnetic field for the diode of Fig. 1. The application of 66.5 kgauss increased the resistance of the dummy to less than 0.7 ohms for longitudinal and less than 0.8 ohms for transverse magnetic field. On the other hand in Fig. 1 the effective diode resistance is increased to over 200 ohms by application of 66.5 kgauss. For all the diodes described below, except the [100] diode, magnetoresistance can be neglected compared to magnetotunneling effects. The [100] diode, for which no dummy was made, did show small magnetoresistance effects at high currents as described below. At normal currents the results are in agreement with those obtained on the [100] oriented diodes fabricated by us, for which magnetoresistance can be neglected.

EXPERIMENTAL PROCEDURE

The tunnel diode current was measured as a function of voltage at constant values of magnetic field and as a function of both the magnitude and direction of magnetic field at constant values of diode voltage. All data were taken on an X-Y recorder. Because the source resistance and associated lead resistance could not easily be made small compared to the resistance of the diodes, in recording the current-magnetic field characteristics the voltage across the diode was maintained constant by a closed loop servo-system. The error between the diode voltage, as measured across voltage probes, and a constant reference voltage was reduced to less than 50 μv by using the servo to vary the source voltage. In order to rotate the diodes with respect to the magnetic field (directed along the bore of the Bitter magnet) the diodes were mounted in a right-angle gear drive mechanism capable of rotation through 320°. The angular displacement was determined from the voltage across a linear potentiometer which was linked to the gear assembly. The zero point angle was set visually introducing a possible error of approximately 10°. The angular error relative to the zero is within 5°. Each diode and the entire gear mechanism were immersed in liquid helium to insure a constant temperature. All the data presented in this paper were taken with the diodes at 4.2°K.

EXPERIMENTAL RESULTS

[100] Diode

As is evident from Fig. 1, large magnetotunneling effects have been observed for the PbTe tunnel diodes. Also, rotation of large magnetic fields in a plane perpendicular to the direction of the tunneling current produces a periodic behavior of this current. Figure 2 shows this behavior for a diode in which the junction

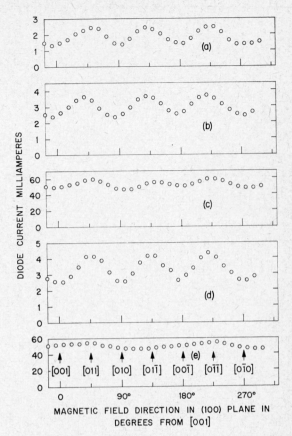

FIG. 2. Diode current as a function of the direction of a 60-kgauss magnetic field in the (100) plane perpendicular to the direction of junction current. The behavior of the current is shown at different fixed diode voltages: (a) 10 mv forward in tunneling region; (b) 10 mv reverse in tunneling region; (c) 80 mv reverse in tunneling region; (d) 80 mv forward in valley region; (e) 175 mv forward in thermal current region. The zero magnetic field currents are respectively: (a) 6.1 ma; (b) 8.3 ma; (c) 112 ma; (d) 9.9 ma; (e) 78 ma.

current is parallel to the [100] axis and a 60-kgauss field is rotated in the (100) plane perpendicular to the current. The 90° periodic anisotropy expected from the presently accepted cubically symmetrical four valley model[7] is obtained for the tunneling current [Fig. 2(a), (b), (c)] as well as for the excess current [Fig. 2(d)]. In all PbTe diodes investigated the periodic anisotropy disappears as the forward bias is increased in the thermal portion of the current characteristic [see Fig. 2(e)]. Thus there seems to be two components of current as one goes from the excess current region into the thermal region, the excess-type current which decreases in magnitude as the bias is increased and the thermal current which seems of different origin. The excess current on the other hand does have the same anisotropy as the tunneling current.

In Fig. 2 at high currents [see Fig. 2(c) and (e)] a 180° periodic anisotropy is observed. This 180° periodic anisotropy (which just means that the effect is independent of the sign of the magnetic field) can be explained in terms of magnetoresistance. While the current at the n-p junction is parallel to the [100] axis, for this particular diode the bulk current is not since the junction and the ohmic contact are not concentric; rather both contacts are on the same face of the wafer. As indicated above this diode is the only one to be discussed made[9] using vapor grown PbTe. In all other diodes considered, the junction and ohmic contact are concentric and only very small periodic anisotropy is observed at high thermal currents.

The 90° anisotropy can be explained by investigating the effects of the magnetic field on the four valleys. Since the electric field is in the [100] direction, all valleys contribute equally to the $H=0$ tunneling current. The minimum current (which is the maximum effect) occurs when the magnetic field is parallel to the [010] and [001] directions. In these directions the effective cyclotron resonance masses m_c^* for all four ⟨111⟩ oriented ellipsoids are equal. As the magnetic field is rotated from these directions, two of these effective masses increase and two decrease. At 60 kgauss, since the heavier m_c^* valleys are significantly lower in energy they contribute more to the tunneling current than the smaller m_c^* valleys. Thus on the average one would expect the smallest "average" effective cyclotron mass in the [010] and [001] directions and the largest effects in these directions.

The minimum effect occurs when the magnetic field is parallel to the [011] and [0$\bar{1}$1] directions. In these directions the effective cyclotron mass for two of the ellipsoids is at its maximum for magnetic fields in the (100) plane. Thus at 60 kgauss where the heavier m_c^* valleys contribute more to the tunneling, the minimum effect occurs in these directions. At lower magnetic fields where the light and the heavy m_c^* valleys contribute nearly equally to tunneling the anisotropy may be small as will be described below.

[111] Diode

Figure 3 shows the 60° periodic anisotropy in the tunneling current as 53.2 kgauss is rotated in the (111) plane perpendicular to the current. The variation in amplitude of the current oscillation as the magnetic field is rotated is believed due to slight misalignment of the magnetic field with respect to the (111) plane. For tunneling from an ellipsoidal band with principal effective masses m_1^*, m_2^*, m_3^*, the number of electrons leaking from the valence to conduction band is[12]

$$n = \frac{F^2}{18\pi\hbar^2 E_g^{\frac{1}{2}}} \left(\frac{m_1^* m_2^* m_3^*}{m_f^{*3}} m_f^* \right)^{\frac{1}{2}} \exp\left(-\frac{\pi m_f^{*\frac{1}{2}} E_g^{\frac{3}{2}}}{2\hbar F} \right), \quad (1)$$

where

$$(m_f^*)^{-1} = \sum_{i=1}^{3} \left[\cos^2\gamma_i / m_i^* \right],$$

and γ_i are the angles between the direction of the electric field and the principal axes of the effective mass tensor. For tunneling in the [111] direction three

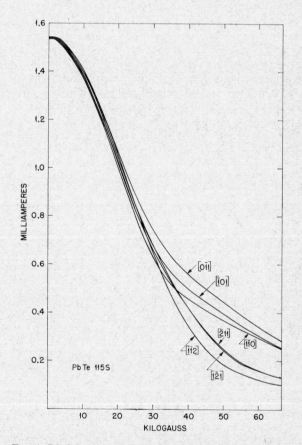

FIG. 4. Diode current as a function of magnetic field for the diode of Fig. 3 and for the magnetic field directions corresponding to the three maxima and three minima of Fig. 3. The diode is biased 20 mv forward.

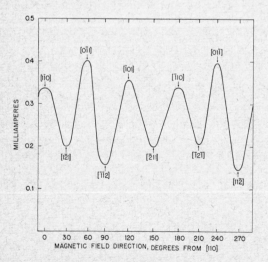

FIG. 3. Diode current as a function of the direction of a 53.2-kgauss magnetic field in the (111) plane perpendicular to the direction of junction current. The diode is in the tunneling region and is biased 20 mv forward.

[12] L. V. Keldysh, Sov. Phys.—JETP 6, 763 (1958).

of the valleys have an effective force mass[13] m_f^* $=0.0137m_0$ while the fourth valley has the large effective force mass $m_f^*=0.06m_0$.

Substituting in Eq. (1) these values for effective mass, $E_g=0.2$ v and $F/e=1.5\times10^5$ v/cm (as determined above) it is seen that less than 0.1% of the tunneling current is carried by carriers from the heavy valley. The heavy force mass valley also exhibits a large mass to the magnetic field in the (111) plane. This cyclotron resonance mass $m_c^*=0.027m_0$ does not change as the magnetic field is rotated in the (111) plane and the current carried by this valley should show no anisotropy. For the other three valleys m_c^* varies with the magnetic field direction varying from a minimum of $0.013m_0$ to a maximum of $0.027m_0$. At high magnetic field the maximum effect (minimum current) occurs as expected in the $[1\bar{2}1]$ direction where one m_c^* is a minimum and the other two are equal. The minimum effect occurs in the $[1\bar{1}0]$ direction as expected where one m_c^* is a maximum and the other two are equal although smaller than the two equal masses in the $[1\bar{2}1]$ direction.

While at high magnetic fields the anisotropy is dictated by the heavy m_c^* valleys this is not true at low magnetic fields where both light and heavy m_c^* valleys must be considered since they all make sizable contributions to the tunneling current. In Fig. 4 the currents for the magnetic field in the directions corresponding to three maxima and three minima of Fig. 3 are shown as a function of magnetic field. As seen from the figure the anisotropy at low fields is small and is different from that at high fields. Also at low fields where contributions from the light m_c^* valleys are more important, the magnetic field effects are as predicted larger.

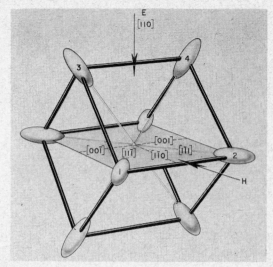

FIG. 6. An artist's view of the relationship of the ellipsoidal band structure of PbTe to the (110) plane. The valleys 1 and 2 have a smaller mass as seen by the $[1\bar{1}0]$ directed electric field than valleys 3 and 4.

$[110]$ Diode

Figure 5 shows the behavior of the tunneling current for a $[110]$ diode as a 53.2-kgauss field is rotated in the (110) plane perpendicular to the current. For tunneling in the $[1\bar{1}0]$ direction two of the valleys have an effective force mass $m_{f1}^*=m_{f2}^*=0.0125m_0$ while the other two have a heavier effective force mass, $m_{f3}^*=m_{f4}^*=0.029m_0$. Thus the contribution, using Eq. (1) and the associated assumptions, to the tunneling current by the last two valleys is about 10% of the total current. Figure 6 shows the relation of the four-valley ellipsoidal band structure of PbTe to the (110) plane. In Fig. 7 four cyclotron resonance masses are plotted as a function of the direction of the magnetic fields in the (110) plane. While this figure is for the conduction band in germanium and the absolute values for the masses are incorrect for PbTe, the general shape of the curves is correct.

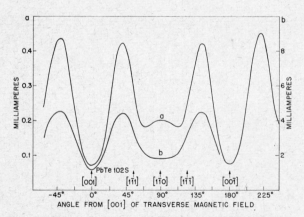

FIG. 5. Diode current as a function of the direction of a 53.2-kgauss magnetic field in the (110) plane perpendicular to the direction of junction current. The top curve (and corresponding left ordinate) are for a diode reverse bias of 30 mv and the bottom curve (and corresponding right ordinate) are for a diode reverse bias of 85 mv.

FIG. 7. Effective mass of electrons in germanium at 4°K for magnetic field directions in a (110) plane. (After Lax, Zeiger, Dexter, and Rosenblum[14].) While the absolute values do not apply, the general shape of the curve is the same as that for PbTe.

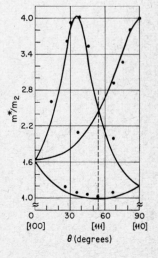

[13] All effective mass values used are reduced values derived from the values in reference 7 assuming identical conduction and valence bands.

[14] B. Lax, H. J. Zeiger, R. N. Dexter, and E. S. Rosenblum, Phys. Rev. 93, 1418 (1954).

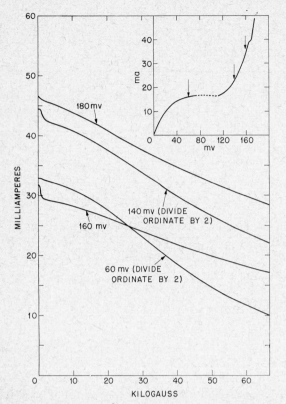

FIG. 8. The effect of magnetic field on hump current. The arrows in the inset, which shows the $H=0$, I-V forward characteristic, indicate the fixed voltages at which the magnetic field effects on the diode current are shown. The magnetic field is parallel to the [111] direction of junction current.

Looking at Fig. 7 and following the rules developed above, at 53.2 kgauss one expects the maximum effect (minimum current) in the [001] direction where all the masses are equal and minimum effects in the direction perpendicular to the [111] direction and in the [1$\bar{1}$0] direction where the heaviest masses occur. The minimum effect, however, may not occur in the [1$\bar{1}$0] direction because the two heaviest m_c^* valleys (valleys 3 and 4) are the heavy m_j^* valleys and only produce 10% of the $H=0$ tunnel current. If one neglects these two valleys, one expects a maximum effect in the [1$\bar{1}$0] direction because the masses of the valleys 1 and 2 are again equal. What one sees in the [1$\bar{1}$0] direction at high magnetic fields and low reverse biases is a minimum effect within a maximum effect (top curve Fig. 5). Increasing the reverse bias as shown in the figure tends to accentuate the maximum effect associated with valleys 1 and 2 as does reducing the magnetic field. Increasing the magnetic field, reducing the bias or going into forward bias accentuates the minimum effect associated with valleys 3 and 4. The experimental results can be explained by noting that the magnetic field raises the energy of valleys 1 and 2 with respect to valleys 3 and 4. Increasing the reverse bias increases the relative tunneling current due to valleys 1 and 2 (which are the easy tunneling valleys) just as increasing the reverse bias in a germanium tunnel diode increases the relative tunneling current due to the more probable transitions to the $k=0$ conduction minimum. This explanation is consistent with the Fermi level not being too far removed from the $\langle 111 \rangle$ valley band edges.

Effect of Magnetic Field on Hump Current

Two PbTe tunnel diodes showed a hump in the current in the "thermal" region at about 165 mv (see inset Fig. 8). In Fig. 8 the effect of longitudinal magnetic field on this hump current is illustrated. The current as a function of magnetic field is plotted at fixed forward biases both larger and smaller than 165 mv. Less than 3 kgauss is necessary to reduce the hump current so it no longer can be distinguished. One explanation of hump current is that it is due to tunneling through the intermediary of a discrete energy level in the forbidden gap. If this explanation is correct, our results indicate that at least for PbTe this energy level is extremely sensitive to magnetic field.

SUMMARY

Experimental results on magnetotunneling in PbTe confirm in detail, although not quantitatively, the multivalley band structure having prolate energy ellipsoids oriented along the $\langle 111 \rangle$ crystalline axes. Heavier mass bands close in energy to these ellipsoids are necessary to explain the apparent position of the Fermi level with respect to the band edges. Quantitative interpretation in terms of the band structure awaits the development of magnetotunneling theory applicable to PbTe. The experiments have brought out differences between the excess current and the "thermal" current and have shown a large magnetic effect on hump current.

ACKNOWLEDGMENTS

The authors wish to thank T. M. Quist, C. R. Grant, and J. M. McPhie for help in taking the data, and H. H. Bessler for assistance in fabricating the diodes. We should also like to thank J. H. Racette and R. N. Hall of the General Electric Research Laboratory who graciously supplied us with 2 PbTe tunnel diodes. We are indebted to B. Lax, J. G. Mavroides, and P. N. Argyres for many helpful discussions. The authors are grateful to F. Smith and W. Mosher of the M.I.T. National Magnet Laboratory for providing us time in the Bitter magnet and helping run the facility for us.

Oxides: SiC; Bi₂Te₃

D. POLDER, *Chairman*

Halides, Oxides, and Sulfides of the Transition Metals

F. J. MORIN

Bell Telephone Laboratories, Incorporated, Murray Hill, New Jersey

The electrical conductivity and optical absorption data for many compounds of the transition metals suggest that there is a trend from ionic insulators, to metals, to covalent semiconductors as the overlap of atomic orbitals is increased. In this paper it is shown that a correlation exists between the electrical behavior of insulators and metals and the magnitude of the overlap integrals $S(d\epsilon d\epsilon)$ and $S(d\epsilon p\pi)$.

THE electrical conductivity and optical absorption data for many compounds of the transition metals suggest that there is a trend in these compounds from insulators, to metals,[1] to covalent semiconductors as the overlap of atomic orbitals is increased. Electrical and optical data are reviewed in the light of this idea and a screening parameter is extracted from optical data which allows an estimate to be made of the overlap integrals $S(d\epsilon d\epsilon)$ and $S(d\epsilon p\pi)$. It is shown that the electrical behavior correlates with the magnitude of these integrals.

The idea of a trend in electrical properties among the transition metal compounds is outlined in Table I. In the insulators NiF_2 and NiO, for example, the partially occupied d levels are relatively isolated. This is indicated by the electron transport, which occurs by thermally activated diffusion of carriers with mobilities much less than 10^{-2} cm²/v sec, and by optical absorption spectra which are similar to that of the hydrated metal ions in water solution. In the metals NiS and MoO_2, for example, transport appears to occur in a narrow, partially-filled band because carrier mobility is in the range 10^{-2} to 1 cm²/v sec with very little temperature dependence. However, some of these metals switch with decreasing temperature into the insulating state with the corresponding low mobility. Compounds[2,3] such as MoS_2 and FeS_2, for example, appear to be covalent semiconductors with wide-energy bands because carrier mobility is in the range 10 to 1000 cm²/v sec and optical absorption[4] indicates band-to-band transition.

TABLE I. Trend in electrical properties with overlap of atomic orbitals increasing.

Electrical behavior:	Ionic insulator $\longrightarrow$ Metal $\longrightarrow$		Covalent semiconductor
Mobility cm²/v sec	$\ll 10^{-2}$	$10^{-2}-1$	10–1000
Example	NiF_2, NiO	NiS, MoO_2	MoS_2, FeS_2
Optical transition:	Isolated ion	?	Band-band

[1] F. J. Morin, Phys. Rev. Letters **3**, 34 (1959).
[2] R. Mansfield and S. A. Salem, Proc. Phys. Soc. (London) **66B**, 377 (1953).
[3] G. Fischer, Can. J. Phys. **36**, 1435 (1958).
[4] J. Lagrenaudie, J. phys. radium **15**, 299 (1954).

TABLE II. Periodic table of transition metal oxides showing carrier mobilities in cm²/v sec estimated from conductivity or Hall effect.

TiO 10^{-1}	VO 10^{-2}	(Insulators)		MnO $\ll 10^{-2}$
Ti_2O_3 10^{-2}	V_2O_3 10^{-2}	Cr_2O_3 $\ll 10^{-2}$		Mn_2O_3 $\ll 10^{-2}$
	VO_2 10^{-1}	CrO_2 10^{-2}		MnO_2 $\ll 10^{-2}$
		MoO_2 10^{-2}		
(Metals)		$Na_{0.75}WO_3$ 1–10		ReO_2 10^{-2}

Some indication of the dependence of electrical properties upon orbital overlap is given in Table II. A periodic arrangement of some lower oxides is shown together with order-of-magnitude mobility estimated (for the metals) from the electrical conductivity of single crystals, assuming that all d electrons contribute to conduction. Metallic behavior is found at the beginning of the transition metal series where nuclear charge is relatively low, with the consequence that the d orbitals extend out the farthest with the greatest possibility of overlap. As nuclear charge is increased and d orbitals contracted by moving across the $3d$ series, there is an abrupt change from metals to insulators. The metallic region becomes extended (indicating an increase in orbital overlap) as the charge and quantum number of the cation are increased. It should also be pointed out here that the extent of the anion orbitals is also important: all of the fluorides appear to be insulators while many sulfides are metals or covalent semiconductors. These factors which influence orbital overlap are more easily seen by examination of optical absorption spectra. To understand these spectra, however, and apply the information to the estimation of overlap integrals, it is necessary to consider in some detail the metal ion in its environment of anions.

In most of the compounds the cation is surrounded by six anions arranged in an octahedron. In this con-

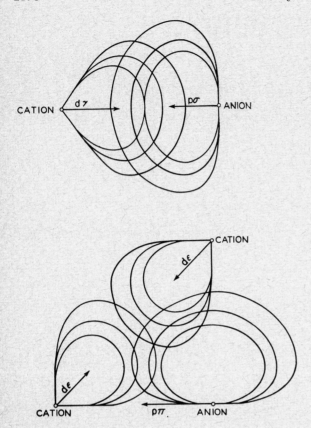

figuration, shown in Fig. 1, two of the five d orbitals (the $d\gamma$) are directed toward the anions and three d orbitals (the $d\epsilon$) are directed away from the anions and toward cation neighbors. Thus, there are three types of overlap as indicated by arrows: (1) $d\gamma p\sigma$ overlap directed along a common axis (σ bonds); (2) $d\epsilon d\epsilon$ overlap directed along a common axis; and (3) $d\epsilon p\pi$ overlap not directed along a common axis (π bonds). Electrostatic interaction between $d\gamma$ and $p\sigma$ electrons (the crystal field) raises the energy of the $d\gamma$ with respect to the $d\epsilon$. This energy separation, characterized by the crystal-field parameter Dq, is obtained from optical spectra. The magnitude of Dq depends upon orbital overlap, because the greater the overlap the greater the electrostatic interaction of electrons in the overlapping orbitals. Dq increases most conspicuously with the formal charge on the cation. As the charge is increased, the electrons in the anion orbitals are pulled toward the cation, increasing orbital overlap. This effect is shown in Fig. 2. Here it can also be seen that Dq varies with the number of d electrons because of the symmetry of the occupied orbitals, and increases with the quantum number of the cation. With large overlap, $d\epsilon p\pi$ interaction becomes appreciable and Dq decreases with further increase in overlap. The relatively low Dq values

found for $(MoF_6)^-$, ReF_6, and the cations in ZnS show this effect.

Another consequence of dp overlap is that the p electrons screen the d electrons from each other and from the cation nucleus. This interaction can be measured by comparing a term separation in the free ion which arises from mutual repulsion of d electrons, such as a $^3F \rightarrow {}^3P$ transition, to the same term separation in the compound. It is found that due to the additional screening by p electrons, the term separation in the compound E is always less than that in the free ion E_0. In Fig. 3 the screening parameter E/E_0 is plotted against Dq for a number of compounds, including a few halides[5] and sulfides.[6] This plot indicates a great decrease in d electron energy and, consequently, a great increase in the extent of d orbitals, with increasing dp overlap. It can be seen that Dq first increases, then decreases with decreasing E/E_0 (increasing overlap). Overlap appears to increase in the order of anions: F^-, $O^=$ (and H_2O), Cl^-, Br^-, I^-, $S^=$.

To determine overlap integrals from the tables,[7] it is necessary to know the effective nuclear charge and the separation of the ions in question. In the present calculation, the effective charge on the cation as determined using Slater's rules was multiplied by $(E/E_0)^{\frac{1}{2}}$ (the energy of an electron being proportional to the square of the charge) to correct it for the additional screening due to overlap. Determination of $d\epsilon$, $d\epsilon$ overlap was restricted by the tables to orbitals whose axes lie in the same plane. Thus, overlap in the rutile structure was calculated only along the C axis, and in hexagonal structures perpendicular to the C axis. In dis-

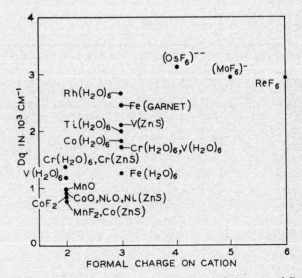

FIG. 2. Energy separation of $d\gamma$ and $d\epsilon$ orbitals in terms of Dq as it depends upon valence state, electronic configuration, and quantum number of the cation and electronegativity of the anion.

[5] C. K. Jorgensen, thesis, Copenhagen, 1957.
[6] R. E. Dietz (to be published); (The octahedral Dq was obtained by multiplying the tetrahedral Dq of ZnS by 9/4.)
[7] H. H. Jaffe, J. Chem. Phys. **21**, 258 (1953).

torted structures, average interatomic distances were used in the calculation. Overlap of $4d\pi$ orbitals, for which tables were not available, was taken as the mean of $3d\pi$ and $5d\pi$ overlap. The results for a number of compounds are shown in Fig. 4 where the integrals $S(d\epsilon,d\epsilon)$ and $S(d\epsilon,p\pi)$ are plotted against each other to indicate their relative importance. For the compounds in parentheses no electrical information is available, and the plot may be taken as a prediction of their metallic or insulating character. The most striking result is that metals and insulators fall into two distinct areas, as indicated by the dotted line. There appears to be a sharp break between the metals and insulators. The clustering of results shows the $d\epsilon$, $d\epsilon$ and $d\epsilon$, $p\pi$ overlap to be of about equal importance in contributing to the conduction band. The insulators TiO_2 and WO_3 are included on the side with the metals because Hall effect measurements indicate that they have the conduction band in agreement with the overlap prediction, although it is normally empty. $Ba_{0.5}TaO_3$ is also in this category. One can see the trend from metal to insulator in an isostructural series as the band attenuates with increasing nuclear charge, in the series TiO_2, VO_2, CrO_2, MnO_2, or TiO, VO, MnO, or Ti_2O_3,

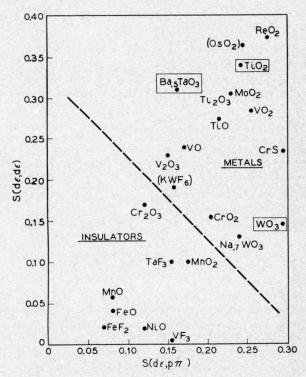

FIG. 4. Comparison of the overlap integrals $S(d\epsilon,d\epsilon)$ and $S(d\epsilon,p\pi)$ for a number of compounds. Dotted line is an arbitrary one which, in fact, separates metals from insulators. Compounds in parentheses are predictions. Compounds in boxes are insulators which have a band in agreement with overlap prediction, although it is normally empty.

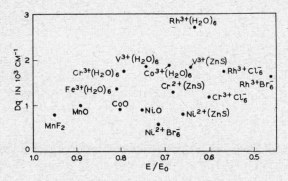

FIG. 3. Relation between Dq and the ratio E/E_0 of term separation of the cation in the crystal to same term separation in the free ion. Term referred to is first term above ground state having same spin multiplicity as the ground state.

V_2O_3, Cr_2O_3. This decreasing bandwidth also is indicated roughly by the trend in electron mobility.

WO_3 is of particular interest. It has the conduction band which is normally empty. When sodium is added to the lattice, electrons are introduced into the band and WO_3 becomes sodium tungstate bronze, a metal. The nuclear magnetic resonance of Na^{23} shows no Knight shift. This suggests that the sodium $3s$ wave functions contribute nothing to the conduction band. Therefore, it appears that sodium is a donor state lying above the band. Compounds having the tetrahedral tungsten bronze structure have been made with

tantalum and niobium oxides.[8] In these the tantalum and niobium ions are in the 5+ state, so the compounds are insulators. However, their conductivity behavior on slight departure from stoichiometry suggests that a conduction band exists, in agreement with the overlap result shown in Fig. 4 for $Ba_{0.5}TaO_3$. These results and the location of Nb and Ta in the periodic system strongly suggest that their lower oxides are metals.

The sulfides TiS, VS, CrS, FeS, and NiS are reported to be metals, and MnS a semiconductor. The overlap integrals for the monosulfides have been estimated in a crude way from optical data taken on the cation in ZnS. The sulfides TiS, VS, and CrS fall in the metallic region of Fig. 4 but the other monosulfides do not. NiS would not be expected to be a metal on the basis of $d\epsilon$ overlap, since these orbitals are full. Therefore, higher lying orbitals must be important in band formation in NiS and probably also in the other monosulfides.

Nothing has been reported on the electrical properties of the chlorides, bromides, and iodides. Overlap considerations suggest that some of them are metals.

[8] F. Galasso, L. Katz, and R. Ward, J. Am. Chem. Soc. 81, 5898 (1959).

JOURNAL OF APPLIED PHYSICS   SUPPLEMENT TO VOL. 32, NO. 10   OCTOBER, 1961

Recent Studies of Bismuth Telluride and Its Alloys

H. J. GOLDSMID

The General Electric Company Limited, Central Research Laboratories, Hirst Research Centre, Wembley, England

This paper reviews the work which has been carried out in these Laboratories during the past two years on single crystals of bismuth telluride and its alloys.

The combination of experiments on Faraday rotation with those performed previously on galvanomagnetic effects has established that there are 3- and 6-valley band structures associated with p- and n-type Bi_2Te_3, respectively. However, observations of the anisotropy ratio for the electrical conductivity and of the galvanomagnetic coefficients for heavily doped n-type material have shown that the shape of the equal-energy surfaces is dependent on carrier concentration. Similar conclusions have been drawn from the behavior of the Seebeck coefficient at low temperatures.

Measurements on Bi_2Te_3 at low temperatures and on alloys of Bi_2Te_3 have shown that the lattice thermal conductivity is particularly sensitive to the substitution of atoms of I, Se, or S for those of Te. It has also been shown that the anisotropy ratio for the lattice thermal conductivity is almost the same for the alloys as for pure Bi_2Te_3.

INTRODUCTION

IN 1959, when the work on Bi_2Te_3 in these Laboratories was last reviewed,[1] many of the characteristics of this semiconductor seemed clear. Measurements of the galvanomagnetic coefficients had been successfully interpreted in terms of 6-valley models for both n-type and p-type material.[2,3] Reasonably good agreement with the results of experiments on, for example, the thermal conductivity[4] and the thermomagnetic effects[5] had been reached on the assumption of acoustic-mode lattice scattering. The room-temperature energy gap of 0.13 ev obtained from the position of the infrared absorption edge[6] agreed well with the value predicted from electrical and thermal measurements.

There were, however, a number of details which defied really satisfying explanations. For example, the variations with temperature of both the Hall coefficient[7] and Seebeck coefficient[8] differed from the theoretical predictions.

Experiments carried out more recently have confirmed or clarified several of the features observed previously. They have also made it increasingly obvious that, while the simplifying assumptions which were originally adopted gave qualitative explanations of most of the phenomena, the real picture of conduction in Bi_2Te_3 must be rather complex.

This paper reviews these more recent developments and also some work on the thermal conductivity of single crystals of Bi_2Te_3 alloys.

INFRARED FARADAY ROTATION

According to the Drude-Zener theory, the Faraday rotation of polarized infrared radiation in a semiconductor is simply related to the carrier concentration and an appropriately-averaged effective mass. The effect is complicated near the absorption edge by interband transitions but, in his experimental work on Bi_2Te_3, Austin[9] was able to separate the free carrier rotation from the interband rotation. He found, as expected, that the free carrier Faraday rotation is proportional to the square of the wavelength. The constant of proportionality allowed him to calculate the product of carrier concentration and the square of the appropriate effective mass. By measuring the Hall coefficient also, an expression for a certain combination of the band parameters was obtained.

The Drude-Zener theory predicts that the same combination of band parameters should occur in the expression for free carrier absorption. Austin's measurements on the free carrier absorption of Bi_2Te_3 gave good agreement with his Faraday rotation results. It was therefore expected that there should have been equally good agreement with the band parameters which fitted the galvanomagnetic measurements of Drabble *et al.*[2,3]

Although the galvanomagnetic measurements had been interpreted in terms of 6 valleys for both conduction and valence bands, they could equally well have been satisfied by the more specialized case of 3 valleys. Austin found, in fact, that the optical and electrical results could be linked together reasonably well if it were assumed that there are 3 and 6 valleys for p- and n-type Bi_2Te_3, respectively.

ANISOTROPY OF THE ELECTRICAL CONDUCTIVITY

It is well known that the electrical conductivity σ_a along the cleavage planes in Bi_2Te_3 is greater than the conductivity σ_c in the perpendicular direction. Previous measurements[10] had also shown that the anisotropy ratio r, equal to σ_a/σ_c, is greater for n-type than for p-type material. However, these results had been obtained using only a single sample of each conductivity type.

[1] D. A. Wright, Research **12**, 300 (1959).
[2] J. R. Drabble, R. D. Groves, and R. Wolfe, Proc. Phys. Soc. (London) **71**, 430 (1958). (This paper will be referred to as DGW.)
[3] J. R. Drabble, Proc. Phys. Soc. (London) **72**, 380 (1958).
[4] H. J. Goldsmid, Proc. Phys. Soc. (London) **72**, 17 (1958).
[5] A. E. Bowley, R. T. Delves, and H. J. Goldsmid, Proc. Phys. Soc. (London) **72**, 401 (1958).
[6] I. G. Austin, Proc. Phys. Soc. (London) **72**, 545 (1958).
[7] B. Yates, J. Electronics and Control **6**, 26 (1959).
[8] H. J. Goldsmid, Proc. Phys. Soc. (London) **71**, 633 (1958).
[9] I. G. Austin, Proc. Phys. Soc. (London) **76**, 169 (1960).
[10] H. J. Goldsmid, Ph.D. thesis, University of London (1957).

The difficulties in making a reliable determination of the anisotropy ratio are twofold. It is first of all unusual to find a sample which is sufficiently thick in the c direction to allow σ_c to be found by a conventional method. Moreover, it is not easy to establish that a superficially sound sample is, in fact, free from cracks between the cleavage planes.

Bowley has overcome these difficulties using a four-probe electrical conductivity apparatus as suggested by Airapetyants and Bresler.[11] The plane containing the four probes could be rotated about an axis perpendicular to the surface of the specimen. When the probes were brought down on one of the cleavage faces the ratio of the potential drop ΔV between the inner probes to the current I did not, of course, change on rotation. However, there was considerable variation of $\Delta V/I$ with orientation when the probes were brought down on a surface which was normal to the cleavage planes. It can be shown that in this case

$$\frac{\Delta V}{I} = \frac{(r^{-1}\cos^2\psi + \sin^2\psi)^{-\frac{1}{2}}}{2\pi\sigma_a s}, \qquad (1)$$

where s is the inter-probe spacing and ψ is the angle between the plane containing the probes and the cleavage planes. Samples could be shown to be crack-free if measurements made on them followed the behavior indicated by Eq. (1). A typical set of results for a specimen of n-type Bi_2Te_3 is given in Fig. 1. Since the inter-probe spacing was only 0.08 cm it was possible to make useful measurements on quite small single crystals.

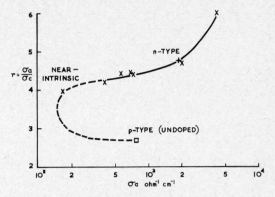

FIG. 2. Anisotropy ratio as a function of the electrical conductivity.

$\times$ 4-probe measurements,
$+$ n-type} previous measurements.
$\square$ p-type}

In Fig. 2, the anisotropy ratio for the electrical conductivity at room temperature is shown for a number of n-type samples of different iodine concentrations. It is clear that the anisotropy ratio is by no means constant but tends to rise as the carrier density increases.

GALVANOMAGNETIC EFFECTS IN HIGHLY-DOPED Bi₂Te₃

The observation that the anisotropy ratio for the electrical conductivity is a variable parameter suggested that it would be worthwhile repeating the galvanomagnetic measurements of Drabble et al.[2] on more heavily doped n-type material. For this purpose a single crystal containing nominally 0.4% by weight of iodine was selected. Its electrical resistivity along the cleavage planes at 77°K, at which temperature the galvanomagnetic coefficients were determined, was no more than 6.57×10^{-5} ohm cm.

The same experimental arrangement as that of DGW was employed. The results are shown in Fig. 3 in which the observed variation of Hall and magnetoresistance effects with orientation of the magnetic field is compared with the theoretical behavior:

case I $\begin{cases} \Delta E_J/JB^2 = \rho_{1111}\cos^2\phi + \rho_{1122}\sin^2\phi & (2a) \\ \\ E_z/JB = -\rho_{312}\sin\phi & (2b) \end{cases}$

case II $\begin{cases} \Delta E_J/JB^2 = \rho_{1111}\cos^2\phi + \rho_{1133}\sin^2\phi \\ \qquad\qquad\qquad + \rho_{1131}\sin2\phi & (3a) \\ \\ E_{z\Delta J}/JB = \rho_{123}\sin\phi. & (3b) \end{cases}$

In these equations the nomenclature is that of DGW and it has been assumed that the Hall effect is linear in magnetic field and that the current passes parallel to one of the three rotation planes of the crystal. Both these points have been checked under the conditions of the experiment.

The measured galvanomagnetic coefficients are given in the second column of Table I. The ratios between

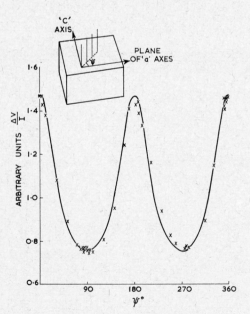

FIG. 1. Typical variation of $\Delta V/I$ with angle on rotation of 4-probe head. $\times$ Experimental results, — Theoretical curve,

$$[(\Delta V/I \propto \{(\cos^2\psi/r) + \sin^2\psi\}^{\frac{1}{2}}].$$

[11] S. V. Airapetyants and M. S. Bresler, Soviet Physics—Solid State 1, 134 (1959).

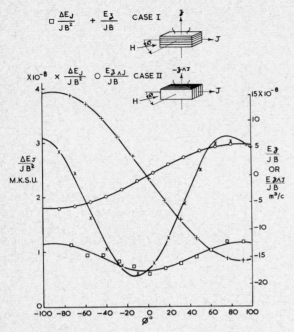

FIG. 3. Galvanomagnetic effects in n-type Bi_2Te_3 with a very high iodine concentration. The curves have the shape predicted by Eqs. (2) and (3).

TABLE I. Galvanomagnetic coefficients of highly-doped Bi_2Te_3 at 77°K.

Coefficient	Experimental value	Calculated values	
		(a)	(b)
ρ_{11}(ohm m)$\times 10^6$	0.657		
ρ_{123}(m³ c⁻¹)$\times 10^6$	0.058	$m_1/m_2 = 1.21$	$m_1/m_2 = 1.0$
ρ_{312}(m³ c⁻¹)$\times 10^6$	0.156	$m_3/m_2 = 0.093$	$m_3/m_2 = 0.05$
ρ_{1111}(m.k.s)$\times 10^8$	0.66	$\cos^2\theta = 0.0546$	$\cos^2\theta = 0.06$
ρ_{1122}	1.18		
ρ_{1133}	3.03		
ρ_{1131}	0.51		
$\rho_{123}{}^2/\rho_{11}\rho_{1111}$	0.78	1.67	0.70
$\rho_{123}{}^2/\rho_{11}\rho_{1122}$	0.43	1.04	0.39
$\rho_{123}{}^2/\rho_{11}\rho_{1133}$	0.17	0.48	0.23
ρ_{312}/ρ_{123}	2.69	2.06	2.70

the coefficients, shown in the last four rows of the table, may be compared with the values (a) predicted for the 6-valley band parameters obtained from the more lightly doped material of DGW. It is obvious that these band parameters do not represent the behavior of the heavily doped Bi_2Te_3. As shown in the last column (b), considerably better agreement is given by assuming that $m_1/m_2 = 1.0$, $m_3/m_2 = 0.05$, and $\cos^2\theta = 0.06$, where m_1, m_2, and m_3 are the principal effective masses of the valleys and θ represents the orientation of the axes of one of the valleys with respect to the crystal axes.[2] It is, however, impossible to select any set of band parameters to yield the very close agreement which is characteristic of DGW. It must be supposed that the equal-energy surfaces are no longer perfectly ellipsoidal but it is apparent that the principal effect of the high doping level has been to compress the surfaces even more strongly in the direction nearly parallel to the 3-fold axis of rotation of the crystal.

It is interesting to note that the anisotropy ratio r, calculated from the band parameters given above, is 4.6 as compared with a value of 4.1 for the set of parameters in DGW. Bearing in mind the different temperatures of the galvanomagnetic and electrical conductivity measurements, it does seem that the former provide a qualitative explanation of the change of anisotropy ratio with doping level observed in the latter.

It is worth mentioning that the latest galvanomagnetic experiments indicate a carrier concentration of 2×10^{19} cm⁻³ and an electron mobility at 77°K of 4600 cm²/v sec compared with 8900 cm²/v sec for the material

of DGW. Bearing in mind the different energies of the electrons in the two cases (the Seebeck coefficients being -22 and $-80\mu v$/°C) the change of mobility is just about as great as would have been predicted for acoustic-mode lattice scattering. While the apparent warping of the energy surfaces prevents any more quantitative interpretation, it seems as if lattice scattering is predominant even at a very high doping level. This agrees with previous observations, though Yates[7] has shown that point-defect scattering becomes important at very low temperatures.

THERMAL PROPERTIES AT LOW TEMPERATURES

Walker[12] has extended the measurements of Seebeck coefficient and thermal conductivity down to liquid helium temperatures. Previously it had been found[8] that the Seebeck coefficient, in the saturation region between 150° and 300°K, increases more rapidly with temperature than expected theoretically. The low-temperature measurements showed a still more complex behavior, certain samples having a minimum for the Seebeck coefficient at about 10°K. There was no real evidence for any contribution from the phonon-drag effect; instead the results could be interpreted on the basis of a changeover from lattice scattering to point-defect scattering of carriers between liquid nitrogen and liquid helium temperatures.

By combining the results of Hall[7] and Seebeck coefficient measurements Walker was able to show that either the density-of-states effective mass, or the relation between Hall coefficient and carrier concentration, or both, must be a function of the doping level. This is, of course, not surprising in view of the change of shape of the equal-energy surfaces which has now been found for heavily doped material.

As the temperature is reduced the lattice thermal conductivity κ_L becomes more predominant over the electronic component so that its value can be estimated from the total thermal conductivity with greater precision. Walker found that, while κ_L is approximately proportional to $1/T$ above 50°K, at lower temperatures

[12] P. A. Walker, Proc. Phys. Soc. (London) 76, 113 (1960).

it rises more rapidly than expected from the $1/T$ law. However, no exponential dependence was found even for the purest samples and it was supposed that phonon scattering due to the abundance of Te isotopes is responsible for reducing the thermal conductivity below its theoretical value.

The increased precision in the determination of κ_L at low temperatures enables the large phonon-scattering effect of iodine, predicted from the high temperature measurements,[4] to be checked. Walker confirmed this effect and showed that it is much greater than expected from the atomic mass fluctuations particularly if, as demonstrated by density measurements,[13] the iodine atoms take up Te sites in the Bi_2Te_3 lattice.

THERMAL CONDUCTIVITY OF Bi_2Te_3 ALLOYS

Figure 4 shows the variation with composition of the lattice thermal conductivity along the cleavage planes for solid solutions between Bi_2Te_3 and Sb_2Te_3, Bi_2Se_3, and Bi_2S_3. Most of the results lie between those of Birkholz[14] and Rosi et al.[15]

It is instructive to compare the observed behavior of κ_L with that expected theoretically. By extending Klemens'[16] theory of point-defect scattering at high temperatures, assuming the scattering of phonons to be due entirely to the mass variation in the alloys, it has been shown[17] that

$$\kappa_L = \kappa_0 \tan^{-1}\Omega/\Omega, \qquad (4)$$

where κ_0 is the lattice thermal conductivity in the absence of point-defect scattering and Ω is given by

$$\Omega^2 = \frac{\pi h}{2k^2}\left(\frac{6\pi^2}{V}\right)^{\frac{2}{3}}\frac{\kappa_0\epsilon}{N\theta_D}, \qquad (5)$$

where θ_D is the Debye temperature, V the volume per atom, and N and ϵ are, respectively, the concentration of unit cells per unit volume and a mass fluctuation parameter.

It has been possible to make reliable estimates of all the parameters which are involved for the region close to Bi_2Te_3. The resulting theoretical predictions are also shown in Fig. 4. It will be noticed that Eq. (4) describes the behavior of $(Bi–Sb)_2Te_3$ very well, but it greatly overestimates the thermal conductivity of $Bi_2(Se–Te)_3$ and $Bi_2(S–Te)_3$. It is concluded that, in the latter two cases, the most important extra contribution to the phonon scattering comes from a disturbance in the forces between the atoms. Considering this in conjunc-

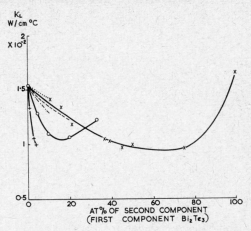

FIG. 4. Lattice thermal conductivity of Bi_2Te_3 alloys along the cleavage planes at 300°K.

	Experimental	Theoretical
$Bi_2Te_3 – Sb_2Te_3$	×	------
$Bi_2Te_3 – Bi_2Se_3$	○	—·—·—
$Bi_2Te_3 – Bi_2S_3$	+	—··—··—

tion with the scattering due to iodine additions which has already been mentioned, it is evident that the interatomic forces are much more sensitive to changes in the electronegative sublattice than in the electropositive sublattice.

It had been found previously[10] that the lattice thermal conductivity of Bi_2Te_3 at 300°K is 2.1 times greater along the cleavage planes than in the perpendicular direction but it was thought that the anisotropy might become less for the alloys. The thermal conductivity perpendicular to the cleavage planes has now been measured for single crystals of $Bi_2Se_{0.3}Te_{2.7}$ and $Bi_{1.6}Sb_{0.4}Te_3$. In both alloys the lattice thermal conductivity for a given direction is appreciably lower than for pure Bi_2Te_3 but the anisotropy ratio remains almost unchanged; experimental values of 2.3 and 2.2 have been obtained for the anisotropy ratio in the two alloys, respectively. An analysis of these results leads one to the conclusion that the mean free path of phonons in Bi_2Te_3 (and its alloys) must be substantially isotropic and that the anisotropy of the lattice thermal conductivity is primarily due to anisotropy of the velocity of sound. This is also suggested by the work of Blitz et al.[18] on the propagation of longitudinal ultrasonic waves in Bi_2Te_3.

CONCLUSIONS

The more recent measurements are consistent with the assumption of 3-fold and 6-fold sets of energy surfaces for p- and n-type Bi_2Te_3, respectively. However, at least for n-type material, it must be supposed that the shape of the energy surfaces is a function of the

[13] H. J. Goldsmid, *Proceedings of the International Conference on the Physics of Semiconductors, Prague, 1960* (Publishing House of the Czechoslovak Academy of Sciences, Prague, 1961), p. 1015.

[14] U. Birkholz, Z. Naturforsch. **13a**, 780 (1958).

[15] F. D. Rosi, B. Abeles, and R. V. Jensen, J. Phys. Chem. Solids, **10**, 191 (1959).

[16] P. G. Klemens, Phys. Rev. **119**, 507 (1960).

[17] J. R. Drabble and H. J. Goldsmid, *Thermal Conduction in Semiconductors* (Pergamon Press, London, to be published).

[18] J. Blitz, D. M. Clunie, and C. A. Hogarth, *Proceedings of the International Conference on the Physics of Semiconductors, Prague, 1960* (Publishing House of the Czechoslovak Academy of Sciences, Prague, 1961), p. 641.

carrier concentration. This affects the anisotropy ratio for the electrical conductivity as well as the galvano-magnetic coefficients and the Seebeck coefficient.

Point-defect scattering of the charge carriers only becomes important well below liquid nitrogen temperatures. However, point-defect scattering of phonons can be observed at much higher temperatures, particularly on the addition of I, S, or Se which substitute for Te. In these cases the theory of phonon scattering by mass fluctuations, which can explain the behavior on the addition of Sb, is inadequate. The fact that the anisot-ropy of the lattice thermal conductivity is little different for Bi_2Te_3 and its alloys can be understood in terms of the anisotropy of the velocity of sound.

ACKNOWLEDGMENTS

The author has been assisted by Miss A. E. Bowley and Mr. L. E. J. Cowles in carrying out those measurements which are reported here for the first time. Single crystal material has been prepared by Mr. R. T. Delves and Mr. D. W. Hazelden.

JOURNAL OF APPLIED PHYSICS   SUPPLEMENT TO VOL. 32, NO. 10   OCTOBER, 1961

Model for the Electronic Transport Properties of Mixed Valency Semiconductors

R. C. MILLER, R. R. HEIKES, AND R. MAZELSKY

Westinghouse Research Laboratories, Beulah Road, Churchill Borough, Pittsburgh 35, Pennsylvania

It is shown that the simple "hopping model" for the transport processes of mixed valency semiconductors is inadequate for impurity concentrations $\gtrsim 1\%$. In particular, it is necessary to redefine (1) the number of free charge carriers, and (2) the density of available states because of the dominant role played by the impurities in the high concentration range.

I. INTRODUCTION

IT has been recognized for some time that the electron transport properties of transition metal chalko-genides, in particular, the oxides, are of an essentially different nature than those of semiconductors such as Ge and Si. In the latter, the use of the concept of energy bands has allowed a convenient interpretation of most of the observed phenomena. On the other hand, in many of the transition metal compounds, it appears more correct to consider the electron as localized on a given lattice site and requiring an activation energy to move to an adjacent site. Such a model was reasonably successful in interpreting the electrical conductivity of the Li substituted transition metal oxides.[1-3] Also, Van Houten[4] gave an interpretation of the Seebeck coefficient of $Li_xNi_{1-x}O$ on the basis of the localized model. Still, the model was incapable of explaining many details of the properties of these mixed valency semi-conductors (MVS). For example, it is frequently found experimentally that the Seebeck coefficient increases with temperature in much the same manner as is found in standard semiconductors, whereas the simple theory[5] of the Seebeck coefficient for MVS indicates that it should be essentially independent of temperature in the temperature region where all carriers are free of impurity centers. In addition, as we will show in the perovskite systems (e.g., $La_xCa_{1-x}MnO_3$), one finds behavior attributable to both the band and the localized model. In an attempt to remove these difficulties, it was decided to restudy the problem, in particular, the $La_xCa_{1-x}MnO_3$ system. This study has led to a model capable of encom-passing most of the known experimental facts. In many respects, it will be seen that, in the region of localized behavior, the present model is essentially a limiting case of the so-called impurity conduction in semiconductors in the low concentration "hopping" limit.

II. EXPERIMENTAL

All of the compounds were prepared by mixing the proper quantities of MnO_2, La_2O_3, and $CaCO_3$ and reacting this material at $\sim 1100°C$. The resulting mate-rial was then ground and pressed into a pellet and fired in an appropriate atmosphere to give material suitable for measurement.

All of the electrical measurements, with the exception of those on $LaMnO_3$ were performed in air. The meas-urements on $LaMnO_3$ were performed in a nitrogen atmosphere. The Seebeck coefficients were measured relative to platinum and corrected to give the absolute values.

III. EXPERIMENTAL RESULTS

Although transport data on perovskite systems have been given before,[6,7] the electrical resistivity and See-

[1] R. R. Heikes and W. D. Johnston, J. Chem. Phys. **26**, 582 (1957).

[2] R. C. Miller and R. R. Heikes, J. Chem. Phys. **28**, 348 (1958).

[3] R. C. Miller, R. R. Heikes, and W. D. Johnston, J. Chem. Phys. **31**, 116 (1959).

[4] S. Van Houten, J. Phys. Chem. Solids **17**, 7 (1960).

[5] R. R. Heikes, *Proceedings of the Rare Earth Conference.*

[6] J. H. Jonker, J. H. Van Sauten, Physica **16**, 599 (1950).

[7] J. Volger, Physica **20**, 49–66 (1954).

beck coefficient of $La_xCa_{1-x}MnO_3$ have not previously been presented in detail. Figures 1 and 2 display the transport properties for small values of x while Figs. 3 and 4 show the remainder of the concentration range. There are several interesting features to be noted from these curves. First, it is seen that for $x=0.005$ and 0.04, the Seebeck coefficient α initially decreases in magnitude with increasing temperature and then begins to increase at about $100°K$. At higher temperatures, α again decreases. This final decrease is attributed to the onset of intrinsic conduction. (If one takes the gap width as twice the high temperature activation energy, one finds a value of 1.6 ev) For $x=0.08$, the initial decrease is not observed. The resistivity curves for these same compounds strongly resemble those of a normal extrinsic semiconductor which goes intrinsic in the high temperature region ($\sim1000°K$). It should be especially noted that the activated portions at low temperatures have nearly the same slope for the 0.005, 0.04, and 0.08 compositions. Second, the temperature dependence of the resistivity of the $x=0.65$ and 0.75 samples appear to be of the metallic type at low temperatures (below the ferromagnetic Curie temperature, T_c of $300°K$) and semiconducting at high temperatures. This transition occurs at the Curie temperature. It should also be noted that α changes sign above T_c for $x=0.75$. Further, the constancy of α at higher temperatures for this composition is rather striking. The remaining compositions will

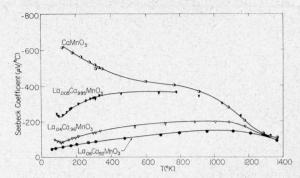

FIG. 2. Seebeck coefficient of Ca-rich Mn perovskites.

not be discussed further in this paper. They are presented only for completeness. The principle reason for avoiding the region $0.10 > X > 0.65$ is that Wollan and Koehler[8] have shown that ordering of the Mn^{3+} and Mn^{4+} is quite predominant here. This would introduce an unnecessary complication into the present model.

IV. DISCUSSION

The first question to be answered is whether the charge carriers in these materials are to be treated on a band model or on a localized model. This question can be perhaps discussed with the help of the magnetic information. In the Mn compounds considered here, it is found[9] from magnetic susceptibility data that the ions have the appropriate moment for spin only with all spins within the ion aligned parallel. This implies that Hund's rule applies to the electrons in the solid, which in turn means that the d electrons within an ion are strongly coupled to one another. Consider $LaMnO_3$. This material is an insulator ($\rho \sim 10^5 \Omega$ cm) in spite of the fact that the d shell is not completely filled. This indicates that the d electrons should be treated as localized. Now each Ca substituted into the $LaMnO_3$ introduces a hole on a Mn^{3+} ion. If the Mn^{3+} lattice is ferromagnetic, all sites in the lattice are equivalent and the hole may occupy any one of them. By taking the appropriate linear combinations of the equivalent states, one can form a band having one state per Mn atom. However, because of a small overlap of the Mn ions, the possibility also exists for a self-trapping of the Landau type due to self-induced polarization. Which of these two situations prevails depends upon the relative free energy of the two states. If bands are formed, they should be expected to be narrow.

Consider next a paramagnetic array of ions. In this case, the translational symmetry is destroyed. And furthermore, as previously noted, the magnetic interactions are not simple perturbations on the d electrons, but instead are the dominant energy terms. Thus, in this case, one should treat the charge carriers as localized. In the interpretation of the data of Sec. III, a

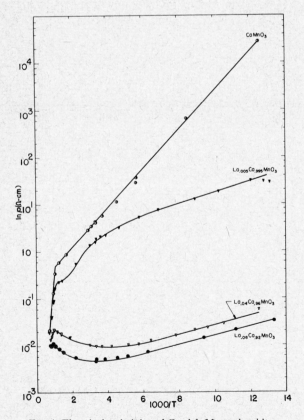

FIG. 1. Electrical resistivity of Ca-rich Mn periovskites.

[8] E. O. Wollan and W. C. Koehler, Phys. Rev. **100**, 545 (1955).
[9] J. H. Jonker and J. H. Van Sauten, Physica **16**, 337 (1950).

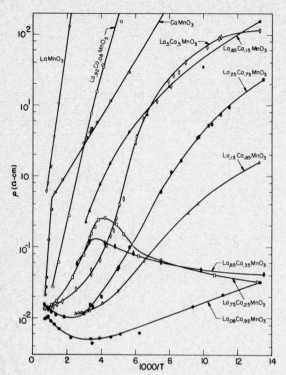

FIG. 3. Electrical resistivity of $La_xCa_{1-x}MnO_3$.

localized model will be used for all except the ferromagnetic cases.

The localized model to be used will first be described and then applied to the perovskite system $La_xCa_{1-x}MnO_3$.[10] The following assumptions will be made:

(1) An activation energy is required to allow the charge carrier to jump from Mn site to Mn site.
(2) The energy of a charge carrier is a function of the number of nearest neighbors of La, the higher the number of La neighbors, the lower the energy.

Let us now discuss the implications of these assumptions. Assumption (2) means that the charge carrier will prefer to move through the crystal in such a way as to remain a nearest neighbor or as near as possible to a La atom. As the temperature is increased, other paths become available. Further, if there exists an isolated La atom, that is, one which is not a sufficiently near neighbor to another La atom, a charge carrier can be bound to it as the temperature is lowered. Thus, there are at least three different energies involved in the transport process. (1) the energy required to remove a charge carrier from an isolated La atom, (2) the energy required to allow the electrons to occupy Mn sites not immediately adjacent to La, and (3) the energy required to overcome the Landau self-trapping.

It is convenient to divide the impurity concentration range into three regions: (1) $x \gtrsim 0.001$, (2) $0.001 \gtrsim x \gtrsim$

---

[10] In the following discussion, we will consider solutions of $LaMnO_3$ in $CaMnO_3$. The same arguments could be applied to the La rich end of the system or to other impurities.

0.04, and (3) $0.04 \gtrsim x \gtrsim 0.10$. Let us first examine the region of highest impurity concentration. This region will be characterized by the fact that the charge carrier can, essentially, always travel through the sample on Mn sites which are nearest neighbors to a La atom. Normally, there will be some sites which are not nearest neighbors to a La atom and thus are not accessible except at increased temperatures. Also, as mentioned above, some of the La atoms may be sufficiently isolated so as to be able to trap an electron as the temperature is lowered. Let us first ask how many such isolated impurities there would be in $La_xCa_{1-x}MnO_3$. Here, the Mn atoms form a simple cubic lattice with the La or Ca atoms distributed randomly at the body centers. A La site will be considered isolated if there is not another La at a distance of $2\frac{1}{2}$ lattice constants or less. Physically, this means that the charge carrier cannot move from Mn site to Mn site and remain nearest neighbors to at least one La atom. In a sphere of such a radius, there would be ~ 60 possible sites for the La. Thus, the fraction, designated by c', of isolated La would be

$$c' = (1-x)^{60}. \qquad (1)$$

Next, we must ask what fraction γ of Mn sites are not nearest neighbors to a La. Clearly, this is

$$\gamma = (1-x)^8. \qquad (2)$$

The exponent 8 results from the fact that each Mn has 8 nearest neighbors, any one of which may be occupied by a La.

It should be recognized that the two concepts just introduced are merely the translation into localized language of the semiconductor concept of impurity levels and the concept that the availability of states increases with temperature. [In semiconductor theory, the number of available states per unit volume is $2(2\pi m^* kT)^{\frac{3}{2}}/h^3$.] We now write down expressions for

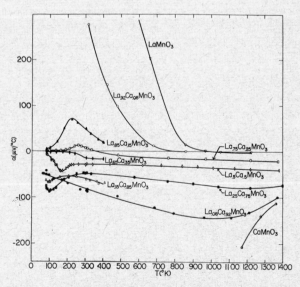

FIG. 4. Seebeck coefficient of $La_xCa_{1-x}MnO_3$.

α and ρ. From reference 1, the electrical conductivity can be written

$$\sigma = \text{const (number of carriers)}$$
$$\text{(number of available sites)} \, e^{-\Delta H/kT}/T, \quad (3)$$

where ΔH is the activation energy for jumping from site to site. The Seebeck coefficient can be written

$$\alpha = -\frac{k}{e}\left\{ -\frac{\Delta S_R}{k} - \ln\frac{\text{(number of carriers)}}{\text{(number of available sites)}} \right\}, \quad (4)$$

where ΔS_R is the entropy change of the system due to the trapping of a charge carrier. Mahanty, Maradudin, and Weiss[11] have shown that $\Delta S_R/k$ should be of the order of unity. Using (1) and (2), one can write

$$\sigma = (\text{const})[(c_0 - c') + c'e^{-Q_1/kT}]$$
$$[(1 - c_0 - \gamma) + \gamma e^{-Q_2/kT}]e^{-\Delta H/kT}/T, \quad (5)$$

where c_0 is the impurity ion concentration, Q_1 is the energy required to free a charge carrier from the vicinity of an isolated La, and Q_2 is the energy required to make sites accessible which are not nearest neighbors to a La atom. The Seebeck coefficient can be written

$$\alpha = \frac{k}{q}\left\{ -\frac{\Delta S_R}{k} - \ln\frac{(c_0 - c') + c'e^{-Q_1/kT}}{(1 - c_0 - \gamma) + \gamma e^{-Q_2/kT}} \right\}, \quad (6)$$

where q carries the sign of the charge carrier.

Equations (5) and (6) give a semiquantitative interpretation of the main features of this perovskite system in the region of *high impurity concentration*. Consider first Eq. (6) giving the Seebeck coefficient. We will assume that $Q_2 > Q_1$; that is, it is more difficult to remove a charge carrier from an ensemble of impurities than from a single impurity. Such an assumption is reasonable from simple electrostatic considerations. Starting at temperatures small compared to Q_1, the exponential terms in Eq. (6) are negligible. As the temperature is increased, charge carriers are freed from the isolated La, thereby increasing the numerator and decreasing α. As the temperature continues to rise, the increase in the denominator of the logarithmic term in Eq. (6) begins to dominate and α begins to increase. If this model is to be seriously considered, it must contain an understanding of (a) the magnitude of the increase in α from the minimum at 100°K to the maximum at ~1000°K, and (b) the absence of a turn-up in α at low temperatures for $x = 0.08$.

Consider $x = 0.04$. At temperatures above the minimum in α at 100°K, the concentration of free carriers may be considered as c_0. If we take the denominator of the logarithmic term in Eq. (6) as $(1 - c_0 - \gamma)$ at 100°K and $(1 - c_0)$ at 1000°K, it is easily seen, using (6), that α should increase by about 110 μv/°C. It should be remarked that the normal semiconductor theory could

give an increase of 300 μv/°C due to the term $(\frac{3}{2} \ln T)$ found in the density of states. Further, to get agreement with the absolute magnitude of α, it is necessary to set $\Delta S_R \simeq k$. This is in keeping with the estimates of reference 10.

Let us now examine the absence of a minimum in α for $x = 0.08$. Such an absence is a natural consequence of the present model. At $x = 0.04$, it is seen from Eq. (1) that $c' = 0.1$; that is, 10% of the La ions are sufficiently isolated so that charge carriers may be bound at low temperatures. However, at $x = 0.08$, $c' = 0.001$; that is, there are essentially no isolated La sites available to trap charge carriers. Thus, one would not expect a turn up at low temperatures.

Further, in the concentration range where Eqs. (1) and (2) apply, it is easily seen that the temperature dependence caused by Q_1 and Q_2 does not mask the exponential behavior due to $\exp(-\Delta H/kT)$. At very small concentrations, however, (region 1, as specified above) where impurity atoms are separated on the average by many lattice distances, c' and γ are meaningless quantities and should be set equal to zero. The fact that $\gamma = 0$ simply means that once freed from a La atom, all sites are accessible. The number of carriers in this concentration region will be of the form $c_0 e^{-Q_1/kT}$. Thus, the measured activation energy for electrical conductivity would be equal to $(Q_1 + \Delta H)$. As already mentioned, in the high concentration region, the measured activation energy is ΔH.

The so-called "pure" CaMnO$_3$ is a case in point for region 1. The "impurity atoms" in this case of probably oxygen vacancies.

This shows why MVS normally have an activation energy which decreases rapidly with increasing impurity concentration until $x \simeq 0.02$ and then tends to level off. The intermediate region (2) is too complicated to handle with the present naive model and will not be discussed further here.

In order to carry through the present ideas in a more rigorous fashion, it will be necessary to consider that the jump frequency of the electron is not only strongly dependent on the local surroundings but is also anisotropic. In order to handle this situation, a Monte Carlo type calculation is being initiated.

The ferromagnetic compounds must now be considered. As noted previously in this section, the energy band formed under these conditions is expected to contain only one state per Mn ion. If this energy band were symmetric (that is, if the point of zero curvature in the energy vs wave number curve occurs at $\frac{1}{2}$ the bandwidth), one would expect the Seebeck coefficients for the ferromagnetic region $(0.6 \gtrsim x \gtrsim 0.8)$ to be positive. This comes about simply because the band would be more than half-filled. Unfortunately, the ferromagnetic compound at $x = 0.65$ has a negative α below T_c. This would indicate either a lack of symmetry in the band or an inadequacy in the present model.

The extremely small Seebeck coefficient, as well as

[11] J. Mahanty, A. A. Maradudin, and G. H. Weiss, Progr. Theoret. Phys. (Kyoto) 24, 648 (1960).

the metallic-type temperature dependence of the electrical resistivity at low temperatures, is evidence in favor of a band interpretation.

As the Curie temperature is approached, the electrical resistivity increases anomalously. This increase can be attributed to the added scattering by the disaligned d shells. Such anomalies are quite common in ferromagnetic substances at the Curie temperature. Above T_c, the resistivity decreases rapidly with increasing temperature. In line with the arguments presented at the beginning of this section, it is expected that the charge carriers are localized above T_c. The resistivity decreases rapidly with increasing temperature. In line with the arguments presented at the beginning of this section, it is expected that the charge carriers are localized above T_c. The activation energy above T_c is probably composed of two parts: (1) the energy required to jump from site to site and (2) a disordering term. The latter would be expected from the work of Wollan and Koehler[8] in which they indicated that there was possibly some order in this composition region. This is somewhat borne out by the fact that the resistivity of $x=0.08$ is less than $x=0.65$, which indicates that the mobility of the charge carrier in the ferromagnetic region ($x=0.65$)

is much less than for low concentrations. An ordering of Mn^{3+} and Mn^{4+} ions could account for this difference.

Finally, the negative value of α at high temperatures for $x>0.5$ must be examined. From Eq. (6), it is seen that α may go negative for positive carriers if $\Delta S_R/k > 1.1$. In fact, for $\Delta S_R/k = 1.3$, one gets the proper α at high temperatures. Because of the uncertainties involved in the calculation of ΔS_R, one can only state that this is a reasonable magnitude.

V. CONCLUSIONS

It is seen that in the nonferromagnetic range, the $La_xCa_{1-x}MnO_3$ system may be described by a model which, in many respects, can be considered as the localized limit of impurity conduction in the hopping region.

Further, if the proposed transition from band to localized behavior does occur at T_c, it gives additional weight to Mott's[12] arguments concerning the transitions from Bloch to Heitler-London behavior.

It is felt that the present model is applicable to all transition metal chalcogenides where the electron-electron coupling within the atoms is strong.

[12] H. F. Mott, Phil. Mag. **6**, 287 (1961).

JOURNAL OF APPLIED PHYSICS   SUPPLEMENT TO VOL. 32, NO. 10   OCTOBER, 1961

Polaron Band Model and Its Application to Ce-S Semiconductors*

J. APPEL AND S. W. KURNICK

John Jay Hopkins Laboratory for Pure and Applied Science, General Atomic Division of General Dynamics Corporation, San Diego, California

The low-lying eigenstates of the "large polarons" have been calculated by several authors for arbitrary strengths of the electron lattice interaction α. However, if $\alpha>1$ the large polaron picture becomes questionable, since for finite temperatures the polaron eigenstates may be strongly affected by the presence of thermal phonons; then a new approach to the polaron theory is applicable which takes into account the atomicity of the lattice and the presence of thermal phonons and which results in the "small polaron" picture. The eigenstates of small polarons depend on T. If the eigenstates form a band, the bandwidth is a function of T, and the eigenstates near the band extremum can be expressed in terms of a T-dependent effective mass. From measurements of the high- and low-frequency dielectric constants, of the temperature dependence of the Seebeck coefficient, and of the electronic mobility, it appears that the eigenstates of the electronic charge carriers in Ce-S semiconductors may be adequately described by the small polaron picture.

I. INTRODUCTION

IN this paper we are concerned with the interpretation of some experimental results which have been obtained on Ce-S[1] semiconductors in the light of the

modern polaron band theory.[2–4] To this end we shall first recall the significant features of the polaron problem; then, the experimental situation for Ce-S semiconductors will be discussed in some detail. It will be shown that, although we cannot unambiguously prove that the charge carriers in Ce-S semiconductors are polarons which are significantly different from slow

* This work is supported by the Advanced Research Projects Agency under a Bureau of Ships contract.

[1] There are several phases present in the cerium-sulfur system; CeS, the unstable phase CeS_2, and the γ-phase semiconductors $Ce_{2+x}S_{3+x}$, where $0<x<1$. The last is the stable high-temperature phase and is easily frozen in on cooling from the melt. In this discussion the designation Ce-S is used to indicate the γ-semiconductor phases.

[2] G. L. Sewell, Phil. Mag. **3**, 1361 (1958).
[3] T. Holstein, Ann. Phys. **8**, 325 and 343 (1959).
[4] J. Yamashita and T. Kurosawa, J. Phys. Soc. Japan **15**, 802 (1960).

crystal electrons moving in the rigid lattice, the experimental results favor the polaron band picture. We mention that so far the only attempt to verify polaron theories rely on mobility measurements.[5] However, since a large polaron is not much different from a slow conduction electron, the corresponding mobilities show the same T-dependencies; they differ only by an effective mass ratio. Some qualitative features of the mobility of small polarons whose eigenstates form a polaron band have been pointed out by Fröhlich and Sewell.[6]

II. NATURE OF THE POLARON PROBLEM

The coulomb field of a slow conduction electron moving through a polar crystal displaces the lattice particles. The displacement polarization, in turn, has some influence on the motion of the conduction electron. The electron together with the accompanying self-consistent polarization field can be thought of as a quasi-particle.[7] This quasi-particle is the polaron. The mathematical problem is to find the polaron states as low-lying eigenstates of the one-electron Hamiltonian H, which consists of three parts: the energy of the electron in the rigid lattice, which we assume to be characterized by an effective mass m_0; $H(\mathrm{el})$, the kinetic and potential energy of the lattice particles $H(\mathrm{lat})$; and the interaction of the electron with the potential change due to lattice displacements $H(\mathrm{int})$. The most important aspect of any polaron theory is that $H(\mathrm{int})$ cannot be treated as a small perturbation, as in nonpolar semiconductors. There, $H(\mathrm{int})$ is treated with perturbation theory in order to calculate the electrical resistivity. In polar substances the displacement interaction can sometimes (when $\alpha < 1$) be divided into two parts. The first part must be considered in zeroth order; it represents the interaction of an electron with the unscreened displacement polarization and determines the low-lying polaron eigenstates which are characterized by two parameters, the lowest eigenvalue E_0 and the polaron mass m^*, i.e., the inertia of a polaron in an external electrical field. The second part of $H(\mathrm{int})$ determines the polaron mobility. However, when the interaction between electron and lattice is strong (i.e., when $\alpha > 1$), $H(\mathrm{int})$ cannot be divided into two independent parts, and for this case no quantitative mobility theory has been worked out.

The Large Polaron Picture

The large polaron theories[8] are characterized by one common feature: the theories of the polaron eigenstates are based on a model in which an excess electron interacts with a dielectric continuum instead of with a crystal lattice. Thus, the interaction part of the one-electron Hamiltonian is

$$H(\mathrm{int}) = e \int \frac{(\mathbf{r}-\mathbf{r}_1) \cdot \mathbf{P}(\mathbf{r}_1)}{|\mathbf{r}-\mathbf{r}_1|^3} d\mathbf{r}_1, \qquad (1)$$

where $\mathbf{P}(\mathbf{r}_1)$ is the polarization vector at $\mathbf{r}_1$ due to an electron at point $\mathbf{r}$. The vector field $\mathbf{P}(\mathbf{r})$ can be expressed in terms of the Fourier amplitudes of longitudinal and transverse polarization waves as convenient dynamical variables. With the usual quantization procedure and the assumption that the energy $\hbar\omega$ of the quantized longitudinal polarization waves is wave-vector independent ($\omega = \mathrm{const}$), the interaction depends on Fröhlich's coupling constant

$$\alpha = \tfrac{1}{2}[(1/\epsilon_\infty) - (1/\epsilon_0)](e^2 u/\hbar\omega) \qquad (2a)$$

with

$$u = (2m_0\omega/\hbar)^{\frac{1}{2}}, \qquad (2b)$$

and where ϵ_∞ and ϵ_0 are the optical and the static dielectric constants, respectively.

So far, the weak and strong coupling theories for the large polaron, $\alpha < 1$ and $\alpha > 1$, respectively, are concerned with the ground state of the system and the low lying excited states. Thus, the polaron states are determined by virtual emission and reabsorption of longitudinal polarization phonons. Consequently, the two characteristic parameters E_0 and m^* of any polaron theory are constants (T-independent) and depend on α only. The same is true, roughly speaking, with respect to the mean number of phonons in the cloud around the electron and with respect to the size of a polaron which is given by u^{-1} [Eq. (2b)].

An obvious weakness[9] of the strong coupling theories for large polarons lies in the continuum approximation. More recent papers overcome this difficulty, as will be pointed out in the next section.

The Small Polaron Picture

According to the weak coupling theory for large polarons ($\alpha \ll 1$) the polaron mass $m^* = m_0(1 + \alpha/6)$. Thus, for small coupling constants the polaron mass is not much larger than the rigid lattice mass m_0, unless m_0 itself is large, in which case the continuum approximation breaks down since it leads to a polaron dimension u^{-1} which is comparable with the lattice constant a. If $u^{-1} \sim a$, a different kind of approach results in what is called the small polaron picture. Recently, a number

[5] T. D. Schultz, Phys. Rev. 116, 526 (1959).

[6] H. Fröhlich and G. L. Sewell, Proc. Phys. Soc. (London) 74, 643 (1959).

[7] L. D. Landau, Soviet Phys.—JETP 3, 920 (1957); see also J. J. Quinn and R. A. Ferrell, J. Nuclear Energy, Part C, 2, 18 (1961).

[8] H. Fröhlich, Advances in Phys. 3, 325 (1954); H. Haken, Halbleiterproblem II, edited by W. Schottky (Vieweg Verlag, Braunschweig, Germany, 1955).

[9] Another weakness of the large polaron theories arises as follows: The mobility of large polarons is calculated by applying perturbation theory to their interactions with free thermal phonons. Such a procedure is not consistent if $\alpha > 1$. In such a case, the presence of thermal phonons will also effect the eigenstates of a conduction electron at moderate and, particularly, at high temperatures; i.e., the induced polarization around an electron consists of virtual and (bound) thermal phonons. The remaining (free) thermal phonons interact with the polaron and determine its mobility.

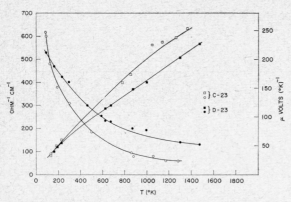

FIG. 1. Conductivity σ and Seebeck potential α as a function of temperature. Sample C-23 is undoped, and its conductivity and Seebeck potential are represented by (○) and (□), respectively. Sample D-23 is doped with BaS; its conductivity is represented by (●) and its Seebeck potential by (■).

of authors have been concerned with the small polaron problem.[2–4] The pertinent feature of the new approaches is that the atomicity of the lattice is taken into account. The calculation of the polaron eigenstates is based on Bloch's tight-binding approximation. However, the displacement of the lattice particles from their equilibrium positions is considered in zeroth order so that the interaction energy of an electron with the lattice is written as

$$H(\text{int}) = e\left[\sum_s \{V(\mathbf{r} - \mathbf{R}_s) + V'(\mathbf{r}' - \mathbf{R}_s')\}\right. \\ \left. - \text{rigid lattice potential}\right], \quad (3)$$

where s is the cell index for the cation and anion sublattices and V and V' are the potentials due to the positive ion at $\mathbf{R}_s$ and the negative ion at $\mathbf{R}_s'$.

With the usual assumptions of Bloch's theory the calculation of band-type solutions is straightforward. The characteristic feature of the small polaron eigenstates is their dependence on the eigenstate of the lattice. This dependence arises because with Eq. (3) the two center overlap integrals, which play a central role in the tight-binding approximation, depend on electron and phonon quantum numbers. By taking an appropriate thermal average over the eigenvalues, finally the low-lying eigenvalues of the small polaron can be defined in terms of two temperature-dependent parameters[2] $E_0(T)$ and m^*. The polaron effective mass increases with rising temperature because the thermal motion of the lattice particles opposes the transfer of the mean positions connected with the motion of a polaron. The polaron band width is a decreasing function of temperature until eventually, when the energy uncertainty associated with the finite lifetime of the polaron is of the order of the band width, the band model breaks down.[3,4] Thus, a crucial test of the small polaron band theory would be possible if one could find an experiment which allows an accurate measurement of the temperature dependence of the effective mass or of the bandwidth.

III. EXPERIMENTAL RESULTS

General Properties of the Ce-S Semiconductors

The γ phase of the Ce-S system[1] is a semiconductor whose crystal structure is that of the Th_3P_4 deficiency lattice[10] and is characterized by a cubic unit cell with 16 sulfur atoms and from $10\frac{2}{3}$ to 12 cerium atoms, depending on the stiochiometry. In the range where cation vacancies are present the sites are distributed randomly. The Ce-S semiconductors range stoichiometrically from Ce_2S_3, a bright glassy red insulator, to Ce_2S_4, which is lustrous black crystalline and a semiconductor. Other physical properties and the chemistry of synthesis are described elsewhere.[11–15] As an indication of the stability of these semiconductors, the high melting points are noted, which range from 1900°C (Ce_2S_3) to 2100°C (Ce_3S_4). It is a remarkable property of this semiconductor that conductivities range from 10^{-10} ohm^{-1} cm^{-1} at the Ce_2S_3 composition up to 2×10^3 ohm^{-1} cm^{-1} at the Ce_3S_4 extreme of composition, a relative change of the order of 10^{13}.

The material used in these investigations was spongy Ce_2S_3 melted down in molybdenum crucibles. The meltdown involved some loss of sulfur. When sulfurrich material was required it was produced by heat treating the cast ingots at about 1200°C in a H_2S atmosphere. The cast samples were impervious to deterioration by ambient gases at room temperature.

Electrical Conductivity and Seebeck Coefficient

Figure 1 shows the electrical conductivities and Seebeck coefficients[16] of both undoped Ce-S and BaS-doped Ce-S from 100° to 1500°K. We have chosen BaS and Ce_2S_3 solutions to be compared to the normal CeS $+Ce_2S_3$ solutions because the Ba^{++} ion is remarkably effective in stretching the lattice.[17] Furthermore, unlike the added Ce which is capable of going into the Ce^{3+} by releasing an electron to the conduction band, the closed shell of the Ba^{++} cannot give off an electron and its effect is only to alter the nearest neighbor distance of two cerium ions. This has been found to alter the electrical properties. The conductivity in both cases is n type. From the conductivity curves as shown in Fig. 1, the behavior is such as to strongly suggest a mobility which varies as T^{-n} (where n is between 0.5 and 1.5)

[10] W. H. Zachariasen, Acta Cryst. 2, 57 (1949).
[11] E. D. Eastman et al., J. Am. Chem. Soc. 72, 2248 (1950); J. Am. Ceram. Soc. 34, 128 (1951).
[1-] J. Flahaut and M. Guittard, Compt. rend. 242, 1318 (1956).
[13] J. Flahaut and M. Guittard, Compt. rend. 243, 1419 (1956).
[14] M. Picon et al., Soc. Chimique de France Bull. 2, 221 (1960).
[15] E. Banks et al., J. Am. Chem. Soc. 74, 2450 (1952).
[16] The different usage of α for both the dimensionless coupling constant and the Seebeck potential is regretted but can easily be interpreted from the accompanying text.
[17] S. W. Kurnick, M. F. Merriam, and R. L. Fitzpatrick (unpublished data on the electrical and physical properties of BaS-doped Ce-S).

and a carrier concentration which is independent of T. The Seebeck potential of Ce-S semiconductors shows an almost linear rise with increasing temperature. The temperature dependence of the mobility and the Seebeck potential is quite different[18] from that of other semiconductors such as NiO and $\gamma - Fe_2O_3$.

Hall Coefficient and Mobility

For the Ce-S semiconductors, Hall measurements by dc techniques have yielded no measurable response, partially due to the unavailability of high magnetic fields and heating associated with large driving currents. These responses were then measured by an ac technique[19] utilizing low-impedance transformers, high-gain (narrow-bandpass) amplifiers and magnetron magnets. The samples required low-resistance contacts, and the best procedure was found in the following recipe. The Hall blanks were first copper-plated electrolytically in a $CuSO_4$ solution and the lead wires were next soldered with indium. The excess copper coating was then removed by sand blasting. The samples prepared this way gave reproducible Hall voltages which corresponded to concentrations between 10^{21} and 10^{22} cm^{-3} and mobilities between 0.2 and 3 cm^2 $volt^{-1}$ sec^{-1}, depending on the electrical conductivity (see Fig. 2). Important features of these measurements may be summarized as follows: (1) Since both Hall coefficient and mobilities are low, accuracies to 20% are acceptable for interpretation. (2) A comparison of "undoped" $Ce_{2+x}S_{3+x}$ with some samples doped with barium, $Ce_{2+x}Ba_yS_{3+x+y}$, was made to investigate the role of distorting barium ions on the electronic properties. For the BaS-doped material, the Hall measurements were taken on samples with $y=0.7$. Conductivities were altered by pumping off sulfur from the melt, i.e., by causing a change in x; it is assumed that BaS was not pumped off appreciably. (3) Inasmuch as Hall voltages are quite low and the change in conductivity with temperatures slight from 78° to 1600°K as compared with the large conductivity

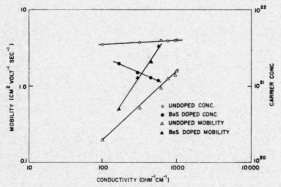

FIG. 2. Hall concentrations and mobilities for both doped and undoped Ce-S as a function of conductivity; $T = 300°$K.

[18] F. J. Morin, Phys. Rev. **93**, 1195, 1199 (1954).
[19] S. W. Kurnick and R. L. Fitzpatrick, Rev. Sci. Instr. **32**, 452 (1961).

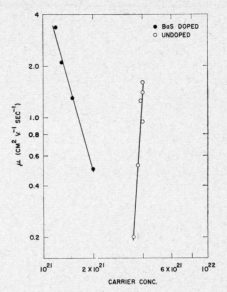

FIG. 3. Hall mobility vs carrier concentration; $T = 300°$K.

variation of 10^{13} due to the sulfur cerium ratio alone, it was decided to restrict the preliminary measurements to room temperature.

The results of the measurements at room temperature are shown in Fig. 3, in which carrier concentration is plotted against mobilities for both doped and undoped Ce-S. In the interpretation of Hall mobilities and concentrations the simplest interpretations have been used, the Hall coefficient being $R_h = 1/n_0e$, where n_0 is the carrier concentration.

Optical and Static Dielectric Constants

According to Eq. (2a) the coupling constant α depends on four parameters, three of which can be measured directly. They are: the frequency ω of the longitudinal polarization waves, the static dielectric constant ϵ_0, and the optical dielectric constant ϵ_∞, which is equal to the square of the index of refraction n. The frequency ω is related to the reststrahlen frequency ω_r by Fröhlich's relation $\omega = (\epsilon_0/\epsilon_\infty)^{\frac{1}{2}}\omega_r$; experiments to measure ω_r are underway.

1. Static Dielectric Constant ϵ_0

The dielectric constant of several thin plates of Ce_2S_3 was determined by placing them between and in contact with the plates of a condenser. Measurements were made on a bridge at audio frequencies (1 kc). The dielectric constant was then determined from the change in capacity when the sample was introduced. The dielectric constant was measured as $\epsilon_0 = 19$.

2. Optical Dielectric Constant ϵ_∞

The reflectivity R was measured on Ce_2S_3 cast samples. The procedure followed was to sulfurize optical blanks of Ce-S to the red insulating form Ce_2S_3 and then optically repolish them. The reflectivity measurements were made before and after sulfurization. This processing eliminates the excess carriers and consequently gives a reflectivity more like that of an insulator than that associated with a strong absorption coefficient, as a result of the many excess conducting electrons. It is interesting to note that, although the original conductivities may have differed—primarily due to the difference in the cerium sulfur ratio—the differences were cancelled out, and all the measurements are consistent with one another in giving a reflectivity to which we can apply the simple relation $R = [(n-1)/(n+1)]^2$. Solving for n gives an index of refraction of 2.5, thus $\epsilon_\infty = 6.25$.

There was little dispersion in the region of measurement between 1 and 13 μ.

IV. DISCUSSION

From the experimental results obtained so far, the question arises as to whether the charge carriers in Ce-S semiconductors are polarons which are significantly different from conduction electrons. Our conclusion is that the electronic charge carriers are small polarons which can be adequately described by Bloch-type eigenstates and which travel through the crystal not by phonon-activated jump processes, but by frequent interchange between one cerium ion and the next nearest cerium ion by tunneling. The probability for such a tunneling process is directly proportional to the value of the appropriate overlap integral, which depends on the lattice displacements and, eventually, on the temperature. We think this model is favored for the following considerations which are based on the experimental results presented in Sec. III.

A large difference has been found between the static and optical dielectric constants ϵ_0 and ϵ_∞. This implies that the effective charge of the lattice particles is about the same as in crystalline compounds characterized by predominant polar binding. The difference $(1/\epsilon_\infty - 1/\epsilon_0)$ is proportional to the strength of the electron-lattice interaction. However, Fröhlich's coupling constant α [Eq. (2a)] depends on two additional parameters, the frequency ω of the longitudinal optical modes and the rigid lattice effective mass m_0. With our values of ϵ_0 and ϵ_∞ one finds

$$\alpha = 1.55 \times 10^7 (m_0/m_e)^{\frac{1}{2}} \omega^{-\frac{1}{2}}. \qquad (4)$$

At this time we neither know an appropriate average value for ω nor do we know the quantitative value of the effective mass m_0; but certainly m_0 is greater than m_e, the mass of the free electron. This conclusion is implied by: (1) the relation between the polaron effective mass and the rigid lattice effective mass for coupling strengths $\alpha \sim 1$; and (2) the polaron effective mass obtained from the results of the measured Seebeck voltage vs temperature. The combined measurements of both the Seebeck effect and the Hall effect have the important advantage that one can measure the charge carrier con-

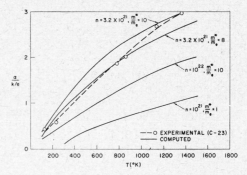

FIG. 4. Normalized Seebeck potential $\alpha e/k$ vs temperature. Dashed curve and open circles represent experimental data. Room-temperature Hall concentration is 3.2×10^{21} cm^{-3}. Solid curves are computed from Eq. (5) with the scattering index $r=0$.

centration as well as their effective mass m^*. If the Hall effect is interpreted on the basis of the conventional formula $R_h = (n_0 e)^{-1}$, one can determine the concentration n_0 and μ, the mobility. From Fig. 2 one can see that the mobilities at $T = 300°$K are of the order of one as to be expected for small polarons.[6] With the assumption that the carrier concentration is independent of temperature, one can obtain the polaron effective mass by analyzing the measured temperature dependency of the Seebeck coefficient with Pisarenko's formula,[20]

$$Q = \pm \frac{k}{e} \left[\frac{r+2}{r+1} \frac{F_{r+1}(\eta)}{F_r(\eta)} - \eta \right], \qquad (5)$$

where $k/e = 86 \mu v (°C)^{-1}$, $\eta = \zeta/kT$ is the reduced Fermi energy, and $F_r(\eta)$ is the integral

$$F_r(\eta) = \int_0^\infty \frac{x^r dx}{\exp(x-\eta)+1}. \qquad (6)$$

The scattering index r is taken equal to zero. Since the electrical conductivity according to Fig. 1 depends only slightly on temperature, it may well be justified to assume that at high temperatures the free path of the carriers is nearly energy independent. With that value $(r=0)$, which corresponds to an energy independent free path, the curves of Fig. 4 have been calculated to best fit the experimental data.

ACKNOWLEDGMENTS

We wish to acknowledge the cooperation of Dr. J. Apfel in the measurements on the static and optical dielectric constants. The authors also wish to acknowledge the aid of R. L. Fitzpatrick in the electrical measurements and L. LaGrange in the chemical synthesis associated with the experimentation. Finally, it is a pleasure to thank Dr. W. B. Teutsch for a critical reading of the manuscript.

[20] A. F. Ioffe, *Semiconductor Thermoelements and Thermoelectricity* (Infosearch Ltd., London, 1957), p. 102.

JOURNAL OF APPLIED PHYSICS   SUPPLEMENT TO VOL. 32, NO. 10   OCTOBER, 1961

Recent Studies on Rutile (TiO₂)*

H. P. R. FREDERIKSE

National Bureau of Standards, Washington, D. C.

A review is made of the work on reduced and "doped" rutile performed since the appearance of Grant's survey article in the Reviews of Modern Physics (1958). Measurements of electrical and optical properties, and of electron spin resonance spectra are discussed. A model of electronic bound states and conduction levels is suggested that is compatible with the results of these experiments. There is strong evidence that the defects in reduced rutile are interstitial Ti^{3+} ions. At very low temperatures, nearly all electrons are self-trapped on cation sites (polarons). As the temperature increases, some of these trapped electrons will be excited into the conduction band. The activation energy for this process is approximately 0.007 ev below 50°K, and about one order of magnitude higher around room temperature. It is concluded that conduction takes place in a narrow $3d$ band associated with Ti ions; the effective mass at the bottom of this band is $\sim25m_0$. If one assumes that the polaron binding energy can be described with a hydrogenic model, one calculates an effective dielectric constant close to the static value. This result is at variance with the commonly accepted ideas concerning electron lattice coupling.

1. REVIEW OF RECENT WORK

FEW materials have such a wide range of interesting physical properties as TiO₂ (rutile). The long list of publications includes investigations of the crystal structure and the dielectric constant of the pure substance, as well as studies of the semiconductivity, magnetic behavior, and optical absorption of reduced and "doped" specimens. An extensive survey of the literature on rutile up till 1958 has been made by Grant.[1] A considerable portion of this work deals with the defect structure and electronic properties of reduced TiO₂. Most investigators come to the conclusion that the imperfections in non-stoichiometric rutile are oxygen vacancies possibly associated with Ti^{3+} ions. These centers would act as doubly charged donors giving rise to two observable activation energies in the electrical resistivity[2] and in the optical absorption.[3]

A number of investigations of recent years, however, have cast some doubt on these interpretations. Hurlen[4] has discussed the defect structure of non-stoichiometric rutile using data on lattice parameters, weight change, electrical conductivity (as a function of oxygen pressure), and oxidation studies. He concludes that of the two alternatives, oxygen vacancies or titanium interstitials, the evidence is somewhat in favor of the latter. The oxygen pressure dependence of conductivity and thermoelectric power of rutile (and other oxides) has been investigated by Rudolph.[5] His results indicate an intrinsic energy gap of 3.12 ev, in good agreement with optical data. A very interesting part of this work is the behavior of iron doped rutile: In a certain pressure range at high temperatures, these samples show p-type conduction (positive thermo emf).

Very important information has been obtained from electron spin resonance experiments. This research has been stimulated by the suitability of rutile for maser applications. Chester[6] has studied reduced crystals as well as Ta- and Nb-doped samples. A large number of paramagnetic centers has been observed by Young et al.[7] in gamma or ultraviolet irradiated rutile. Crystals doped with Cr, V, and Ni have been investigated by Gerritsen et al.[8] The paramagnetic resonance spectrum of iron in rutile has been measured and analyzed by Carter and Okaya[9] and by Low et al.[10]

A pattern of narrow lines found in both reduced and irradiated TiO₂ can be attributed to an electron located on Ti ions. The weak hyperfine structure observed by Young (apparently due to the small number of Ti nuclei with spins $\frac{5}{2}$ and $\frac{7}{2}$) strongly confirms this interpretation. Chester has found that the spectrum of the center in reduced rutile depends on the method and degree of reduction. With the magnetic field parallel to the a plane, he observes a set of four lines that might be caused by interstitial Ti^{3+} or by perturbations of other defects on the environment of regular Ti^{3+} ions.

In all doped rutile samples (with the possible exception of Ni-doped crystals), the spectrum shows that the foreign ion replaces the Ti ion on a regular lattice site and causes a hyperfine splitting in accordance with the nuclear spin of the substituting element. There is, however, still considerable confusion about the interpretation of the anisotropy of the different g factors.

The optical absorption of medium and strongly reduced rutile crystals has been measured by Cronemeyer (at 300°K).[3] It appears that the absorption is proportional to the room temperature conductivity of the samples. The author's interpretation is given in terms of the model mentioned in the first paragraph of this paper. This model is untenable if one considers the

---

* This work was supported by the Office of Naval Research.

[1] F. A. Grant, Revs. Modern Phys. **31**, 646 (1959).
[2] R. G. Breckenridge and W. R. Hosler, Phys. Rev. **91**, 793 (1953).
[3] D. C. Cronemeyer, Phys. Rev. **113**, 1222 (1959).
[4] Tor Hurlen, Acta Chem. Scand. **13**, 365 (1959).
[5] J. Rudolph, Z. Naturforsch. **14a**, 727 (1959).
[6] P. F. Chester, Bull. Am. Phys. Soc. **5**, 73 (1960); J. Appl. Phys. **32**, 866 (1961); Westinghouse Research Rept. 403FD449–R5; J. Appl. Phys. **32**, 2233 (1961).
[7] C. G. Young, A. J. Shuskus, and O. R. Gilliam, Bull. Am. Phys. Soc. **6**, 248 (1961).
[8] H. J. Gerritsen, S. E. Harrison, H. R. Lewis, and J. P. Wittke, Phys. Rev. Letters **2**, 153 (1959); H. J. Gerritsen and H. R. Lewis, Phys. Rev. **119**, 1010 (1960); H. J. Gerritsen and E. Sabisky, Bull. Am. Phys. Soc. **6**, 235 (1961).
[9] D. L. Carter and A. Okaya, Phys. Rev. **118**, 1485 (1960).
[10] W. Low (private communication).

TABLE I. Characteristics of reduced and Nb-doped rutile samples.

Sample	Treatment			m^*/m_0 (400–500°K)	E_2(ev)[a] (~20°K)	E_3(ev)[a] (~300°K)
(1) $l \perp c$	8 hr	1000°C	~10^{-3} mm Hg	22	7×10^{-3}	8×10^{-2}
(2) $l \perp c$	1 hr	1050°C	~10^{-3} mm Hg	32	10×10^{-3}	7×10^{-2}
(3) $l \perp c$	3 hr	1200°C	<10^{-5} mm Hg	20	···	$\geqslant 5 \times 10^{-2}$
(4) $\langle lc \rangle 45°$	$\frac{1}{2}$ hr	1000°C	~10^{-3} mm Hg	12	5×10^{-3}	4×10^{-2}
(5) $l \perp c$	Nb-doped—re-oxidized at 1 atm of O_2 pressure and 700°C			23	25×10^{-3}	8×10^{-2}

[a] E_2 and E_3 are defined as $(-k/e) \times$ slope of $\ln(RT^{3/2})$ vs $1/T$.

evidence presented by the ESR measurements. In this connection, it is interesting to report that the absorption band observed in Nb-doped specimens at room temperature is very similar to that of reduced samples.[11]

Soffer[12] has shown that a small but very sharp absorption band at 0.4 ev observed in all rutile single crystals is caused by OH. Considering that the crystals are grown in an oxygen-hydrogen flame, the presence of such a group is not surprising. Although the concentration of this impurity is appreciable, the OH group has only recently been taken it into consideration when interpreting any of the physical properties of TiO_2.[6]

Work on the dielectric constant and dielectric losses is somewhat outside the scope of this conference. The interested reader is referred to the literature.[13–18] We will limit our comments to the following remark: So far no satisfactory absorption mechanism has been suggested to explain observed dielectric losses when certain impurities were introduced.

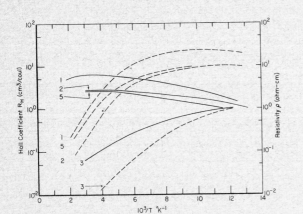

FIG. 1. Resistivity (——) and Hall coefficient (– – –) of reduced and Nb-doped rutile single crystals above 78°K. The numbers refer to the samples listed in Table I.

Measurements and analysis of electrical conductivity, Hall coefficient, and thermoelectric power of medium reduced rutile single crystals were reported briefly last year by Frederikse et al.[19] From the transport data and the ESR results, a model of electronic conduction in rutile was derived. According to this model, electrons are self-trapped on Ti ions (polarons); the observed conduction is due to electrons that are thermally excited from the polaron state into the conduction band. The second part of this paper describes this work in detail.

Results of experiments concerning the piezoresistivity of rutile single crystals have been reported by Hollander.[20] The same author has also investigated the anisotropy of the resistivity in rutile.[21]

2. ELECTRONIC TRANSPORT IN REDUCED RUTILE

The resistivity ρ, Hall coefficient R, and thermoelectric power Q of a number of monocrystalline rutile samples were determined. The reduction or doping processes to which these specimens were exposed are indicated in Table I. Their behavior is representative for rutile with resistivities in the range 0.1–10 ohm cm. The Nb-doped crystal was grown with Nb addition at MIT[22]; samples cut from this boule were oxidized at 700°C in order to prevent the presence of titanium interstitials or oxygen vacancies. Considering the results of spin resonance experiments performed on an identical sample from this boule, we believe that this goal was achieved.

The sample holder and dewar arrangement for low-temperature determinations of conductivity and Hall coefficient have been discussed in earlier publications.[23] Leads were soldered to the sample with indium. Thermoelectric power measurements were made in an apparatus similar to that described by Scanlon.[24]

[11] R. F. Blunt (private communication).

[12] B. H. Soffer, Techn. Rept. 140, Lab. for Insul. Research, MIT, August, 1959.

[13] Rebecca A. Parker and J. H. Wasilik, Phys. Rev. 120, 1631 (1960).

[14] L. Nicolini, Nuovo cimento 13, 257 (1959).

[15] Rebecca A. Parker, Phys. Rev. (to be published).

[16] K. G. Srivastava, Phys. Rev. 119, 516 (1960).

[17] J. van Keymeulen, Naturwissenschaften 45, 56 (1958).

[18] N. P. Bogoroditskii and I. D. Fridberg, Elec. Technol. U. S. S. R. (English translation) 2, 259 (1959); Zhur. Tekh. Fiz. 26, 1890 (1956) [translation, Soviet Phys.—Tech. Phys. 1, 1826 (1957)].

[19] H. P. R. Frederikse, W. R. Hosler, and J. H. Becker, Proceedings of the International Conference on Semiconductor Physics, Prague, 1960 (Publishing House of the Czechoslovak Academy of Sciences, Prague, 1961), paper R7, p. 868.

[20] L. E. Hollander, T. J. Diesel, and G. L. Vick, Phys. Rev 117, 1469 (1960).

[21] L. E. Hollander and Patricia L. Castro, Phys. Rev. 119, 1882 (1960).

[22] A. Linz, Jr. kindly made this crystal available to us.

[23] H. P. R. Frederikse, W. R. Hosler, and D. E. Roberts, Phys. Rev. 103, 67 (1956).

[24] H. P. R. Frederikse, V. A. Johnson, and W. W. Scanlon, Methods of Experimental Physics, edited by L. Marton (Academic Press, Inc., New York, 1959), Vol. 6B, p. 121.

Welded joints of the Pt-Pt (Rh) thermocouples were jammed into holes drilled into the samples with the use of a Cavitron. All measurements were made in vacuum. As the samples show additional reduction above 500°C, no measurements above this temperature are reported in this paper.

Results of these measurements are shown in Figs. 1–4 and Table I. One can easily distinguish two regions: below 50°K where the activation energies are $(5\text{-}10) \times 10^{-3}$ ev (except for the Nb-doped sample), and around room temperature where higher activation energies of the order $(4\text{-}8) \times 10^{-2}$ ev are observed. Hall mobilities at 300°K are approximately 0.2 and 1.0 cm²/v sec in the a and c directions, respectively. At low temperatures, however, values of 10^2–10^3 cm²/v sec are obtained. Above 150°K, the temperature dependence of the mobility approaches exp $(0.10/kT)$.

Using the well-known expressions for the resistivity, Hall coefficient, and thermoelectric power of a nondegenerate electron gas,[25] one calculates an effective mass m^* of 12-32 m_0. This figure and the order of magnitude of the mobility indicate that we are dealing with electronic transport in a narrow conduction band. We identify this band with the lowest $3d$ band of the titanium ions. The exponential temperature variation of the mobility at higher temperatures leads to the conclusion that the scattering is due to optical lattice modes.

The Hall coefficient and the conductivity of the Nb-doped sample show a considerably larger activation energy at low temperatures. Above 60°K, the temperature dependence of these parameters is identical with that of the reduced samples; the effective mass and mobility at room temperature are also the same.

Now the question is to find a suitable model of

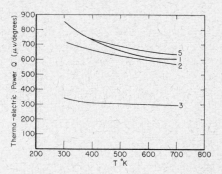

FIG. 3. Thermoelectric power of reduced and Nb-doped rutile single crystals above room temperature. The numbers refer to the samples listed in Table I.

electronic energy states that will explain—at least qualitatively—the temperature behavior of the conductivity and Hall coefficient. The electron spin resonance experiments indicate that, in irradiated and in reduced rutile, the electrons are located on Ti³⁺ ions at low temperatures. The centers observed in irradiated rutile are presumably Ti³⁺ ions on regular lattice sites. In reduced rutile, at 4°K (and below), the resonance spectrum seems to indicate more than one center. Chester[6] has suggested the following possibilities: Ti³⁺ ions on regular lattice sites, interstitial Ti³⁺ ions, or "regular" Ti³⁺ ions perturbed by neighboring defects. At somewhat higher temperature (8°K), the spectrum is compatible with electrons moving between "regular" Ti⁴⁺ ions. The energy difference between these different sites will probably be very small. One can visualize that, at very low temperatures, the electrons are distributed over interstitial and regular lattice sites with a preference for the former according to a Boltzmann factor. On raising the temperature, electrons from interstitial (or perturbed) lattice sites will move among regular cation positions.

It is furthermore assumed that the extra electrons produced by the reduction are bound to the Ti ions by self-polarization of the surrounding lattice (self-trapped

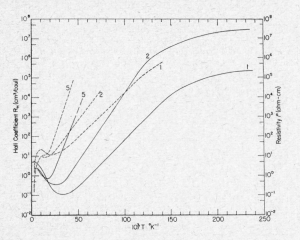

FIG. 2. Resistivity (——) and Hall coefficient (– – –) of reduced and Nb-doped rutile single crystals at low temperatures. The numbers refer to the samples listed in Table I.

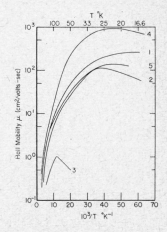

FIG. 4. Hall mobility of electrons in reduced and Nb-doped rutile single crystals. The numbers refer to the samples listed in Table I.

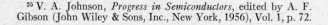

[25] V. A. Johnson, *Progress in Semiconductors*, edited by A. F. Gibson (John Wiley & Sons, Inc., New York, 1956), Vol. 1, p. 72.

electron[26] or polaron[27,28]). We believe that the potential binding such an electron can be approximated by $-e^2/\epsilon^* r$.[29] The energy of the ground state in such a well will be given by

$$E_s = (e^4 m^*)/(2h^2 \epsilon^{*2}). \qquad (1)$$

There are 3 different mechanisms that allow the electron to conduct[30-32]: (a) it can jump by thermal activation from one site to the next, (b) the different states have enough overlap to give rise to a polaron band and the electron *with* its polarization can move in such a band, or (c) through thermal activation (phonons) the electron can escape from its polarization and is able to travel large distances as a "bare" electron in the conduction band until it is stopped and becomes self-trapped again.

Considering the ability to measure the Hall coefficient and the very appreciable Hall mobilities, it is our feeling that the observed conduction is due to "bare" electrons. Hence the binding energy E_s is identified with the activation energy of 0.01 ev observed in reduced rutile samples below 50°K. Using the value of 25 for m^*, one calculates an effective dielectric constant of 185. It appears now that ϵ^* is of the same order as the static dielectric constant ($\epsilon_{st}{}^c = 260$ and $\epsilon_{st}{}^a = 140$ at $T = 0°K$).[15] The commonly accepted value of ϵ^* used in the polaron theory[27,28] is $[(1/\epsilon_{op}) - (1/\epsilon_{st})]^{-1}$. As $\epsilon_{op} \approx 7$ for TiO_2, this expression reduces to $\epsilon^* \approx \epsilon_{op}$, and the binding energy would be very much larger.

There are, however, two major objections to the derivation of the electron-lattice coupling as presented in the literature. First, the calculation uses the macroscopic electric and polarization fields which are applicable only if the interaction is long range. Secondly, it is assumed that the polarization can be sharply divided into an ionic and an electronic part. However, as the frequency region of interest is nearly always the dispersion range, such a division is unrealistic.

In the case of TiO_2, the electron energy ($\hbar\omega_{el} = 0.01$ ev) is somewhat smaller than the lattice energy ($\hbar\omega_L = 0.04$-0.06 ev). The ions will be able to follow the "slow" motion of the electron rather closely; the *dynamic* interaction between the oscillating electron and the oscillating lattice will be weak, and hence the binding energy will be small.

As stated before, we believe that the electron can occasionally be freed from its self-induced polarization and become a "bare" electron (or conduction electron). Such a process will be the result of a fluctuation of the

potential well caused by optical phonons of frequency $\nu_i (\sim 10^{13} \text{ sec}^{-1})$, where i indicates different lattice modes. (Even one phonon has enough energy to evict the electron from its well.) The frequency with which this process will take place is given by

$$1/\tau_s = \nu = \sum_i \nu_i e^{-E_s(\nu_i)/kT}. \qquad (2)$$

τ_s is the mean time of stay of an electron in its well. This time changes rapidly, $\tau_s \sim 10^{-2}$ sec at 4°K and $\sim 3 \times 10^{-11}$ sec at 40°K. At about 100°K, the frequency of eviction will approach ν_i.

The temperature dependence of the Hall coefficient indicates that above 100°K the activation energy E_s increases by about one order of magnitude. We do not believe that this activation energy is connected with any unknown impurity in the crystal. Considering that the conduction behavior of Nb-doped rutile is very similar to that of reduced TiO_2 (Figs. 1-4), one is led to the conclusion that the "donor" is Ti^{3+} in both cases.

Although the increase of E_s is poorly understood, several qualitative arguments can be presented that may elucidate the problem. First, the static dielectric constant decreases with rising temperature[15]; if we accept our conclusion that $\epsilon^* \approx \epsilon_{stat}$, then a decrease of ϵ_{stat} will result in a higher activation energy. However, this effect is too small to explain the total change of E_s.

Another way of looking at this phenomenon is the following. The use of Eq. (1) more or less implies that we consider the self-trapped electron as a donor center. In that case, m^* should be equated with the polaron mass, m_{pol}.[33] Sewell[34] has shown that m_{pol} increases with temperature as $\exp[\coth\gamma_l{}^0(\hbar\omega_L/2kT)]$ (where $\gamma_l{}^0 \approx 3$) leading to the observed result. This explanation would certainly be applicable to the binding energy of an interstitial Ti^{3+} ion.

Yet another approach is based on the fact that, in the neighborhood of 100°K, the frequency of excitation becomes comparable with the fundamental lattice frequency. When the mean time of stay of an electron on a cation is very short, the average polarization of its environment is small. Hence the effective dielectric constant which describes the local situation above 100°K will be much smaller than the low temperature ϵ^*, and the binding energy will be correspondingly larger. One can also formulate this by saying that the dielectric shielding will diminish when the electron does not remain localized long enough to produce a large polarization.

To explain the temperature behavior of the transport properties of Nb-doped rutile, the following model is suggested. In the temperature range below 60°K, the donor levels are Nb^{4+} ions; the observed Nb hfs leads immediately to this conclusion.[6] Once the electrons are

[26] L. Landau, Physik. Z. Sowjetunion **3**, 664 (1933).

[27] S. I. Pekar, *Studies of the Electron Theory of Crystals* (Gostekhizdat, 1951). Translated into German.

[28] H. Fröhlich, H. Pelzer, and S. Zienau, Phil. Mag. **41**, 221 (1950).

[29] N. F. Mott and R. W. Gurney, *Electronic Processes in Ionic Crystals* (Clarendon Press, Oxford, England, 1948), p. 87.

[30] G. H. Wannier, *Elements of Solid State Theory* (Cambridge University Press, New York, 1959), p. 169.

[31] T. Holstein, Ann. Phys. **8**, 325, 343 (1959).

[32] Jiro Yamashita, J. Appl. Phys. **32**, 2215 (1961).

[33] Conyers Herring, *Photoconductivity Conference*, edited by Breckenridge, Russell, and Hahn (John Wiley & Sons, Inc., New York, 1956), p. 93.

[34] G. L. Sewell, Phil. Mag. **3**, 1361 (1958).

thermally excited from these donors, they can still be self-trapped on a Ti ion. It requires then the higher activation energy (0.08 ev) of the Ti^{3+} sites to transform them into "bare" electrons in the 3d band of the cations.

The model of conduction described in this paper differs considerably from the diffusion theory developed for certain oxides by Morin,[35] Heikes and Johnston,[36] van Houten,[37] etc. For all the oxides studied by these workers, only the conductivity (and thermoelectric power) were measured; no data on Hall coefficient could be obtained. An exponentially increasing conductivity can be explained in two ways: either the mobility or the number of carriers is thermally activated. The assumption of a diffusion mechanism automatically leads to the former. However, if one is able to measure the Hall coefficient and finds that this parameter decreases exponentially with temperature, the evidence is strongly in favor of carrier excitation.

ACKNOWLEDGMENTS

Thanks are due W. R. Hosler for taking the data on transport properties presented in part 2 of this paper. The author expresses his appreciation to Dr. P. F. Chester for transmission and discussion of his data prior to publication. It is a pleasure to acknowledge the fruitful discussions with Dr. A. H. Kahn.

[35] F. J. Morin, Bell System Tech. J. **37**, 1047 (1958).
[36] R. R. Heikes and W. D. Johnston, J. Chem. Phys. **26**, 582 (1957).
[37] S. van Houten, J. Phys. Chem. Solids **17**, 7 (1960).

JOURNAL OF APPLIED PHYSICS   SUPPLEMENT TO VOL. 32, NO. 10   OCTOBER, 1961

Heitler-London Approach to Electrical Conductivity

JIRO YAMASHITA
The Institute of Solid State Physics, University of Tokyo, Tokyo, Japan

A Heitler-London approach to electrical conductivity is proposed in order to discuss conduction in semiconductors with incomplete d shells, in which the mobility of carriers is very low. The physical conditions which make the hopping motion of the electrons predominant is examined in detail.

I. INTRODUCTION

IN this paper we shall consider the problem of conduction in semiconductors with incomplete d shells, in which the mobility of carriers is very low. A typical example of these materials is nickel oxide. It seems well established experimentally that pure NiO has a very high electrical resistance and the observed electrical conductivity in less pure NiO is associated with the presence of Ni^{3+} ions in the lattice. It is well known that alloying of Li$_2$O with NiO in a homogeneous solid solution leads to semiconducting material with a remarkable range of conductivity depending on the lithium content. The temperature coefficient of conductivity is also strongly dependent on the lithium concentration. A number of other crystals, for examples, LaMnO$_3$, Fe$_2$O$_3$, magnetites, and some ferrites, exhibit similar conduction properties. We notice that some organic semiconductors also show similar properties.

It does not seem appropriate to apply the band theory to conduction electrons of these materials. The band theory will give a very narrow band and a very heavy effective mass. The speed of an electron in such a band is so small, that the time to traverse an interatomic distance is comparable with the electronic mean free time between collisions. If an electron is scattered in a time comparable with its time of flight for one lattice period, then it cannot experience the effect of the lattice periodicity, and **k** is not a good quantum number. Since very low mobility is observed in the materials mentioned previously, another approach may be necessary to discuss the problem of conduction in these materials. In the following we shall mention one of these possible approaches, which is, in fact, based on a rather oversimplified model of the very complicated real situations.

II. TIGHTLY BOUND ELECTRON IN PERFECT LATTICE

Let us imagine a perfect periodic lattice, which has insulating properties in a usual condition. Now, we insert an extra electron in the lattice, or take out an electron from the lattice, and suppose that the extension of the wave function of the electron is fairly small as compared with the lattice spacing. If the extra electron stays around an atom in the lattice for a sufficiently long time as compared with the period of the lattice vibration, then the electrical polarization, in a polar lattice, or the elastic deformation, in a nonpolar lattice, will be induced around the atom so as to minimize the total energy. If the energy of this localized state is lower than that of the usual Bloch state, then we might suppose that a kind of the self-trapping of an electron is realized. Here we notice that Holstein[1] has introduced a very simple, but very convenient model. Let us imagine an one-dimensional lattice which is

[1] T. Holstein, Ann. Phys. **8**, 343 (1959).

composed of hydrogen molecules with a certain lattice constant. If we take out one electron from the lattice, then we may imagine two possible states for the positive hole. One is the state represented by a localized wave function of the hole to a special hydrogen molecule, that is, a hydrogen molecular ion in the otherwise perfect lattice. In this case the internuclear distance of the molecular ion is changed so as to minimize the energy. Then, the system gains the energy corresponding to this process. The other state is represented by a running wave through the lattice and the hole gains the energy from a band formation. On the other hand, the motion of the hole is so fast, that any hydrogen molecules remain in the initial state irrespective of the motion of the hole, and there is no energy-gain caused by the deformation of the molecule. Which of the states has a lower energy? It depends upon the lattice spacing. If the lattice spacing is sufficiently large, the localized state may have a lower energy, and vice versa, if the lattice spacing is small. The localized state, however, is not the state of the lowest energy, if the lattice potential is perfectly periodic, because there exist matrix elements, though small, for transition of the hole between nearest neighbor orbits and the linear combination of such localized states gives a lower energy. Then, the electron makes a translational motion drawing the lattice distortion with it and its effective mass becomes heavier.

Since the mathematical treatment of the problem in three-dimensional cases is very complicated,[2-5] we shall adopt the one-dimensional molecular crystal model and follow Holstein's treatment. We assume that the wave function of the system has the form

$$\psi(\mathbf{r}, x_1, \cdots x_N) = \sum_n a_n(x_1, \cdots x_N)\phi(\mathbf{r}-n\mathbf{a}, x_n), \quad (1)$$

where $\varphi(\mathbf{r}-n\mathbf{a}, x_n)$ is a one-electron wave function, localized about the nth molecular site, x_i is the deviation of an internuclear separation of an ith molecule from the equilibrium position. The wave function of the lattice vibration $a_n(x_1, x_2, \cdots)$ is determined by the following equation:

$$i\hbar\frac{\partial a_n(x_1, \cdots x_N)}{\partial t}$$

$$= \left[\sum_{m=1}^{N}\left(-\frac{\hbar^2}{2M}\frac{\partial^2}{\partial x_m^2}+\frac{1}{2}M\omega_0^2 x_m^2\right.\right.$$

$$\left.\left.+\frac{1}{2}M\omega_1^2 x_m x_{m+1}\right)-Ax_n\right]a_n - J(a_{n+1}+a_{n-1}). \quad (2)$$

Here J is the transition matrix of the electron which is

defined by

$$J(x_n, x_{n\pm1}) = \int \phi^*(\mathbf{r}-n\mathbf{a}, x_n)U(\mathbf{r}-n\mathbf{a}, x_n)$$

$$\times\phi[\mathbf{r}-(n\pm1)\mathbf{a}, x_{n\pm1}]dv, \quad (3)$$

and finally $(-Ax_n)$ is the electron-lattice interaction energy. Introducing a set of the normal mode coordinates q_k we have

$$i\hbar\frac{\partial a_n(\cdots q_k\cdots)}{\partial t} = \sum_k\left[\left(-\frac{\hbar^2}{2M}\frac{\partial^2}{\partial q_k^2}+\frac{1}{2}M\omega_k^2 q_k^2\right)\right.$$

$$\left.-\left(\frac{2}{N}\right)^{\frac{1}{2}}Aq_k\sin(nka+\pi/4)\right]$$

$$\times a_n(\cdots q_k\cdots) - J(a_{n+1}+a_{n-1}), \quad (4)$$

where $\omega_k^2 = \omega_0^2 + \omega_1^2\cos(ka)$. Then, we see that, if we put $J=0$, the eigenstate is easily determined as

$$a_n^0(\cdots q_k\cdots) = \prod_k \Phi_{N_k}[(M\omega_k/\hbar)^{\frac{1}{2}}(q_k-q_k^{(n)})], \quad (5)$$

where Φ is a harmonic oscillator eigenfunction with an equilibrium point

$$q_k^{(n)} = (A/M\omega_k^2)(2/N)^{\frac{1}{2}}\sin(nka+\pi/4). \quad (6)$$

The wave function

$$\Psi^0 = \varphi(\mathbf{r}-n\mathbf{a})\prod_k \Phi_{N_k}[(M\omega_k/\hbar)^{\frac{1}{2}}(q_k-q_k^{(n)})] \quad (7)$$

represents the localized state around the nth molecule. Then, we find easily that the wave function of the total system is represented by a superposition of the isolated wave function (7):

$$\Psi_\sigma = \sum_n \exp(i\sigma na)\varphi(\mathbf{r}-n\mathbf{a}, x_n)$$

$$\times\prod_k \Phi_{N_k}[(M\omega_k/\hbar)^{\frac{1}{2}}(q_k-q_k^{(n)})], \quad (8)$$

and the total energy of the system is given by

$$E_\sigma = E_b + \sum_k (N_k+\tfrac{1}{2})\hbar\omega_k - 2J\cos(\sigma a)\exp(-S_T), \quad (9)$$

where E_b is the deformation energy of a molecule and S_T is defined by

$$S_T = \sum_k (\gamma_k/N)\coth(\hbar\omega_k/2kT),$$

$$\gamma_k = A^2(1-\cos ka)/2M\hbar\omega_k^3. \quad (10)$$

The factor $\exp(-S_T)$ in (8) comes from the matrix element between the oscillator part of the wave functions, and it makes the bandwidth quite small when the lattice deformation around the electron is considerably large. The bandwidth is maximum at absolute zero and diminishes rapidly with increasing temperature.

[2] S. V. Tjablikov, Zhur. Eksptl. i Theoret. Fiz. 23, 381 (1952).
[3] G. L. Sewell, Phil. Mag. 3, 1361 (1958).
[4] J. Yamashita and T. Kurosawa, J. Phys. Chem. Solids 5, 34 (1958).
[5] J. Yamashita and T. Kurosawa, J. Phys. Soc. Japan 15, 802 (1960).

III. HOPPING MOTION OF THE ELECTRON

The ideal situation, that is, the perfectly periodic lattice potential, does not seem to be realized in crystals with which we are concerned in this paper. We expect rather that the potential is fluctuating from lattice point to lattice point owing to the very existence of many impurity ions. If we denote the transition probability of an extra electron from one atom to its nearest neighbor as W, then the energy level of the state has a natural width of the order of $\hbar W$. If the amount of the potential difference between nearest neighbor lattice points is larger than $\hbar W$, the electron cannot make a translational motion from ion to ion, that is, the motion of the electron as propagating waves is prohibited. We note that if the transition probability W is large as in usual semiconductors, a small amount of fluctuation in potential is regarded as a small perturbation for propagating motion of electron waves. The situation is, however, completely different when W is very small.

Although the propagating motion of electron waves is impossible, the localized state around an ion is a stationary state only in the first approximation. The electron will jump from an ion to its nearest neighbor ions with a certain probability by the effect of other ions, which is regarded as a perturbation. This transition of the electron must be accompanied by emission and absorption of many phonons, because the energy of the system must be conserved on the transition. Hereafter we shall distinguish two kinds of the matrix element of transition; one in which the vibration quantum numbers (N_k) remain unchanged, the other in which some of these numbers change by one unit. We call the former diagonal and the latter nondiagonal. Here we are concerned with the nondiagonal matrix elements. We need the thermal average of the transition probability:

$$W_T(n\,|\,m)=Z^{-1}\sum_{\cdots N_k\cdots} W_{N_k}(n\,|\,m)$$
$$\times\exp[-\sum_k(N_k+\tfrac{1}{2})\hbar\omega_k/kT], \quad (11)$$

where

$$W_{N_k}(n\,|\,m)=\sum_{\cdots N_k\cdots} W(n,\cdots N_k\cdots\,|\,m,\cdots N_k'\cdots),$$
$$W(n,\cdots N_k\cdots\,|\,m,\cdots N_k'\cdots)$$
$$=\frac{2}{\hbar^2}|(n,\cdots N_k\cdots\,|\,U\,|\,m,\cdots N_k'\cdots)|^2\frac{\partial}{\partial t}\Omega(x) \quad (12)$$

and $\Omega(x)$ is

$$\Omega(x)=2\sin^2(xt/2\hbar)/(x^2/\hbar^2). \quad (13)$$

Here x is defined by

$$x=\sum_k\hbar\omega_k(N_k-N_k')+\Delta E, \quad (14)$$

where ΔE is the difference of the potential energy be-
tween two cells before and after transition. In principle the matrix element of the transition $(n,\cdots N_k\cdots\,|\,U\,|\,m,\cdots N_k'\cdots)$ is evaluated by the standard perturbation theory. Here we assume that the matrix element is negligibly small, unless the lattice position m is one of nearest neighbors of the position n. Then, we see that the motion of the electron is represented by a series of random hopping motions.

Although the possibility of such a hopping motion seems to be self-evident under the circumstances mentioned previously, it is also possible to prove it by a more fundamental approach.[6] The problem is to prove that such a wandering motion of the electron obeys the transport equation

$$\frac{dP_n(t)}{dt}=\sum_m\left[W(n,m)P_m(t)-W(m,n)P_n(t)\right] \quad (15)$$

under certain conditions. Here $P_n(t)$ is the probability that an electron is situated at the ion n at the time t, and $W(n,m)$ is the probability of the transition of the electron from the ion n to the ion m. In fact, Van Hove[7] has proved that the statistical behaviors of a sufficiently large system started from an appropriate initial state obey such an equation as (15), if the perturbing Hamiltonian is sufficiently small and has characteristic singular properties. In our case, however, the perturbing Hamiltonian, which causes the hopping motion of the electron from an ion to another, has extra singularities in addition to those required by Van Hove. They work to maintain the wavelike propagating motion of the electron, so that Eq. (15) does not hold exactly, if they remain to work. If the crystal potential, however, fluctuates from cell to cell owing to randomly distributed impurity ions and consequently the electron can find no place of the equal potential energy among the nearest and second nearest neighbor ions, then following the Van Hove method we can prove that the extra singularities cause no wavelike motion and Eq. (15) will be valid.

For the one-dimensional molecular crystal model the transition probability is evaluated as follows:

$$W_T(n\,|\,n\pm1;\,\Delta E)=(J/\hbar)^2\exp(-2S_T)\exp(-\Delta E/2kT)$$
$$\times\int_{-\infty}^{\infty}\left\{\exp\left[\sum_k\frac{2\gamma_k}{N}\operatorname{csch}\left(\frac{\hbar\omega_k}{2kT}\right)\cos\omega_k\tau\right]-1\right\}d\tau. \quad (16)$$

For our present purpose the necessary quantity is not the function $W(n\,|\,n+1;\,\Delta E)$ itself, but the average of it with respect to the probable distribution of the final states $\rho(\Delta E)$:

$$\bar{W}=\int_{-\infty}^{\infty}\rho(\Delta E)W_T(n\,|\,n\pm1;\,\Delta E)d(\Delta E). \quad (17)$$

[6] T. Kurosawa, J. Phys. Soc. Japan 15, 1211 (1960).
[7] L. Van Hove, Physica 21, 517 (1955).

For the three-dimensional ionic crystals, where the optical frequency is given by $\omega_k = \omega_0 + \Delta\omega_k$, the transition probability $W(nm; \Delta E)$ is evaluated as

$$W_T(\Delta E) = \sum_{p=-\infty}^{\infty} \exp\{-S(2N_0+1)\} I_p\{2S[N_0(N_0+1)]^{\frac{1}{2}}\}$$

$$\times \exp(-p\hbar\omega_0/2kT) f_p(\Delta E - p\hbar\omega_0), \quad (18)$$

where N_0 is the number of phonons in thermal equilibrium, the function $I_p(z)$ is defined by

$$I_p(z) = \frac{1}{2\pi} \int_0^{2\pi} \exp\{ipx + 2\cos x\} dx, \quad (19)$$

and the function $f_p(\Delta E - p\hbar\omega_0)$ is, in general, a very complicated function which tends to the δ function: $\delta(\Delta E - p\hbar\omega_0)$ when $\Delta\omega_k$ tends to zero. The quantity S is defined by

$$S = \frac{1}{\hbar\omega_0} \{E^{(m)}(X_0^{m+1}) - E^{(m)}(X_0^m)\}. \quad (20)$$

Here $E^{(m)}(X_0^m)$ is the energy of the system when the electron is at the lattice site m and the lattice is in equilibrium after inducing polarization around m, while $E^{(m)}(X_0^{m+1})$ is the energy of the system when the lattice takes the equilibrium configuration as if the electron were at the lattice site $(m+1)$, but the electron is in fact at the site m.

When the density of the donor-impurity is very small, most of the electrons are bound to the vicinity of the impurity centers and only a small fraction of the electrons is activated from the centers. In this case we may approximately divide ΔE into two parts: one is the activation energy of the impurity center, the other is a fluctuation of the potential energy at the lattice sites far from the impurity centers. Thus, we can divide the electrons into two parts: electrons which are bound to the centers and electrons which participate the conduction. On the other hand, when the concentration of donor impurities is high, we can no more distinguish between conduction and bound electrons. We may say that each electron moves by hopping motion through some efficient path and participates in the conduction each with a certain probability.

Next, let us consider the case in which the lattice potential is not much perturbed by impurities, but the temperature is so high, that the nondiagonal transition is much more dominant than the diagonal transition. In this case there exists a wavelike motion of the electron, but the normal transit of the electron from ion to ion by the wavelike motion is so slow that it cannot compete with the transit by thermally activated process. Then, the wavelike motion is damped out so quickly, that it gives only a small contribution to the electric conduction. By using the one-dimensional molecular crystal model and the usual perturbation treatment Holstein evaluated the diagonal and nondiagonal

matrix elemets of the transition: (Here we put $\Delta E = 0$.)

$$W_T(n|n\pm 1) = W_T^d(n|n\pm 1) + W_T^{nd}(n|n\pm 1), \quad (21)$$

where

$$W_T^d(n|n\pm 1) = 2(J/\hbar)^2 \exp(-2S_T)t, \quad (22)$$

and

$$W_T^{nd}(n|n\pm 1) = (J/\hbar)^2 \exp(-2S_T)$$

$$\times \int_{-\infty}^{\infty} \left\{ \exp\left[\sum_k \frac{2\gamma_k}{N} \operatorname{csch}\left(\frac{\hbar\omega_k}{2kT} \right) \right. \right.$$

$$\left. \left. \times \cos(\omega_k\tau) \right] - 1 \right\} d\tau. \quad (23)$$

We see that the diagonal contribution increases indefinitely with time. This difficulty is only apparent, because the probability of the initial state decreases very rapidly by the nondiagonal transitions. By considering damping of the initial state through nondiagonal transitions by the well-known Weisskopf-Wigner perturbation treatment we find the total transition probability is approximately given by,

$$W_T(n|n\pm 1) = W_T^{nd}(n|n\pm 1)$$

$$\times \left(1 + \frac{2(J/\hbar)^2 \exp(-2S_T)}{W_T^{nd}(n|n\pm 1)} \right). \quad (24)$$

Here we assume that

$$W_T^{nd}(n|n\pm 1) \gg (J/\hbar)^2 \exp(-2S_T).$$

We may also understand the situation by the idea of wave packet. Let us imagine that we construct a wave packet localized in a very narrow region around an ion from a set of Bloch functions. The packet has a natural tendency to spread out with time through the crystal. The rate of diffusion of the packet, however, depends upon the mass of the particle. If the bandwidth is very narrow and the effective mass is very large, the rate of diffusion is so small, that it cannot compete with the transit through activated states. Then, the latter processes will break the coherent motion of the packet.

IV. DISCUSSIONS

At present we are not very convinced that we have observed a real hopping current, although there are many cases which are likely to be so. Therefore, it seems to be worth while to mention some characteristics of the current which are expected from the theory. (a) The mobility is quite small. Since it takes a time of the order of the lattice vibration to induce lattice polarization, or lattice distortion around the electron, the lifetime of the localized state should be sufficiently long as compared with a period of the lattice vibration. Therefore, the condition that the H-L approach is good is given by: $\nu \gg W$. According to the Einstein relation we can determine the mobility from the transition probability W:

$$\mu = (ea^2/kT)W,$$

where a is the distance between nearest neighbor ions and τ is the relaxation time. Inserting the value $T=300$, $a=4\times10^{-8}$ cm, and $W=10^{13}$ sec^{-1} we have: $\mu=0.6$ cm^2/v sec. Next, let us investigate a condition for the Bloch theory of the conductivity. We set a condition for it:

$$h/\text{bandwidth} \ll \tau.$$

If we estimate the bandwidth by $\hbar^2 k_0{}^2/2m$, where k_0 is nearly equal to π/a we have: $\mu = e\tau/m (\gg 4/\pi)ea^2/\hbar \sim 2$ cm^2/v sec. This condition gives the lower bound of the mobility for the usual Bloch theory. Thus, we may roughly determine the borderline of the two alternative approaches. When the mobility is much larger than 1 cm^2/v sec, the usual Bloch theory is applicable, while the H-L approach may be good when the mobility is much smaller than 1 cm^2/v sec. (b) The mobility shows a positive temperature coefficient. (c) There is a connection between the absolute value and the activation energy of the mobility. The mobility is smaller, then the activation energy becomes larger. (d) There is a connection between the density of the impurity ion and the activation energy of the mobility. If the density of donors is low, the main part of the activation energy comes from the work to separate an electron from vicinity of the donor to a distant point in the lattice. In usual situations in the polar lattice this energy amounts to 1 ev. On the other hand, if the density of the donors is high, an electron which is at one of the nearest neighbors of a donor can transfer directly to another position which is again one of the nearest neighbors of another donor with nearly the same energy. Therefore, we expect that the activation energy of the current changes rather abruptly at some density of the donors. The critical density of donors may be 0.1. According to these characteristics we may pick up some examples which are likely to be a Heitler-London current.

The typical example is the NiO crystal doped with lithium,[8] where we can observe all of the four characteristics of the current mentioned previously. Another good example is the alloy of La$_{1-x}$Ca$_x$MnO$_3$.[9] When x is much smaller than 0.1, the crystal shows a semiconducting character with a large activation energy. When x becomes larger than 0.1, the conductivity increases rapidly with increasing x and moreover it shows a metallic temperature dependence. Recently similar characters of the current are found in some organic

semiconductors.[10] At present it is not certain whether the current is carried by the hopping process in these materials. But we may expect that some of them really have a hopping current. Here we do not discuss this interesting field in detail, but we only mention that the current in vioanthrene-iodine and similar materials has very interesting characters, which may be interpreted by some kind of the hopping process.[11] Another interesting example is M$^+$(TCNQ)$^-$(TCNQ)^0 which was found by Kepler and others.[12] We suppose that the current is carried by the exchange of charge among two kinds of anion radicals (TCNQ)$^-$ and (TCNQ)^0 rather than by the usual translational motion of charges.

We must notice, however, that we have no definite theoretical scheme to analyse the experimental results about the low-mobility materials. At present we do not know whether the Hall voltage can be observed in the H-L current, although it is quite certain that a part of the current which is accompanied with the wavelike motion of the electrons shows the usual Hall effect. Perhaps we might find some complicated hopping processes which cause the Hall voltage. However, it may be hard to expect that we can draw some useful information to determine the number of carriers from the observed Hall voltage. On the other hand, thermoelectric power of these materials has been observed for determining the sign and the number of carriers. We are not quite certain, however, that the thermoelectric power is really useful to give the desired information, because it does not seem permissible to disregard the contribution from the kinetic energy part of the hopping motion.[13] If it is necessary to consider this factor, the observed thermoelectric power is connected with a sum $(E-\zeta+Q)$ instead of $(E-\zeta)$, where Q is the activation energy of the hopping motion. As is well known, $(E-\zeta)$ is connected with the number of carriers by the relation

$$n = N \exp[-(E-\zeta)/kT],$$

but in this case we cannot determine $(E-\zeta)$ from the observed data, unless Q is very small as compared with $(E-\zeta)$. Another complication comes from the fact that the conductivity is not always expressed by a simple formula $\sigma = ne\mu$, but it may be expressed by a more complicated one $\sigma = e \sum n_i \mu_i$. If the density of donors is high, the observed current is a sum of many partial currents which follow efficient paths with different activation energies.

[8] F. J. Morin, Phys. Rev. **93**, 1199 (1954).

[9] E. J. W. Verwey, *Semiconducting Materials*, edited by H. K. Henish (Butterworths Scientific Publications, London, 1951), p. 151.

[10] C. G. B. Garett, *Semiconductors*, edited by N. B. Hannay (Reinhold Publishing Corporation, New York, 1959), p. 634.

[11] H. Akamatu and H. Inokuchi, *Proceedings of the Third Conference on Carbon* (Pergamon Press, New York, 1959), p. 51; H. Inokuchi, Bull. Chem. Soc. Japan **28**, 570 (1955).

[12] R. G. Kepler, P. E. Bierstedt, and R. E. Merrifield, Phys. Rev. Letters **5**, 503 (1960).

[13] M. Tuji, J. Phys. Soc. Japan **14**, 1640 (1959).

Investigations on SnS

W. ALBERS, C. HAAS, H. J. VINK, AND J. D. WASSCHER

Philips Research Laboratories, N. V. Philips' Gloeilampenfabrieken, Eindhoven, Netherlands

The p, T, x diagram of the Sn-S system was determined especially in the region of the compound SnS. The pressure of S_2 in equilibrium with SnS and a liquid phase was found to extend over several decades up to 25-mm Hg at the "Sn-rich" side, whereas at the "S-rich" side the S_2 pressures in equilibrium with solid SnS and a liquid phase lie between 25-mm Hg and 100-mm Hg. It was shown that the existence region of solid SnS very probably lies entirely at the excess sulfur side. The hole mobility in a plane perpendicular to the c axis, ≈ 90 cm²/v sec at room temperature, was proportional to $T^{-2.2}$ for higher temperatures. The mobility in the direction of the c axis was about five times smaller. Reversible annealing effects were found for temperatures above 200°C which could be explained by assuming association of neutral Sn vacancies. Absorption measurements showed that the edge absorption is due to indirect transitions. The bandgap was 1.08 ev at 300°K and 1.115 ev at 77°K. Interband transitions in the valence band were also found. The effective charge of the atoms ($e^* = 0.7e_0$) and the effective masses of the holes in the three principal crystal directions ($m_a^* = m_b^* = 0.20m_0$; $m_c^* \approx m_0$) were determined from reflection measurements in the infrared. From these values and the value for the density of states mass obtained by means of the Seebeck effect ($m_d^* \geqslant 0.95m_0$), the number of equivalent maxima of the valence band was found to be at least four.

I. INTRODUCTION

IN view of the extensive research carried out on the semiconducting compounds PbS, PbSe, and PbTe, it seemed worthwhile to study somewhat more closely the compounds SnS and SnSe which so far have attracted relatively little attention, the investigations of Asanabe and his co-workers[1,2] on SnSe being the most recent. An additional incentive for the study of SnS and SnSe is the fact that, where as PbS, PbSe, PbTe, and SnTe crystallize in the NaCl structure, the compounds SnS and SnSe have an anisotropic crystal structure. It was thought that, in view of the great number of cubic semiconductors presently under investigation, it would be interesting to study a strongly anisotropic one.

II. CRYSTAL STRUCTURE

According to the investigations of Hofmann,[3] SnS and SnSe have an orthorhombic structure that may be described as pseudotetragonal (Fig. 1). Every Sn atom is surrounded by three S atoms as nearest neighbors (drawn lines). These three bonds form angles with each other which are nearly 90°. Therefore in general one can say that the bonding type, apart from its ionic part, is covalent of the p^3 type.

Many of the semiconductors now under investigation show a bonding that, in its covalent part, is of the sp^3 type. This difference in type of bonding is another reason for studying these substances. Fig. 1 also shows clearly that SnS has a layer structure with double layers perpendicular to the c axis.

III. p, T, x DIAGRAM OF Sn-S

In order to grow single crystals under controlled conditions it is necessary to have a knowledge of the p, T, x diagram of the system Sn-S, particularly in the region of the compound SnS. Of special interest in this

connection is the study of the three-phase (solid-liquid-vapor) equilibrium lines of the system.

The relation between the composition of the liquid, the temperature, and the corresponding sulfur pressure along these three-phase lines have been determined by a method already used for other substances.[4–7] The T-x projection of this "liquidus" line[8] is shown in Fig. 2, together with some data of other investigators. The existence of the compounds SnS and SnS_2 is clearly indicated. The maximum melting points of these compounds are 881.5° and ≈ 870°C, respectively. At the Sn-rich side of SnS there is a large region of liquid immiscibility extending down to the quadruple point at 860°C, where solid SnS, a liquid phase having a composition close to SnS, a second liquid phase rich in Sn, and a vapor phase are in equilibrium.

The relation between the S_2 pressure and the temperature along the various three-phase lines is shown in Fig. 3. It is seen that the S_2 pressures in equilibrium with SnS and SnS_2 at their maximum melting points are 25 mm Hg and about 40 atm, respectively. From these two graphs a plot of Ps_2 vs composition of the liquid can be constructed for the three-phase equilibria (Fig. 4). The existence of the liquid immiscibility gap is clearly demonstrated by the horizontal part of the graph at the Sn-rich side of SnS. From the foregoing, it is clear that at the Sn-rich side of solid SnS the pressure of S_2 in equilibrium with SnS and a liquid phase extends over several decades up to 25 mm Hg, whereas at the S-rich side the corresponding S_2 pressure extends over a very limited range, viz. from 25 mm Hg to 100 mm Hg.

The S_2 pressure as a function of temperature of another three-phase equilibrium, viz. solid SnS at the S-rich side of its existence region, a solid phase of a

[1] S. Asanabe and A. Okazaki, Proc. Phys. Soc. (London) **A73**, 824 (1959).

[2] S. Asanabe, J. Phys. Soc. Japan **14**, 281 (1959).

[3] W. Hofmann, Z. Krist. **92**, 161 (1935).

[4] J. Bloem, Philips Research Repts. **11**, 273 (1956).

[5] J. Bloem and F. A. Kröger, Z. physik. Chem. (Frankfurt) **7**, 1 (1956).

[6] J. van den Boomgaard and K. Schol, Philips Research Repts. **12**, 127 (1957).

[7] D. de Nobel, Philips Research Repts. **14**, 357, 430 (1959).

[8] W. Albers and K. Schol, Philips Research Repts. (to be published).

higher S content (SnS$_2$), and vapor, is indicated by the dotted line in Fig. 3 but has not been determined experimentally. The highest S$_2$ pressure (100 mm Hg) of this three-phase equilibrium is given by the quadruple point at 740°C in Fig. 3, where a liquid phase is also present, this point being the lowest temperature (and highest S$_2$ pressure) where the three-phase equilibrium, solid SnS (at the S-rich side of its existence region), liquid, and vapor is possible.

The limited range of equilibrium S$_2$ pressures at the sulfur-rich side of the three-phase equilibria of SnS is unusual compared to substances like PbS,[4,5] CdTe[7] and CdS.[9]

With regard to the solidus line of the three-phase equilibria, i.e., the composition of solid SnS as a function of temperature and corresponding S$_2$ pressure, it must be borne in mind that the existence region of solid compounds is often very small.[5,7,10] It has thus often proved to be impossible to determine the compositions of the boundaries of the existence region of the solid by purely chemical means. However, if measurements of the Hall effect can be interpreted in terms of excess or defect concentrations of the constituents, it offers a very sensitive, though indirect, method of determining the composition of the solid along the three-phase equilibria. A necessary condition of the ability to do this is the certainty that no chemical impurities are present in a concentration sufficiently high to make the interpretation of the Hall measurements in terms of excess of

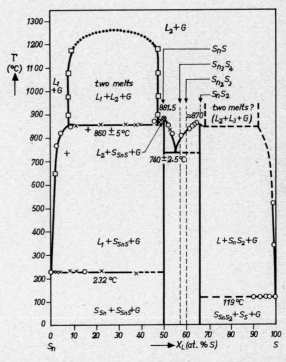

FIG. 2. T-x projection of the three-phase equilibria in the phase diagram of the system Sn-S, x_L giving the composition of the liquid phase. The composition of the vapor phase has been omitted. Our data are indicated by the circles.

one of the constituents doubtful. Therefore, in order to try to determine the solidus line, pure SnS is needed. This was obtained by a double distillation at 960°C in vacuum of SnS prepared from the purest Sn and S available. Spectrochemical analysis of this purified SnS showed that the concentration of all detectable elements was below the limits of detectability with the exception of Pb, which was still present at a concentration of $7 \cdot 10^{16}$ at./cm^3.

Sometimes the distillation itself gave crystals of sufficient size to carry out electrical or optical measurements; sometimes it was necessary to zone melt the material in a horizontal boat to get single crystal bars out of which samples could be cut.

Small crystals of pure SnS obtained in this way were brought into equilibrium at various temperatures with the pressure of S$_2$ corresponding to the three-phase equilibrium at those temperatures. The S$_2$ pressures needed were established either by the two-temperatures furnace method,[5-8] where the sulfur pressure in question must be known, or by the elegant method of Brebrick and Allgaier,[10] where exact knowledge of the sulphur pressures is not necessary. Hall measurements carried out on those crystals after quenching from the temperature of equilibrium to room temperature showed that the two methods mentioned above did not give results entirely consistent with each other, especially with regard to the dependence of the carrier concentration on the S$_2$ pressure. This question is still under study.

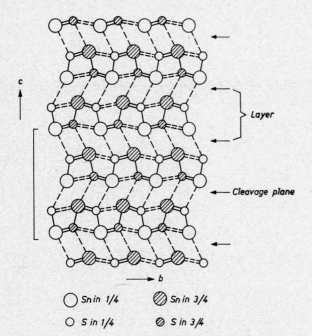

FIG. 1. Crystal structure of SnS according to Hofmann, drawn lines indicating bonds that are at least partly covalent.

[9] F. A. Kröger, H. J. Vink, and J. van den Boomgaard, Z. physik. Chem. (Leipzig) 203, 1 (1954).
[10] R. F. Brebrick and R. S. Allgaier, J. Chem. Phys. 32, 1826 (1960).

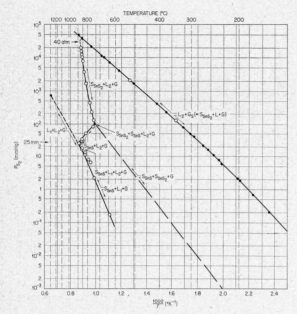

FIG. 3. The S_2 pressure, as a function of temperature, over various three-phase equilibria in the system Sn-S.

However, for both methods p-type conductivity was found with a carrier concentration of about 1–$3 \cdot 10^{18}$ holes per cm³ at both boundaries of the existence region of SnS. In view of this result, it seems reasonable to assume that the existence region of solid SnS at the temperatures concerned lies entirely at the excess sulfur side. However, with regard to the inconsistencies just

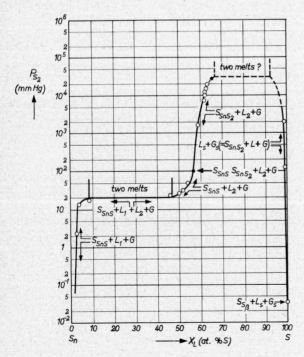

FIG. 4. The S_2 pressure, as a function of the composition of the liquid phase, over the various solid-liquid-vapour three-phase equilibria in the system Sn-S.

mentioned, this result must be considered with some reserve.

IV. ELECTRICAL PROPERTIES

Hall-effect measurements have been carried out on a great number of crystals. Resistance measurements have also been carried out.

The Hall mobility perpendicular to the c axis $\mu_\perp = (8/3\pi)R_H\sigma_\perp$, where $\sigma_\perp = (\sigma_a\sigma_b)^{\frac{1}{2}}$, was measured as a function of temperature from 600°K down to 20°K (Fig. 5). At high temperatures the hole mobility is proportional to $T^{-2.2}$, indicating lattice scattering. The low-temperature branch could be explained by assuming impurity scattering of the Conwell-Weisskopf type and a concentration of charged centers of the order of $3 \cdot 10^{17}$

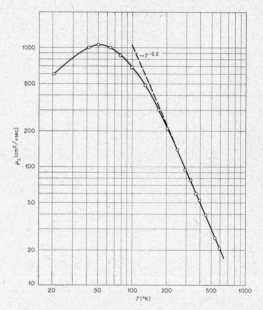

FIG. 5. The Hall mobility of the holes in SnS perpendicular to the C axis $[(\mu_a\mu_b)^{\frac{1}{2}}]$ as a function of temperature.

at./cm³. This concentration agrees well with the number of acceptor levels ($4 \cdot 10^{17}$ at./cm³) found for this crystal by Hall measurements.

With regard to the anisotropy of the crystal it was found that the hole-mobility at room temperature differs for the different crystal axes, viz. $\mu_a/\mu_c = 5.5 \pm 0.5$ and $\mu_a/\mu_b = 1.15 \pm 0.1$, if μ_a and μ_b are the mobilities measured in the directions of the two nearly equivalent directions of the pseudo-tetragonal structure and μ_c is the mobility measured in the direction perpendicular to the a,b plane ($a < b < c$).

To calculate the concentration p of the free holes from the Hall effect, the well-known relation

$$p = (3\pi/8) \cdot (1/R_He) \qquad (1)$$

was used. It is of course doubtful whether this simple relation is entirely valid over the range of temperatures used.

A typical log p versus $1/T$ curve obtained in this way from Hall measurements is shown in Fig. 6. It can be seen that even at 20°K the concentration of holes shows only a small tendency to drop. This seems to indicate that the activation energy of the corresponding level is virtually zero. At higher temperatures there is a slight increase in the concentration of holes as calculated by formula (1). This could be caused either by the fact that (1) is not valid over the entire range of temperatures or by an incipient ionization of another deeper lying level. To decide between these two possibilities Hall measurements at higher temperatures are needed. However, when measured at higher temperatures the Hall effect changed with time in a complicated way, reminiscent of the similar effects found for SnSe by Asanabe.[1,2]

V. EFFECTS OF ANNEALING

This effect of the Hall effect changing with time at higher temperatures was thought to be interesting enough for closer investigation. This was done in the following way. At a series of temperatures lying be-

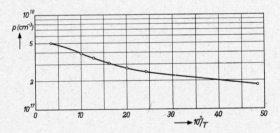

FIG. 6. The concentration of free holes in SnS, as determined by the Hall effect, as a function of temperature.

tween 230–500°C the crystals were allowed to "anneal" for a time sufficient to establish equilibrium over the entire crystal. Then the crystals were rapidly cooled to room temperature, and the concentration of free holes at room temperature measured by means of the Hall effect. The result is shown in Fig. 7. It can be seen that when the concentration of holes measured at room temperature is plotted logarithmically against the reverse of the absolute temperature of annealing, a straight line is obtained. The temperatures of annealing could not be chosen higher than about 500°C because for these temperatures the rate of cooling was very probably not high enough to prevent some degree of annealing from taking place during the cooling also. The experiments further showed that at the temperatures used, the crystal does not change its composition, because the experiments could be repeated several times going up and down the temperature scale without evincing any change in the Hall effect at room temperature for a given annealing temperature.

Starting from the assumption, mentioned in Sec. III, that solid SnS always contains excess sulfur, these phenomena could be explained by the formation of

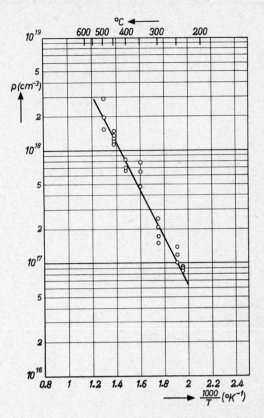

FIG. 7. The concentration of free holes in SnS measured at room temperature, after quenching, as a function of the temperature of annealing.

associates at lower temperatures:

$$2h^{\cdot}+2V_{Sn}{}' \rightleftharpoons (V_{Sn}V_{Sn})^{\times}, \qquad (2a)$$

or

$$4h^{\cdot}+2V_{Sn}{}'' \rightleftharpoons (V_{Sn}V_{Sn})^{\times}, \qquad (2b)$$

with $\cdot$, $\times$, and x denoting the effective charges with respect to the crystal of the positive hole, the negative Sn vacancy, and the neutral associate. Such an association is not improbable[11] because one way of visualizing such a $(V_{Sn}V_{Sn})$ center is to consider it as a S_2 molecule dissolved in the SnS crystal.

Now, assuming that the $(V_{Sn}V_{Sn})$ center neither acts as a donor nor as an acceptor, and taking into account that the concentration of holes nearly equals that of the nonassociated V_{Sn} centers, for high degrees of

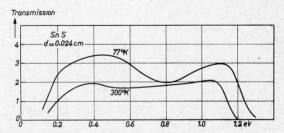

FIG. 8. Transmission of SnS as a function of photon energy.

---

[11] F. A. Kröger (private communication).

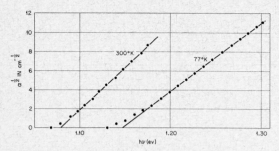

FIG. 9. The square root of the absorption coefficient as a function of the photon energy.

association, for the concentration of holes measured at room temperature after quenching the following is obtained:

$$p = (N/2K)^{1/4}, \qquad (3a)$$

or

$$p = (2N/K)^{1/6}, \qquad (3b)$$

K being the reaction constant of reaction (2) and N the total number of Sn vacancies, associated and non-associated.

If now $\log p$ is plotted against $1/T$, as in Fig. 7, $\frac{1}{4}$ of the dissociation energy of reaction (2) is found from the slope, neglecting the temperature dependence of the pre-exponential factors or, if one assumes that a V_{Sn} center can take up two electrons from the valence band, $\frac{1}{6}$ of that dissociation energy.

From Fig. 7 1.4 ev or 2.1 ev, respectively, can be calculated for this energy.

VI. SEEBECK EFFECT

Over an extensive range of carrier concentrations the thermoelectric power α at room temperature could be described[12] by the well-known formula:

$$\alpha = k/q[r + 2 + \ln(N_v/p)], \qquad (4)$$

with $e^r(m_d{}^*)^{\frac{3}{2}} = 1.5 m_0{}^{\frac{3}{2}}$. The highest reasonable value of r is $\frac{1}{2}$ (polar scattering below the Debye temperature), which would give a value 0.95 m_0 for the density of states effective mass $m_d{}^*$. This is a lower limit for $m_d{}^*$.

VII. OPTICAL MEASUREMENTS

Infrared transmission measurements were carried out[13] at 77° and 300°K as a function of wavelength from $\lambda = 1\mu$ to $\lambda = 10\mu$. Some of the results have already been given in a previous paper.[14] Figure 8 gives curves of the transmission versus photon energy for SnS having about $5 \cdot 10^{17}$ free holes per cm³. In the region of the absorption edge, the square root of the absorption coefficient turned out to be proportional to the photon energy (Fig. 9). This means that absorption in this

region is due to indirect transitions. From this it may be concluded that the maximum of the valence band and the minimum of the conduction band occur at different values of the crystal momentum. A further analysis showed the bandgap to be 1.080 and 1.115 ev at 300° and 77°K, respectively. Absorption due to free carrier absorption was observed as well. This absorption increases almost linearly with the concentration of holes and is proportional to the square of the wavelength λ in a large spectral region, starting at about $\lambda = 5\mu$. In addition an absorption band was found at 77°K at 0.8 ev (absorption starting at 0.4 ev). This absorption, too, was proportional to the hole concentration. It is thought that this absorption band is due to interband transitions of holes from the highest valence band to a valence band lying about 0.4 ev lower.

Reflection measurements have also been carried out in the infrared as a function of wavelength between 2 and 25μ.[15] These measurements were done at different temperatures and different concentrations of free holes, varying from $p < 10^{14}$ cm⁻³ (Sb-doped), to $p = 1.1 \cdot 10^{19}$ cm⁻³ (Ag-doped). Figure 10 gives two typical curves of the reflection coefficient as a function of wavelength for a SnS crystal containing a concentration p of holes equal to $p = 1.06 \cdot 10^{19}$ cm⁻³. The results were carefully analyzed with regard to the contributions of lattice vibrations of a polar crystal, and of the free carriers, taking into account the fact that the crystal structure of SnS is rhombic. Therefore, sometimes polarized infrared light was used. An analysis of the lattice vibration part of the complex refractive index gave the following results: The extrapolated value of the index of refraction n_0 ($\lambda = 0$) = 3.6 ± 0.1, effective charge on the atoms $e^* = 0.7 e_0$, dielectric constant $\epsilon = 19.5 \pm 2$. The values of n_0 and e^* are essentially the same for the three principal axes in the crystal. The value found for the dielectric constant, could be confirmed by capacitance measurements of layers of high-ohmic SnS ($\epsilon = 19 \pm 1$). Moreover, the refractive index found in this way could be corrobated by an analysis of interference fringes in thin plates of high-ohmic SnS for wavelengths lying between 4 and 10μ, also leading to a value of $n_0 = 3.6$.

The contribution to the complex refractive index due to free carriers was analyzed in order to obtain the

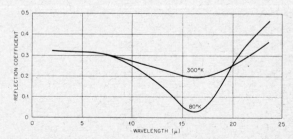

FIG. 10. Reflection coefficient of SnS as a function of wavelength ($p = 1.06 \cdot 10^{19}$ cm⁻³).

[12] Our thanks are due Dr. F. N. Hooge of this Laboratory for kindly carrying out these experiments.

[13] W. Albers, C. Haas, H. Ober, G. R. Schodder, and J. D. Wasscher (to be published).

[14] W. Albers, C. Haas, and F. van der Maesen, J. Phys. Chem. Solids 15, 306 (1960).

[15] C. Haas and M. M. G. Corbey, J. Phys. Chem. Solids (to be published).

effective mass of the holes. It was found that the effective mass in the plane perpendicular to the c axis is approximately isotropic and has a value $m_a^* \simeq m_b^* = 0.20\,m_0$. The effective mass parallel to the c axis is much larger and was estimated to be $m_c^* \approx m_0$.

It is remarkable that the ratio m_c^*/m_a^* is almost equal to the ratio μ_a/μ_c. From this it can be concluded that, although the effective masses differ considerably for the two crystal directions, the relaxation time is isotropic.

In Sec. VII a lower limit of $0.95\,m_0$ was found for the density of states effective mass m_d^*. Now one has

$$(m_d^*)^{\frac{3}{2}} = N(m_a^* \cdot m_b^* \cdot m_c^*)^{\frac{1}{2}}, \qquad (5)$$

where N is the number of equivalent maxima of the valence band. Substitution of the various masses in this equation shows that N must be at least 4. The maximum of the valence band therefore certainly does not lie at $K = 000$. A value $N = 4$ is not inconsistent with the crystal symmetry.

JOURNAL OF APPLIED PHYSICS　　SUPPLEMENT TO VOL. 32, NO. 10　　OCTOBER, 1961

Investigations on Silicon Carbide

H. J. van Daal, C. A. A. J. Greebe, W. F. Knippenberg, and H. J. Vink

Philips Research Laboratories N. V. Philips' Gloeilampenfabrieken, Eindhoven, Netherlands

Measurements of Hall effect and resistivity up to 1300°K on p-type hexagonal SiC showed an acceptor level for aluminium of 0.27 ev at zero donor concentration and a not yet identified acceptor level of 0.39 ev. The spin multiplicity of this unknown center appears to be four times smaller than that of the aluminium center, so that we may conclude that this unknown center in non-ionized state has paired electrons. Taking a temperature dependence of the level depths proportional to that of the bandgap, the density-of-states effective mass of the holes amounts to 0.59 m_0. The Hall mobility shows at high temperatures the same temperature dependence as that ascribed to scattering of holes by optical phonons. Assuming that optical phonons really come into effect, the behavior of the Hall mobility in the temperature range from 1300° to 300°K can be explained taking also into account the effect of scattering by acoustical phonons and charged impurities. By a study of I–V characteristics of grown junctions in αSiC and also by applying Roosbroeck-Shockley's theory to the spectral distribution of the p-n luminescence under forward bias, inhomogeneities were found over the junction area. By means of pyrolysis of gaseous compounds of Si and C pure crystals ($4 \times 2 \times 2$ mm³) of "cubic" βSiC were obtained. With the aid of polarized light the existence of a skeleton of a hexagonal twinning system was found in these crystals, the cubic SiC filling up the pores of this skeleton structure.

INTRODUCTION

IN this paper a short review will be given of current investigations on silicon carbide. Therefore what is given here is an interim report on several investigations still going on. This being so, no rounded-off picture of the various problems can be expected. Nevertheless it seemed worthwhile publishing some results obtained so far. Three different problems will be dealt with. In the first section an account is given of Hall effect measurements carried out on p-type hexagonal SiC prepared in the Lely furnace from technical grade material using argon gas either pure or doped with $AlCl_3$, Al, and CCl_4. The concentration of centers not added deliberately is rather high. Nevertheless it is believed that some conclusions might be drawn with regard to the properties of an acceptor center due to aluminium. Other data too could be obtained in this connection.

Quite a different problem is treated in Secs. 2, 3, and 4.

Grown p-n junctions in α SiC were studied. A brief description of their preparation is given, followed by a discussion of their forward characteristics as a function of temperature. The p-n luminescence emitted from these junctions under forward bias was also studied. The I–V characteristics could in principle be explained

assuming a p-i-n structure of the diode. With regard to the p-n luminescence it appeared that in the case of light having photons of an energy higher than the bandgap all phenomena could be explained by applying Roosbroeck-Shockley's theory. In both cases, however, the existence of inhomogeneities over the junction area had to be assumed. There was some indication that some of these inhomogeneities had to do with the polytypism of SiC.

The last part of the paper (from Sec. 5 onwards) deals with the problem of preparing pure SiC by thermal reduction, on a hot wire, of gaseous compounds of Si and C. It was found that crystals of a reasonable size ($4 \times 2 \times 2$ mm³) of "cubic" SiC could be obtained with a large efficiency by the "hot-wire" method, if only the right conditions were applied. A remarkable feature of these "cubic" crystals is the fact that they consist actually of a "sponge-like" matrix of thin colorless platelets of anisotropic hexagonal SiC, intersecting each other at definite angles, filled up with the isotropic cubic material. Some results of chemical analysis are also given, indicating that these crystals are very pure indeed. Yet the crystals showed n-type conductivity with a concentration of free electrons of $n = 10^{17}$ cm^{-3}.

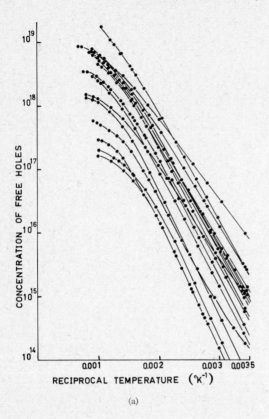

(a)

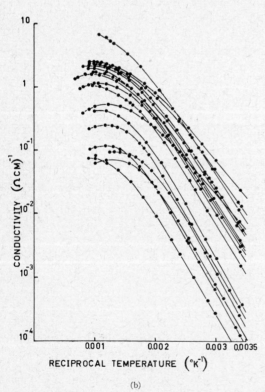

(b)

FIG. 1. (a) The concentration of free holes in SiC as determined by the Hall effect as a function of temperature. The samples were grown using argon as an ambient. (b) The conductivity as a function of temperature. The same samples as in (a).

1. ELECTRICAL PROPERTIES OF HEXAGONAL SiC

Hall effect and resistivity measurements were carried out on a great number of hexagonal crystals prepared in the Lely furnace.[1,2] The starting material for the preparation of these crystals consisted of technical grade silicon carbide and in some cases out of purer silicon carbide as prepared by pyrolysis of chlorosilanes. In the first instance we wanted to examine p-type SiC,[3] because very little was known about the electrical behavior of this conductivity type.[4] To that purpose we started on a series of batches prepared in firings using as an ambient argon charged in each case with a certain amount of AlCl₃.

From chemical analysis it appeared that apart from Al these crystals also contained large quantities of Cl. In connection herewith we started to find out the role played by chlorine. Therefore, measurements were carried out on crystals originating from batches made under varying CCl₄ pressures. Also batches made under varying CO pressures have been looked into, because oxygen might play the same role as chlorine.

On the other hand, it was our aim to study material exclusively doped with aluminium. This was done on batches fired in argon under certain dissociation pressures of aluminium carbide. Finally, we submitted batches made solely in argon to an examination. Figures 1(a) and (b) show respectively the number of holes and the conductivity as a function of temperature for a number of crystals originating from batches prepared using argon as an ambient. These curves are evidence that these argon batches contain many impurities that can differ from charge to charge.

This can be understood since in the preparative method used for these crystals, the different possible sources of impurities as there are ambient, the furnace, and the starting material had not yet been brought up to a constant low-value impurity content.

It is obvious that on the basis of the material described above it is difficult to obtain reliable information on a special center. All samples turned out to be p-type and were taken to be partly compensated, containing in any case nitrogen as a donor. To make reliable analyses possible, measurements at high temperatures, preferably up to 1300°K, appeared to be indispensable. This, in connection with the high activation energies, from 0.15 to 0.35 ev, and the great numbers of impurities present. The number of holes p was deduced from the Hall constant R_H according to the formula

$$p = 3\pi/8(R_H e)^{-1}.$$

[1] J. A. Lely, Colloquium I.U.P.A.C., Münster, September 2–6, 1954.

[2] J. A. Lely, Ber. Deutsch. Keram. Ges. 32, 229 (1955).

[3] H. J. van Daal, W. F. Knippenberg, and J. D. Wasscher, J. Phys. Chem. Solids (to be published).

[4] J. A. Lely and F. A. Kröger, *Halbleiter und Phosphore Vorträge des internationalen Kolloquiums 1956 in Garmisch-Partenkirchen*, edited by M. Schön and H. Welker (Friedrich Vieweg & Sohn, Braunschweig, 1958), p. 525.

The Hall curves have been analyzed by means of the formula

$$p \cdot (p + N_d)/(N_a - N_d - p) = N\theta^{-\frac{3}{2}} \exp(-E_a/kT),$$

in which $\theta = 10^3 \, T^{-1}$ and $N = N_v \cdot g^{-1} \cdot \exp(\alpha/k)$. In this formula N_v represents the density of states at 1000°K, g is a factor arising from the spinmultiplicity of the acceptor center, and α is the temperature coefficient of the acceptor level concerned. Figure 2 gives the relation between N_a and N_d for all samples. Roughly, the samples can be divided into two groups. The first group fairly scatters along the line $N_d = 0.5 \, N_a$. In this group of crystals a sofar unknown center is dominating and apparently its building in takes place according to a special until now uncontrolled chemical reaction. The second group, scattering along by the line $N_d = 0.1 \, N_a$, represents probably the building in of Al and B via carbides.

In Fig. 3(a) the activation energies for all samples have been plotted versus the average reciprocal distance of the donors. Irrespective of the scattering, two parallel lines can be drawn, the slope of which can be deduced from existing theory[5,6] using the dielectric constant value 10.2 for SiC. Comparison with chemical and spectrochemical analyses shows that the lower lying level must be ascribed to aluminium and that chlorine does not play an active electric role. It is still an open question which center gives the upper level. In some cases, marked in the figures, the number of the deep

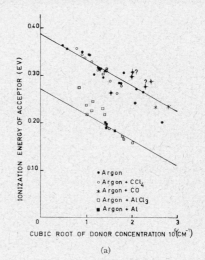

(a)

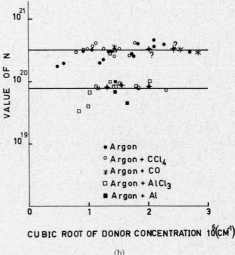

(b)

Fig. 3. (a) The ionization energy of the acceptors, for the same samples as in Fig. 2, as a function of the average reciprocal distance of the donors partly compensating the acceptors. (b) The value of N vs the cubic root of the donor concentration for the same samples as in Fig. 2.

lying acceptors corresponded with the number of boron atoms present in these samples. This boron concentration was determined by chemical and paramagnetic resonance spectra analyses.

Figure 3(b) gives the value of N as a function of the average reciprocal distance of the donors. Here again two groups can be distinguished of almost the same samples as in Fig. 3(a). Each group is characterized by more or less a constant value of N. (The ratio between the average N values amounts to 4.06.) This ratio can be explained by difference in spin multiplicity for the two levels. For an acceptor center with unpaired electron applies $g = 2$ and for a center with paired electrons $g = \frac{1}{2}$, so that a ratio 4 is involved.[7,8] For the aluminium level, having an unpaired electron, the lower N value is to be expected and was indeed found. A difference in tem-

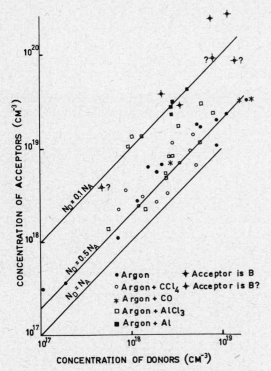

Fig. 2. A comparison between the acceptor- and donor-concentration of the p-type crystals examined.

[5] P. P. Debye and E. M. Conwell, Phys. Rev. **93**, 693 (1954).
[6] M. L. Chetkarov, Sov. Phys.—Tech. Phys. **3**, 895 (1958).

[7] P. T. Landsberg, Proc. Phys. Soc. (London) **B69**, 1056 (1956).
[8] F. W. G. Rose, Proc. Phys. Soc. (London) **B70**, 801 (1957).

perature dependence between both levels can also lead to different values of N in case of the same spinmultiplicity for these levels. Taking this dependence proportional to the temperature coefficient of the bandgap gives a contribution of about 1%. Calculating the effective mass of the holes, m_h^*, from N, taking into account the proper g-value and an estimated linear temperature dependence of the level concerned gives $m_h^* = 0.79\ m_0$.

Note added in proof. Taking into account the possible decrease of activation energy with increasing temperature due to increasing concentration of free holes gives a correction for both mean N values with the same factor 0.63. The effective mass then amounts to $m_h^* = 0.59\ m_0$.

In Fig. 4(a) the reciprocal Hall mobility is seen as a function of temperature for a few crystals. It appears that the mobility at high temperatures over a range of 800°K fits very well with the theoretical curve for scattering of holes by optical phonons.[9] In this theoretical curve a Debye temperature of 1400°K was used. This Debye temperature was deduced from infrared reflection measurements on SiC by Spitzer *et. al.*[10] In Fig. 4(b) a combination of three scattering-mechanisms, viz., scattering by optical and accoustical phonons and by charged centers,[11] is shown. The full lines indicate the theoretical curves assembled out of these three components. These lines coincide fairly well with the measured points down to room temperature. At lower temperatures, evidently, additional factors come into play. From the other side, an explanation of the absolute value of the experimentally found mobility, using the formula for scattering by optical modes, leads to an unprobably high value for the effective mass of the holes. Probably for SiC quite another theoretical approach is necessary.

2. PREPARATION OF GROWN *p-n* JUNCTIONS

By changing the dope of argon during the growth of SiC crystals in the Lely furnace *p-n* junctions can be made. This can be easily done by first starting with "pure" argon using technical grade SiC as starting material. As has been shown in the previous section the crystals made in this way are *p* type and contain about $3 \times 10^{18} - 3 \times 10^{19}$ acceptor centers cm^{-3} with an activation energy of about 0.3 ev. After the crystals grown under these circumstances have reached appreciable dimensions, the argon flowing through the furnace suddenly was doped with N_2 to a partial pressure of 0.01 atm. Consequently the parts of the crystals growing during the second stage were *n* type with some 10^{19} donors cm^{-3} having an activation energy of 0.08 ev.[4]

[9] D. J. Howarth and E. H. Sondheimer, Proc. Roy. Soc. (London) **A219**, 53 (1953).

[10] W. G. Spitzer, D. A. Kleinman, and D. J. Walsh, Phys. Rev. **113**, 127 (1959).

[11] H. Brooks, Advances in Electronics and Electron Phys. **8**, 85 (1955).

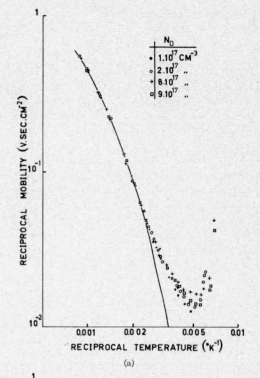

(a)

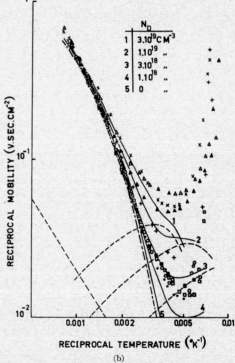

(b)

FIG. 4. (a) The solid line represents the theoretical dependence of the reciprocal mobility on temperature for scattering by optical phonons. Markings represent experimentally found values. (b) The broken lines represent theoretical curves for optical and accoustical modes of scattering and for scattering by charged impurities. For the solid lines—numbered 1 up to 5 inclusive—these three contributions have been taken into account for the range of donor-concentrations indicated at the top of the figure. The markings indicate measured points for crystals with a donor-concentration in the range from $3 \cdot 10^{19}$ to $1 \cdot 10^{17}$ cm^{-3}.

The resulting p-n junction is clearly visible due to the difference in color between p-type (colorless) and n-type (green) regions. Bars of about $0.02 \times 0.05 \times 0.5$ cm³ were sawn out of the crystals containing a junction with an area of about 0.02×0.05 cm². Ohmic contacts were made using a Au-Ta alloy on the n-type and an Au-Ta-Al alloy on the p-type parts.[12]

3. FORWARD CHARACTERISTICS

The direct-current forward characteristics had the general appearance of the curved lines in Fig. 5. The slope of the exponential part of these ln I versus V curves was obtained for low current densities by extrapolating the low-voltage part of the curve (which shall be referred to as the "foot" of the characteristic and is shown only partly in Fig. 5) to higher voltages and subtracting the extrapolation from the experimental curve. This was done on the assumption that the foot of the characteristic constituted the only current path parallel to the junction proper. This procedure seems reasonable, because the foot appeared to be the only part of the characteristic which could be influenced externally. For high current densities the curvature in the ln I versus V lines is caused by the increasing influence of a series resistance. It was possible to choose a definite

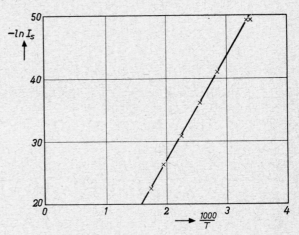

FIG. 6. The saturation current of a grown junction in α-SiC as a function of temperature. The slope gives an activation energy of 1.4 ev.

value of this series resistance in order to correct this part of the characteristic in such a way that a linear ln I versus V relationship was obtained over a large range of current densities. A series of such corrected ln I versus V curves for a certain diode at different temperatures is also shown in Fig. 5. As has been shown in a previous paper,[12] for every junction a minimum temperature T_0 could be found, above which the slope of this corrected linear ln I versus V line is given by

$$\alpha = q/2kT.$$

Also, for every $T > T_0$ the point of intersection of the straight ln I line with the current axis (which will be called the saturation current I_s) behaved similarly for

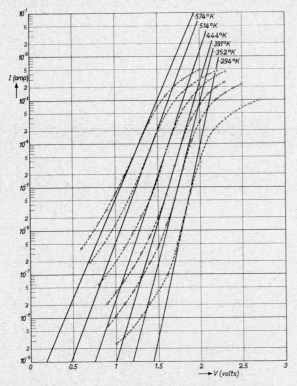

FIG. 5. The direct-current forward characteristics (dashed lines) of a grown junction in hexagonal SiC as a function of temperature. The drawn straight lines are the characteristics corrected for series resistance and "foot."

[12] C. A. A. J. Greebe and W. F. Knippenberg, Philips Research Repts. **15**, 120 (1960).

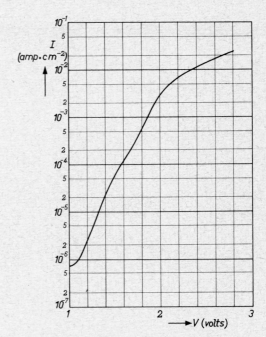

FIG. 7. The direct-current forward characteristic of a grown junction in SiC, showing an additional component.

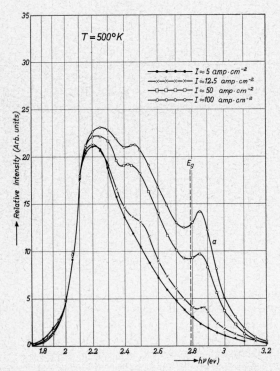

FIG. 8. *p-n* luminescence in a grown junction of hexagonal SiC under forward bias at 500°K as a function of photon energy for various current densities. Below 2.1 ev the curves are determined by the efficiency of the multiplier tube.

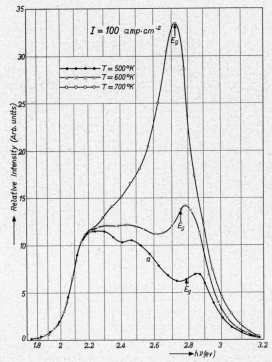

FIG. 9. *p-n* luminescence of the same junction as of Fig. 8 for 100 amp cm⁻² as a function of photon energy for various temperatures. Below 2.1 ev the curves are determined by the efficiency of the multiplier tube.

every junction: (a) The dependence of I_s on temperature (Fig. 6) could be described by an activation energy of about 1.4 ev, that is half the bandgap of 6H-SiC and (b) its order of magnitude ($I_a \approx 10^{-19}$ amp cm⁻² at room temperature) satisfies the relation:

$$I_s = qn_i(L/\tau), \qquad (3)$$

with n_i the intrinsic carrier concentration for the given temperature ($T > T_0$), L the diffusion recombination length, and τ the lifetime, taken to be 10^{-8} sec, a reasonable value for SiC.

For temperatures below T_0 the slope of this ln I versus V line was smaller than $q/2kT$, and also, the value of I_s was too large to fit (3) and did not show an exponential dependence on $1/T$.

For different junctions different T_0's were found, but T_0 never exceeded 700°K. A lower limit of T_0 cannot be given because for temperatures below room temperature the series resistance of the junction increased so rapidly that no accurate measurements of I could be obtained.

These data strongly suggest that, apart from the foot, for $T > T_0$, the junctions behave essentially as *p-i-n* junctions, but that for $T < T_0$ an additional parallel current component, that can be neglected for $T > T_0$, is important. This might be an indication that the current density is distributed inhomogeneously over the junction area apart, of course, from surface effects indicated by the presence of the "foot." In some junctions which were prepared more recently, such an additional com-

ponent could be observed directly.[13] Figure 7 shows a ln I versus V characteristic for such a junction. At a current of about 5×10^{-5} amp cm² one clearly sees a

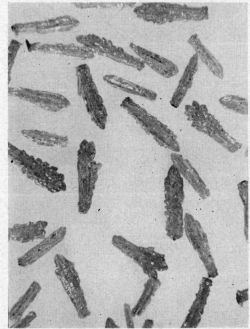

FIG. 10. Crystals of SiC obtained by thermal reduction of Si and C compounds. Their length is about 2–3 mm.

[13] C. A. A. J. Greebe and W. F. Knippenberg, Philips Research Repts. (to be published).

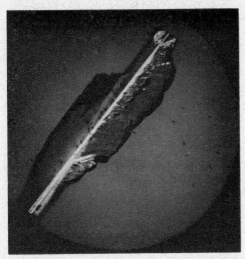

FIG. 11. A cross section parallel to the long axis of a crystal as shown in Fig. 10 viewed through crossed nicols showing a double refracting stem in 45° position.

shoulder in the characteristic. One characteristic was found suitable for analysis with some accuracy because of an appropriate position of the shoulder between the foot and the high current part. This analysis was based on the assumption of two parallel p-i-n junctions. The ln I versus V graphs of these two junctions both gave slopes equal to $q/2kT$, whereas the activation energies of the respective saturation currents were equal to 1.2 ev and 1.55 ev. These values are exactly half the bandgaps of 3C- and 4H-SiC, respectively.[13]

4. p-n LUMINESCENCE UNDER FORWARD BIAS

Under forward bias the spectral distribution of the p-n luminescence was measured for junctions that showed an activation energy of the saturation current equal to 1.43 ev. The current density used was of the order of 100 amp cm^{-2}, that is much higher than that used for the investigation of the forward characteristics.

Figures 8 and 9 give such spectral distribution curves. In Fig. 8 one sees the spectral distribution curve as a function of current density at 500°K and in Fig. 9 at a current density of 100 amp cm^{-2} for different temperatures. Curve a is the same in both figures only reproduced on different intensity scales. It is seen that the band-band emission increases with increasing current density and with increasing temperature. E_g indicates the position of the bandgap found by Choyke and Patrick.[14] After suitable corrections for the geometrical arrangements and for self-absorption, it must be expected that that part of the spectral distribution corresponding to photon energies larger than the bandgap can be related to the absorption spectrum by means of the theory of Roosbroeck and Shockley. As has already been found for technical grade material[15] this relation seemed to be valid only over a small region of the spectrum concerned. Detailed analysis[13] of the spectral distribution, however, showed that these data could be made to fit the Roosbroeck-Shockley theory, if a certain inhomogeneity of the emission over the junction area was admitted. This assumption could be strengthened by grinding away parts of the junctions and in doing so, changing the emission spectra in a predictable way.

One must conclude, therefore, that even at these high current densities, inhomogeneity effects do occur. Whether the inhomogeneities found on analyzing the forward characteristics and those found by means of a study of the p-n luminescence are of the same nature is unclear at the moment.

FIG. 13. A cross section perpendicular to the long axis of a crystal as shown in Fig. 10 viewed through crossed nicols in the 0° position showing a cross section of the double refracting primary plates. The angles between the light emitting lines are about 70° and 40°.

[14] W. J. Choyke and L. Patrick, Phys. Rev. **105**, 1721 (1957).
[15] L. Patrick and W. J. Choyke, J. Appl. Phys. **30**, 236–248 (1959).

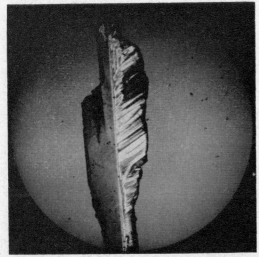

FIG. 12. The same cross section as of Fig. 11 showing the side branches, under a 60° angle with the stem, in the 45° position.

FIG. 14. The same cross section as of Fig. 13 in the 45° position.

5. PREPARATION OF SiC BY THERMAL REDUCTION

The pyrolysis of gaseous silicon and carbon compounds in hydrogen resulting in the formation of SiC is a possible way of preparing very pure SiC crystals in a one-step procedure.

Hot-filament techniques similar to the procedure applied in the preparation of pure metals, were used by several authors.[16] The crystals produced were mostly very small.

Tube furnaces were also used for the decomposition of the silicon and carbon compounds.[17–19] They were introduced with the intention of having the growing surface of the crystals at a constant temperature.

We have experimented with both techniques.[20] The hot-filament technique was found to give the best results. As the temperature of the filament is constant over its whole length, undesirable side reactions could be avoided entirely, leading in this way to maximum possible yield, in contrast to the case of tube-furnaces where a temperature gradient along the axis is inevitable. By using thick filaments (2 mm in diameter) the temperature of the filament could be lowered down to about 1650°C. This is important because at low filament temperatures the crystals do not fall off the filament. In this way crystal growth can last for a much longer time

resulting in ellipsoidal crystals of 2–3 mm in length or even longer.

The cracking apparatus consisted of a cylindrical quartz jar with an axial carbon rod with a diameter of 2 mm. This rod was heated by resistance heating. It was necessary to use the gaseous silicon and carbon compounds in a high dilution. Hydrogen was used as the dilutent, containing about 1% of the silicon and carbon compounds.

Of course, the purity of the hydrogen gas is of great importance. Use was made either of mixtures, e.g., $SiHCl_3$ and like $SiH_2Cl_2CH_3$ or halogen free silicons like SiH_3CH_3.

Figure 10 shows a picture of the crystals obtained, having an elongated ellipsoidal form.

6. PURITY OF THE CRYSTALS

No impurity could be detected with the aid of normal spectrographs. The nitrogen content was analyzed by means of a vacuum fusion technique. This method does not allow N contents less than 20 ppm to be found. A nitrogen concentration in excess of this value was not found. In the crystals made out of compounds containing chlorine, a chlorine content of 100–300 ppm was found.

The crystals always showed n-type conductivity, with a charge-carrier concentration of 10^{17} cm^{-3}. The chemical nature of the donor giving rise to these free electrons has so far not been established.

No fluorescence was observed for these crystals, when irradiated by ultraviolet, at room temperature or liquid

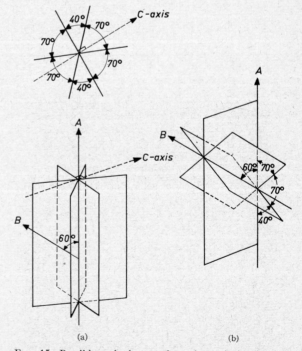

FIG. 15. Possible twinning configurations of the hexagonal platelets (a); a c axis is shown. Around axes like B the same twinning system is possible. One such system is drawn in (b).

[16] J. T. Kendall, J. Chem. Phys. **31**, 821 (1953).

[17] K. M. Merz in *Proceedings of the Conference on Silicon Carbide, Boston, Massachusetts*, edited by J. R. O'Connor and J. Smiltens (Pergamon Press, London, 1960), p. 73.

[18] V. E. Straughon and E. F. Mayer in *Proceedings of the Conference on Silicon Carbide, Boston Massachusetts*, edited by J. R. O'Connor and J. Smiltens (Pergamon Press, London, 1960) p. 85.

[19] S. Susman, R. S. Spriggs, and H. S. Weber in *Proceedings of the Conference on Silicon Carbide, Boston, Massachusetts*, edited by J. R. O'Connor and J. Smiltens (Pergamon Press, London, 1960), p. 94.

[20] W. F. Knippenberg, *Preparation of Silicon Carbide Crystals*, Philips Research Repts. (to be published).

N_2 temperature; a red flourescence appearing, however, when traces of aluminium were added.

7. OPTICAL OBSERVATIONS

Plan-parallel platelets were polished out of the crystals parallel and perpendicular to the crystal axis. When viewed through a polarization microscope with crossed nicols, sections parallel to the axis showed a main stem of double refracting material, often accompanied by side branching, at an angle of 60° with the stem, on one or both sides. Figure 11 gives a picture of such a cross section with a double refracting stem in the 45° position and in Fig. 12 the same sample shows the side-branches in 45° position. Figures 13 and 14 show a cross section perpendicular to the axis of the crystal in the 0° and 45° position, respectively. Here the angles between the light transmitting lines are about 70 and 40 deg. This twinning over a 70° angle of hexagonal platelets in SiC was reported by Thibault[21] and recognized as the tetraeder angle of the cubic symmetry. (See also refer-

ence 22.) From the foregoing the texture of the crystals may be visualized as consisting primarily of a skeleton of hexagonal platelets intersecting each other along the main axis of the crystal at angles of 70 deg, according to Fig. 15 (compare Figs. 13 and 14).

A same twinning configuration is possible around side axes, lying in the primary plates like B in Fig. 15 and forming an angle of 60° with the crystal axis. One of these secondary systems is also shown in Fig. 15. Of course this twinning can be repeated. The space between these colorless hexagonal twinning is filled up by yellow isotropic material.

By etching the crystals, the cubic material was taken away preferentially, leaving behind a skeleton of double refracting colorless material. In some cases this twinning system of hexagonal platelets could also be seen on crystals, without any etching. It looked like a mass of colorless platelets intersecting each other and protruding from that part of the crystal already filled up with cubic material.

[21] N. W. Thibault, Am. Mineralogist **29**, 269 (1944).

[22] N. N. Padurow, Tschermak's mineral. u. petrol. mitt. (3) **3**, 1, 10 (1952).

JOURNAL OF APPLIED PHYSICS   SUPPLEMENT TO VOL. 32, NO. 10   OCTOBER, 1961

Electron Spin Resonance in Semiconducting Rutile*

P. F. CHESTER

Westinghouse Research Laboratories, Pittsburgh 35, Pennsylvania, and Central Electricity Research Laboratories, Leatherhead, Surrey, England†

The ESR spectra of oxygen-deficient and doped rutile have been investigated at liquid helium temperatures. Niobium and tantalum are shown to give rise to the donors Nb^{4+} and Ta^{4+} rather than Ti^{3+}. The spectrum of reduced rutile depends on the method of reduction. Reasons for this are discussed. Under certain circumstances, involving hydrogen reduction, a particularly simple spectrum is observed. This degenerates and is replaced by a single line as the temperature is raised. A similar effect is obtained by increasing the concentration of centers. Possible assignments are discussed. Vacuum reduction results in distinctly different spectra which persist to higher temperatures. Resistivity measurements indicate a higher activation energy, by a factor of three, for vacuum-reduced samples.

ALTHOUGH a great deal of work has been carried out on the semiconductivity of oxygen-deficient and doped rutile,[1-5] we cannot yet say with certainty which of the models proposed for the donor center is the correct one, nor are we sure of the role of Ti^{3+} in the conduction process. Recently,[6,7] the further possibility of polaron states and/or bands has been recognized. Since the Ti^{3+} ion and some of the proposed donor centers are paramagnetic, it is to be expected that measurements of electron spin resonance (ESR) would give additional, and perhaps decisive, information.

This expectation has been fulfilled in the case of doped rutile. Results published in another connection on niobium and tantalum doping[8] show that, at low temperatures, the effect is not the stabilization of Ti^{3+}, as had previously been assumed. Instead, the donors Nb^{4+} and Ta^{4+} are formed. Gerritsen and Lewis[9] similarly have shown that vanadium enters as V^{4+}, but with a much larger ionization energy. It is now apparent that, at low temperatures, a distinction must be drawn between oxygen-deficient rutile and rutile doped with pentavalent impurities.

* This work was supported in part by the U. S. Air Force under contract.

† Present address.

[1] References to earlier work are given in the review article by F. A. Grant, reference 2.

[2] F. A. Grant, Revs. Modern Phys. **31**, 646 (1959).

[3] D. C. Cronemeyer, Phys. Rev. **113**, 1222 (1959).

[4] H. P. R. Frederikse and W. R. Hosler, N.B.S. Report 6585, November 1, 1959.

[5] L. E. Hollander and Patricia L. Castro, Phys. Rev. **119**, 1882 (1960).

[6] H. P. R. Frederikse, Conference on Compound Semiconductors, Schenectady, New York, June, 1961, and International Semiconductor Conference, Prague, 1960.

[7] T. Holstein (private communication).

[8] P. F. Chester, J. Appl. Phys. **32**, 866 (1961).

[9] H. J. Gerritsen and H. R. Lewis, Phys. Rev. **119**, 1010 (1960).

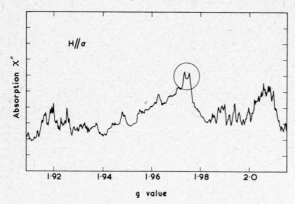

FIG. 1. Spectrum of Linde "as grown" rutile, $H \parallel a$, 2.0°K. The circled portion corresponds with the A spectrum.

The situation in oxygen-deficient rutile is not so simple. In the present work,[10] a number of samples have been examined with an X-band spectrometer in the liquid helium temperature range. ESR spectra are observed which, under certain conditions, are particularly simple in form.[11] An unexpected complication, however, was the discovery that the spectra observed, and by implication the defect structures responsible for them, depend strongly on the method of sample preparation.

Starting materials for the present work were commercial single-crystal boules[12] grown by flame fusion using an oxy-hydrogen flame. In the "as grown" condition, without deliberate oxidation, the Linde material had a resistivity of a few tenths ohm cm[13] and was an opaque blue-black color in a thickness of 1 mm. The spectrum of this material, for one orientation, is shown in Fig. 1. It extends over more than 250 oe and clearly indicates a high degree of disorder in the crystal. This material was not investigated in detail. Attention was concentrated on samples of higher resistivity and greater internal order as outlined below.[10]

Samples with resistivities in the range 10 to 40 ohm cm were prepared in two ways. The first was to heat Linde "as grown" material in air at about 625°C for $2\frac{1}{2}$ hr. The second was to cut directly from National Lead "as supplied" boules which had been partially reoxidized in air at 1100°C after growth. All the samples so prepared were a pale greyish-blue color in a thickness of 1 mm. They showed the same ESR spectrum which was investigated in detail. This spectrum, which we shall denote "A," corresponds closely with the circled portion in Fig. 1. It appears that, of all the centers contributing to the resonance in Linde "as grown" material,

only the A centers survive the process of partial reoxidation. It should be noted however that the A centers themselves can be removed, as far as ESR is concerned, by complete oxidation.

At 4.2°K with the magnetic field H in the (001) plane, the A spectrum consists of two, three, or four lines depending on the angle ϕ between H and an a axis. The appearance of these lines, which are individually about 1 oe in width, is shown for a Linde specimen in Fig. 2. With $H \parallel c$, a single line, 1 oe wide, is observed. No other lines are observed at any orientation in the range 500 to 8400 oe. The observed g values are well described by the relation: $g^2 = g_x^2 \cos^2\theta_x + g_y^2 \cos^2\theta_y + g_z^2 \cos^2\theta_z$, where θ_x, θ_y, and θ_z are the angles made by H with the magnetic axes x,y, and z. There are four sites per unit cell, each having $S=\frac{1}{2}$, $g_x=1.974$, and $g_z=1.941$.[14] The four z axes coincide and lie along the crystal c axis. The x axes make angles of $\pm 26°$ with the crystal a axes. This fourfold multiplicity rules out an assignment to Ti^{3+} ions on normal titanium sites. Attempts were made, with H carefully aligned along the c axis, to detect hyperfine structure. Although a pattern of satellite lines, accounting for a few percent of the total absorption, was observed, they were asymmetrically disposed about the center line and could not be interpreted in terms of hyperfine interaction with single nuclei of Ti^{47} or Ti^{49}. The spin-lattice relaxation time of the A spectrum with $H \parallel c$ was found to be 0.012 sec at 2.16°K in a Linde sample.

It is apparent from Fig. 2 that the intensities of the lines are not symmetric about the midpoint of the pattern, as would be expected from four crystallographically equivalent sites. This appears to be due to the presence of another line, on the low-field side of the A spectrum, whose g value is independent of ϕ. The asymmetry was somewhat less at 1.3°K and increased markedly as the temperature was raised above 4.2°K, as shown in Fig. 3.

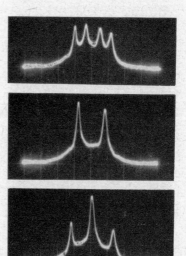

FIG. 2. The A spectrum observed in partially reoxidized Linde rutile, $H \perp c$, 2.16°K. Sweep 18.5 oe, field increasing from left to right. Top: $\phi=9°$. Center: $\phi=0°$. Bottom $\phi=18°$.

[10] A full account is contained in Westinghouse Scientific Paper No. 908 C 901-P3.

[11] P. F. Chester, Bull. Am. Phys. Soc. **5**, 72 (1960).

[12] One from the Linde Company, 30 East 42nd Street, New York 17, New York, two from the National Lead Company, P. O. Box 58, South Amboy, New Jersey.

[13] Unless otherwise stated, all resistivities refer to the c axis and room temperature.

[14] This value was confirmed by a measurement at 19.66 kMc.

It appears that the A spectrum degenerates with increasing temperature and that a new line, to be denoted "B" and which is present to some extent at 1.3°K, grows out of it.

It was observed that the National Lead samples showed a greater asymmetry and less sharp lines than the Linde sample. This may be connected with the fact that the latter contained about 100 times as much aluminium impurity (0.1%).

The A spectrum was also found to be present, although it accounted for only a small fraction of the total absorption, in a sample[15] prepared by reduction in a mixture of hydrogen and argon to a resistivity of about 60 ohm cm.

Since the temperature dependence of the A spectrum and the emergence of the B spectrum were suggestive of a thermally activated conduction process, attempts were made to increase the concentration of centers and thereby to increase the "hopping" rate. Four samples having a deep clear blue color were prepared, one by reoxidation of Linde "as grown" material in air for 3 hr at 450°C (giving $\rho_c = 0.82$ ohm cm) and three by heating National Lead "as supplied" material in oxygen at a few mm pressure for 8 hr at 825°C (giving in two cases $\rho_c = 0.76$ and 0.78 ohm cm, respectively). At 4.2°K these samples showed a single line spectrum with axial symmetry about the crystal c axis. The parameters of this spectrum, which will be denoted "C," are: $g_{\parallel} = 1.941$ and $g_{\perp} = 1.976$. These values correspond closely with those of the B spectrum. The C spectrum was not

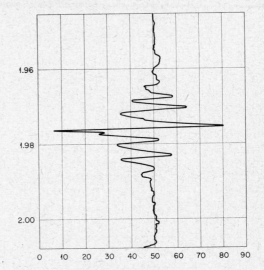

FIG. 4. Derivative spectrum of vacuum-reduced, 10-ohm cm sample, $H//[110]$, 4.4°K.

equally "clean" in all the samples. In the best case ($\rho_c = 0.76$ ohm cm), the linewidth was measured as a function of temperature and orientation. With $H \parallel c$ it was 1.9 oe and nearly independent of temperature, while with $H \perp c$ it decreased from 3.8 oe at 2.3°K to 1.8 oe at 17.8°K. The spin-lattice relaxation time was found to be about 100 μsec at 2°K. The center responsible for the C spectrum appears, therefore, to be partly mobile. Its hopping frequency at 10°K is of the order of 10^8 sec^{-1}. A careful comparison shows that the C spectrum is not centered on the A spectrum. This rules out simple exchange between the four stationary sites of the A spectrum.

At the present stage, a definitive interpretation of the above spectra cannot be made. The most obvious assignment of the A spectrum is to an interstitial site occupied by Ti^{3+} (with or without a lattice polarization) or by a self-trapped electron. There are four interstitial sites in the unit cell, each enclosed by a distorted octahedral arrangement of oxygen atoms, which differ only in the orientations of their principal axes in the (001) plane. These axes make angles of about ±31° with the a axes. Another possible assignment, already discussed,[11] is that of Ti^{3+} (with or without a lattice polarization) on a titanium site that is perturbed by an oxygen vacancy in a next-nearest neighbor position in the (001) plane. A third possible assignment is to an unidentified center whose presence is due to the incorporation of hydrogen in the rutile lattice. It is known[16] from infrared measurements that significant quantities of hydrogen can be taken up to form O–H bonds which then constitute defects in the rutile structure. It may be significant that the A spectrum has so far been observed only in samples prepared in hydrogen. The absence of hyperfine structure rules out atomic hydrogen or the

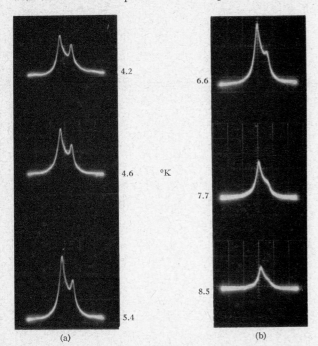

(a)   (b)

FIG. 3. Temperature dependence of A spectrum in "as supplied" National Lead rutile, $H \parallel 110$. Sweep ~30 oe, field increasing from left to right. The scale of absorption at different temperatures is not constant.

[15] Kindly loaned by L. E. Hollander.

[16] B. H. Soffer, M.I.T. Laboratory for Insulation Research, Tech. Report 140, August, 1959.

O–H group itself as the source. This is supported by the observation[16] that complete oxidation does not remove the O–H groups while it does remove the A spectrum.

The B spectrum, because of its axial symmetry, must correspond either to a center whose wave function extends over several unit cells or to a mobile center. If the A spectrum is indeed due to a weakly trapped polaron, then the B spectrum could well correspond to a polaron band. In this case, the C spectrum might be assigned to 'impurity' conduction—perhaps involving the A sites. Another possibility is that the transition from the A to the C spectrum is linked with the change in optical absorption observed by Cronemeyer[3] between 3 and 4 ohm cm (a axis). The a axis resistivities of the samples showing the C spectrum were about 2.7 ohm cm, while those of the samples showing the A spectrum were in excess of 40 ohm cm.

In an attempt to produce a series of samples of graduated resistivity to explore the transition from the A to the C spectrum, a series of samples of National Lead material were first oxidized in pure oxygen at 600°C for 8 hr (which removed the A spectrum) and then reduced in vacuum for 24 hr at temperatures in the range 600° to 1150°C. Resistivities ranged from 80 to 2 ohm cm (c axis). Unexpectedly, none of the samples showed the simple A or C spectra. The 600° samples showed only weak resonances. With $H \perp c$, the others showed similar spectra, each consisting of a central peak at $g = 1.975$, whose position was roughly independent of ϕ, together with an irregular pattern of large satellites, whose appearance was markedly dependent on ϕ, extending over about 70 oe. No measurements were made with $H \| c$. The structure and extent of these spectra remained unchanged up to at least 30°K in a 5-ohm cm sample and to at least 20°K in a 10-ohm cm sample. Although there is not enough data at present for an accurate description of these centers, it is clear that they differ markedly from the A center produced by partial reoxidation of "as grown" material and are more strongly bound. This latter point was confirmed by resistivity measurements below room temperature on one partially reoxidized sample (National Lead "as supplied," 10 ohm cm) and two vacuum-reduced samples (10 and 5 ohm cm). Below 30°K the activation energies observed were 0.01, 0.028, and 0.034 ev, respectively.

The central line in Fig. 4, with $H \perp c$, is very similar to the B spectrum of Fig. 3. If they can in fact be proved to be the same, the B spectrum would have to be regarded as the fundamental spectrum of reduced rutile, and logically attributed to mobile or spatially extensive polaron states.

The results outlined above show clearly that "oxygen-deficient" rutile of a given resistivity has a defect distribution which is dependent on the method of reduction. Although the defect structure of rutile is by no means well understood, one can see how the method of preparation can enter. Straumanis et al.[17] have recently shown that the TiO_2 phase boundary is at $TiO_{1.983}$. Beyond this point, according to Hurlen[18] and Andersson et al.,[19] one can expect the condensation of oxygen vacancies and their elimination in the formation of planes of interstitial titanium. The defect structure remaining after partial reoxidation will then depend on the relative diffusivities of the defects. This, in turn, is likely to depend on impurities (e.g., nitrogen) in the reoxidizing atmosphere[20] or in the crystal (aluminium is likely to be particularly important here). The possible role of hydrogen has already been discussed.

Since the magnetic and electrical properties of the various possible defects are by no means identical, it is essential for the understanding of the low temperature conductivity of "reduced rutile" that the defect structure be known and controlled. Room temperature resistivity is not a sufficient specification. In this connection the diagnostic value of ESR has been demonstrated.

ACKNOWLEDGMENTS

The experimental work in this paper was carried out at the Westinghouse Research Laboratories and it is a pleasure to thank T. Holstein and E. I. Blount for many informative and stimulating discussions there. The interpretation was carried out in part at the Central Electricity Research Laboratories, where the work is continuing, and the author is grateful to the C.E.G.B. for the opportunity to present this paper.

[17] M. E. Straumanis, T. Ejima, and W. J. James, Acta Cryst. 14, 493 (1961).
[18] Tor Hurlen, Acta. Chem. Scand. 13, 365 (1959).
[19] S. Andersson, B. Collén, U. Kuylenstierna, and A. Magnéli, Acta Chem. Scand. 11, 1641 (1957).
[20] V. I. Arkharov and G. P. Luckin, Doklady Akad. Nauk. S.S.S.R. 83, 837 (1952).

JOURNAL OF APPLIED PHYSICS   SUPPLEMENT TO VOL. 32, NO. 10   OCTOBER, 1961

2-5 and 2-6 Compounds

H. Brooks, *Chairman*

Optical Properties of Free Electrons in CdS

W. W. Piper and D. T. F. Marple

General Electric Research Laboratory, Schenectady, New York

The contribution of free electrons to the refractive index and extinction coefficient of Ga-doped CdS has been measured in a series of samples with carrier concentrations ranging from about 10^{17} to 2×10^{19} electrons/cm^3. The effective mass m^* for electrons near the bottom of the conduction band was calculated from the free electron contribution to the refractive index and from the carrier concentration as determined from the Hall coefficient. The result, $m^* = (0.22\pm0.01)m_e$ is in satisfactory agreement with previous studies by other workers. The magnitude and wavelength variation of the absorption coefficient observed with about 10^{17} carriers/cm^3 are in fair agreement with theoretical results calculated for a polar-mode lattice scattering mechanism. Although similar data for 2×10^{18} carriers/cm^3 are in good quantitative agreement with the impurity scattering theory at room temperature, the absorption for high carrier concentration is observed to decrease with temperature, in contradiction with the theory for a nondegenerate population. A theory for the degenerate concentration range is needed. Reflection and transmission of radiation polarized parallel and perpendicular to the crystal C axis were studied for one sample. These data show $(m_\perp{}^*/m_{||}{}^*) = 1.08\pm0.05$.

INTRODUCTION

CADMIUM sulfide is of considerable interest at present in the development of the theory of compound semiconductors. Large crystals are available and transport properties have been studied experimentally to the extent that they can now be used to test the theory in some detail. Since extensive studies of photoconductivity and luminescence have also been made, it is hoped that a more unified and detailed theoretical understanding of CdS and other II-VI compounds may eventually emerge aided by the growth of understanding of the band structure and transport properties.

Emphasis in this paper is on the optical properties of free electrons, the interpretation of the data in the light of the existing theory, and comparison of the results with previously reported transport properties and exciton spectra in CdS. Spitzer and Fan[1] have shown that the effective mass may be estimated from the contribution of free electrons to the optical constants by a method that is independent of the electron scattering mechanisms. The main contribution of the present work is to apply this method to CdS. Additionally, the spectral dependence of the absorption coefficient has also been determined at several temperatures, and the results are compared with the most recent theories for free carrier absorption.

EXPERIMENTAL

Single crystals of CdS were grown in this Laboratory by a method described by Piper and Polich.[2] These crystals were doped in a controlled fashion during crystal growth. They were large enough that rectangular sections could be cut 2 to 4 mm by 15 mm, a size range appropriate for the optical and electrical measurements. The crystals were doped with Ga to make them n type and heat treated in Cd vapor to minimize compensation by cation vacancies. Polished faces were obtained with conventional lapping techniques. Before reflectance measurements most crystals were also chemically polished by immersion for 30 sec to 2 min in HPO_3 at 280°C followed by a rinse in distilled water. Although real differences in the reflectance could be detected between mechanically and chemically polished crystals these differences were not large enough to limit the precision of the experiment.

The free carrier concentration was determined from measurements of the Hall coefficient in a magnetic field of 7400 gauss. A cryostat of conventional design was used and the voltages were measured with an L & N microvoltmeter. Electrical contact was made to the crystal by fusing small In dots to the surface at 280°C.[3] Reflectance was measured at about 10° angle of incidence with a double-pass CsI prism monochromator in the wavelength range 3 to 35 μ and in some cases with a NaCl prism in the 1 to 15 μ range. With CsI, spectral resolution was better than 0.3 μ in the 3 to 10 μ range, decreasing to better than 0.7 μ in the 25 to 35 μ range. A globar source was imaged on the sample by an optical system that included a scatterplate that reduced the short wavelength energy incident on the crystal and thus reduced sample heating. The sample and source were imaged on the spectrometer slits with a mirror train designed to minimize abberations in the image. Tests with filters showed that at all wavelengths studied radiation at undesired wavelengths contributed 2% or less to the signal obtained

[1] W. G. Spitzer and H. Y. Fan, Phys. Rev. **106**, 882 (1957).
[2] W. W. Piper and S. J. Polich, J. Appl. Phys., **32**, 1278 (1961).

[3] For more detailed information on the apparatus and techniques used for the electrical measurements, see W. W. Piper and R. E. Halsted, *Proceedings of the International Conference on Semiconductor Physics, Prague, Czechoslovakia, 1960* (Publishing House of the Czechoslovak Academy of Sciences, Prague, 1961), p. 1046.

with the evaporated-aluminum reflectance standard. Low temperature measurements were made with a small single-wall cryostat with KBr windows. The samples were cemented to copper plates cooled by the cryostat and were enclosed in a metal-and-KBr chamber also cooled by the cryostat to minimize condensation of films on the crystal. For transmission measurements a CaF_2 prism was substituted for CsI to give greatly improved spectral resolution.

RESULTS AND DISCUSSION

A. Reflectance

Spitzer and Fan[1] have pointed out that the carrier effective mass may be deduced in a very direct way from the contribution of free carriers to the optical properties of an extrinsic semiconductor. Starting from the Boltzmann transport equation they show that for spherical surfaces of constant energy the free carriers contribute to the electric susceptibility an amount

$$\chi_c = -Ne^2/\omega^2 m^* \qquad (1)$$

if $\omega\tau \gg 1$, where N is the carrier concentration, m^* is the carrier effective mass, and ω is the angular frequency. This result is independent of the scattering mechanism. Measurement of χ_c makes possible a determination of the effective mass without a detailed knowledge of the scattering processes.

The real part of the dielectric constant may be written as

$$\epsilon(\lambda) = n^2 - k^2 = \epsilon_0(\lambda) + 4\pi\chi_c, \qquad (2)$$

where $\epsilon_0(\lambda)$ is the contribution of the intrinsic crystal and n and k are the refractive index and extinction coefficient of the doped crystal. Solving (1) and (2)

$$m^* = \frac{Ne^2}{\pi c^2} \frac{\lambda^2}{(\epsilon_0 - \epsilon)}. \qquad (3)$$

The normal incidence reflectance is related to n and k by

$$R = \frac{(n-1)^2 + k^2}{(n+1)^2 + k^2} \qquad (4)$$

and if $(n-1)^2 \gg k^2$, n may be determined from R alone.

$$n = (1+\sqrt{R})/(1-\sqrt{R}).$$

Figure 1 summarizes the reflectance data on several n-type crystals doped to different carrier concentrations ranging from 10^{18} cm^{-3} to 2×10^{19} cm^{-3}, as well as the reflectance of an undoped crystal. The undoped crystal shows the sharp dip and rapid rise in reflectance at $31\,\mu$ which are characteristic of the reststrahlen reflection.

$\epsilon_0(\lambda)$ was determined from the reflectance of the undoped crystal by (5). The results were checked against the relatively accurate values published by Czyzak

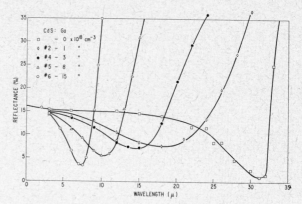

FIG. 1. Reflectance of Ga-doped cadmium sulfide as a function of wavelength for several different free electron concentrations.

et al.[4] for short wavelengths and were found to be in satisfactory agreement. As a further check, interference fringes were observed in transmission on each of a set of plane-parallel samples polished to thickness ranging from 8 to $35\,\mu$. Measurements were extended out to 30-μ wavelength to find the absolute interference order and obtain an accurate dispersion curve. All of these results were consistent within the experimental errors, and were adequate to determine ϵ_0.

Figure 1 shows that the reflectance of the doped samples passes through a minimum near the wavelength at which the free carrier contribution to χ_c causes n to fall to unity. If $k \ll n$, the reflectance drops quite sharply to a minimum value near zero,[5] and m^* may be obtained accurately if the wavelength of the minimum and $\epsilon = 1$ are substituted into (3). Figure 1 shows that for n-type CdS the minima are rather broad and shallow. This indicates that the extinction coefficient already contributes significantly to the reflectance at the minimum. However, since k drops quite rapidly with decreasing wavelength there is in the present cases a wavelength region short of the minimum where n may be obtained accurately from (5), and $(\epsilon_0 - \epsilon)/\lambda^2$ may be calculated from the refractive indices for the doped and undoped crystals. With decreasing wavelength $(\epsilon_0 - \epsilon)/\lambda^2$ at first increases (the effect of $k > 0$) but flattens out to a constant value before $(\epsilon_0 - \epsilon)$ becomes so small that it is obscured by experimental errors. The value of $(\epsilon_0 - \epsilon)/\lambda^2$ for each crystal shown in Fig. 1 is tabulated in Table I along with an estimate of the experimental error.

Table I also lists the carrier concentration for each crystal. The formula

$$R_H = -1/Ne \qquad (6)$$

was used to compute the carrier concentration from the Hall coefficient. The error limits on the concentra-

[4] S. J. Czyzak, R. C. Crane, W. M. Baker, and J. B. Howe, J. Opt. Soc. Am. 47, 240 (1957).

[5] For example, see the results on InSb (reference 1) and on GaAs; W. G. Spitzer and J. M. Whelan, Phys. Rev. 114, 59 (1959).

tion reflect the estimated error in the Hall voltage measurements and crystal dimensions. A further source of error is the spatial fluctuation of impurity concentration. Crystal specimens were selected which showed no evidence of grain boundaries, a linear potential drop along the length, and a constant Hall voltage at different points along the length. Nevertheless a 5 to 10% variation in impurity concentration within the sample is not precluded. The Hall mobility μ is tabulated in column 3 of Table I. R_H and μ were measured between room temperature and liquid nitrogen temperature and found to be substantially constant.

For the more lightly doped samples the carrier concentration is probably not accurately given by (6), and this probably explains the tabulated decrease in effective mass at low concentration. For $m^* = 0.2\ m_e$ the degeneracy concentration at room temperature is 2×10^{18} cm^{-3}. Thus for samples 2, 3, and 4, the Fermi level is less than $2kT$ above the bottom of the conduction band, and scattering theory appropriate for the degenerate region may not apply. If one assumes that N is proportional to $(\epsilon_0 - \epsilon)/\lambda^2$ and that $N = a/R_H e$, then $a = 1.5 \pm 0.3$ for sample 2 and $a = 1.3 \pm 0.2$ for sample 3. These values for a are not inconsistent with values expected for charged impurity scattering.[6] A more precise check on this point requires measurements of the Faraday effect in CdS.

The most probable conclusion to be drawn from the data in Table I is that the effective mass for electrons near the bottom of the conduction band is (0.22 ± 0.01) m_e. Three other estimates of this effective mass have been made. Piper and Halsted[3] measured the activation energy of an isolated donor level. They assumed a hydrogenic model to be a valid approximation and found $m^* = 0.19\ m_e$. They also obtained a good fit to the theoretical expression[7] for the temperature dependence of mobility with $m^* = 0.16\ m_e$. Hopfield and Thomas[8] analyzed the Zeeman effect on the exciton absorption spectrum of CdS and found $m^* = 0.21\ m_e$ for the conduction band.

TABLE I. Analysis of the reflectance and electrical data for several samples of n-type CdS.

Sample number	$(\epsilon_0 - \epsilon)/\lambda^2$ (10^4 cm^{-2})	N (10^{18} cm^{-3})	μ (cm^2/v-sec)	m^*/m_e
1	⋯	0.09±0.01	295	⋯
2	85±15	1.29±0.06	256	0.14±0.05
3	115±15	2.01±0.06	259	0.15±0.04
4	130±2	3.06±0.06	228	0.213±0.01
5	300±20	7.84±0.15	229	0.235±0.02
6	625±15	14.6±0.3	234	0.211±0.01

[6] E. Conwell and V. F. Weisskopf, Phys. Rev. **77**, 388 (1951).

[7] D. J. Howarth and E. H. Sondheimer, Proc. Roy. Soc. (London) **A219**, 53 (1953).

[8] J. J. Hopfield and D. A. Thomas, Phys. Rev. **122**, 35 (1961).

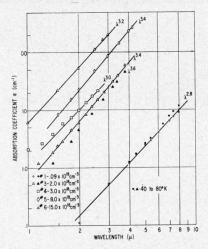

FIG. 2. Absorption coefficient of Ga-doped cadmium sulfide as a function of wavelength at room temperature for several different free electron concentrations. Samples 1 and 3 were also measured at 40°K and at 80°K, and these points are plotted with solid symbols.

B. Transmission

Transmission measurements were made on the samples used for the reflectance study, Part A above. For this work the thickness of the samples was reduced by grinding and instead of being roughened on one side (desirable in the reflectance study to minimize multiple reflections) the samples were polished on both sides. Sample 1, not studied in Part A, was also included in the transmission study. The absorption coefficient α was calculated[9] from the transmission with

$$\alpha = d^{-1} \log_e [T(1-R)^{-2}].$$

Thicknesses used ranged from 0.03 to 0.1 cm. Reflectance directly measured on the transmission sample was used in the calculation except in a few instances where extrapolated results of Czyzak et al.[4] were used.

Figure 2 summarizes the results. With the exception of sample 6, data were taken on only one sample thickness and the most accurate values lie near the middle and upper end of each set of points. The wavelength dependence of the absorption coefficient for the degenerate samples groups around a power law with exponent somewhat greater than 3, but a cubed dependence is not entirely beyond experimental error.

Sample 1 was measured on two occasions some three months apart and a significant decrease in the absorption coefficient was found. The wavelength dependence was $\lambda^{2.7}$ at the earlier time and at 2 microns the absorption coefficient was also about twice as large. This change is not understood and has not been explored further. One possible explanation is a slow change in the charge state of relatively deep traps and a contribution to the absorption from these bound states. Figure 3 shows the dependence of the absorption coe-

[9] This approximate formula neglects multiple reflections, but is a good approximation for fairly small transmission.

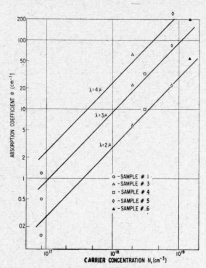

FIG. 3. Absorption coefficient of Ga-doped cadmium sulfide as a function of carrier concentration for three different wavelengths. The straight lines have unit slope.

fficient on the carrier concentration for three wavelengths. The absorption coefficient is seen to be proportional to the concentration, although some deviation is noted at each end of the curve. A significant point is that in the degenerate region the data is not consistent with an N^2 dependence.

For two samples the absorption was also measured at temperatures in the 40° to 80°K range. The temperature dependence is not strong for either crystal 1 or 3; in the first case there is a 20% increase in absorption and in the second a somewhat larger *decrease*.

Visvanathan has carried out the most recent theoretical analysis of free carrier absorption. He computed the absorption coefficient for the nondegenerate region due to scattering by the optical modes of a polar crystal[10] and due to scattering by charged impurities.[11] Assuming $m^* = 0.2\ m_e$, room temperature, and photons much more energetic than the reststrahlen energy, his theory for phonon scattering predicts

$$\alpha = 3.2 \times 10^{-19} N \lambda^{2.5}\ \text{cm}^{-1}$$

for all samples except 6, if λ is measured in microns and N in cm^{-3}. The prediction for charged impurity scattering in the same samples at room temperature, assuming the Elwert approximation, and that the charged impurity concentration is the same as the free electron concentration, is

$$\alpha = 4.5(10^{-19} N)^2 \lambda^3\ \text{cm}^{-1}.$$

Due to the relatively large dispersion in the refractive index in sample 6 as compared to the others, the wavelength dependence exponent for this crystal is increased by about 0.2 in both cases in the region of our measurements.

For sample 1 the impurity scattering is much too weak to account for the observed absorption. At 4-μ wavelength, the phonon scattering theory can account for three fourths, and at 6 μ two thirds of the observed absorption. The theory predicts a 10 to 20% decrease in the absorption at 40° to 80°K as compared to room temperature, whereas a significant increase is observed in the 5 to 8-μ region. However, in view of the anomalous change in absorption over a period of time mentioned above, we conclude that the theory for scattering by optical phonons is not contradicted by the data and that most of the absorption at the lowest concentration studied can be accounted for by this mechanism.

The sum of the predicted contributions from both mechanisms together is less than the observed absorption for sample 4 by about 10% at 2 and at 3 μ. For sample 6, which has the highest carrier concentration, the data lie 25% below the theoretical prediction. The contribution calculated for impurity scattering is three to four times as large as the phonon scattering at this concentration.

A theoretical analysis appropriate to the degenerate region is needed for absorption due to charged impurity scattering in order to make a meaningful comparison with the experimental results. The Maxwellian distribution assumed by Visvanathan obviously does not apply to most of the samples measured. A crude attempt to modify the analysis for the degenerate case indicates that the carrier concentration N in Eq. (7) should be replaced by the effective density of states. If a more careful analysis bears this out, the linear dependence of absorption on the carrier concentration and the temperature dependence which are observed experimentally might be better understood.

C. Anisotropy

Measurements of reflectance and transmission were made on one crystal with radiation polarized parallel and perpendicular to the crystal C axis. An infrared polarizing film[12] was used in the transmission measurements (1.5 to 2.1 μ) and a polyethylene sheet polarizer[13] in the reflectance measurements. The analysis used in Part A to determine the effective mass could not be carried through because the polyethelene had strong absorption bands which prevented measurements at wavelengths shorter than the minimum in the reflectance. However, it was possible to determine the wavelength for the minimum for both polarizations. The reflectance was less than 5% at the minimum for both polarizations. Ignoring the contribution from the change in absorption it is possible to estimate the effective mass anisotropy from the two minimum-reflectance wavelengths. With $\epsilon = 1$ inserted into (3) above, one

[10] S. Visvinathan, Phys. Rev. **120**, 376 (1960).
[11] S. Visvinathan, Phys. Rev. **120**, 379 (1960) (previous theories are reviewed in references 10 and 11).
[12] "Polaroid" film HR38.
[13] A. Mitsuishi, Y. Yamada, S. Fujita, and H. Yoshinaga, J. Opt. Soc. Am. **50**, 433 (1960).

obtains

$$(m_\perp{}^*/m_{11}{}^*) = (\lambda_\perp/\lambda_{11})_{min}{}^2 = 1.08 \pm 0.04.$$

The absorption coefficients observed for the two polarizations were found to have a constant ratio between 1.5 and 2.0 μ with

$$(\alpha_{11}/\alpha_\perp) = 1.13.$$

The effective mass dependence of the absorption coefficient is uncertain in the case of impurity scattering in the degenerate range. For the nondegenerate case the absorption coefficient is predicted to be proportional to $(m^*)^{-\frac{3}{2}}$ for charged impurity scattering and proportional to $(m^*)^{-1}$ for optical phonon scattering.

With the absorption ratio given above, these theoretical results point to an effective mass tensor with $m_\perp{}^*$ 8 to 13% greater than $m_{11}{}^*$. This conclusion is in satisfactory agreement with the result from reflectance, above, and with the exciton study of Hopfield and Thomas[8] who found less than 5% anisotropy in the effective mass.

ACKNOWLEDGMENTS

R. S. MacDonald measured infrared transmission in several thin CdS samples. S. J. Polich carried out most of the crystal-growing experiments and prepared most of the samples for electrical and optical measurements. A. F. Razzano made most of the optical measurements. These contributions are acknowledged with thanks.

JOURNAL OF APPLIED PHYSICS   SUPPLEMENT TO VOL. 32, NO. 10   OCTOBER, 1961

Electrical and Optical Properties of the II–V Compounds

W. J. TURNER, A. S. FISCHLER,* AND W. E. REESE
IBM Research Center, Yorktown Heights, New York

The noncubic II-V semiconductors have been studied recently by several workers. A review will be given of the present situation. The energy gaps of these materials range from 0.13 to over 1 ev. Room temperature mobilities of 10–15 000 cm²/v sec have been observed. Anisotropy of electrical and optical properties have been reported for several of the compounds. For CdAs₂ it has been possible to explain the anisotropy of Hall mobility by a simple energy band model.

GENERAL

THE authors have recently published a paper on the general physical properties of several II-V semiconductors.[1] The purpose of the present paper is to update that work with respect to our own data and to call attention to recent publications of other workers. While in the previous paper, 6 compounds were discussed *viz* Zn_3As_2, $ZnAs_2$, $ZnSb$, Cd_3As_2, $CdAs_2$, $CdSb$, the bulk of this paper will be devoted to the three cadmium compounds.

Table I outlines some of the general properties of the materials. Since none of the materials is cubic, it is necessary to make measurements as functions of crystal orientation. In the determination of the galvanomagnetic coefficients, it is highly desirable to measure as many parameters as possible on a single sample. This minimizes effects of impurity gradients.

The electrical and optical properties of $ZnAs_2$[1,2] and $CdAs_2$[1,3] show marked anisotropy. Anisotropy has been reported in $ZnSb$[4,5] and $CdSb$.[5,6] For $CdAs_2$ it is pos-

sible to interpret the results of a detailed study of the anisotropy in terms of a simple band picture.[3]

The energy gaps shown were determined from optical absorption data since irreversible changes in Hall coefficient and resistivity are observed at relatively low temperature, i.e., $T > 400°K$.[1] These changes result from thermal dissociation of the compounds.[7] Reproducible electrical data at temperatures up to 700°K have been reported recently on Zn_3As_2[8,9] and Cd_3As_2.[10,11] Pigon[9] has calculated a 0°K electrical gap of 0.91 ev for Zn_3As_2 and Zdanowicz[11] has found 0.14 ev for Cd_3As_2. According to these workers,[12] upon cooling from 700 to 400°K, the Hall coefficient changed its value and slope; and above 700°K, perceptible and irreversible changes were observed. The same value of Hall coe-

* Present Address: Watson Scientific Computing Laboratory at Columbia University.

[1] W. J. Turner, A. S. Fischler, and W. E. Reese, Phys. Rev. **121**, 759 (1961).

[2] W. J. Turner, A. S. Fischler, and W. E. Reese, Bull. Am. Phys. Soc. **3**, 379 (1958).

[3] A. S. Fischler, Phys. Rev. **122**, 425 (1961).

[4] M. V. Kot and I. V. Kretsu, Soviet Phys.—Solid State **2**, 1134 (1960).

[5] L. Stourac, J. Taub, and M. Zavetova, *Proceedings of the*

International Conference on Semiconductor Physics, Prague, 1960 (Publishing House of the Czechoslovak Academy of Sciences, Prague, 1961), p. 1091.

[6] I. K. Andronik and M. V. Kot, Soviet Phys.—Solid State **2**, 1022 (1960).

[7] G. A. Silvey, V. J. Lyons, and V. J. Silvestri, J. Electrochem. Soc. **108**, 653 (1961).

[8] W. Trzebiatowski, K. Pigon, and J. Rozyczka, Bull. acad. polon. sci. ser. Chim. **8**, 197 (1960).

[9] K. Pigon (to be published).

[10] W. Trzebiatowski and W. Zdanowicz, Bull. acad. polon. sci., ser. Chim. **8**, 511 (1960).

[11] W. Zdanowicz, *Proceedings of the International Conference on Semiconductor Physics, Prague, 1960* (Publishing House of the Czechoslovak Academy of Sciences, Prague, 1961), p. 1095.

[12] W. Zdanowicz (private communication).

TABLE I. General properties of several II-V semiconductors.[a]

Compound	Crystal system	Melting point (°C)	ΔE_{op} (ev) at 297°K	Type	Carrier concentration (cm^{-3})	Hall mobility at 297°K (cm^2/v sec)	Effective mass
ZnSb	orthorhombic	546	0.53	p	5×10^{16b}	350–575[b]	$m_l/m_0 = 0.175 \pm 0.010$ $m_t/m_0 = 0.146 \pm 0.010$ electrons[d]
ZnAs$_2$	monoclinic	768	$E \parallel C$, 0.90 $E \perp C$, 0.93	p	10^{16}	50	
Zn$_3$As$_2$	tetragonal	1015	0.93	p	10^{18}	10	
CdSb	orthorhombic	456	0.46	p	3×10^{15c}	900–2000[c]	$m_l/m_0 = 0.140 \pm 0.010$ $m_t/m_0 = 0.159 \pm 0.010$ electrons[d]
CdAs$_2$	tetragonal	621	$E \parallel C$, 1.00 $E \perp C$, 1.04	n	5×10^{14}	a axis 100 c axis 400	$m_l/m_0 = 0.150 \pm 0.030$ $m_t/m_0 = 0.580 \pm 0.040$ electrons $m_l/m_0 = 0.094 \pm 0.030$ $m_t/m_0 = 0.346 \pm 0.025$ holes
Cd$_3$As$_2$	tetragonal	721	0.13	n	2×10^{18}	15 000	$m^*/m_0 = 0.10 \pm 0.05$ electrons

[a] For references see W. J. Turner, A. S. Fischler and W. E. Reese, Phys. Rev. **121**, 759 (1961) unless specifically noted.
[b] M. V. Kot and I. V. Kretsu, Soviet Phys.-Solid State **2**, 1134 (1960).
[c] I. K. Andronik and M. V. Kot, Soviet Phys.-Solid State **2**, 1022 (1960).
[d] M. J. Stevenson, Phys. Rev. (to be published).

fficient was again obtained when the sample was repolished.

Irreversible changes in electrical properties have been established for ZnSb, ZnAs$_2$, CdSb, and CdAs$_2$ at temperatures above 400–450°K by several workers including the authors. There does not appear to be any simple relationship between energy gap and melting points as has been observed for other materials, e.g., III-V compounds. This is not unreasonable since the III-V compounds all have the same crystallographic modifications whereas this is not true for the II-V compounds.

The approximate room temperature Hall mobilities are shown. The six compounds exhibit a wide range of mobilities. For ZnSb, Kot and Kretsu[4] report hole mobilities of between 350–575 from the conductivities of samples cut in three crystallographic directions. They assumed an average carrier concentration of 5×10^{16} holes/cm^3 from Hall data without regard to anisotropy factors. Only p-type material was studied and no Hall reversal was observed. For the zinc arsenides the hole mobilities are very low.

In the cadmium compounds, hole mobility has been measured only for CdSb. No Hall reversal is observed for p-type CdSb. We will discuss CdSb in more detail below. Both the cadmium arsenides are n type as grown. Judging from the cyclotron resonance effective mass data, the holes in CdAs$_2$ might be expected to be more mobile than electrons.[13,14]

For CdAs$_2$, the electron mobility in the c direction is four times that in the a direction.[1,3,14,15] For Cd$_3$As$_2$, the

electron mobility is comparable to the highest observed in semiconductors with greater than 10^{18} carriers /cm^3.

Despite the low room temperature mobilities, cyclotron resonance has been observed in CdAs$_2$,[13,14] CdSb, and ZnSb.[14] For CdAs$_2$, both carriers have been shown to give resonance; while for CdSb and ZnSb, the minority (electrons) gave rise to the resonance.[16] For Cd$_3$As$_2$ the average value of the electron effective mass has been obtained from optical and thermoelectric data.[1,17]

ZnAs$_2$, ZnSb, CdAs$_2$, and CdSb all show a wide range of high optical transmission, while Cd$_3$As$_2$ shows free carrier absorption and dispersion in reflectivity due to free carrier resonance.[1,17]

CdSb

The material as grown is p type with an optical energy gap of 0.465 ev at 300°K.[1,18] Several workers have obtained samples showing n-type conductivity over a limited temperature range using various dopants, e.g., In,[15,18] Pb.[6] To date, no Hall reversal has been observed in p-type material. This may be caused by the formation of acceptors at elevated temperatures due to thermal decomposition or because the hole mobility is greater than the electron mobility. If the latter were true it is easy to show that the intrinsic slope of the curves lnρ and ln$R$ versus $1/T$ would differ.[18] Recently, Stevenson[16] has identified the sign

[13] M. J. Stevenson, Phys. Rev. Letters **3**, 464 (1959).
[14] M. J. Stevenson, *Proceedings of the International Conference on Semiconductor Physics, Prague, 1960* (Publishing House of the Czechoslovak Academy of Sciences, Prague, 1961), p. 1083.
[15] W. J. Turner, A. S. Fischler, and W. E. Reese, J. Electrochem. Soc. **106**, 206C (1959).

[16] M. J. Stevenson, Phys. Rev. (to be published).
[17] W. J. Turner, A. S. Fischler, and W. E. Reese, J. Electrochem. Soc. **107**, 189C (1960).
[18] W. J. Turner, A. S. Fischler, and W. E. Reese, *Proceedings of the International Conference on Semiconductor Physics, Prague, 1960* (Publishing House of the Czechoslovak Academy of Sciences, Prague, 1961), p. 1080.

of the carriers giving rise to the cyclotron resonance absorption in both p-type ZnSb and CdSb as negative. Although this does not exclude the presence of a more mobile hole in CdSb at elevated temperatures, it does indicate care is needed in the interpretation of galvanomagnetic data above 400°K.

The data of Andronik and Kot[6] on CdSb and of Kot and Kretsu[4] on ZnSb have been interpreted as indicating mobility anisotropy in both compounds. As no attempt was made in these publications to determine values for the same order resistivity tensor (as a function of the applied magnetic field) on the same sample, it is difficult to decide whether the observed effects are entirely due to anisotropy of the electrical coefficients or partly due to inhomogeneities. Andronik and Kot also report thermoelectric power data which they interpret as indicating a hole mobility slightly higher than the electron mobility.

Stourac et al.[5] have also reported Hall and resistivity anisotropy for CdSb as well as for mixtures of CdSb and ZnSb. These workers have found that the absorption of polarized light depended on the direction of polarization with respect to the crystallographic direction of the sample. The absorption coefficient was different for 3 directions of polarization; however, for two directions the dependence of K on photon energy was similar. The third direction was slightly different. Turner and Reese[19] have found similar results.

For CdSb as well as ZnSb, it would appear that these materials have anisotropic properties. Since the anisotropy factors for these materials have not been determined as has been done for CdAs$_2$,[3] it is not possible to describe the anisotropy of properties accurately, e.g., give Hall mobility ratios.

CdAs$_2$

As grown CdAs$_2$ is n type. The purest material which the authors have studied had a net carrier concentration of 5×10^{14} cm^{-3}. To date, no p-type CdAs$_2$ has been reported. Since CdAs$_2$ is tetragonal,[20] both Hall and resistivity are tensor quantities. Fischler[3] has investigated the galvanomagnetic properties of oriented single crystals of n-type CdAs$_2$. From these measurements, the two independent Hall components are known to be equal within experimental error. The resistivity in the a direction is four times that in the c direction. Resistivity anisotropy with isotropic Hall coefficients indicate the surfaces of constant energy to be ellipsoids of revolution located along the symmetry axis (c axis) of the tetragonal crystal system. An isotropic relaxation time is indicated by the same temperature variation of the mobilities.

Using a single ellipsoidal model and an isotropic relaxation time, the ratio of electron effective mass in the a direction to that in the c direction is equal to the

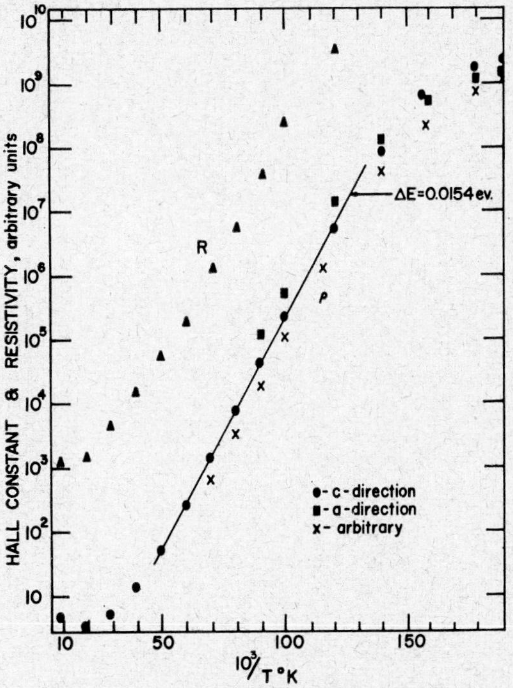

Fig. 1. Hall coefficient (R) and resistivity (ρ) in arbitrary units for indium doped CdAs$_2$.

resistivity ratio which is 4. This is in good agreement with Stevenson's cyclotron resonance value of 3.87. Magnetoresistance measurements confirm this conduction band model and show the scattering to be primarily acoustical lattice mode scattering with some degree of impurity scattering present.

Fischler and Koenig[21] have recently studied the electrical properties of n-type CdAs$_2$ doped with known amounts of indium. Figure 1 shows their preliminary data which indicate an impurity activation energy of 0.015 ev, independent of crystallographic direction. Below 6°K, these workers noted what might be the onset of impurity band conduction. The analysis of Keyes[22] of impurity states in axially symmetric crystals predicts an activation energy of 0.047 ev for hydrogenlike impurities in n-type CdAs$_2$. The discrepancy may be due to the fact that the value for the dielectric constant which Keyes used in his computation was obtained from high frequency reflectivity data while the appropriate parameters may be the static dielectric constant. An increase in the dielectric constant from 10 to 17.5 would bring the theoretical value of the activation energy into agreement with the experimental value. For a substance with some ionic nature, it is not unreasonable to expect the static dielectric constant to be larger than the high frequency value by this amount.

CdAs$_2$ shows high optical transmission from the intrinsic edge to the cutoff at approximately 37 μ. The

[19] W. J. Turner and W. E. Reese (unpublished data).
[20] N. R. Stemple and M. E. Senko, Acta. Cryst. (to be published).
[21] A. S. Fischler and S. Koenig (to be published).
[22] R. W. Keyes, IBM J. Research Develop. **5**, 65 (1961).

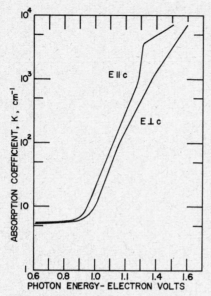

FIG. 2. Optical absorption coefficient of $CdAs_2$ near the intrinsic edge for two directions of polarization relative to the c axis.

transmission is a maximum for nonpolarized light propagated along the c axis of the crystal.[1,15] From polarized light studies near the intrinsic edge, the authors have reported an apparent optical gap of 1.04 ev for the E vector perpendicular to c and 1.00 ev for E parallel to c.[1] These measurements have been extended to much higher values of the absorption coefficient in order to determine the nature of the optical transitions in $CdAs_2$. The results of these measurements are seen in Fig. 2. The light is propagated perpendicular to a sample containing the c axis in the plane of the sample. The energy dependence for the light polarized with E vector parallel to c is quite different than for E vector perpendicular to c. For a given photon energy, the value of the absorption coefficient is higher for the parallel polarization than for the perpendicular. This may result from the effects in the transition probabilities of the different effective masses for the c and a directions.[1] It would appear, for both polarizations, that there are indirect transitions. However, for E parallel to c, a direct transition seems to dominate at about 1.3 ev. This indicates the existence of selection rules which makes this transition allowed in the parallel polarization and not in the perpendicular case. Further analysis of the optical data will be given in a future publication.[23]

Cd_3As_2

Cd_3As_2 is a tetragonal[24,25] semiconductor with a solid-solid phase transition at 578°C[26] and only n-type

Cd_3As_2 has been reported to date. The electron mobility of this compound is extremely high[1,10,11,27] but, due to the phase transition, it is difficult to obtain perfect single crystal samples for detailed studies of the electrical properties. In an effort to avoid or minimize the effects of cracking in the material during cooling from above the phase transition, several methods of crystal growth were studied.[7] Table II shows a summary of electrical data obtained by the authors on samples prepared by different methods. In all cases, the material was n type as determined by thermal probing or the sign of the Hall coefficient. The results clearly indicate that the best method of crystal preparation, from a mobility point of view, is pulling. In this case, a modified Gremmelmaier technique was used.[7] Other workers[10,11] have obtained comparable crystals using a modified Bridgman technique.

The intrinsic activation energy of Cd_3As_2 has been determined by several workers by electrical and optical means. An electrical gap at 0°K of 0.14 ev was first given by Moss[28] from the slope of resistance versus temperature measurements on evaporated films of Cd_3As_2. More recently, other workers have confirmed this value from Hall data on single crystals.[10,11,29]

The best high temperature work to date has been done by Zdanowicz[11,29] who has studied the electrical and thermoelectric properties of Cd_3As_2 up to 700°K. From that work, the mobility is proportional to $T^{-2.5}$ in the intrinsic range of 475–700°K, $T^{-1.5}$ in the range of 350–475°K and finally T^{-1} between 95–350°K. This last value is in agreement with the authors value of $T^{-0.88}$ between 100–300°K. Zdanowicz computed a value of 1.3 for the electron to hole mobility ratio. As he remarked, this seems too low. From the temperature dependence of thermoelectric power between 300 and 700°K, he concluded that the Fermi level is in the conduction band even at the highest temperatures for material of carrier concentration greater than 10^{18} cm^{-3}. From the thermoelectric power he also deduced the effective mass of electron using an analysis appropriate

TABLE II. Summary of electrical data for Cd_3As_2 grown by several methods.

Method of growth	Temp (°K)	ρ (μ-Ω cm)	R (cm^3/coul)	Hall mobility (cm^2/v sec)
Directional freeze (polycrystal)	297	111	0.40	3 600
	77	31	0.20	6 450
Vapor deposition (single crystal)	297	556	3.20	5 750
	77	135	3.20	23 700
Vapor deposition 4 sublimations (single crystal)	297	334	3.7	11 000
	77	102	3.1	31 000
Pulled (single crystal)	297	233	3.18	13 650
	77	66	3.10	47 300

[23] W. J. Turner (to be published).
[24] M. V. Stackelberg and R. Paulus, Z. physik. Chem. **B28**, 427 (1935).
[25] N. R. Stemple (private communication).
[26] M. Hanson, *Constitution of Binary Alloys* (McGraw-Hill Book Company, Inc., New York, 1958).

[27] A. J. Rosenberg and T. C. Harmon, J. Appl. Phys. **30**, 1621 (1959).
[28] T. S. Moss, Proc. Roy. Soc. (London) **B63**, 167 (1950).
[29] W. Zdanowicz, Acta Phys. Polon. (to be published).

for a degenerate semiconductor.[30],[31] At room temperature, the value is $m^* = 0.046\ m_0$.

The room temperature optical energy gap of Cd_3As_2 is 0.13 ev as determined from a plot of the optical absorption coefficient versus photon energy. At nitrogen temperature, the gap increases to 0.17 ev. These values were measured by making careful transmission and reflection measurements on samples of Cd_3As_2 containing approximately 2×10^{18} electron/cm³. Moss[28] has earlier reported an optical gap of 0.6 ev as determined from a photo response curve on evaporated films. Zdanowicz has also obtained an optical gap at 297°K of 0.6 ev from a photosensitivity curve of an evaporated layer of Cd_3As_2. He explained the large optical gap in terms of the Burstein effect,[32] i.e., the increase of optical gap in a semiconductor with small density of states or low effective mass and large number of carriers. In this case, an electron is excited from the valence band to the lowest available states in the conduction band. This may require an appreciably larger energy than the true optical gap for undoped sample. It is not clear to the authors whether Zdanowicz has observed optical transmission at the higher energies or just a photoresponse. Although the Burstein effect may well exist in Cd_3As_2 with higher carrier concentration, it was not evident in the present authors' absorption data.

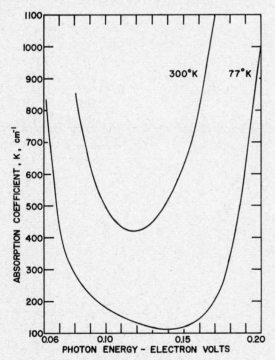

FIG. 3. Optical absorption of n-type Cd_3As_2 at 300° and 77°K as a function of photon energy.

[30] J. Tauc and M. Matyas, Czechoslov. J. Phys. **5**, 369 (1955).
[31] A. G. Samoilovich and L. L. Korenblit, Uspekhi Fiz. Nauk **57**, 577 (1955).
[32] E. Burstein, Phys. Rev. **93**, 652 (1954).

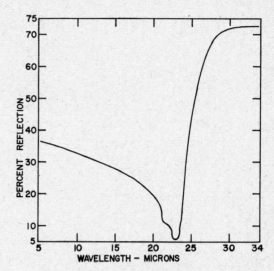

FIG. 4. Reflection of n-type Cd_3As_2 at long wavelengths.

Cd_3As_2 shows free carrier absorption which increases as approximately the square of the wavelength. See Fig. 3.

The reflectivity of Cd_3As_2 shows a dispersive behavior at long wavelengths as is seen in Fig. 4. This results from the electronic contribution to the dielectric constant when the electrons move like a plasma. Using classical dispersion theory, Moss[33] shows that the condition for minimum of reflection occurs when the index of refraction, $n \approx 1$, i.e. when $Ne^2/(m^* \epsilon_0 \omega_{min}^2) \approx n_c^2 - 1$. The equation above is valid if $\omega^2 \tau^2 \gg 1$ and $k > 1$, where the symbols have the following meaning:

N = electron concentration/m³,
e = electronic charge in coulomb,
m^* = effective mass of the electrons in kg,
ω_{min}^2 = angular frequency of the light at the minimum,
ϵ_0 = dielectric constant of free space = 8.85×10^{12} coul/newton m²,
n_c = index of refraction in the nondispersive region,
ω = angular frequency of light in sec⁻¹,
τ = collision time in sec,
k = extinction coefficient.

The value of N is obtained from the Hall effect. From reflectivity at 12 μ, $n_c^2 = 12.1$. Therefore, it is possible to determine m^* from the position of the reflection minimum. An estimate of the electron effective mass of 0.1 $m_0 \pm 0.05\ m_0$ was obtained from: (1) the value of the absorption coefficient in the free carrier region, (2) the wavelength of the minimum in reflectivity, and (3) the value of the thermoelectric power on several different samples. This compares with the value of $m^* = 0.046\ m_0$ which Zdanowicz finds from thermoelectric power measurements.

[33] T. S. Moss, *Optical Properties of Semiconductors* (Academic Press, Inc., New York, 1959).

JOURNAL OF APPLIED PHYSICS   SUPPLEMENT TO VOL. 32, NO. 10   OCTOBER, 1961

Electrical and Optical Properties of Mercury Selenide (HgSe)

H. Gobrecht, U. Gerhardt, B. Peinemann, and A. Tausend

II. Physikalisches Institut der Technischen Universität Berlin, Germany

HgSe single crystals are grown by zone melting. The compound crystallizes in the zinc-blende structure and splits into (100) planes. For temperatures ranging from 90° to 500°K conductivity, Hall effect and thermoelectric power are measured; above 500°K evaporation of HgSe begins. The lowest carrier concentration of the crystals at 300°K is 3.5×10^{17} cm^{-3}. Only n-type conduction is found. The highest mobility at 300°K is 18 500 cm^2/v sec. Magnetoresistance shows that the longitudinal effect is very small compared with the transverse. From the photo emf of the p-n junction Se/HgSe crystal and from the absorption edge of layers the energy gap of 0.5 to 0.75 ev is obtained. Using the temf and the absorption an estimation of the effective mass leads to 0.04 to 0.07 m_0.

I. INTRODUCTION

ONLY a few facts about the semiconductor HgSe are available. Polycrystalline material and single crystals are known to have a high electron mobility.[1-3] The mobility in HgSe layers is also remarkable. The conductivity of such layers practically does not depend upon temperature.[4] There are contradicting experimental results about the energy gap and little work has been done on the scattering mechanism. We have tried to obtain more information about the HgSe.

II. PREPARATION OF SAMPLES

Even at room temperature, Hg and Se join at the surface to form HgSe. The compound is produced by annealing stoichiometric amounts of the components in evacuated quartz crucibles. The melting point of 960°K is given by Blum and Regel.[1] Rodot,[5] however, finds the point at 1035°K. In spite of the high melting point, the evaporation of the material already begins at 500°K. In a vacuum, it sublimates again as HgSe at the cold sides of the vacuum bell jar. This is pointed out by x-ray investigation. To prevent evaporation, zone melting is carried out under high partial pressure of both components. The purity of Se is 10^{-5} and of Hg 10^{-4}. Before preparing the compound, Hg is distilled three times in a vacuum.

Single crystals obtained by splitting the ingot have an electron concentration of about 5×10^{18} cm^{-3}. By annealing at 500°K for several hours the carrier concentration is lowered; a constant value as low as 3.5×10^{17} cm^{-3} has been reached. Figure 1 shows a sample with a lowest carrier concentration of 1.2×10^{18} cm^{-3}. Annealing

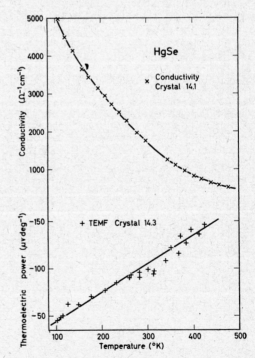

FIG. 1. Variation of electron concentration and electron mobility of a HgSe crystal with annealing time.

FIG. 2. Conductivity and thermoelectric power of HgSe crystals plotted against temperature.

[1] A. J. Blum and A. R. Regel, J. Tech. Phys. (U.S.S.R.) **21**, 316 (1951).

[2] T. C. Harman, J. Electrochem. Soc., **106**, 205C (1959).

[3] C. H. L. Goodman, Proc. Phys. Soc. (London) **B67**, 258 (1954).

[4] C. D. Elpat'evskaja, J. Tech. Phys. (U.S.S.R.) **28**, 2676 (1958).

[5] M. Rodot and H. Rodot, Compt. rend. **250**, 1, 1447 (1960).

in air or in a vacuum without any Hg leads to the same effect. However, by a small partial pressure of Hg the electron concentration increases. After annealing, the crystals show a gray coating. It is quite probable that this might be excessive Hg.

We have tried to dope HgSe by adding other elements and by deviations of stoichiometry; but we failed in getting p-type conduction. The method given by de Nobel[6] for CdTe has also achieved no success.

HgSe crystallizes in the zinc-blende lattice, and in splitting generally favors (100) planes. This is confirmed by x-ray investigation. Crystals of about $0.5 \times 1 \times 3.5$ mm³ suitable for electrical measurements can be split from the ingot. Cd contacts on the ends of the crystals are created by evaporation. Cd shows practically no influence on the carrier concentration contrary to Ag, Bi, and Sn.

III. EXPERIMENTAL RESULTS

Conductivity and Hall effect are measured on HgSe single crystals. Figure 2 shows that the conductivity decreases as the temperature increases. In Fig. 3 the electron concentration is plotted against reciprocal temperature. These curves supply an activation energy of 0.03 ev. At lower temperatures the concentration, which often varies from test to test, remains at a constant level. The carrier concentration is determined from the Hall coefficient; the mobility is the Hall mobility (Fig. 4). A T^{-a} law holds approximately in two temperature areas. Between 100° and 200°K $a = 0.6$ to 0.9; however, between 300° and 500°K $a = 3.5$ for all crystals. The temf shown in Fig. 2 increases roughly linearly with temperature and is -100 μv/deg at 300°K. A similar dependence has been observed with HgSe layers.[4]

For three crystals with different orientation, the magnetoresistance is determined at 300°K. One crystal is split from the ingot, the direction of current is [100] and the mobility is 16 200 cm²/v sec. The other two are

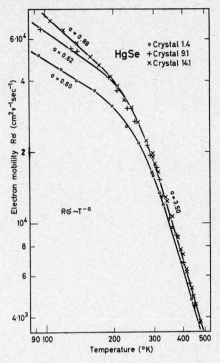

FIG. 4. Hall mobility of the same crystals versus temperature.

grounded with current directions of [110] with mobility of 17 900 cm²/v sec and [111] with mobility of 17 200 cm²/v sec, respectively. The electrodes at the ends are built up by evaporation of Cd. In Fig. 5 magnetoresistance is plotted against the angle between current and field. The rotation axes are [010], [1$\bar{1}$0], and [11$\bar{2}$] for current directions [100], [110], and [111], respectively.

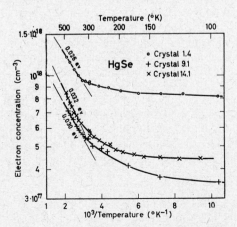

FIG. 3. Electron concentration of HgSe crystals versus reciprocal temperature.

[6] D. de Nobel, Philips Research Repts. **14**, 361 (1959).

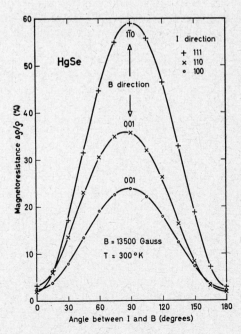

FIG. 5. Variation of magnetoresistance as H is rotated.

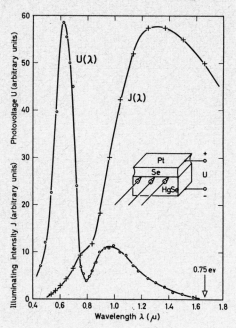

FIG. 6. Dependence of photovoltage and illuminating intensity on wavelength.

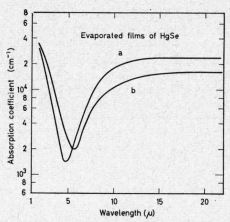

FIG. 7. Absorption coefficient of a HgSe layer, before and after annealing, versus wavelength. Carrier concentrations: curve a, $n=1.55\times10^{19}$ cm^{-3}; curve b, $n=1.13\times10^{19}$ cm^{-3}.

The transverse magnetic field depends on the strength of the field and shows a slight parabolic increase from 0 to 5000 gauss and then a linear dependence.

The contact of HgSe with most of the metals shows no barrier layer. A p-n junction can be obtained by smearing Se upon a HgSe crystal followed by annealing at 480°K. Se is contacted with Pt to avoid a barrier layer at this side. This cell (Fig. 6) can be used as a rectifier; upon illuminating one side of the junction, a photo emf is observed. These effects can only be obtained when the thickness of the Se layers is more than 50 μ. There is no photo emf at photon energies smaller than 0.75 ev.

The polished slice surfaces of KBr receive HgSe and Cd electrodes through evaporation. (Size of layers 4.5×1.5 cm^2.) The thickness of the layers varies from 0.1 to 2 μ. At 300°K the electrical measurements give values for conductivity ranging from 300 to 1000 $(\Omega$ cm$)^{-1}$ and carrier concentrations of 1 to 5×10^{19} cm^{-3}.

We get the absorption coefficient from measurements of transmission and reflectivity of light at these layers over the wavelength range of 1 to 22 μ (Fig. 7). A single crystal with a thickness of 80 μ shows no transmission. A range exists for all layers in which the absorption decreases linearly as the wavelength increases. The center of this linear range corresponds to an absorption coefficient of 10^4 cm^{-1}. If this value is defined as the energy gap ΔE, we calculate for films with different carrier concentrations $\Delta E=0.5$ to 0.7 ev. In Fig. 7 the absorption coefficient of one layer with the concentration of 1.5×10^{19} cm^{-3} (curve a) is shown. The concentration is lowered to 1.13×10^{19} cm^{-3} by annealing for

several hours (curve b). The absorption coefficient is proportional to the concentration in the long wavelength tail. Between 5 to 10μ α increases with the law α prop. $\lambda^{3.5}$. The behavior of α at short wavelength (1 to 5 μ) is remarkable. At a constant wavelength the layer with the higher carrier concentration shows the smaller absorption coefficient. The refractive index has a constant value of 2.4 from 5μ to long wavelengths.

IV. DISCUSSION

There are important relations between the elements and the compounds which crystallize in the diamond and zinc-blende structure within the periodic table of the elements. HgSe can be regarded as an imitation of the gray tin and of the InSb, respectively, as Hg and Se are positioned symmetrically to this element and to the components of the latter compound in the periodic table. InSb has an extremely high electron mobility and indeed, HgSe shows a high mobility also. However it is reduced by the increasing share of the ionic bonding which leads to a stronger interaction of the electrons with the vibrating polar lattice. The change in the nature of the chemical bond can clearly be found in the different splitting of the crystals. The group IV elements split most easily into the (111) planes, as here the lowest number of C–C bonds per unit plane have to be cut. As it was pointed out by Pfister,[7] III-V compounds split into the (110) planes, while with HgSe we always find splitting results into the (100) planes. This change in splitting is caused by the electrostatic power becoming more effective. The energy gap also changes regularly when moving from the group IV elements to the compounds with zinc-blende structure. Goodman[3] predicted $\Delta E=0.7$ ev for HgSe, using the scale of electronegativities. This corresponds well with the value of Sorokin,[8] who calculated $\Delta E=0.7\pm0.10$ ev from the photoelectron emission. Because of the evaporation above

[7] H. Pfister, Z. Naturforsch. **10a**, 79 (1955).
[8] O. M. Sorokin, J. Tech. Phys. (U.S.S.R.) **28**, 1413 (1958).

500°K and the comparable high carrier concentration, we failed in getting intrinsic HgSe. Regarding the spectral distribution of the photovoltage of the junction Se/HgSe crystal (Fig. 6), there is a great probability that the effect in the infrared is due to band-band transitions within the HgSe. The photovoltage vanishes when the energy of the photons decreases. Thus $\Delta E = 0.75$ ev is obtained. The photovoltage shows no distinct threshold, as it is observed with p–n junctions within the same lattice. This is so because there is a continuous junction of two bands with different energy gaps. A photodiffusion voltage in the Se is thought to be responsible for the minimum at $0.8\,\mu$. In this region, the photovoltage consists of a short and a long time effect. Some seconds after illuminating the cell, this long time effect even produces a negative sign of the photovoltage in the region between 0.7 and $0.85\,\mu$. The energy gap, given by the photovoltage, corresponds to a certain degree to ΔE which is obtained from the slope of the absorption coefficient of HgSe layers (Fig. 7). Here, a region of $\Delta E = 0.5$ to 0.7 ev has been found, as there is a dependence of ΔE on the electron concentration. At higher concentrations the energy gap increases. This effect is also found in InSb. It is explained by the degeneracy of the electrons in the conduction band. The lower levels of the conduction band are filled progressively by the electrons, so that absorption transitions can only go the higher, empty conduction band levels.

The activation energy of 0.03 ev which was found for the crystals above 300°K, is thought to be due to a donor. It is not yet clear whether it is caused by strange impurities or by excessive Hg. The exact scattering mechanism in HgSe is not known; therefore the electron concentration is calculated by the simple formula $n = f/eR$; $f = 1$. This introduced error should be of little significance; f goes to 1 when approaching degeneracy of the electron gas and in the case of increased scattering by optical vibration of the lattice.

Using the experimental data of the thermoelectric power (Fig. 2) for different temperatures, m_n^* is evaluated from the formula

$$\Theta = -k/e[2 + \ln 2(2\pi m_n^* kT)^{\frac{3}{2}}/nh^3];$$

$m_n^* = 0.04$ to $0.07\,m_0$ is obtained. The assumptions of the formula, i.e., nondegenerate electron gas and spherical energy surfaces are not fulfilled here. On the other hand, the thermoelectric power is not a sensitive indicator for the scattering mechanism. Calculations based on different scattering mechanisms do not differ significantly. Also in InSb $m_n^* = 0.013\,m_0$ is obtained by cyclotron resonance experiments. Concerning the mentioned relations to this compound, the value for m_n^*, (about four times larger than in InSb), is quite reasonable, especially since the mobility in InSb is about four times larger than in the HgSe.

Another estimation of m_n^* in HgSe layers can be made by using the infrared absorption between 5 and $10\,\mu$. The increase of the absorption coefficient follows the law α prop. $\lambda^{3.5}$. This means ionized impurity scattering, when you take into account the enlarged theory of Fröhlich.[9] Presuming this mechanism, m_n^* may be evaluated using the data from optical and electrical experiments. These are conductivity, electron concentration, absorption coefficient, and refractive index. The evaluation is carried out as it is shown in[10] and leads to $m_n^* = 0.04\,m_0$. This corresponds well with m_n^* obtained from the temf, although crystals and layers with a different scattering mechanism are observed. If ionized impurity scattering really would be the exact mechanism, the mobility of the layers should vary prop. $T^{1.5}$. As has been pointed out,[4] there is virtually no dependence on temperature. This is exactly what we find for the mobility in layers. In the theory of Fröhlich, nondegeneracy is supposed, but degeneracy is indicated by the dependence $\Delta E\ (n)$. When calculated by this method m_n^* does not offer a high degree of accuracy.

Rodot[5] notes that magnetoresistance in HgSe crystals with current in [111] direction is even larger in a longitudinal field than in a transverse field. Our experiments cannot confirm this matter. Three crystals with current in [111], [110], and [100] show a similar dependence of magnetoresistance on the angle between field and current. In the case of a longitudinal field, there is only a very small change in resistance in the three crystals compared with the transverse field. Perhaps there may be a certain deviation from an isotropic conduction, because in the case of current in [111] direction, magnetoresistance is larger than in the other crystals. This is not due to the different mobilities since there is no significant difference between them. Contrary to Rodot we have to assume that the simple spherically symmetrical band centered at $k = 0$ gives a better approach than his model. For a transverse field, the dependence of magnetoresistance upon the field is linear already at low fields. This is quite reasonable, when you consider the high mobility in HgSe. The simple theory, which leads to a parabolic dependence, is valid only for $\mu B \ll 1$.

Rodot[5] presents the dependence of the mobility in HgSe single crystals with $n = 3.5 \times 10^{18}$ cm^{-3}. He finds μ prop. T^{-a}, $a = 2$. In polycrystalline HgSe with $n = 2 \times 10^{18}$ cm^{-3}, $a = 2.5$ was found.[4] In a crystal with $n = 2.3 \times 10^{18}$ cm^{-3}, we determined $a = 2.5$ between 200° and 500°K. Purer crystals (5 to 9×10^{17} cm^{-3}) always show $a = 3.5$ in this region (Fig. 4). Consequently, the scattering mechanism depends upon the electron concentration. The exponent increases from 2 to 3.5 when the concentration is lowered from 3.5×10^{18} to 5×10^{17} cm^{-3}. At lower temperatures the curve of the mobility is flattened, indicating that a different scattering mecha-

[9] Y. H. Fan and M. Becker, *Proceeding of the Reading Conference* (Butterworths Scientific Publications, Ltd., London, 1951), p. 132.
[10] K. J. Planker, and E. Kauer, Z. angew. Physik **12**, 425 (1960).

nism might have become effective. Because of the relation to InSb, similar scattering mechanisms (as pointed out for InSb by Ehrenreich[11,12]) might be strong in HgSe. The screened polar interaction is thought to be of particular importance since the ionic part of the bond should be larger in HgSe than in InSb and screening effects are likely because of the high density of conduction electrons. On the other hand, intrinsic InSb with a considerably lower carrier concentration has been observed.

[11] H. Ehrenreich, J. Phys. Chem. Solids **2**, 131 (1957).
[12] H. Ehrenreich, J. Phys. Chem. Solids **9**, 129 (1959).

JOURNAL OF APPLIED PHYSICS   SUPPLEMENT TO VOL. 32, NO. 10   OCTOBER, 1961

Edge Emission in Zinc Selenide Single Crystals

D. C. REYNOLDS
Aeronautical Research Laboratory

AND

L. S. PEDROTTI AND O. W. LARSON
Air Force Institute of Technology, Wright-Patterson Air Force Base, Ohio

Edge emission in single crystals of ZnSe subjected to ultraviolet radiation at low temperatures has been examined in the temperature interval from 4.2° to 77°K. Two distinct edge emission spectra have been found indicating that two different types of single crystals exist. For type I crystals the edge emission spectrum at 4.2°K contains 10 lines located between 4400 A and 4800 A; at 77°K the emission spectrum contains two lines. For type II crystals the edge emission spectrum at 4.2°K contains 14 lines located between 4400 A and 4900 A; at 77°K the emission spectrum contains three lines, one of which is located at the fundamental absorption edge of the crystal. Both crystal emissions show evidence of phonon interaction with the ZnSe lattice and both emissions undergo significant reductions in intensity as the crystal temperature increases from 4.2° to 77°K.

I. INTRODUCTION

SINGLE crystals of ZnSe irradiated with ultraviolet light at low temperature emit a characteristic fluorescence known as edge emission.[1] This emission, appearing on the long wavelength side of the absorption edge of the crystal, is made up of a series of lines and bands whose wavelengths vary between 4400 A and 4900 A. In appearance the edge emission spectrum in ZnSe is similar to the edge emission spectrum in CdS.[2-5] This might be expected since ZnSe, which has the zinc-blende structure, and CdS, which has the wurzite structure, are similar II–VI semiconductor compounds. The two compounds have fundamental absorption edges in the blue region of the spectrum. A value of 2.554 ev has been reported by Thomas and Hopfield[6] for CdS and the value for ZnSe is 2.83 ev, both values being obtained at 4.2°K. In CdS the emission is made up of a series of lines and bands between 4800 and 5600 A which reflects the longer wavelength absorption edge.

Edge emission in ZnSe has been examined for several samples of single crystals taken from different crystal growth runs. All of the crystals examined were grown from the vapor phase by the same method used to grow single crystals of cadmium sulfide.[7] The samples of ZnSe crystals used for this investigation were made up of a series of yellow rod-shaped crystals. The two distinct edge emission spectra observed for the different samples indicate that two different types of single crystals are present. It is possible that the difference in the two types of crystals may be due to different host lattice defects.

It is the purpose of this paper to report on the nature and structure of edge emission observed in the two

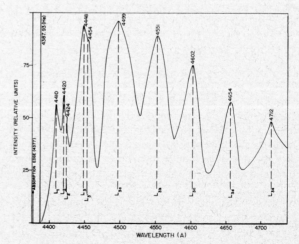

FIG. 1. Edge emission spectrum for type I ZnSe single crystals at 4.2°K.

[1] F. A. Kroger, Physica **7**, 1 (1940).
[2] L. R. Furlong and C. F. Ravilious, Phys. Rev. **98**, 954 (1955).
[3] J. J. Lambe and C. C. Klick, Phys. Rev. **98**, 909 (1955).
[4] M. Bancie-Grillot, E. F. Gross, E. Grillot, and B. S. Razbirin, Optika i Spektroskopiya **5**, 461 (1959).
[5] L. S. Pedrotti and D. C. Reynolds, Phys. Rev. **119**, 1897 (1960).
[6] D. G. Thomas and J. J. Hopfield, Phys. Rev. **116**, 573 (1959).
[7] L. C. Greene, D. C. Reynolds, S. J. Czyzak, and W. M. Baker, J. Chem. Phys. **29**, 1375 (1958).

TABLE I. Wavelengths of spectral lines in edge emission spectrum of type I crystals at 4.2°K and 77°K.

Wavelengths at 4.2°K (A)	Wavelengths at 77°K (A)
L_1—4410	N_1—4421
L_2—4420	N_2—4436
L_2'—4424	
L_3—4448	
L_3'—4454	
L_{3a}—4499	
L_{3b}—4551	
L_{3c}—4602	
L_{3d}—4654	
L_{3e}—4712	

different types of ZnSe single crystals at temperatures of 4.2° and 77°K, and to describe the changes in the structure of this emission for each type of crystal between 4.2° and 77°K. Some attempt will be made to account for the origin of some of the observed lines and to explain some of the changes observed in the spectra as the crystal is heated from 4.2° to 77°K.

II. EXPERIMENTAL RESULTS

The two samples of crystals, hereafter referred to arbitrarily as type I and type II crystals, were stimulated to fluoresce by irradiating the crystals with a concentrated beam of ultraviolet light while the crystals were immersed in either liquid nitrogen or liquid helium. In each instance the emission was photographed with a Bausch and Lomb two meter grating spectrograph equipped with a 4 in.×4 in. grating blazed at 3000 A with 55 000 lines/in. giving a dispersion of 2 A/mm in the first order. Each of the photographs was analyzed with a NSL Spec-Recorder to obtain densitometric traces.

The change in the spectral structure of the edge emission for both type I and type II crystals resulting from a change in crystal temperature between 4.2° and 77°K was investigated in the following manner. The specimens were cemented with silver paint to an aluminum rod which was immersed in liquid helium, all of which was contained in the inner Dewar of a double Dewar arrangement. The crystal was irradiated with a 100 WSP4 mercury vapor lamp filtered to eliminate the visible. The emission was detected with a Bausch and Lomb grating monochromator equipped with a IP28 photomultiplier tube. The temperature was measured with a copper-constantan thermocouple imbedded in the aluminum rod at a position near the crystal. The temperature rise between 4.2° and 77°K occurred naturally as the liquid helium evaporated from the inner Dewar. The temperature rise occurred in approximately 2 hr. No attempt was made to achieve temperature equilibrium and for this reason a small temperature lag between the indicated thermocouple temperature and the actual crystal temperature may have existed. Temperatures between 4.2° and 12°K are based on an

extrapolation of the copper-constantan calibration curve between these two temperatures.

A. Edge Emission Spectrum in Type I ZnSe Crystals

Figure 1 shows a densitometric trace of the edge emission spectrum for type I crystals at 4.2°K. The series of 10 lines is located just to the long wavelength side of the absorption edge at 4377 A. In addition, there are three very broad bands (not shown in Fig. 1) which vary in width from 500 to 1000 A with centers near 4950, 5350, and 6110 A. Several interesting features are to be noted in Fig. 1. Lines L_2' and L_3' appear as less intense lines located a few angstroms on the long wavelength side of L_2 and L_3. The separation energy between lines L_3, L_{3a}, L_{3b}, L_{3c}, L_{3d}, and L_{3e} is approximately 0.03 ev which is about equal to the longitudinal optical phonon energy determined from the measured value of the transverse optical phonon energy 0.027 ev,[8] indicating that the five broad lines L_{3a}—L_{3e} result from phonon interactions.

Fig. 2 shows a densitometric trace of the edge emission spectrum for the same crystal at 77°K. The overall intensity of the emission is significantly reduced and at 77°K only two lines, N_1 and N_2, remain in the edge emission spectrum. The three very broad bands mentioned above have become less intense and somewhat more diffuse at this temperature. The wavelengths of all of the lines shown in Fig. 1 and Fig. 2 are listed for convenience in Table I.

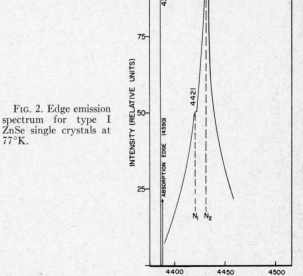

FIG. 2. Edge emission spectrum for type I ZnSe single crystals at 77°K.

[8] D. T. F. Marple, M. Aven, and B. Segall, Bull. Am. Phys. Soc. 2, 6 (1961).

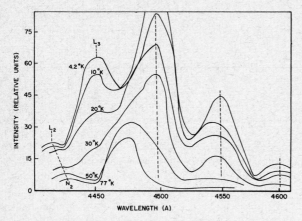

FIG. 3. Relative intensity of edge emission lines vs wavelength for type I ZnSe single crystals at several crystal temperatures between 4.2° and 77°K.

Figure 3 shows the gradual change in the intensity and position of the lines in the edge emission spectrum at 4.2°K as the crystal temperature increases from 4.2 to 77°K. The six curves, obtained with the Bausch and Lomb grating monochromator as explained above, indicate that line L_2 at 4.2°K moves toward longer wavelengths at 77°K and eventually becomes line N_2,

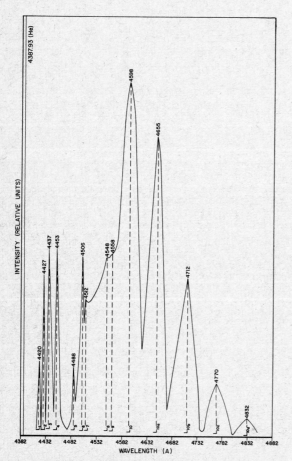

FIG. 4. Edge emission spectrum for type II ZnSe single crystals at 4.2°K.

while the remaining lines decrease steadily in intensity and vanish at 77°K. (Other evidence based on wavelength calculations indicates that line L_1, not shown in Fig. 3, changes into line N_1 at 77°K). The broad hump between 4450 and 4500 A is due to the presence of the 4470 A mercury line in the reflected radiation. This line made resolution of line L_3 with temperature change quite difficult. As a result, it was not possible to say how L_3 shifted as the temperature increased.

B. Edge Emission Spectrum in Type II ZnSe Crystals

Type II crystals were examined in the same manner as type I crystals. Figure 4 shows the edge emission spectrum for a type II crystal at 4.2°K. In this spectrum there are 14 lines. The first seven lines are sharply

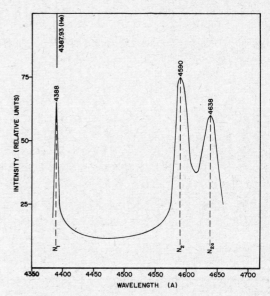

FIG. 5. Edge emission spectrum for type II ZnSe single crystals at 77°K.

defined and closely spaced compared with the lines at longer wavelengths, which are quite broad. The last five lines, L_{10} through L_{10d}, have a constant energy separation of approximately 0.03 ev, the measured phonon energy, indicating again that lines L_{10a} through L_{10d} result from phonon interactions.

Figure 5 shows the edge emission spectrum for the identical crystal at 77°K. Again, as was observed for type I crystals, the edge emission intensity is noticeably reduced at 77°K and much of the structure evident at 4.2°K is missing. As seen in Fig. 5, three lines, N_1, N_2, and N_{2a}, remain and these three lines are not related to the two remaining lines at 77°K for type I crystals. Note that line N_{2a} is separated from line N_2 by the phonon energy 0.03 ev. The lines in Figs. 4 and 5 are listed with their respective wavelengths in Table II.

In an attempt to effect some correlation between the edge emission spectrum for type II crystals at 4.2°K with the edge emission spectrum at 77°K, the change in structure of edge emission with change in crystal temperature was examined. The intensity of lines L_1—L_9 was too low to be resolved with the spectrometer and phototube detector; therefore, only the broad blue lines were examined. Figure 6 shows the structural change for lines L_{10}—L_{10c}. Several interesting results are evident. It is seen that as line L_{10} diminishes in intensity and shifts gradually toward longer wavelengths with an increase in temperature, lines L_{10a}, L_{10b}, and L_{10c} do likewise in an almost step by step manner, thereby strengthening the assumption that lines L_{10a}—L_{10d} result from the parent line L_{10} via multiple phonon interactions. In addition, it is seen that at 13°K, with L_{10} already decidedly reduced in intensity, another set of lines appears on the short wavelength side of the L_{10} set. As the temperature increases to 28°K, the new set of lines increase in intensity while the L_{10} set continues to decrease, and with additional temperature increase

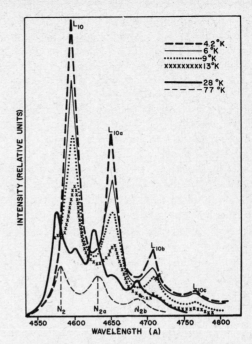

FIG. 6. Relative intensity of edge emission lines vs wavelength for type II ZnSe single crystals at several crystal temperatures between 4.2° and 77°K.

TABLE II. Wavelengths of spectral lines in edge emission spectrum of type II crystals at 4.2°K and 77°K.

Wavelengths at 4.2°K (A)	Wavelengths at 77°K (A)
L_1—4420	N_1—4388
L_2—4427	N_2—4590
L_3—4437	N_{2a}—4638
L_4—4453	
L_5—4488	
L_6—4506	
L_7—4512	
L_8—4548	
L_9—4558	
L_{10}—4598	
L_{10a}—4655	
L_{10b}—4712	
L_{10c}—4770	
L_{10d}—4832	

both begin to decrease in intensity and shift slightly toward longer wavelengths. These lines are identified by their wavelengths to be lines N_2 and N_{2a} in Fig. 5. Thus, it is clear that lines N_2 and N_{2a} at 77°K do not arise from the shifting of lines which are prominent at 4.2°K. This observation is exactly analogous to the phenomenon observed for similar lines in the green emission of CdS between 4.2 and 77°K.[4]

III. DISCUSSION OF RESULTS

In the absence of satisfactory energy models which must eventually account for the observed lines in the spectra of the two types of crystals, it is possible to discuss some of the significant features of the edge emission spectra presented in the previous section. For type I crystals one is able to conclude that lines L_1 and L_2 at 4.2°K change gradually into lines N_1 and N_2 at 77°K. Line L_3 and the series of lines L_{3a}—L_{3e}, which are related to L_3 by multiple phonon interactions, decrease in intensity and dissappear at 77°K. The behavior of lines L_2' and L_3' with a change in crystal tem-

perature is unknown and their proximity to lines L_2 and L_3 make their resolution with temperature changes quite difficult. It is clear that for type I crystals, as a minimum requirement, a satisfactory energy model must account for the five lines L_1, L_2, L_2', L_3, and L_3'.

The structure at 4.2°K for type II crystals is somewhat more complicated than for type I crystals. Nevertheless, some conclusions can be drawn. It has been shown in Fig. 6 that line L_{10}, (and the related lines L_{10a}—L_{10c}) prominent at 4.2°K, vanish as the crystal temperature is increased to 77°K, and two different lines N_2 and N_{2a} appear in the process. Since line N_{2a} is related to line N_2 via a phonon interaction, one need only account for the origin and behavior of line N_2.

Based on a similar phenomenon for the green emission in CdS,[4] it is possible to suggest an energy scheme which accounts for the disappearance of line L_{10} and the appearance of line N_2 as the crystal is warmed from 4.2° to 77°K. Such an energy scheme is shown in Fig. 7. An impurity center located 0.015 ev below the conduction band edge, and another located 0.12 ev above the valence band edge, are necessary to account for the

FIG. 7. Suggested energy scheme accounting for edge emission line L_{10} at 4.2°K and line N_2 at 77°K in type II crystals of ZnSe.

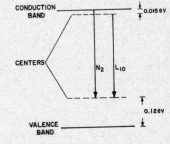

energy of the transitions L_{10} and N_2. According to this energy scheme L_{10} is a center-to-center transition which is dominant at 4.2°K; as the crystal temperature is increased the 0.015 ev center becomes depopulated, and the dominant transition N_2 now occurs between the conduction band edge and the 0.12 ev center. It should be pointed out that a mirror reflection of these centers about the center of the band gap would serve equally well to explain the transitions.

Another interesting characteristic of the edge emission observed in type II crystals is evident in Fig. 5. Line N_1 at 77°K, which has no counterpart at 4.2°K, occurs at the absorption edge of the crystal and may result from exciton emission.

One need only compare the complete dissimilarity between lines N_1 and N_2 at 77°K for type I crystals with lines N_1, N_2, and N_{2a} at 77°K for type II crystals to appreciate fully the fact that type I and type II crystals are indeed distinct. The true nature of the difference between the two types of crystals is of funda-

mental importance in the understanding of the edge emission spectrum observed. As mentioned previously, it is felt that the difference may be due to differences in host lattice defects. In order to investigate the character of this difference, electron bombardment experiments, similar to thos carried out by Kulp and Kelley for CdS,[9] are planned for ZnSe. Kulp and Kelley established that the sulphur interstitial atom was responsible for the green edge emission bands in CdS. It is hoped that similar experiments will determine whether or not the selenium atom plays an analogous role for the blue emission characterized by line L_{10} in ZnSe.

ACKNOWLEDGMENT

The authors are indebted to C. W. Litton of the Solid State Research Branch of the Aeronautical Research Laboratory for the generous and willing donation of his time and technical skills in photographing the emission spectra.

[9] B. A. Kulp and R. H. Kelley, J. Appl. Phys. **31**, 1057 (1960).

JOURNAL OF APPLIED PHYSICS   SUPPLEMENT TO VOL. 32, NO. 10   OCTOBER, 1961

Some Properties of HgSe-HgTe Solid Solutions

M. RODOT, H. RODOT, AND R. TRIBOULET
Laboratoire de Magnétisme et de Physique du Solide CNRS. Bellevue, S. et O., France

HgTe$_{1-x}$Se$_x$ solid solutions have been prepared, with x varying from 0 to 1. The samples are n type near $x=1$ and p type near $x=0$, but, due to the high electron-to-hole mobility ratio, electronic conudction is dominant in the range 100–400°K in all samples. The concentration of free electrons lies between $5 \cdot 10^{16}$ and $3 \cdot 10^{18}$ cm^{-3}. Measurements of the Hall mobility μ_H and magnetothermoelectric effect ΔQ show that, for Se-rich samples, lattice scattering is dominant in the range 77–400°K and that, near room temperature, $\mu_H \propto T^{-2}$. For Te-rich samples, lattice scattering is dominant in the range 200–400°K and, near room temperature, $\mu_H \propto T^{-1}$. Effective masses have been calculated and it is seen that the conduction band is not parabolic. The detailed band structure and the exact value of the mobility seem to depend little upon structural factors. For $x=0.5$ and $x=0.9$, the electron mobility can reach 12 000 cm^2 v^{-1} sec^{-1} at 293°K and 30 000 cm^2 v^{-1} sec^{-1} at 77°K.

I. INTRODUCTION

MERCURY selenide and telluride are two semiconducting compounds with high electron mobility and low energy gaps. We have studied the HgSe$_x$Te$_{1-x}$ solid solutions, in order to determine whether their electrical properties can be derived from those of the compounds or are dependent upon the disorder of the lattice, that is the coexistence of Se and Te on the anion sites.

Some electrical properties of HgTe have been previously studied[1-7] and a value of 0.025 ev has been

estimated for the energy gap ΔE.[4,5] The conduction band is isotropic and nonparabolic; the electron effective mass has been found to increase with temperature and is[7] $m_n = 0.030 m_0$ at 300°K, for a concentration of free electrons equal to $n = 5.10^{17}$ cm^{-3}. Near room temperature, the electron mobility μ_H is limited by the interaction of electrons with acoustical phonons[2,7] and varies as $T^{-\frac{1}{2}}$ according to Tsidil'kovski[2] and as T^{-1} according to Harman et al.[4]; μ_H has been found[7] as high as $2.2.10^4$ cm^{2}v^{-1}sec^{-1} at 300°K; the ratio of electron-to-hole mobility is[6,7] of the order $b \simeq 70$. While the optical properties of HgTe–CdTe solid solutions have been studied, those of pure HgTe seem yet undescribed.

HgSe is not so well known.[1,3,7-9] Its band structure is apparently similar to that of HgTe; $m_n = 0.044 m_0$ for $n = 4 \cdot 10^{18}$ cm^{-3} (references 7,8). For the same purity the mobility μ_H can reach $0.6 \cdot 10^4$ cm^2 v^{-1} sec^{-1} at

[1] A. I. Blum and A. R. Regel, Zhur. Tekh. Fiz. **21**, 316 (1951).
[2] I. M. Tsidil'kovski, Zhur. Tekh. Fiz. **27**, 1744 (1957).
[3] R. O. Carlson, Phys. Rev. **111**, 476 (1958).
[4] T. C. Harman, M. J. Logan, and H. L. Goering, J. Phys. Chem. Solids **7**, 228 (1958).
[5] J. Black, S. M. Ku, and H. T. Minden, J. Electrochem. Soc. **105**, 723 (1958).
[6] W. D. Lawson, E. Nielsen, E. H. Putley, and A. S. Young, J. Phys. Chem. Solids **3/4**, 325 (1959).
[7] H. Rodot and M. Rodot, C. R. Acad. Sci. **248**, 934 (1959); M. Rodot, Czechoslov. J. Phys. (to be published, 1961).

[8] G. B. Wright and B. Lax, Bull. Am. Phys. Soc. **5**, 177 (1960).
[9] D. Redfield, Bull. Am. Phys. Soc. **2**, 121 (1957).

TABLE I. Properties of the $Hg(Te_{1-x}Se_x)$ solid solutions.

Sample	x	$T=77°K$				$T=293°K$			$T=373°K$			
		n^a	μ_H^b	r	m_n/m_0	n^a	μ_H^b	K_{ph}^c	n^a	μ_H^b	r	m_n/m_0
HSZ3	1.0	26	2.07	-0.23	0.044	22.5	0.73	$\cdots$	22.5	0.43	-0.11	0.044
F9B	0.9	4.9	2.86	-0.04	0.022	7.2	1.22	11.4	8.4	0.75	-0.39	0.033
F8B	0.8	2.4	3.25	$+0.19$	0.038	7.1	1.38	7.4	8.7	0.99	-0.19	0.0
F7B	0.7	5.2	2.25	$+0.05$	0.034	8.6	1.13	9.3	10.0	0.85	-0.17	0.034
F6B	0.6	1.8	2.86	$\cdots$	$\cdots$	7.1	1.18	7.4	8.5	0.91	$\cdots$	$\cdots$
F5C	0.5	4.8	3.35	-0.07	0.022	9.1	1.05	$\cdots$	11.0	0.76	-0.26	0.032
F5D	0.5	2.1	3.5	$+0.29$	0.012	7.3	1.27	5.0	8.8	0.99	-0.19	0.032
F4B	0.4	2.1	3.04	$\cdots$	$\cdots$	7	1.22	5.8	7.8	1.00	$\cdots$	$\cdots$
F3B	0.3	$\cdots$	$\cdots$	$\cdots$	$\cdots$	$\cdots$	$\cdots$	6.0	$\cdots$	$\cdots$	$\cdots$	$\cdots$
F2B	0.2	3.3	1.37	$\cdots$	$\cdots$	7	1.15	6.6	8.1	0.89	$\cdots$	$\cdots$
F1B	0.1	9.0	1.28	$\cdots$	$\cdots$	8.3	1.12	8.0	9.1	0.92	$\cdots$	$\cdots$
HTZ8	0.0	5.9	0.81	$\cdots$	$\cdots$	5.9	1.37	21	7.3	0.99	$\cdots$	$\cdots$
HTQ5	0.0	0.7	2.25	$+0.31^d$	0.012^d	4.0	1.97	$\cdots$	5.4	1.55	-0.26	0.029

a 10^{17} cm^{-3}; b 10^4 cm^2 v^{-1} s^{-1}; c 10^{-3} w cm^{-1} deg^{-1}
d at 150°K, where $n=1.9\cdot10^{17}$ cm^{-3} and $\mu_H=3.1\cdot10^4$ cm^2 v^{-1} s^{-1}.

300°K, is proportional[7] to T^{-2}, and is limited by the electron-acoustical phonon interaction[7] in the whole range $77-300°K$. The study[9] of the optical absorption indicates that $\Delta E < 0.08$ ev.

For HgSe–HgTe solid solutions, Nikolskaya and Regel[10] have already observed a maximum of mobility for a fifty-fifty composition, but most of their samples had low mobilities.

II. PREPARATION OF SAMPLES

About forty samples, with selenium content x varying by 0.1, were melted in sealed quartz vessels and quickly cooled. They are polycrystalline, with coarse grains ($\simeq3$ mm). Their x-ray diagram (Debye-Scherrer method) shows one phase, which has the zinc-blende structure. For the samples listed in Table I, this diagram showed no extra lines, and the deduced lattice parameter is indicated in Fig. 1. Electrical and thermal properties of these samples have been measured.

III. ELECTRICAL PROPERTIES

The Hall constant R_H has been measured in the range 77–400°K, for an induction $H=6600$ gauss (Fig. 2). One can see that there are two kinds of samples: n-type Se-rich samples, for which R_H varies little with T, and Te-rich samples, with temperature-dependent R_H and sometimes a maximum of R_H for an intermediate temperature. The latter are probably p type at very low temperature. Near 77°K, they present mixed conduction, as is seen by the behavior of R_H, the thermoelectric power Q, and its variation with a transverse magnetic field, i.e., the MTE effect ΔQ (see Fig. 4). At higher temperature, the transport properties in all samples are essentially determined by the electrons, due to the high electron-to-hole mobility ratio.

The Hall mobility has been calculated from the formula $\mu_H=R_H\sigma$ ($\sigma=$ electrical conductivity). Typical results are plotted in Fig. 3. The slope of the log μ_H (log T) curves near room temperature is negative, varying from -1 for HgTe and Te-rich samples to -2 for Se-rich samples and HgSe. This negative value of the slope suggests lattice scattering to be a dominant mechanism. This is again true for Se-rich samples, down to 77°K. For the Te-rich samples, the slope is positive at low temperatures, due to the onset of impurity scattering or hole conduction (or both).

Measurements of ΔQ (Fig. 4) support these conclusions. It is known[11] that a positive MTE effect is ob-

FIG. 1. $HgTe_{1-x}Se_x$: Lattice parameter a versus composition.

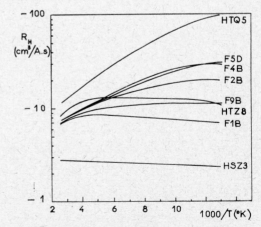

FIG. 2. Hall constant ($H=6600$ gauss) versus temperature (definition of samples: see Table I).

[10] E. Nikolskaya and A. R. Regel, Zhur. Tekh. Fiz. 25, 1352 (1955).

[11] M. Rodot, Ann. Phys. 5, 1085 (1960); M. Rodot, C. R. Acad. Sci. 252, 2626 (1961).

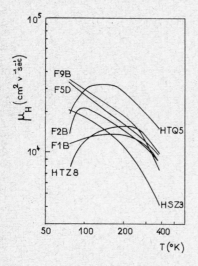

FIG. 3. Hall mobility versus temperature (definition of samples: see Table I).

served in the case of ambipolar conduction (electrons + holes) or in the case of electronic conduction limited by ionized impurity scattering. On the other hand, a negative MTE effect is associated either with electronic conduction limited by acoustical phonon scattering or with "ambielectronic" conduction (i.e., by two kinds of electrons, belonging to two distinct bands). The latter situation is not probable for HgTe and HgSe. As ΔQ is negative for all samples near room temperature—and for $x \simeq 1$, down to 77°K—we can conclude that, in these conditions, acoustical-phonon scattering is dominant and that the mobility is characteristic of the crystal lattice, although impurity scattering might be also present.

The thermoelectric power Q has also been measured; from Q, R_H, and ΔQ, one can use a previously described method[7] to deduce the effective mass m_n, the reduced Fermi level η, and the scattering index r (i.e., the exponent of the formula $\tau = \tau_0 \epsilon^r$ relating the relaxation time τ to the electron energy ϵ, assuming only one

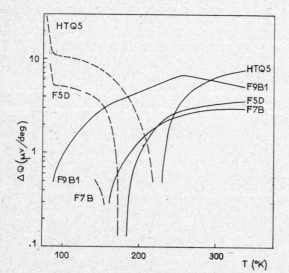

FIG. 4. Magnetothermoelectric effect ΔQ versus temperature (definition of samples: see Table I. Full lines, $\Delta Q < 0$; broken lines, $\Delta Q > 0$).

scattering mechanism to be effective). The results are also given in Table I.

IV. DISCUSSION

The calculated effective masses, listed in Table I, are only accurate to $\pm 15\%$. Nevertheless, it can be seen that, at 77°K, m_n is an increasing function of n, whereas at 373°K, at which temperature most samples have approximately the same purity, m_n does not vary significantly for the whole set of samples. This means that the conduction band is not parabolic.

Near room temperature, the lattice electron mobility varies as T^{-1} for $x < 0.8$ and as T^{-2} for $x \geqslant 0.8$. The reason for this difference is not known and cannot be explained by considering the variation of m_n with T. As for the exact value of μ_H at a given temperature, a satisfactory correlation has not been obtained between the electron mobility and the values of n and m_n. Generally speaking, the electron mobility in semiconducting compounds depends upon intrinsic properties (ionicity of bonds, band scheme, effective mass, dielectric constant, etc.) and also upon the electron concentration. In the alloys, the electron mobility is also affected by structural factors. Because of the atomic disorder, perturbations of the band scheme arise ("band tails" studied by Parmenter[12]). The effect of these perturbations on the transport properties is probably important, but is not well understood at the moment. For our samples, the possible influence of these perturbations on the mobility is masked by the effect of the density of electron through impurity scattering. The quite-independence of m_n upon x at room temperature seems to indicate that the influence of lattice disorder on the band structure is not important. However, it is possible that the study of optical properties would reveal such an influence.

It is expected that atomic disorder might have more influence upon the lattice thermal conductivity K_{ph} than upon the electrical properties. Weill[13] has measured the total thermal conductivity K, and obtained K_{ph} by subtracting an electronic contribution determined from the general Wiedemann-Franz-Lorenz formula (with $r = 0.25$). The values of K_{ph} (Table I) are, on the whole, in agreement with previous measurements.[14] The mass difference between Te and Se atoms causes the expected decrease of thermal conductivity for the solid solutions.

V. CONCLUSIONS

We wish to stress the following points. In the HgTe$_{1-x}$Se$_x$ solid solutions, the electron mobility is essentially characteristic of the lattice, and proportional to T^{-1} for $x \simeq 0$ and to T^{-2} for $x \simeq 1$. The conduction band is not parabolic. The values of the electron effective mass and lattice mobility seem to be little sensitive to the disorder of the lattice.

[12] R. H. Parmenter, Phys. Rev. 97, 587 (1955).
[13] G. Weill (unpublished).
[14] A. V. Ioffe and A. F. Ioffe, Fiz. Tverdogo Tela 2, 781 (1960).

JOURNAL OF APPLIED PHYSICS   SUPPLEMENT TO VOL. 32, NO. 10   OCTOBER, 1961

Righi-Leduc Effect in Mercuric Selenide

CHARLES R. WHITSETT

Parma Research Laboratory, Union Carbide Corporation, Parma 30, Ohio

A preliminary study has been made of the Righi-Leduc effect in mercuric selenide (HgSe). The Righi-Leduc magnetothermal effect is the thermal analog of the Hall effect wherein temperature plays the role of voltage and heat flow replaces electric current. The effect is particularly large in HgSe because of the coincidence of large electron mobilities (as high as 1.5 m²/v sec at 300°K) and low lattice thermal conductivity (about 0.02 w/°K cm at 300°K). Results are presented of measurements of the Righi-Leduc coefficient S as a function of temperature, magnetic field strength, and electron concentration. Classically, for a one-carrier material, $S \cong (\kappa_E/\kappa)\mu$, where κ_E is the electronic, κ is the total thermal conductivity, and μ is the electron mobility. This expression is in qualitative accord with the experimental results. At room temperature S ranged between 0.27 and 0.34 m²/v sec for samples that had between 55×10^{17} and 5.6×10^{17} electrons/cm³.

I. INTRODUCTION

MERCURIC selenide (HgSe) has an extremely large ratio of electron mobility to lattice thermal conductivity and for this reason exhibits large magneto-thermal effects. In this paper are presented results of measurements on HgSe of the Righi-Leduc coefficient as a function of temperature, magnetic field strength, and electron concentration. Simple analysis shows that the behavior of the Righi-Leduc effect is in qualitative accord with theory.

The Righi-Leduc effect is the production of a transverse temperature gradient by the action of a perpendicular magnetic field upon a sample carrying a longitudinal heat current. It is the thermal analog of the Hall effect wherein temperature plays the role of voltage and heat flow replaces electric current. Treatments of the Righi-Leduc effect are included in treatises by Jan[1] and Putley.[2] Its measure is the Righi-Leduc coefficient S, which for metals should be equal to the Hall mobility[3]; in materials which have an appreciable lattice thermal conductivity, S is diminished by the ratio of the electronic thermal conductivity to the total thermal conductivity.

In mercuric selenide the electron mobility is large (ranging up to 1.5 m²/v sec at 300°K in the samples reported here) and the lattice thermal conductivity, while appreciable, is low (about 0.02 w/°K cm at 300°K). Therefore, the Righi-Leduc coefficient in HgSe is quite large (ranging up to 0.4 m²/v sec at 300°K in purer samples) and easily measurable.

II. ELEMENTARY THEORY

If the axis of the sample and the longitudinal heat flow are taken to be in the x direction, and if the magnetic field H_z is directed along the z axis, then the induced transverse temperature gradient component parallel to the y axis is the Righi-Leduc temperature gradient. Refer to Fig. 1 for the geometrical relationships between the sample, heat currents, and magnetic field as they are to be considered here. The Righi-Leduc coefficient S is defined by the relation

$$S = \frac{1}{H_z} \frac{\partial T/\partial y}{\partial T/\partial x}. \tag{1}$$

This definition assumes that there are no electric currents and that there is no net transverse heat flow (i.e., adiabatic conditions prevail). If only electrons contribute to the effect then the warm side will be that indicated in Fig. 1 and the Righi-Leduc coefficient will be negative.

An elementary analysis of the effect is based upon the assumption that the magnetic field induces a transverse electron heat current $Q_{R.L.}$ proportional to the longitudinal electron heat current, with the proportionality constant being $\mu_H H_z$ (μ_H=Hall mobility):

$$Q_{R.L.} = \mu_H H_z \kappa_E (\partial T/\partial x), \tag{2}$$

where κ_E is the electronic thermal conductivity and $\partial T/\partial x$ is the longitudinal temperature gradient. For there to be no net heat flow this must be exactly cancelled by the opposing heat flow

$$Q_y = \kappa(\partial T/\partial y), \tag{3}$$

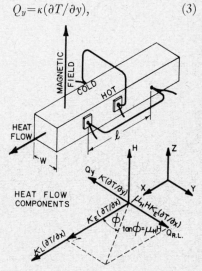

FIG. 1. Geometrical relationship between sample, magnetic field, and heat currents active in the Righi-Leduc effect. The indication of the hot and cold sides is for conduction by electrons.

[1] J.-P. Jan in *Solid State Physics*, edited by F. Seitz and D. Turnbull (Academic Press, Inc., New York, 1957), Vol. 5, pp. 1–96.

[2] E. H. Putley, *Hall Effect and Related Phenomena* (Butterworths and Company, London, 1960).

[3] See reference 1, p. 65.

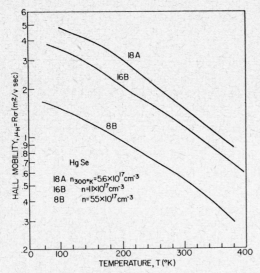

FIG. 2. Hall mobility as function of temperature for HgSe samples with different electron concentrations.

where κ is the total thermal conductivity and $\partial T/\partial y$ is the transverse temperature gradient. Equating (2) and (3) gives

$$\mu_H H_z \kappa_E (\partial T/\partial x) = \kappa (\partial T/\partial y), \qquad (4)$$

and substitution into (1) gives

$$S = \frac{1}{H_z} \frac{(\partial T/\partial y)}{(\partial T/\partial x)} = \frac{\kappa_E}{\kappa} \mu_H. \qquad (5)$$

Putley[2] solved the Boltzmann equation under the assumptions that $\mu_H H_z \ll 1$, that classical statistics apply, and that electron scattering is by acoustic lattice vibrations; he obtained

$$S = -\frac{21\pi}{32} \left(\frac{k^2 T}{e\kappa} \right) n\mu^2, \qquad (6)$$

where $k = $ Boltzmann constant, $e = $ electron charge, and $\mu = $ conductivity mobility. Using the classical expressions

$$\kappa_E = 2(k/e)^2 \sigma T, \qquad (7)$$

where σ is the electrical conductivity, and

$$\mu_H = (3\pi/8)\mu, \qquad (8)$$

Eq. (6) may be rewritten

$$S = (7/8)(\kappa_E/\kappa)\mu_H. \qquad (9)$$

Putley also derived general expressions which may be used to compute S for any degree of degeneracy and for any values of magnetic field for which the Boltzmann equation is valid. Since the HgSe samples studied here were largely degenerate, and further since the zero magnetic field approximation should not apply to them, these more general expressions should be used for a careful analysis of the Righi-Leduc effect in HgSe.

The general expressions, however, are complex functions of Fermi integrals and depend upon the value of the Fermi energy, the effective electron mass, and the energy-dependence of the electron mean free time. The HgSe samples studied were not sufficiently well characterized to justify the more exact analysis.

III. EXPERIMENTAL METHOD

The samples of HgSe used for this study were unoriented single crystals about $1 \times 2 \times 15$ mm^3 in size. The Hall coefficient and electrical resistivity as functions of temperature were measured for each sample during runs preliminary to the Righi-Leduc coefficient measurements. For the Righi-Leduc coefficient measurements the samples were supported between a heat sink at one end and a heater at the other end. Differential Chromel-constantan-Chromel thermocouples, 0.003 in. in diameter, were used to measure the longitudinal and transverse temperature gradients. To satisfy the requirement that no electric currents flow in the sample it was necessary to electrically insulate one junction of each differential thermocouple from the sample; this insulation was provided by a minute layer of Duco cement. The other junction of each differential thermocouple was allowed to contact the sample. A separate thermocouple was used to measure the sample temperature at the location of the transverse differential thermocouple junctions. All thermocouple junctions were pressed against the sample and secured with General Electric 7031 varnish. Measurements were made with the sample in a chamber evacuated to a pressure less

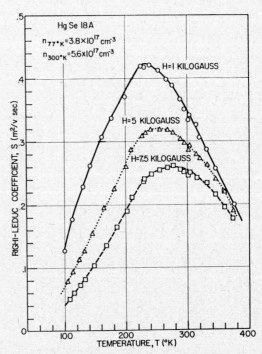

FIG. 3. Righi-Leduc coefficient as function of temperature for HgSe in different magnetic fields.

than 10^{-5} mm of Hg and as the sample warmed slowly from near liquid nitrogen temperature to room temperature. The sample chamber could be heated to obtain data above room temperature.

The sample heater was used to deliver 15 to 30 mw of power to the sample. Longitudinal temperature gradients produced ranged from 5° to 20°K/cm, and the Righi-Leduc temperature gradients ranged from 0.015° to 0.4°K/cm.

The greatest sources of error would be from the failure of the transverse differential thermocouples to accurately sense the Righi-Leduc temperature gradients and from lateral heat flow from the sample due to conduction by the thermocouples and to radiation. These effects would reduce the measured transverse temperature gradient below the true value and lead to a too small value for S. The main indications of reliability of the results were their reproducibility when samples were remounted, the almost negligible time lag of the Righi-Leduc temperature difference as the magnetic field was turned on and reversed, and the lack of dependence of the Righi-Leduc coefficient upon the magnitude of the heat flow through the sample.

IV. RESULTS AND DISCUSSION

The three samples, 18A, 16B, and 8B, reported upon here had room temperature electron concentrations of 5.6×10^{17} cm^{-3}, 11×10^{17} cm^{-3}, and 55×10^{17} cm^{-3}, respectively. The electron concentrations of the latter two samples did not vary with temperature below room temperature, but in the first sample the concentration decreased to 3.8×10^{17} cm^{-3} at liquid nitrogen tempera-

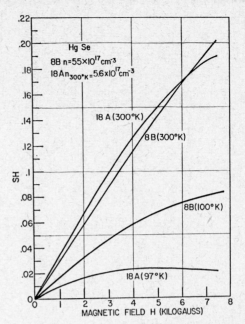

FIG. 5. Plots of SH vs H for HgSe samples having extremes of mobility values. For S to be field independent, the plots must be linear.

ture. The zero magnetic field Hall mobilities as functions of temperature for these samples are shown in Fig. 2.

The Righi-Leduc coefficient of the sample 18A, having the lowest electron concentration, is plotted as a function of temperature and for three values of magnetic field in Fig. 3. Figure 4 gives plots of the Righi-Leduc coefficient at 1000 gauss for all three samples. In every case the Righi-Leduc coefficient peaked between 200° and 300°K. This is as expected since at higher temperatures the decrease of mobility with increase in temperature dominates, and at lower temperatures the strong decrease of the ratio of κ_E/κ with decrease in temperature dominates the behavior of S. Below 100°K the ratio κ_E/κ becomes very small, but it may be that as the absolute zero of temperature is approached this ratio, and thus S, will again become large.

The Righi-Leduc coefficient in most cases was strongly dependent upon magnetic field strength. Since μH ranged from 0.05 to 0.5 for fields between 1000 and 7500 gauss, a decrease of S with increase in magnetic field is expected and is due to the falling off of both κ_E and μ_H as the field strength is increased. The decrease in κ_E in magnetic fields also gives rise to a pronounced thermal magnetoresistance effect in HgSe, as has also been observed by Amirkhanov and co-workers.[4] According to the classical expression for S, Eq. (6), S will vary as μ^2. The higher mobility samples did exhibit a magnetoresistance at liquid nitrogen temperature which implied a mobility decrease large enough to account for the magnetic field dependence of S. Data for quantitatively comparing the magnetic field depend-

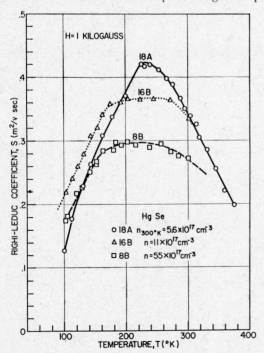

FIG. 4. Righi-Leduc coefficient at 1000 gauss as function of temperature for samples having different electron concentrations.

[4] Kh. I. Amirkhanov, A. Z. Daibov, and V. P. Zhuze, Doklady Akad. Nauk S.S.S.R. **98**, 557 (1954).

TABLE I. Summary of electrical and thermal data for HgSe.

Sample No. and temp.	Electron density, n (cm^{-3})	Hall mobility, μ_H, (m²/v sec)	Estimated total thermal conductivity (w/°K cm)	Electronic thermal conductivity, κ_E (w/°K cm)	Calculated Righi-Leduc coefficient, S (m²/v sec)	S, measured at 1000 gauss (m²/v sec)
8B						
100°K	55×10¹⁷	1.45	0.099	0.019	0.24	0.17
150°K	55×10¹⁷	1.24	0.071	0.024	0.37	0.27
200°K	55×10¹⁷	0.93	0.054	0.024	0.39	0.30
250°K	55×10¹⁷	0.71	0.045	0.023	0.32	0.29
300°K	55×10¹⁷	0.55	0.035	0.021	0.29	0.27
16B						
100°K	11×10¹⁷	3.52	0.109	0.010	0.28	0.22
200°K	11×10¹⁷	2.12	0.050	0.012	0.45	0.36
300°K	11×10¹⁷	1.16	0.027	0.010	0.38	0.34
18A						
100°K	3.8×10¹⁷	4.78	0.086	0.0043	0.21	0.13
200°K	3.9×10¹⁷	3.04	0.045	0.0057	0.35	0.38
300°K	5.6×10¹⁷	1.49	0.025	0.0060	0.30	0.34

encies of S and μ at higher temperatures were not obtained. In most cases it was established that the values for S at 1000 gauss were very nearly the same as the zero field limits. This is evident from examination of Fig. 5 which shows SH vs H at 100° and 300°K for samples with the extremes of mobility values. Only for sample 18A at low temperatures was the value for S appreciably reduced at 1000 gauss.

One unusual characteristic of mercuric selenide is that for the best crystals the electron mobility increases in proportion to the reciprocal square root of the electron concentration. Thus $n\mu^2$ is a constant for all samples. This constancy was valid for samples 8B and 16B, but the mobility in 18A was slightly low for the corresponding carrier concentration which implies imperfection of the crystal. Since, according to Eq. (6), S is proportional to $n\mu^2/\kappa$, the sole variation of S with electron concentration should be because of the variation of the electronic contribution to the total thermal conductivity. As the electron concentration increases the total thermal conductivity also increases. Therefore, S should decrease with an increase in electron concentration and was observed to do so except when the Righi-Leduc effect became very dependent upon magnetic field strength.

At temperatures above 300°K the Righi-Leduc coefficient became less dependent upon magnetic field because of the rapid diminishment of the electron mobility. Data for temperatures above room temperature for samples 8B and 16B were not considered sufficiently reliable for inclusion here because, in a vacuum environment above room temperature, the characteristics of higher carrier concentration samples changed with elapsed time, presumably because of the loss of excess mercury from the material.

To check the validity of the classical expression for S it was necessary to estimate the total thermal conductivity. Such estimates were based upon the measured temperature gradients in the samples, the power supplied by the sample heater, and estimated losses of heat through the sample mount. Above 220°K radiation losses became too large to allow any reasonable estimate of thermal conductivity to be made, but the lower temperature values were extrapolated to room temperature. The thermal conductivity estimates are given, along with a summary of other results, in Table I. The electronic thermal conductivity was calculated from Eq. (7); the calculated values of the Righi-Leduc coefficient were obtained by using the classical expression, Eq. (9). The agreement between the calculated and measured values is as good as could be expected, not only because of the inadequacy of the classical expression, but also because the total thermal conductivity estimates are crude.

The classical expression qualitatively describes the weak field behavior of the Righi-Leduc effect in HgSe. After the band parameters and electron scattering mechanisms in HgSe have been sufficiently well established, a more exact treatment of the Righi-Leduc effect will be attempted. More extensive measurements and analyses are expected to provide essentially a detailed knowledge of the conduction of heat by electrons and of the effect of magnetic fields upon this heat transport.

ACKNOWLEDGMENTS

The assistance of Ralph D. Thomas with the preparation and mounting of the samples and the collection of data was invaluable and is gratefully acknowledged.

JOURNAL OF APPLIED PHYSICS   SUPPLEMENT TO VOL. 32, NO. 10   OCTOBER, 1961

Some Electrical and Optical Properties of ZnSe

M. Aven, D. T. F. Marple, and B. Segall

General Electric Research Laboratory, Schenectady, New York

Single crystals of ZnSe have been prepared by the vapor growth technique and optical and electrical measurements on these crystals are reported. Analysis of the reststrahlen reflection peak gives 0.026 ev for the transverse optical phonon energy. The longitudinal optical phonon energy is 0.031 ev as calculated from the transverse phonon energy, the static dielectric constant, $\epsilon_0 = 8.1 \pm 0.3$, and the high-frequency dielectric constant, $\epsilon_\infty = 5.75 \pm 0.1$. The effective ionic charge calculated from the Szigetti formula is 0.7 ± 0.1. Exciton absorption peaks associated with the valence and conduction bands in the vicinity of Γ were observed at liquid hydrogen temperature with the principal peak at 2.81 ± 0.01 ev. The exciton reduced mass $0.1\ m_0$ combined with the room temperature electron-to-hole mobility ratio of 12, ob-

tained by preliminary transport measurements on n- and p-type ZnSe gives tentative values of $0.1\ m_0$ and $0.6\ m_0$ for the electron and hole masses, respectively.

Reflectance was determined by various methods in the range 0.025 to 14.5 ev photon energy and was analyzed by the Kronig-Kramers inversion method to obtain the optical constants in the 1 to 10 ev range. A number of peaks appear in the imaginary part of the dielectric constant.

The first set of peaks, 2.7 and 3.15 ev, are believed to be due to exciton and interband transitions at Γ with a spin-orbit valence band splitting of 0.45 ev. The second set of peaks, 4.75 and 5.1 ev, are tentatively assigned to transitions at L with a spin-orbit splitting of 0.35 ev. Other peaks are observed at higher energies.

INTRODUCTION

PAST work on ZnSe has been mainly concerned with its photoluminescent[1] and photoconductive[2] properties. Recently several workers[3,4] have reported on the synthesis of large single crystals of ZnSe. Larson, Pedrotti, and Reynolds[5] as well as Halsted and Aven[6] have reported on the edge emission. Marple, Segall, and Aven[4] have described exciton absorption and infrared reflection spectra.

The present paper is an extension of the work described in reference 4. Optical studies now include investigation and partial interpretation of the optical constants at energies higher than the optical band gap, and observations of absorption bands on the high-energy side of the reststrahlen absorption. The electrical work to be reported is a preliminary investigation of the electron and hole transport properties.

SAMPLE PREPARATION

ZnSe crystals were grown by the vapor growth technique first reported by Greene, Reynolds, Czyzak, and Baker[7] and modified by Piper and Polich.[8] The starting material was G.E. Chemical Products Plant ZnSe powder which was first sintered in the presence of excess Se in a hydrogen atmosphere. The sintered material was then passed at the rate of approximately 0.2 mm/hr through a furnace heated to about 1350°C, which sublimed the ZnSe to the cold end of the growing tube. The product was usually a boule of several cm³

containing single crystal regions of a tenth to one cm³. Doped crystals were grown by adding the impurity to the ZnSe powder prior to the sintering. Native imperfections were introduced, when desired, by firing the crystals in Se or Zn. Sample bars approximately 1 cm long and 4 mm² in cross-sectional area were cut from the boules described above. Prior to electroding the bars were chemically polished in hot concentrated NaOH. The electrode material was In for the n-type and In-Ag alloy for the p-type samples. Various sizes and shapes of cleaved or polished crystals, depending on the type of measurement and the spectral range, were used in the optical work.

EXPERIMENTAL

Figure 1 shows the reflectance of ZnSe at room temperature in the range 0.026 to 14.5 ev. Data included here were obtained on a variety of crystals by several different methods. In the region 0.026 to 0.05 ev measurements were made on thick mechanically polished sections cut from three different boules. Conventional arrangements were used with a double-pass CsI prism monochromator. In the 0.05 to 2.5 ev range the reflectance given in Fig. 1 was calculated from the refractive index. In the 1.8 to 2.5 ev range the index was calculated from refraction produced by a ZnSe prism. Interference fringes were observed in transmission through a 5.35μ thick mechanically polished crystal in the range 0.4 to 2.4 ev with a double-pass CaF₂ prism spectrometer[9] and in a $114\ \mu$ thick mechanically polished crystal with a prism-grating spectrometer in the range 0.04 to 0.5 ev. Matching the dispersion and magnitude of the index in the overlapping regions and the reflectance directly measured at 0.04 ev made possible a unique assignment of interference orders in the analysis of the fringe data, and hence a determination of both the index and the crystal thicknesses.

Reflectance in the 2.5 to 3.8 ev range was twice

[1] R. E. Shrader, S. Lasof, and H. W. Leverenz, *Solid Luminescent Materials*, Cornell University Symposium (John Wiley & Sons, Inc., New York, 1946), pp. 215–257.

[2] R. H. Bube and E. L. Lind, Phys. Rev. **110**, 1040 (1958).

[3] A. G. Fisher, Bull. Am. Phys. Soc. **6**, 17 (1961).

[4] D. T. F. Marple, B. Segall, and M. Aven, Bull. Am. Phys. Soc. **6**, 19 (1961).

[5] O. W. Larson, L. S. Pedrotti, and D. C. Reynolds, Bull. Am. Phys. Soc. **6**, 111 (1961).

[6] R. E. Halsted and M. Aven, Bull. Am. Phys. Soc. **6**, 312 (1961).

[7] L. C. Greene, D. C. Reynolds, S. J. Czyzak, and W. M. Baker, J. Chem. Phys. **29**, 1375 (1958).

[8] W. W. Piper and S. Polich, J. Appl. Phys. **32**, 1278 (1961).

[9] The optical arrangement was similar to the equipment described by W. W. Piper, D. T. F. Marple, and P. D. Johnson, Phys. Rev. **110**, 323 (1958).

measured on a cleaved face of a single crystal, first a few hours after cleaving and again after about 500-hr exposure to still air. The equipment described in reference (9) was slightly modified for these measurements, and a freshly evaporated Al mirror was used for a calibration standard.[10] No systematic differences in the two measurements were found, and the average values are given in Fig. 1.

Reflectance in the 3.8 to 14.5 ev range was measured with a vacuum ultraviolet monochromator[11,12] on three crystals. Two of these were measured a few hours after cleaving and the third after about 800-hr exposure to air. This last crystal was remeasured after chemical polishing with NaOH.[13] With the possible exception of the region above 10 ev, no systematic differences were found between these four surfaces, and the average reflectance at each energy for the group, normalized to match the absolute value found at 3.8 ev, is shown in Fig. 1. The angle of incidence in all the reflection measurements was less than 15 deg, and the results were analyzed with the theory for normal incidence.

It is quite difficult to estimate the possible systematic errors in the reflectance measurements in Fig. 1 especially at energies of 2.6 ev and more. Aside from possible systematic errors connected with the condition of the sample surface, the absolute reflectance near 2.6 ev is believed to be correct within $\pm 5\%$ of the value stated, with increasing possible systematic error up to $\pm 15\%$ beyond 7 ev. Random errors, due to detector noise for example, would influence the shape of small details in the curve, and are believed to be about ± 0.005 in the reflectance.

Transmission through three mechanically polished samples cut from two different boules, with thicknesses 14.1, 114, and 2590 μ was measured with the double-pass CsI prism spectrometer. Figure 2 shows the absorption coefficient in the range 0.031 to 0.08 ev as calculated from these data. Corrections for reflection loss were made with the results shown in Fig. 1. Widely spaced interference fringes were observed in the thinnest crystal, and although the results for the absorption peak at 0.037 ev, are qualitatively reliable, they are less accurate than the rest due to the difficulties in the analysis of the fringes.

Optical properties in the absorption edge region, 2.6 to 2.9 ev received further study at room temperature and near liquid hydrogen temperature. Results for the absorption coefficient are shown in Fig. 3. Transmission was measured on two mechanically polished crystals, one 4.1 and the other 5.35 μ thick, for energies where the absorption coefficient was less than about 2×10^4

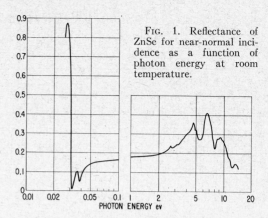

FIG. 1. Reflectance of ZnSe for near-normal incidence as a function of photon energy at room temperature.

cm^{-1}. The optical constants also were measured by reflection of polarized light at 80° angle of incidence.[14] Crystals used in these measurements had reflecting faces about 6×8 mm, were about 2 mm thick and were cemented to copper plates which could be cooled by a cryostat. For the low-temperature measurements the crystal was entirely enclosed in a metal-and-glass shield also cooled by the cryostat. No changes in the optical constants were observed during a period of eight hours near liquid hydrogen temperature. Two cleaved crystal faces and two chemically polished surfaces were studied at low temperatures, but only the data for one crystal was included in Fig. 3. The other crystals gave qualitatively similar results, but at low temperature the large absorption peak was shifted by up to 0.01 ev from the results presented here. The cause of this shift is not known, but is possibly associated with strains or impurities in the crystals. Both the transmission and reflection data in Fig. 3 were obtained on sections of the same boule. The sharp structure shown in Fig. 3 was entirely absent in mechanically polished crystals.

Measurements of transport properties in ZnSe are at present in a preliminary stage. Such measurements are complicated by the tendency of ZnSe (as well as most other wide band-gap II–VI compounds) to compensate the intentionally added donor or acceptor impurities through formation of native imperfections, vacancies or interstitials. An additional complication in p-type sulfides and selenides of zinc and cadmium arises from the relatively large hole ionization energy, between 0.5 and 1.2 ev in these materials. The Hall effect apparatus built for these measurements covered the temperature range from room temperature up to about 600°C, and operated on currents between 10^{-3} and 10^{-11} amp. The measurements were performed by applying a dc potential, usually of the order of 0.1 v, across the long dimension of the crystal. The current through the crystal was ascertained by measuring the potential drop across a standard resistor in series with the crystal. The potential drop across the crystal and the Hall voltage were measured by two sets of appropriately

[10] G. Hass, J. Opt. Soc. Am. **45**, 945 (1955).

[11] P. D. Johnson, J. Opt. Soc. Am. **42**, 278 (1952).

[12] The same procedure, with almost the same numerical value was used for Ge. See H. R. Philipp and E. A. Taft, Phys. Rev. **113**, 1002 (1959).

[13] The mechanically polished crystal was immersed for one minute in boiling 50% NaOH, followed by a rinse in briskly flowing tap water at 70°C followed by a rinse in distilled water.

[14] Apparatus and techniques described by S. Roberts, Phys. Rev. **114**, 104 (1959); **118**, 1509 (1960).

TABLE I. Mobilities and carrier concentrations for doped and undoped ZnSe.

Material	Temp. range (°C)	Mobility (cm²/v-sec) Electrons	Mobility (cm²/v-sec) Holes	Carrier concentration (cm⁻³) 27°C	Carrier concentration (cm⁻³) 200°C
ZnSe:Cu	200–400		11	⋯	10^{10}
ZnSe:Cu, Se-fired	130–260		16	⋯	10^{11}
ZnSe, Se-fired	200–400		15	⋯	10^{10}
ZnSe:Ga	27–400	80		10^{6}	⋯
ZnSe:Ga, Zn-fired	27	150		10^{15}	⋯
ZnSe, Zn-fired	27–250	260		10^{16}	⋯

located electrodes on the faces of the crystal. The measurements were performed with an Applied Physics Corporation vibrating reed electrometer. Magnetic fields between 5 and 6 kgauss were used.

Although measurements were made at a number of different temperatures, the limited range covered does not allow a reliable mobility-temperature relationship to be established at this time. Table I lists the average mobilities in the temperature range covered for three types of samples for each conductivity type. The carrier concentrations are given for room temperature for the n-type samples. The p-type samples had too high resistivity for reproducible Hall measurements at room temperature, and therefore the carrier concentrations for these samples are given for 200°C.

DISCUSSION

The reststrahlen reflection peak at 0.026 ev was fitted with the classical theory for a single resonance absorp-

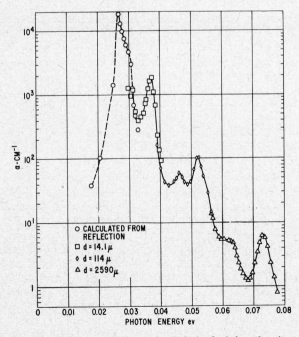

FIG. 2. Absorption coefficient α for ZnSe in the infrared region as a function of photon energy at room temperature.

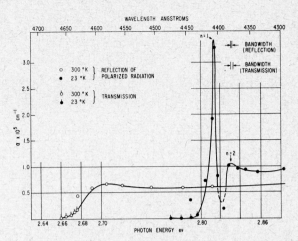

FIG. 3. Absorption coefficient α for ZnSe in the absorption edge region as a function of photon energy.

tion,[15] ignoring the small reflection anomaly at 0.037 ev which is associated with the absorption peak at about the same energy. The static dielectric constant was found to be 8.1 ± 0.3 from capacitance measurements on a single-crystal wafer about 0.015 cm thick and 1.5 cm in diameter. Analysis of the refractive index data gave 5.75 ± 0.1 for the high-frequency dielectric constant. With these constants inserted in the equations at the outset the data could be fitted within the scatter in the data with a resonance energy of 0.026 ev. The absorption coefficient as calculated from the reflection is included in Fig. 2 and agrees satisfactorily with the transmission results where they overlap. The largest absorption peak is due to transverse optical phonons at or near the center of the Brillouin zone. The longitudinal phonon energy calculated from the above resonance energy with the Lyddane-Sachs-Teller formula[16] is 0.031 ev, in agreement with the structure observed in luminescence at energies near the absorption edge by Halsted.[17]

The effective ionic charge sZ has been discussed by Szigetti[18] and Callen[19] among others. With the dielectric constants and reststrahlen frequency reported above the Szigetti formula gives $sZ=0.70$. From the standpoint of effective charge ZnSe is intermediate in "ionicity" compared to the other "fourth-row" compounds, as sZ for CuBr and GaAs are 1.00 and 0.43, respectively.[18,20]

We believe that the series of weaker absorption peaks seen in Fig. 2 are due at least in part to multiple phonon absorption processes involving both the optical

[15] For a convenient summary see T. S. Moss, *Optical Properties of Semiconductors* (Butterworth Scientific Publications Ltd., London, 1959), pp. 15–22.

[16] R. H. Lyddane, R. G. Sachs, and E. Teller, Phys. Rev. **59**, 673 (1941).

[17] R. E. Halsted (private communication).

[18] B. Szigetti, Trans. Faraday Soc. **45**, 155 (1949).

[19] H. B. Callen, Phys. Rev. **76**, 1394 (1949).

[20] G. Picus, E. Burstein, B. W. Henvis, and M. Hass, J. Phys. Chem. Solids **8**, 282 (1959).

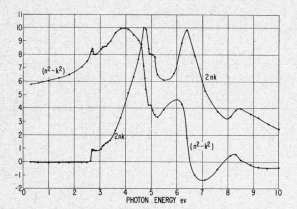

FIG. 4. Real and imaginary parts of the dielectric constant, (n^2-k^2) and $2nk$, respectively, for ZnSe as a function of photon energy at room temperature. The refractive index n and extinction coefficient k were calculated from the results in Fig. 1 by the Kronig-Kramers inversion.

and acoustical branches of the phonon spectrum. A spectrum qualitatively similar to Fig. 2 has been observed in InSb,[21] GaP[22] and other materials. However, some of the structure seen in Fig. 2 may possibly be due to impurity absorption. Further analysis of the results is in progress.

A consistent quantitative interpretation of the low temperature absorption shown in Fig. 3 is obtained by the assumption that the large peak is due to weakly bound direct excitons formed in the ground state and that the small secondary peak[23] is the first excited state. The sharpness and strength indicate that the absorption is a direct transition.

On the basis of our knowledge of the band structure of the compound semiconductors, it is plausible to assume that the smallest "vertical" gap is at Γ (i.e., at $k=0$). Thus we believe that excitons are formed from the conduction band and upper valence band which have respectively[24] Γ_6 and Γ_8 symmetries at $k=0$. If the degeneracy in the Γ_8 valence band is neglected,[25] the hydrogenic formula is correct in the limit of weak binding. The spacing of the peaks gives (0.020 ± 0.004) ev for the exciton binding energy and (0.10 ± 0.03) m_0 for the exciton reduced mass. The oscillator strength for the ground state excitons calculated from the absorption and the refractive index (not presented here) is about 3×10^{-3} in satisfactory agreement with rough theoretical estimates. The first excited state is roughly

a factor of ten weaker in oscillator strength. Theoretically it is expected to be about $\frac{1}{8}$ as strong.

It is possible to obtain an estimate of the effective masses of electrons and holes from the exciton reduced mass given above and the mobility ratio obtained from the transport measurements. Assuming that in the temperature range covered the dominant mobility-limiting mechanism is scattering by longitudinal optical phonons, the ratio of the hole and electron masses is given by[26] $m_h/m_e=(\mu_e/\mu_h)^{\frac{2}{3}}$. By using the averages for the mobility values in Table I for both electrons and holes, the above equation in conjunction with the exciton reduced mass of 0.1 m_0 yields approximately 0.1 m_0 for the effective mass of electrons and 0.6 m_0 for the effective mass of holes. These assignments have to be regarded as tentative at this time, subject to revision after more exhaustive transport measurements and after elucidation of the scattering mechanisms.

In this connection it is worth noting that according to the carrier concentration figures in Table I, native imperfections have a strong influence on the position of the Fermi level. ZnSe:Ga, for example, is but weakly n type, probably because most of the Ga donors are compensated by Zn vacancies. ZnSe fired in Zn, which treatment is expected to produce either selenium vacancies or interstitial zinc, however, is strongly n type even without the addition of chemical donor impurities. In p-type samples this effect is less pronounced, both ZnSe:Cu and Se-fired ZnSe having about equal carrier concentrations.

Figure 4 shows the optical constants, expressed as the real and imaginary parts of the dielectric constant, for the energy range 1 to 10 ev, as calculated from the room temperature reflectance, Fig. 1, by the Kramers-Kronig inversion procedure.[27] In this analysis the reflectance at energies higher than 14.5 ev was extrapolated assuming[12] that $d(\log R)^{\frac{1}{2}}/d (\log h\nu)=-2.21$ in order to bring the phase angle to zero in the transparent region between 0.5 and 2.5 ev.

The small absorption peak at 2.70 ev calculated by the Kramers-Kronig inversion is in satisfactory agreement with the absorption for room temperature given in Fig. 3. It is interpreted as exciton absorption and the onset of direct transitions at Γ. The bump at 3.15 ev in the imaginary part of the dielectric constant Fig. 4 is thought to be due to excitons formed from the "split-off", Γ_7 valence band and the Γ_6 conduction band, and to direct transitions between the bands. That the lifetime of these excitons is appreciably shorter than the lower-energy ones results from the fact that they can undergo autoionization. The reflectance peak associated with this absorption has also been observed at 23°K. At this temperature it is considerably sharper, with peak

[21] S. J. Fray, F. A. Johnson, and R. H. Jones, Proc. Phys. Soc. (London) **76**, 939 (1960).

[22] D. A. Kleinman and W. G. Spitzer, Phys. Rev. **118**, 110 (1960).

[23] Observations of reflectance at normal incidence made near liquid hydrogen temperature on one cleaved crystal with spectral resolution of 0.8 A confirm the existence of this second peak and give a more accurate measurement of the peak spacing than is shown in Fig. 3. The width of both peaks at half-maximum absorption coefficient is 4 ± 1.5 A.

[24] The notation employed here is given in G. Dresselhaus Phys. Rev. **100**, 580 (1955).

[25] This degeneracy leads to errors which are probably small compared to the experimental uncertainty in the line spacing.

[26] H. Ehrenreich, J. Phys. Chem. Solids **8**, 130 (1959).

[27] For an example and summary of the literature, see W. G. Spitzer and D. A. Kleinman, Phys. Rev. **121**, 1324 (1961), or reference 12. We wish to thank D. S. Story who programmed and performed most of the calculations on an IBM 1620 computer.

reflection at 3.22 ev and half-width very roughly 0.04 ev. From the energy difference for the two absorptions at room temperature as evaluated from Fig. 4, one obtains 0.45 ± 0.04 ev for the splitting of the valence band at Γ due to spin-orbit interaction.

We suggest that the absorption peaks in Fig. 4 at 4.75 and 5.10 ev are due to transitions between the two valence band states and the conduction band at L[i.e., $k = 2\pi/a(\frac{1}{2},\frac{1}{2},\frac{1}{2})$]. The valence band splitting at L due to spin-orbit interaction is theoretically expected to be about $\frac{2}{3}$ of that at Γ. Within the experimental error the separation of the two peaks, 0.35 ± 0.08 ev, is in accord with this prediction.

One may speculate that the absorption peak at 6.4 ev is due to transitions between the valence and conduction bands in the vicinity of X[i.e., $k = 2\pi/a$ (1,0,0)]. This ordering of the different direct-transition band gaps suggested for ZnSe is the same as that suggested for Ge by Phillips.[28] The relative magnitude of the absorption at 4.75 and 6.4 ev and the width of the peak at 6.4 ev are comparable to the corresponding structure observed in Ge.[12] We believe that interpretation of the structure above 6.4 ev, as well as confirmation of the assignment of the lower energy structure, will be greatly aided by improved calculations of the band structure of II–VI compounds, and by more accurate optical data.

ACKNOWLEDGMENTS

We wish to thank H. R. Philipp and E. A. Taft for assistance in obtaining the reflectance data in the 3.8 to 14.5 ev energy range, and for advice on the Kronig-Kramers inversion calculations, and Henry Ehrenreich for stimulating discussions.

[28] J. C. Phillips, J. Phys. Chem. Solids 12, 208 (1960).

JOURNAL OF APPLIED PHYSICS   SUPPLEMENT TO VOL. 32, NO. 10   OCTOBER, 1961

Band Structure of HgSe and HgSe-HgTe Alloys

T. C. Harman and A. J. Strauss

Lincoln Laboratory, *Massachusetts Institute of Technology, Lexington 73, Massachusetts*

A detailed analysis of Hall coefficient data obtained at temperatures between 77° and 350°K has been made for HgSe and HgSe$_{0.5}$Te$_{0.5}$ samples containing excess donor concentrations up to 10^{19} cm^{-3}. On the basis of previous magnetoresistance, Seebeck coefficient, and reflectivity data, a spherically symmetric non-quadratic conduction band exhibiting the $\epsilon(k)$ dependence described by Kane was adopted in making the analysis. Calculations based on a conventional two-band model failed to give quantitative agreement with experiment, but good agreement was obtained on the basis of a model in which the conduction band and one valence band overlap in energy. Therefore the materials are semimetals rather than semiconductors. The best fit to the data was obtained with an overlap energy of 0.07 ev for both HgSe and HgSe$_{0.5}$Te$_{0.5}$, with hole density-of-states masses of 0.17 m_0 and 0.30 m_0, respectively. With increasing carrier concentration, the optical absorption edge for heavily doped HgSe exhibits a shift to higher energies which is characteristic of n-type materials with low electron effective masses. Qualitatively, the optical data are consistent with a semimetal band model rather than with a semiconductor model, since the interband absorption edge apparently occurs at photon energies less than the Fermi energy.

INTRODUCTION

THE II-VI compounds HgSe and HgTe, which crystallize in the zinc-blende structure, form a continuous series of pseudobinary solid solutions. Values of 0.01–0.02 ev for the energy gap of HgTe have been obtained by several authors[1-5] who analyzed the variation of Hall coefficient with temperature on the basis of a simple two-band semiconductor model. Zhuze[6] has reported the energy gap of HgSe to be 0.12 ev, without specifying the nature of the experimental data or the band model employed, while Goodman[7] has predicted a gap of 0.7 ev for HgSe on theoretical grounds. No energy gap values for HgSe-HgTe alloys have been reported.

In the present investigation, the variation of Hall coefficient and resistivity with temperature has been determined experimentally for samples of HgSe and the following alloys: HgSe$_{0.75}$Te$_{0.25}$, HgSe$_{0.5}$Te$_{0.5}$, and HgSe$_{0.25}$Te$_{0.75}$. A detailed analysis has been made of the Hall coefficient data for HgSe and HgSe$_{0.5}$Te$_{0.5}$ samples varying widely in net donor concentration. It has not been possible to explain these data on the basis of a simple two-band model. Good agreement with experiment is obtained, however, with a band model in which the conduction band and one valence band overlap

* Operated with support from the U. S. Army, Navy, and Air Force.

[1] I. M. Tsidilkovski, Zhur. Tekh. Fiz. (USSR) 27, 1744 (1957).

[2] T. C. Harman, M. J. Logan, and H. L. Goering, J. Phys. Chem. Solids 7, 228 (1958).

[3] R. O. Carlson, Phys. Rev. 111, 476 (1958).

[4] J. Black, S. M. Ku, and H. T. Minden, J. Electrochem. Soc. 105, 723 (1958).

[5] W. D. Lawson, S. Nielsen, E. H. Putley, and A. S. Young, J. Phys. Chem. Solids 9, 325 (1959).

[6] V. P. Zhuze, Zhur. Tekh. Fiz. (USSR) 25, 2079 (1955).

[7] C. H. L. Goodman, Proc. Phys. Soc. (London) B67, 258 (1954).

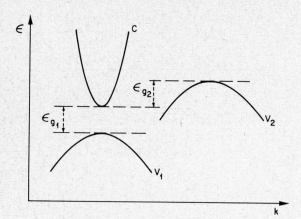

FIG. 1. Schematic $\epsilon(k)$ diagram for the band structure model adopted for HgSe and HgSe$_{0.5}$Te$_{0.5}$.

in energy, as shown schematically in Fig. 1. Data on the infrared absorption of HgSe are also consistent with this model. Thus, these two materials are semimetals rather than semiconductors. This result suggests that HgTe may also be a semimetal and that previous analyses of its electrical properties should be reconsidered from this point of view.

EXPERIMENTAL PROCEDURE

Ingots of HgSe and HgSe-HgTe alloys were prepared by a modified Bridgman method. Commercial high-purity elements (99.999+%) were placed in a quartz tube tapered to a point at one end. The tube was sealed off under vacuum and heated in a horizontal two-zone resistance furnace. After reaction was complete, the furnace was rotated into the vertical position, and the melt was frozen directionally by lowering the tube out of the furnace at the rate of about 4 mm/hr. The vapor pressure of mercury over the melt was kept constant during crystallization by controlling the temperature of the upper zone of the furnace. Single crystals of undoped HgSe up to 2.5 cm in diameter and 10 cm in length were obtained by this method, while doped HgSe ingots and alloy ingots were generally composed of large grains. Where the electrical properties of single crystal and polycrystalline samples could be compared, they were found to be the same within experimental error. Undoped ingots of HgSe were n type, probably due to the presence of excess mercury.[8] The minimum donor concentration which could be obtained was about 1×10^{17} cm^{-3}. Donor concentrations up to 3×10^{19} cm^{-3} were obtained by doping with aluminum, but no acceptor impurity could be found. In the HgSe$_{0.5}$Te$_{0.5}$ alloy, donor concentrations up to 7×10^{18} cm^{-3} and acceptor concentrations up to 3×10^{19} cm^{-3} were obtained by doping with aluminum and copper, respectively.

---

[8] The stoichiometry of HgSe, HgTe, and their alloys, as well as the electrical behavior of various impurities in these materials, will be discussed in a subsequent publication.

The resistivities and Hall coefficients of parallelepiped samples cut from HgSe and alloy ingots were measured by conventional dc potentiometric methods. The magnetic field used for the Hall measurements was approximately 6000 gauss. Measurements at room temperature and liquid nitrogen temperature were made with pressure contacts, while indium-soldered contacts were used for measurements at liquid helium temperature and for those made as a function of temperature. In making the latter measurements, the sample was first cooled to either liquid nitrogen or liquid helium temperature and then allowed to warm up slowly while data were taken automatically with a recording potentiometer.

Infrared reflection and transmission data used to calculate optical absorption coefficients were obtained with Perkin Elmer model 221 (double beam) and model 12C (single beam) spectrophotometers, respectively. The samples were etched before measurement in order to remove the work damage produced by grinding and polishing.

EXPERIMENTAL DETERMINATION OF FREE ELECTRON CONCENTRATIONS

Experimental values of the free-electron concentration (n) in all samples for which $n \geqslant p$, the free-hole concentration, were calculated from the measured Hall coefficients (R_H) according to the usual expression for one-carrier conduction: $n = -1/R_H ec$. This expression is found to be applicable to such samples on the basis of data which show that the ratio of electron mobility to hole mobility $(b = \mu_n/\mu_p)$ is of the order of 100 in the HgSe-HgTe system. For HgSe$_{0.5}$Te$_{0.5}$, a comparison between the Hall mobilities of extrinsic n-type and p-type samples at 4.2°K gives $b = 85$, while analysis of the temperature dependence of R_H for a sample containing an excess acceptor concentration of 3.4×10^{18} cm^{-3} gives $b = 1.1 \times 10^2$. This analysis utilizes the expression $R_{max}/R_{ext} = (1-b)^2/4b$, where R_{max} is the maximum negative value of R_H and R_{ext} is the value of R_H in the extrinsic range. In deriving this expression, it is assumed that only one species of electrons and one species of holes make appreciable contributions to the conductivity, and also that the rate of change of b with temperature is small compared to the rate of change of the intrinsic concentration with temperature. No additional assumptions concerning the details of band structure, statistics, or scattering mechanism are required. In the case of HgTe, the same analysis of R_H as a function of temperature gives $b = 100$,[2,3] but no comparison between the Hall mobilities of extrinsic n-type samples is possible, since extrinsic n-type material cannot be prepared.[8] Neither type of data on the mobility ratio of HgSe can be obtained, since extrinsic p-type material cannot be prepared.[8] It seems reasonable, however, to assume that the mobility ratio in HgSe is also of the order of 100, particularly since the Hall mobilities of electrons in HgSe, HgTe, and their alloys are the same to within 25%.

For a semiconductor with $b \gg 1$ and $n \geqslant p$, the exact expression for R_H reduces to $R_H = -A_n/nec$, where A_n is a parameter whose value depends on degree of degeneracy, scattering mechanism, the product of carrier mobility and magnetic field, and band structure. Since analysis of galvanomagnetic and thermomagnetic data for HgSe indicates that the value of A_n is very nearly unity,[9] n is very nearly equal to $-1/R_H ec$ when $n \geqslant p$.

THEORETICAL CALCULATION OF FREE ELECTRON CONCENTRATIONS

The energy-band model adopted in the theoretical calculations is shown schematically in Fig. 1. Free electrons are present in the conduction band C, as in the simple two-band model, but free holes are present only in valence band V_2, not in valence band V_1 as in the two-band model. In order to calculate n as a function of temperature and donor concentration, theoretical expressions for n and p were first obtained in terms of the band parameters of bands C and V_2, respectively. These expressions were derived on the basis of the general equation for the concentration of free carriers in a band:

$$n = 1/4\pi^3 \int f_0 d^3k, \qquad (1)$$

where f_0 is the equilibrium distribution function and k is the wave vector.

The properties of the conduction band of HgSe required to evaluate n according to Eq. (1) are known with considerable accuracy from previous investigations. Measurements in this laboratory of magnetoresistance as a function of the angle between current and magnetic field indicate that the band is spherically symmetric,[9] although it should be noted that magnetoresistance results described by Rodot and Rodot[10] are inconsistent with spherical symmetry. Both Seebeck coefficient data[11] and infrared reflectivity measurements[12] show that the electron energy relative to the bottom of the conduction band (ϵ) does not exhibit a simple quadratic dependence on wave vector (k), so that the effective mass is not independent of energy. The data are found to be consistent with the expression given by Kane[13] for $\epsilon(k)$:

$$\epsilon = -\epsilon_{g1}/2 + (\epsilon_{g1}^2 + 8P^2k^2/e)^{\frac{1}{2}}/2, \qquad (2)$$

where ϵ_{g1} is the energy gap between the conduction band and valence band V_1, as shown in Fig. 1, and P is a matrix element defined by Kane.[13] When the integral $\int f_0 d^3k$ in Eq. (1) is evaluated on the basis of this expression, using Fermi-Dirac statistics, the equation

obtained for the free-electron concentration is

$$n = \frac{3}{4\pi^2} \left(\frac{3}{2}\right)^{\frac{1}{2}} \left(\frac{kT}{P}\right)^3 \int_0^\infty \frac{X^{\frac{1}{2}}(X+\phi_1)^{\frac{1}{2}}(2X+\phi_1)}{1+\exp(X-\eta)} dX, \qquad (3)$$

where $X \equiv \epsilon/kT$, $\phi_1 \equiv \epsilon_{g1}/kT$, and $\eta \equiv \epsilon_F/kT$; ϵ_F is the Fermi level relative to the bottom of the conduction band.

The following values of P and ϵ_{g1} for the conduction band of HgSe have been used in evaluating Eq. (3): $P = 9 \times 10^{-8}$ ev-cm, $\epsilon_{g1} = 0.1$ ev. These are the values which yield theoretical results for the Seebeck coefficients in quantitative agreement with the experimental data[11]; somewhat different values of the band parameters are derived on the basis of the reflectivity data. It is of interest that P has essentially the same value for HgSe as for the III-V compounds InSb, InAs, GaSb, InP, and GaAs.[14]

Seebeck and reflectivity data for $HgSe_{0.5}Te_{0.5}$, although not as extensive as those for HgSe, appear to be consistent with the Kane model of the conduction band. Therefore, Eq. (3) has been used in the theoretical calculations for the alloy as well as for HgSe. The data indicate that, for electrons of a given energy, the effective mass is considerably greater in $HgSe_{0.5}Te_{0.5}$ than in HgSe. The band parameter values adopted for the alloy are: $P = 9 \times 10^{-8}$ ev cm (the same as for HgSe), $\epsilon_{g1} = 0.2$ ev. For electrons at the bottom of the conduction band, the effective mass calculated from these values is 0.014 m_0, compared with the corresponding mass of 0.007 m_0 for HgSe.

The fact that the conduction band in HgSe and $HgSe_{0.5}Te_{0.5}$ has the form given by Kane indicates that a valence band V_1 is present at an energy ϵ_{g1} below the bottom of the conduction band, as shown in Fig. 1. No other information concerning the valence bands of HgSe or $HgSe_{0.5}Te_{0.5}$ is available from previous investigations. As stated above, the additional valence band V_2 shown in Fig. 1 has been included in the band structure in order to account for the data obtained in the present investigation, since these data could not be explained quantitatively by a structure containing only bands C and V_1.

In obtaining a theoretical expression for p analogous to Eq. (3) for n, it was assumed that valence band V_2 is a simple parabolic band, in which hole energy exhibits a quadratic dependence on wave vector. Evaluation of Eq. (1) for such a band, using Fermi-Dirac statistics, gives the well-known expression

$$p = 4\pi(2kT/h^2)^{\frac{3}{2}}(m_p^*)^{\frac{3}{2}}F_{\frac{1}{2}}[-(\eta+\phi_2)], \qquad (4)$$

where m_p^* is the density-of-states effective mass for holes, $F_{\frac{1}{2}}$ is the Fermi integral, η is the reduced Fermi level defined as in Eq. (3), and $\phi_2 \equiv \epsilon_{g2}/kT$; ϵ_{g2} is the energy separation between bands C and V_2, as shown in Fig. 1. Equation (4) is valid regardless of the number

[9] T. C. Harman and A. J. Strauss (to be published).
[10] M. Rodot and H. Rodot, Compt. rend. **250**, 1447 (1960).
[11] T. C. Harman, Bull Am. Phys. Soc. **5**, 152 (1960).
[12] G. B. Wright, A. J. Strauss, and T. C. Harman, Bull. Am. Phys. Soc. **6**, 155 (1961).
[13] E. O. Kane, J. Phys. Chem. Solids **1**, 249 (1957).

[14] H. E. Ehrenreich, Phys. Rev. **120**, 1951 (1960).

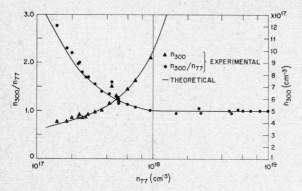

FIG. 2. Free-electron concentration at 300°K (n_{300}) and free-electron concentration ratio (n_{300}/n_{77}) versus free-electron concentration at 77°K (n_{77}) for HgSe.

of valleys associated with band V_2 and is applicable in the presence of degenerate valence bands.

The evaluation of ϵ_{g2} and m_p^* for HgSe and HgSe$_{0.5}$Te$_{0.5}$, which formed an important part of the present investigation, was accomplished by substituting trial values of these parameters into Eq. (4), calculating n as a function of temperature and donor concentration according to the method now being described, and selecting those values which gave the best fit to the experimental data. The value of ϵ_{g2} obtained in this manner is -0.07 ev for both HgSe and HgSe$_{0.5}$Te$_{0.5}$. The values of m_p^* for HgSe and HgSe$_{0.5}$Te$_{0.5}$ are 0.17 m_0 and 0.30 m_0, respectively.

By using the band parameter values listed for HgSe and HgSe$_{0.5}$Te$_{0.5}$, theoretical values of n and p for a specified temperature can be calculated from Eqs. (3) and (4), respectively, provided that the Fermi level is also specified. Since the present calculations were restricted to cases in which $n \geqslant p$, the values of ϵ_f adopted were those which led to values of n and p satisfying the relationship

$$n = p + N_D, \qquad (5)$$

where N_D is the excess donor concentration. This relationship is applicable because the excess donors are fully ionized over the entire temperature range, as shown by the fact that for all samples the Hall coefficient increased to a constant value when the temperature was reduced sufficiently. Complete ionization is predicted theoretically, since for electron effective masses as low as those in HgSe and HgSe$_{0.5}$Te$_{0.5}$ the ionization energy of shallow donors is expected to vanish at concentrations much lower than those studied in the present investigation, due to the overlap of conduction band and donor wave functions.

In order to obtain theoretical values suitable for comparison with experimental data, two sets of calculations were made on the basis of Eqs. (3), (4), and (5). In one case, n was evaluated as a function of N_D for certain fixed temperatures, using a desk calculator, while in the other the variation of n with T for certain fixed donor concentrations was calculated with an

IBM 709 computer. All band parameter values were taken to be independent of temperature.

RESULTS AND DISCUSSION

Experimental data for free-electron concentrations at 300° and 77°K in HgSe samples varying widely in net donor concentration are shown in Fig. 2, where n_{300} and n_{300}/n_{77} are plotted against n_{77}. The qualitative features of the data may be explained in terms of the relationship of n_{300} and n_{77} to N_D, in the same manner as if HgSe were an n-type semiconductor. Samples containing sufficiently high donor concentrations are extrinsic at both temperatures, with $N_D \gg p$; according to Eq. (5), $n_{300} = n_{77} = N_D$, and $n_{300}/n_{77} = 1$. As N_D is decreased below 1×10^{18} cm^{-3}, p_{300}—which is equal to the concentration of intrinsic electrons promoted from the valence band to the conduction band—becomes appreciable compared to N_D. Therefore, n_{300} becomes greater than N_D. On the other hand, n_{77} remains equal to N_D until N_D is reduced to values considerably less than 1×10^{18} cm^{-3}, since the intrinsic carrier concentration at 77°K is lower than at 300°K. Therefore n_{77} decreases more rapidly than n_{300} with decreasing N_D, and n_{300}/n_{77} increases as n_{77} decreases.

Although the general features of the data in Fig. 2 are consistent with a conventional semiconductor model in which there is a positive energy gap between the valence and conduction bands, calculations based on such a model failed to give quantitative agreement with experiment. Calculations based on the semimetal (overlapping band) model of Fig. 1 did give quantitative agreement with the experimental data, however. The results of these calculations are shown as theoretical curves in Fig. 2. In order to obtain the curves, n_{300} and n_{77} were first calculated as functions of N_D, as shown in Fig. 3. As stated previously, the band parameter values used in the calculations were $\epsilon_{g1} = 0.1$ ev, $\epsilon_{g2} = -0.07$ ev, and $m_p^* = 0.17$ m_0. Each of the points

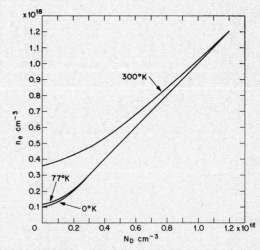

FIG. 3. Theoretical dependence of free-electron concentration (n_e) at 300°, 77°, and 0°K on net donor concentration (N_D) for HgSe.

used in drawing the curves of Fig. 2 was then obtained by comparing n_{77} for a given N_D with n_{300} for the same N_D.

In addition to the results for 300° and 77°K, Fig. 3 shows the theoretical variation of n with N_D at 0°K. Whereas the intrinsic free electron concentration (n_i) is zero for a semiconductor at 0°K, n_i calculated according to the semimetal model has the rather large value of 1.0×10^{17} cm^{-3} at 0°K. As the temperature is increased, n_i changes very slowly, increasing to only 1.2×10^{17} cm^{-3} at 77°K and to 3.6×10^{17} cm^{-3} at 300°K. According to this result, none of the HgSe samples studied in the present investigation was in the intrinsic range even at room temperature.

Experimental results for HgSe$_{0.5}$Te$_{0.5}$ at 300° and 77°K are shown in Fig. 4. These results are more complex than those for HgSe, since they include data for samples containing excess acceptors as well as for those containing excess donors. For samples containing sufficiently high acceptor concentrations, the free-electron concentrations cannot be obtained from the measured Hall coefficients, since R_H is given by the expression for mixed conduction rather than by the single carrier expression $-1/nec$. Therefore, the data in Fig. 4 are presented in terms of R_{77}/R_{300} and $-1/R_{77}ec$; these coordinates are equivalent to n_{300}/n_{77} and n_{77}, respectively, for samples in which $n \geqslant p$.

The experimental curve in Fig. 4 consists of two branches. The part of the upper branch for values of $-1/R_{77}ec$ greater than 2.8×10^{17} cm^{-3} includes data for samples in which $n \geqslant p$. The variation of n_{300}/n_{77} ($=R_{77}/R_{300}$) observed in this region occurs for the same reasons described previously for the case of HgSe, and the theoretical curve shown was calculated in the same way as the corresponding curve for HgSe shown in Fig. 2. As stated above, the band parameter values adopted for HgSe$_{0.5}$Te$_{0.5}$ in order to obtain quantitative agreement with the experimental data were $\epsilon_{g1} = 0.2$ ev, $\epsilon_{g2} = -0.07$ ev, and $m_p^* = 0.30 \ m_0$. The values of n_i calculated for these parameters are 2.8×10^{17} cm^{-3} at 77°K and 7.8×10^{17} cm^{-3} at 300°K.

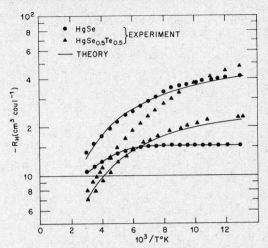

FIG. 5. Dependence of Hall coefficient (R_H) on $1/T$ for two samples of HgSe and two samples of HgSe$_{0.5}$Te$_{0.5}$.

The theoretical part of the curve in Fig. 4 ends at the point where the net donor concentration becomes zero due to the compensation of donor and acceptor impurities and consequently $n_{77} = n_i$. The remainder of the curve is traversed as the net acceptor concentration (N_A) is increased. Initially, R_{77} increases with increasing N_A, both because of the decrease in n_{77} and because of the onset of mixed conduction. Therefore, $-1/R_{77}ec$ decreases, and the experimental points fall along the upper branch of the curve. In this region, the increase in R_{77} causes an increase in R_{77}/R_{300}, since the samples remain very nearly intrinsic at 300°K and R_{300} therefore remains essentially constant. When N_A increases sufficiently, R_{77} begins to decrease toward zero as a result of mixed conduction. Therefore, $-1/R_{77}ec$ increases, and the points fall along the lower branch of the curve. In this region, R_{300} is greater than R_i at 300°K, so that for a given value of R_{77} the value of R_{77}/R_{300} is lower than for the upper branch of the curve.

In addition to the Hall coefficient data obtained at 300° and 77°K for a large number of samples, R_H was measured for several samples as a function of temperature between 77°K and about 350°K. The data for two samples of HgSe and two samples of HgSe$_{0.5}$Te$_{0.5}$ are shown in Fig. 5. The three theoretical curves shown were calculated in the manner described above, using the same band parameters used to calculate the theoretical curves of Figs. 2, 3, and 4. The agreement between theory and experiment is seen to be quite satisfactory. No attempt was made to calculate a theoretical curve for the fourth sample, which contained sufficient excess acceptors to be in the mixed conduction region over the whole temperature range.

Optical absorption coefficients measured at 300°K for three samples of HgSe with free-electron concentrations from 1.5×10^{18} cm^{-3} to 1.8×10^{19} cm^{-3} are shown as a function of wavelength in Fig. 6. The minima in the curves result from the simultaneous occurrence of two absorption processes, one of which increases with in-

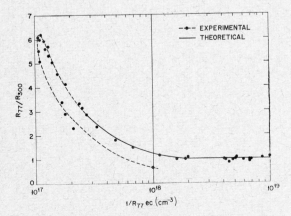

FIG. 4. Variation of Hall coefficient ratio (R_{77}/R_{300}) with $-1/R_{77}ec$ for HgSe$_{0.5}$Te$_{0.5}$.

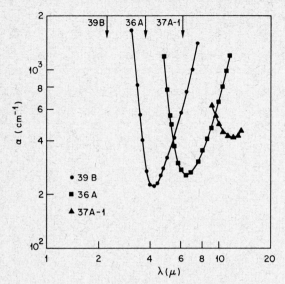

FIG. 6. Absorption coefficient (α) versus wavelength (λ) for samples of HgSe with carrier concentrations of 1.5×10^{18} cm^{-3} (37A-1), 5.0×10^{18} cm^{-3} (36A), and 1.8×10^{19} cm^{-3} (39B). Each vertical arrow indicates the wavelength which corresponds to the Fermi energy for the designated sample.

creasing photon energy while the other decreases. The former process is presumably an interband absorption, while the latter is presumably free-carrier absorption. The usual method of obtaining corrected data for the interband absorption by subtracting out the free-carrier absorption could not be applied, since the latter does not follow a simple power law in the spectral region investigated, and therefore could not be extrapolated accurately to shorter wavelengths. Consequently, no attempt was made to analyze the data in a quantitative fashion. The qualitative features are of considerable interest, however.

The absorption edge for interband transitions exhibits a marked shift to higher energies with increasing free-electron concentration. Such an increase in energy is also observed for n-type InSb,[15] InAs,[16] and CdSnAs$_2$.[17,18] It occurs in materials with low electron effective masses because the Fermi level increases appreciably with increasing concentration; as the lower states in the conduction band become filled, photons of higher energy are required to promote electrons from

the valence band to higher unoccupied states in the conduction band.[19] The shift of the absorption edge in HgSe thus supports the conclusion that this edge is associated with promotion of electrons into the conduction band.

In terms of the present investigation, it is even more significant that intense interband absorption ($\alpha \sim 10^3$ cm^{-3}) occurs in HgSe at photon energies considerably less than the Fermi energy. In order to illustrate this fact, a vertical arrow has been placed in Fig. 6 at the wavelength corresponding to the Fermi level calculated from Eq. (3) for each sample investigated. In each case, the absorption edge occurs at wavelengths significantly greater than the one indicated by the arrow. Qualitatively, this observation is consistent with the semimetal band model proposed for HgSe in the present investigation, rather than with a semiconductor band model. If phonon absorption or emission is neglected, on the basis of the semimetal model the minimum photon energy required to promote an electron into the conduction band would be less than the Fermi energy by the value of the overlap energy (ϵ_{g2}) between the valence and conduction bands. The semiconductor model, on the other hand, requires a minimum energy equal to the Fermi energy plus the energy gap. A detailed experimental and theoretical analysis of the shape of the absorption edge would be required before accepting the optical data as convincing evidence for the semimetal model.

CONCLUSION

Detailed analysis of Hall coefficient data for HgSe and HgSe$_{0.5}$Te$_{0.5}$ leads to the conclusion that these materials are semimetals, in which the valence and conduction bands overlap by approximately 0.07 ev, rather than semiconductors. Optical absorption data for HgSe are consistent with this conclusion.

ACKNOWLEDGMENTS

The authors are grateful to A. E. Paladino for his assistance in preparing the materials and making many of the electrical measurements, to Mrs. M. C. Plonko for making the optical measurements, and to S. Hilsenrath and Mrs. N. B. Rawson for performing most of the theoretical calculations. They are also pleased to acknowledge the helpful comments of Dr. G. B. Wright and Dr. J. M. Honig.

[15] M. Tanenbaum and H. B. Briggs, Phys. Rev. 91, 1561 (1953).
[16] R. M. Talley and F. Stern, J. Electronics 1, 186 (1955).
[17] A. J. Strauss and A. J. Rosenberg, J. Phys. Chem. Solids 17, 278 (1961).
[18] W. G. Spitzer, J. H. Wernick, and R. Wolfe, Solid-State Electronics 2, 96 (1961).

[19] E. Burstein, Phys. Rev. 93, 632 (1954).

2-6 Compounds

S. NIKITINE, *Chairman*

Exciton Structure and Zeeman Effects in Cadmium Selenide*†

J. O. DIMMOCK‡ AND R. G. WHEELER

Sloane Physics Laboratory, Yale University, New Haven, Connecticut

A semi-empirical theory of exciton structure in the presence of an external magnetic field developed from Dresselhaus' effective mass approximation [G. Dresselhaus, J. Phys. Chem. Solids **1**, 15 (1956)] has been used to obtain the band parameters at $K=0, 0, 0$ of CdSe from observed exciton spectra. The theory has been approximated to the case of uniaxial crystals of small anisotropy by the assumption that the effective mass tensor for both the hole and the electron is diagonal, the remaining anisotropy is cylindrical and small and that the anisotropy in the dielectric constant is also cylindrical and small. The exciton spectra of CdSe has been observed and identified by optical reflection, absorption, and Zeeman structure at $1.8°K$. The reflection and absorption spectra indicate the presence of two nonoverlapping exciton series. From observed optical selection rules, the conduction band is identified as having Γ_7 symmetry. The two series correspond to the $\Gamma_9 - \Gamma_7$ valence band crystal field splitting of approximately 200 cm^{-1}. A third series, at higher energies, has been observed in absorption corresponding to a Γ_7 valence band state split by spin-orbit effects from the other two states by approximately 3490 cm^{-1}. The $n_1 = 1, 2, 3,$ and 4 states of the first, Γ_9, series, the $n_2 = 1$ and 2 states of the second (first Γ_7) series, and the $n_3 = 1$ state of the third (second Γ_7) series have been observed and identified in absorption. The series limit of the first series, corresponding to the band gap, has been measured to be 14 850 cm^{-1}. The band parameters have been obtained by comparing theory and experiment. The effect of the finite photon momentum has been observed through changes in the Zeeman structure of the $n_1 = 2$, P states upon 180° rotation of the magnetic field in the plane perpendicular to the crystal C axis.

INTRODUCTION

THE observation and interpretation of the exciton spectra in semiconductors gives detailed information concerning the electronic band structure of the material in question. This is to report on such observations in wurtzite cadmium selenide. The identification and subsequent interpretation was facilitated by measurements of the magneto-optical effects of the spectra. A qualitative understanding of the spectra is obtained from a review of the allowed band symmetries at $K=0, 0, 0$, in crystals of C_{6v}^4 symmetry. The exciton symmetries and selection rules are obtained group theoretically. A quantitative comparison with an anisotropic exciton mass theory, which will be outlined, permits evaluation of the electron and hole mass and g-value parameters.

As pointed out by Birman,[1] Glasser,[2] and others, if one considers the valence band as P like and the conduction band as S like in the II-VI wurtzite class of semiconductors, at $K=0, 0, 0$, the conduction band including spin will have a Γ_7 symmetry, the valence band $\Gamma_9 - \Gamma_7 - \Gamma_7$ symmetries. The valence band splitting is due to the spin-orbit and crystal field effects. The exciton states then will reflect these symmetries, as well as the symmetries of the hydrogenic state of the exciton in the center of mass. As Dresselhaus[3] and others[4] have pointed out, the exciton wave function is a product function of the electron wave function and the hole wave function multiplied by the Fourier transform of the particular hydrogen state in the center of mass. The states observable by dipole radiation will be those whose representation[5] is either Γ_1 or Γ_5 corresponding to polarizations $E\|C$ and $E\perp C$, respectively. These remarks are summarized in Fig. 1 for S and P hydrogenic states.

Using the $\mathbf{K \cdot p}$ method Casella[6] has shown that the energy surfaces for the Γ_9 symmetry give rise to a diagonal mass tensor. For the Γ_7 symmetry, the energy may contain terms linear in wave vector. In what follows the assumption has been made that these linear terms are small, such that a diagonal mass tensor describes the Γ_7 as well as the Γ_9 energy surfaces. The

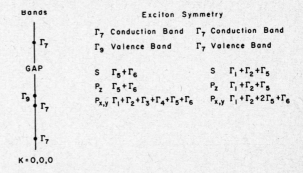

FIG. 1. The band symmetries and their relative energies are indicated as is compatible with the observed optical selection rules for three exciton series in cadmium selenide. With the light polarization parallel and perpendicular to the C axis, dipole transitions to the Γ_1 and Γ_5 states, respectively, are allowed.

* Supported in part by the Office of Scientific Research, U. S. Air Force.

† A preliminary report of this work was given at the meeting of the American Physical Society, March, 1961.

‡ National Science Predoctoral Fellow.

[1] J. L. Birman, Phys. Rev. Letters **2**, 157 (1959).

[2] M. L. Glasser, J. Phys. Chem. Solids **10**, 229 (1959).

[3] G. Dresselhaus, J. Phys. Chem Solids **1**, 15 (1956).

[4] H. Haken, J. Phys. Chem. Solids **8**, 166 (1959).

[5] V. Heine, *Group Theory in Quantum Mechanics* (Pergamon Press, New York, 1960).

[6] R. C. Casella, Phys. Rev. Letters **5**, 371 (1960).

analysis of the data in terms of an anisotropic effective mass formalism will describe the experimental situation if the following assumptions characterize the electronic band structure of cadmium selenide.

(1) The band extrema are at or very near $K=0, 0, 0$. The hole and electron masses may be expressed as diagonal tensors such that the energy surfaces are ellipsoidal.

(2) The resultant exciton mass anisotropy is small, allowing first-order perturbation calculations to be made for the energy states as well as for the magnetic field effects.

(3) The valence band splittings are larger than the exciton binding energies, allowing one to neglect mixing between valence bands.

(4) The energy of the longitudinal optical phonon is much greater than the exciton binding energy, allowing low frequency dielectric constants to be used in the calculations with no corrections for polaron effects.[4,7]

THEORY

In the effective mass approximation the Hamiltonian for an exciton in the presence of an external magnetic field can conveniently be considered as a sum of terms.[3]

$$\hat{H}=\hat{H}_1+\hat{H}_2+\hat{H}_3+\hat{H}_4+\hat{H}_{K1}+\hat{H}_{K2}+\hat{H}_{K3},$$

where

$$\hat{H}_1=-\frac{\hbar^2}{2m}\left\{\frac{1}{\mu_x}\frac{\partial^2}{\partial x^2}+\frac{1}{\mu_y}\frac{\partial^2}{\partial y^2}+\frac{1}{\mu_z}\frac{\partial^2}{\partial z^2}\right\}$$
$$-\frac{e^2}{\epsilon\eta^{\frac{1}{2}}}(x^2+y^2+\eta^{-1}z^2)^{-\frac{1}{2}},$$

$$\hat{H}_2=-2i\beta\left\{\frac{A_x}{\Delta_x}\frac{\partial}{\partial x}+\frac{A_y}{\Delta_y}\frac{\partial}{\partial y}+\frac{A_z}{\Delta_z}\frac{\partial}{\partial z}\right\},$$

$$\hat{H}_3=\frac{e^2}{2mc^2}\left\{\frac{1}{\mu_x}A_x^2+\frac{1}{\mu_y}A_y^2+\frac{1}{\mu_z}A_z^2\right\},$$

$$\hat{H}_4=\tfrac{1}{2}\beta\sum_{\alpha=x,y,z}(g_{e\alpha}s_{e\alpha}+g_{h\alpha}s_{h\alpha})H_\alpha,$$

$$\hat{H}_{K1}=-\frac{i\hbar^2}{2m}\left\{\frac{K_x}{\Delta_x}\frac{\partial}{\partial x}+\frac{K_y}{\Delta_y}\frac{\partial}{\partial y}+\frac{K_z}{\Delta_z}\frac{\partial}{\partial z}\right\},$$

$$\hat{H}_{K2}=\beta\left\{\frac{K_x}{\mu_x}A_x+\frac{K_y}{\mu_y}A_y+\frac{K_z}{\mu_z}A_z\right\},$$

$$\hat{H}_{K3}=\frac{\hbar^2}{8m}\left\{\frac{1}{\mu_x}K_x^2+\frac{1}{\mu_y}K_y^2+\frac{1}{\mu_z}K_z^2\right\},$$

in which m is the free electron mass, μ_α is the reduced

---

[7] T. Muto, Suppl. Progr. Theory Phys. (Japan) **12**, 3 (1959).

effective mass of the exciton in the direction α,

$$\frac{1}{\mu_\alpha}=\left(\frac{m}{m_{e\alpha}{}^*}+\frac{m}{m_{h\alpha}{}^*}\right)$$

with $m_{e\alpha}{}^*$ and $m_{h\alpha}{}^*$, respectively, the electron and hole effective masses in the direction α. In the term $\hat{H}_1$ the possibility of dielectric anisotropy is considered with ϵ being the dielectric constant transverse to the crystal C axis (taken as the Z direction), and $\epsilon\eta$ being the dielectric constant in the Z direction.

$$\beta=e\hbar/(2mc),$$

$$\mathbf{A}=\tfrac{1}{2}(\mathbf{H}\times\mathbf{r}),$$

$$\frac{1}{\Delta_\alpha}=\left(\frac{m}{m_{e\alpha}{}^*}-\frac{m}{m_{h\alpha}{}^*}\right).$$

K_α is the projection in the direction α of the wave vector $\mathbf{K}$ of the light which creates the exciton. Since we assume that the valence and conduction band extrema are essentially at $K=0, 0, 0$, $\mathbf{K}$ represents the displacement of the exciton in K space from $K=0, 0, 0$. One recognizes the origin of the various terms to be as follows. $\hat{H}_1$ is the Hamiltonian for a hydrogenic system in the absence of external fields with the possibility of mass and dielectric anisotropies included. $\hat{H}_2$ is the linear (Zeeman) magnetic field term. $\hat{H}_3$ is the quadratic (diamagnetic) magnetic field term. $\hat{H}_4$ is a convenient representation of the spin energy in the presence of a magnetic field. We note that due to the small reduced mass μ, and large dielectric constant ϵ of CdSe, the "radius" of the exciton states will be much larger than the corresponding hydrogen state "radii." Hence, since spin-orbit coupling is proportional to r^{-3} and is thus quite small, it is legitimate to write the magnetic field perturbations in the above Paschen-Bach limit. The last three terms are the standard $\mathbf{K}\cdot\mathbf{P}$ term, the $\mathbf{K}\cdot\mathbf{A}$ term due to the magnetic field and the standard $\mathbf{K}^2$ term, respectively. In the case under analysis, these last three terms are small enough to be neglected.

It will be convenient to consider the first four terms as consisting of a zero-order Hamiltonian plus a perturbation. The following choice is made:

$$\hat{H}_0=-\frac{\hbar^2}{2m}\left\{\frac{1}{\mu_x}\frac{\partial^2}{\partial x^2}+\frac{1}{\mu_x}\frac{\partial^2}{\partial y^2}+\frac{1}{\mu_z}\frac{\partial^2}{\partial z^2}\right\}$$
$$-\frac{e^2}{\epsilon\eta^{\frac{1}{2}}}\left(x^2+y^2+\frac{\mu_z}{\mu_x}z^2\right)^{-\frac{1}{2}},$$

$$\hat{H}_1'=-\frac{e^2}{\epsilon\eta^{\frac{1}{2}}}\left\{(x^2+y^2+\eta^{-1}z^2)^{-\frac{1}{2}}-\left(x^2+y^2+\frac{\mu_z}{\mu_x}z^2\right)^{-\frac{1}{2}}\right\},$$

$$\hat{H}_2'=\hat{H}_2+\hat{H}_3+\hat{H}_4.$$

The zero-order state functions which satisfy the

equation

$$\hat{H}_0 \Phi_n = E_n{}^0 \Phi_n$$

are related to the hydrogenic state functions Ψ_n by

$$\Phi_n(x,y,z) = (\mu_z/\mu_x)^{\frac{1}{2}} \Psi_n(x,y,(\mu_z/\mu_x)^{\frac{1}{2}}z)$$

on a scale where the first Bohr orbit is

$$a_0' = \frac{\hbar^2}{me^2} \frac{\epsilon}{\mu_x} \eta^{\frac{1}{2}} = a_0 \frac{\epsilon}{\mu_x} \eta^{\frac{1}{2}}.$$

The energy $E_n{}^0$ is related to the energy of the corresponding state in the hydrogen atom by

$$E_n{}^0 = (\mu_x/\epsilon^2 \eta) E_n{}^H.$$

It will be convenient in calculating matrix elements to use the functions $\Psi_n(x,y,z)$ instead of $\Phi_n(x,y,z)$. Consider the matrix element of the operator $\hat{R}(x,y,z)$

$$\langle \Phi_{n_1}(x,y,z) | \hat{R}(x,y,z) | \Phi_{n_2}(x,y,z) \rangle$$
$$= \langle \Psi_{n_1}(x,y,z) | \hat{R}(x,y,(\mu_x/\mu_z)^{\frac{1}{2}}z) | \Psi_{n_2}(x,y,z) \rangle$$
$$\equiv \langle n_1,l_1,m_1 | \hat{R}(x,y,(\mu_x/\mu_z)^{\frac{1}{2}}z) | n_2,l_2,m_2 \rangle.$$

The anisotropy perturbation is obtained from the matrix elements

$$\langle n_1,l_1,m_1 | \hat{H}_1'(x,y,(\mu_x/\mu_z)^{\frac{1}{2}}z) | n_2,l_2,m_2 \rangle.$$

The calculation is facilitated by writing

$$\hat{H}_1'\left(x,y,\left(\frac{\mu_x}{\mu_z}\right)^{\frac{1}{2}}z\right) = \frac{e^2}{\epsilon \eta^{\frac{1}{2}} r}\{1 - (1 - \alpha \cos^2\theta)^{-\frac{1}{2}}\},$$

where

$$\alpha = 1 - \frac{\mu_x}{\mu_z} \cdot \frac{\epsilon_x}{\epsilon_z} = 1 - \frac{\mu_x}{\mu_z \eta}.$$

This perturbation lifts the degeneracy of the exciton system. The observed zero field splitting of identified states gives a measure of α and hence of μ_x/μ_z since η is known.

The energies of the exciton states of principle quantum number $n_1 = 1$, 2, of the first series, in the presence of an external magnetic field both along $z(H\|C)$ and along $x(H \perp C)$ have been obtained to first order. The hydrogenic functions $\Psi_n(x,y,z)$ and the

TABLE I. Eigenfunctions and energies for $H\|C$ first series $n_1 = 1,2$.

State	C_{6v}	Observed	Function	Energy
1S	Γ_5	$E \perp C$	$R_{1s}Y_0{}^0/\alpha_h\beta_e$	$E_1{}^0\left(1 + \dfrac{1}{3}\alpha + \dfrac{3}{20}\alpha^2\right) + \sigma H_z{}^2 + \frac{1}{2}(g_{hz} - g_{ez})\beta H_z$
			$R_{1s}Y_0{}^0/\beta_h\alpha_e$	$E_1{}^0\left(1 + \dfrac{1}{3}\alpha + \dfrac{3}{20}\alpha^2\right) + \sigma H_z{}^2 - \frac{1}{2}(g_{hz} - g_{ez})\beta H_z$
2S	Γ_5	$E \perp C$	$R_{2s}Y_0{}^0/\alpha_h\beta_e$	$E_2{}^0\left(1 + \dfrac{1}{3}\alpha + \dfrac{3}{20}\alpha^2\right) + 14\sigma H_z{}^2 + \frac{1}{2}(g_{hz} - g_{ez})\beta H_z$
			$R_{2s}Y_0{}^0/\beta_h\alpha_e$	$E_2{}^0\left(1 + \dfrac{1}{3}\alpha + \dfrac{3}{20}\alpha^2\right) + 14\sigma H_z{}^2 - \frac{1}{2}(g_{hz} - g_{ez})\beta H_z$
2P_0	Γ_5	$E \perp C$	$R_{2p}Y_1{}^0/\alpha_h\beta_e$	$E_2{}^0\left(1 + \dfrac{3}{5}\alpha + \dfrac{9}{28}\alpha^2\right) + 6\sigma H_z{}^2 + \frac{1}{2}(g_{hz} - g_{ez})\beta H_z$
			$R_{2p}Y_1{}^0/\beta_h\alpha_e$	$E_2{}^0\left(1 + \dfrac{3}{5}\alpha + \dfrac{9}{28}\alpha^2\right) + 6\sigma H_z{}^2 - \frac{1}{2}(g_{hz} - g_{ez})\beta H_z$
2$P_{\pm 1}$	Γ_5	$E \perp C$	$R_{2p}Y_1{}^{-1}/\alpha_h\alpha_e$	$E_2{}^0\left(1 + \dfrac{1}{5}\alpha + \dfrac{9}{140}\alpha^2\right) + 12\sigma H_z{}^2 - \left[\dfrac{1}{\Delta_x} - \dfrac{1}{2}(g_{hz} + g_{ez})\right]\beta H_z$
			$R_{2p}Y_1{}^1/\beta_h\beta_e$	$E_2{}^0\left(1 + \dfrac{1}{5}\alpha + \dfrac{9}{140}\alpha^2\right) + 12\sigma H_z{}^2 + \left[\dfrac{1}{\Delta_x} - \dfrac{1}{2}(g_{hz} + g_{ez})\right]\beta H_z$
2$P_{\pm 1}$	$\Gamma_1 - \Gamma_2$	$E\|C$	$R_{2p}Y_1{}^{-1}/\alpha_h\beta_e$	$E_2{}^0\left(1 + \dfrac{1}{5}\alpha + \dfrac{9}{140}\alpha^2\right) + 12\sigma H_z{}^2 - \left[\dfrac{1}{\Delta_x} - \dfrac{1}{2}(g_{hz} - g_{ez})\right]\beta H_z$
			$R_{2p}Y_1{}^1/\beta_h\alpha_e$	$E_2{}^0\left(1 + \dfrac{1}{5}\alpha + \dfrac{9}{140}\alpha^2\right) + 12\sigma H_z{}^2 + \left[\dfrac{1}{\Delta_x} - \dfrac{1}{2}(g_{hz} - g_{ez})\right]\beta H_z$

where $\sigma = -\dfrac{1}{4}\dfrac{e^2}{mc^2}a_0{}^2 \dfrac{\epsilon^2}{\mu_x{}^3}\eta$

TABLE II. Energies for $H \perp C$ first series $n_1 = 1, 2$.

State	Energy
$1S^+$	$E_1^0\left(1 + \frac{1}{3}\alpha + \frac{3}{20}\alpha^2\right) + \frac{\mu_x}{\mu_z}\sigma H_x^2 + \frac{1}{2}g_{ex}\beta H_x$
$1S^-$	$E_1^0\left(1 + \frac{1}{3}\alpha + \frac{3}{20}\alpha^2\right) + \frac{\mu_x}{\mu_z}\sigma H_x^2 - \frac{1}{2}g_{ex}\beta H_x$
$2S^+$	$E_2^0\left(1 + \frac{1}{3}\alpha + \frac{3}{20}\alpha^2\right) + 14\frac{\mu_x}{\mu_z}\sigma H_x^2 + \frac{1}{2}g_{ex}\beta H_x$
$2S^-$	$E_2^0\left(1 + \frac{1}{3}\alpha + \frac{3}{20}\alpha^2\right) + 14\frac{\mu_x}{\mu_z}\sigma H_x^2 - \frac{1}{2}g_{ex}\beta H_x$
$2P_x^+$	$E_2^0\left(1 + \frac{1}{5}\alpha + \frac{9}{140}\alpha^2\right) + 6\frac{\mu_x}{\mu_z}\sigma H_x^2 + \frac{1}{2}g_{ex}\beta H_x$
$2P_x^-$	$E_2^0\left(1 + \frac{1}{5}\alpha + \frac{9}{140}\alpha^2\right) + 6\frac{\mu_x}{\mu_z}\sigma H_x^2 - \frac{1}{2}g_{ex}\beta H_x$
$2P_y^+$	$E_2^0\left(1 + \frac{2}{5}\alpha + \frac{27}{140}\alpha^2\right) + 12\frac{\mu_x}{\mu_z}\sigma H_x^2 + \frac{1}{2}g_{ex}\beta H_x + \pi$
$2P_y^-$	$E_2^0\left(1 + \frac{2}{5}\alpha + \frac{27}{140}\alpha^2\right) + 12\frac{\mu_x}{\mu_z}\sigma H_x^2 - \frac{1}{2}g_{ex}\beta H_x + \pi$
$2P_z^+$	$E_2^0\left(1 + \frac{2}{5}\alpha + \frac{27}{140}\alpha^2\right) + 12\frac{\mu_x}{\mu_z}\sigma H_x^2 + \frac{1}{2}g_{ex}\beta H_x - \pi$
$2P_z^-$	$E_2^0\left(1 + \frac{2}{5}\alpha + \frac{27}{140}\alpha^2\right) + 12\frac{\mu_x}{\mu_z}\sigma H_x^2 - \frac{1}{2}g_{ex}\beta H_x - \pi$

where

$$\pi = \left\{ \left[E_0^2\left(\frac{1}{5}\alpha + \frac{9}{70}\alpha^2\right)\right]^2 + \frac{a^2\beta^2}{\Delta_x^2}H_x^2 \right\}^{\frac{1}{2}}$$

$$\sigma = -\frac{1}{4}\frac{e^2}{mc^2}a_0^2\frac{\epsilon^2}{\mu_x^3}\eta \qquad a = \frac{1}{2}\left(\frac{\mu_z}{\mu_x}\right)^{\frac{1}{2}}\left(\frac{\Delta_x}{\Delta_z} + \frac{\mu_x}{\mu_z}\right)$$

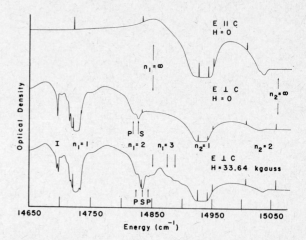

FIG. 2. The spectra observed for a 0.5-μ crystal. The unlabeled lines at energies less than the $n_1 = 1$ are due to impurity excitons. These and other lines observable in other crystals are variable from crystal to crystal. The calibration emission lines are iron lines third order, which overlap the second-order absorption spectra.

to be handled by a perturbation approach. Instead a variational method should be used. However, it is the opinion of the authors that a greater understanding of the situation would not therein be obtained.

EXPERIMENTAL PROCEDURE

The experiments were performed on CdSe single crystal platelets of 0.5–5 μ thickness with the crystal C axis (i.e., z) in the plane of the platelet. The crystals, mounted strain free, were immersed in liquid helium at 1.8°K or lower for all experiments. A concave grating spectrograph in a stigmatic Wadsworth mounting was used in the second order. The dispersion is about 2.1 A/mm with a measured resolving power using 103a-F Kodak spectrographic plates of greater than 70 000. The magnet was calibrated to 0.1%, with current stability of the order of one part in 10^5. Emission lines from various sources were imposed on each plate for calibration purposes. The spectrum was obtained from the plates by a recording densitometer.

EXPERIMENTAL RESULTS AND INTERPRETATION

Typical absorption spectra observed for a single crystal about 0.5 μ thick are shown in Figs. 2 and 3. The data from these very thin crystals give evidence for three exciton series which we associate with a single Γ_7 conduction band and the three valence bands. Within a given exciton series, one not only expects the nearly hydrogen-like energy spacing but also theoretically that the intensities will fall approximately as $1/n^3$.[8,9] With reference to the group theoretical selection rules of Fig. 1, the band symmetries are obtained from the observed optical selection rules of the exciton $1S$

corresponding exciton energies for $H \| C$ are given for observable states in Table I. The energies of observable exciton states for $H \perp C$ are given in Table II. Note that the S states in this table will be observed only in the polarization $E \perp C$. With $H \perp C$, the spin contribution of the hole is zero ($g_{hx} = 0$), since the valence band splitting is large compared to spin magnetic energies effectively quenching the hole spin along the crystal C axis. Only the exciton states of principle quantum number $n_1 = 1, 2$ are treated since an essentially different approach must be used for higher n states. This is because the large exciton radius for these states results in magnetic field perturbations, through the diamagnetic term $\hat{H}_3$ which strongly mix states of different principle quantum number. Already in the $n_1 = 2$ states, this mixing is observed. A deviation between theory and experiment is expected at high fields, since the magnetic field effects are too great

[8] G. Dresselhaus, Phys. Rev. **106**, 76 (1957).
[9] R. J. Elliott, Phys. Rev. **108**, 1384 (1957).

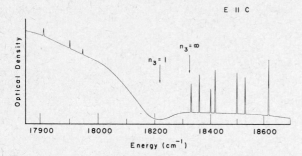

FIG. 3. This spectra indicates the $n_3 = 1$ line of the third series as observed with the same crystal as in Fig. 2. This spectrum however was taken in the first order. We have observed that this line is usually only observable with the very best strain-free extremely thin crystals.

states. Up to the writing of this report we have only analyzed the magnetic effects on the excited states of the first series. The experimental plots for these lines and their identification as made by the selection rules and magnetic splittings are shown in Figs. 4 and 5 for both orientations of the magnetic field with $E\|C$. Similar data exist for $E \perp C$. As is apparent from Fig. 4, $n_1 = 3$ and higher states have energies at high fields larger than the zero field series limit, such lines in the past being usually identified with transitions between "Landau" levels in the conduction and valence bands.

In order to analyze the data and concurrently test the validity of the theoretical approach, the following procedure was adopted. The effective mass μ_x can be obtained from the diamagnetic shift, $H\|C$, of the two states in Table I observed $E\|C$, since the dielectric constants, $\epsilon_x = 9.7$ and $\epsilon_z = 10.65$, are known. Using the zero field position of the $2P_{\pm 1}$ states and the extrapolated zero field position of the $2P_0$ state we can determine μ_z. (μ_z is also obtained from the diamagnetic shift, $H \perp C$, of the $n_1 = 2$ states in Table II.) This information is sufficient to determine the series limit, and the energies of the $n_1 = 1$ and $n_1 = 3$ states. The calculated and experimental values are given in Table III.

Further information can be determined from the linear Zeeman splitting for $H\|C$ on both the Γ_5 and $\Gamma_1 - \Gamma_2$, $2P_{\pm 1}$ states. To date we have been unable to observe the splitting of the Γ_5 $2S$ state which would permit a determination of all g values. From the linewidth, however, we place an upper bound of $|g_{ez} - g_{hz}|$ < 1.0 thus obtaining an estimate of Δ_x and consequently of the transverse hole and electron effective masses. In the $H \perp C$ case, the small energy difference between the two cases of H along $+x$ and H along $-x$ may be ascribed to the transverse electron spin energy. The numbers thus obtained have been used to calculate the energies of the states indicated in Figs. 6 and 7. The parameter π is an experimentally obtainable quantity which when squared and plotted against H^2 yields a straight line with a slope nearly equal to $a^2\beta^2/\Delta_x^2$. The value of this quantity also helps to

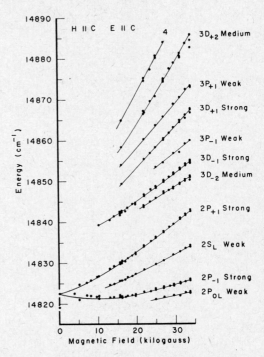

FIG. 4. The energies of the excited exciton states of the first series are plotted against magnetic field with the field and polarization parallel to the crystal C axis. In all cases, the propagation direction of the light was normal to the plane of the platelet and normal to the field. This figure includes data from 3 crystal samples. The line labeled $2S_L$ has the properties as described by Hopfield and Thomas for a longitudinal exciton.[10] That is, its intensity is dependent upon the orientation of the light propagation vector relative to the C axis. When these vectors are exactly perpendicular, the line is unobservable.

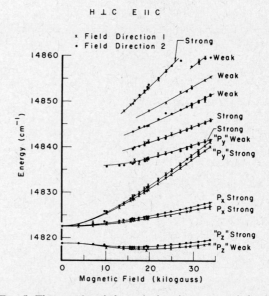

FIG. 5. The energies of the excited exciton states of the first series are plotted against magnetic field with the field perpendicular and the polarization parallel to the crystal C axis. The propagation direction of the light was normal to the plane of the crystal platelet and normal to the field. This figure includes data obtained from 3 crystal samples.

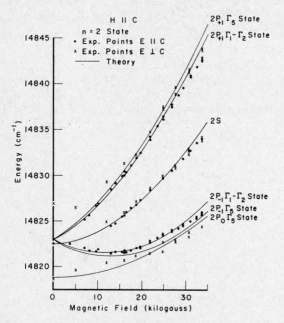

FIG. 6. The $n_1=2$ states with the magnetic field along the C axis are shown in both polarizations as compared with the theory. Only the data of the P states is used to determine the necessary parameters for the theory.

determine the band parameters. The results of the comparison of theory and experiment are given in Table III. The errors indicated reflect not only the experimental uncertainties but also the theoretical limitations. There is a general agreement between theory and experiment to better than 10%. The errors quoted in Table III, however, result from the data

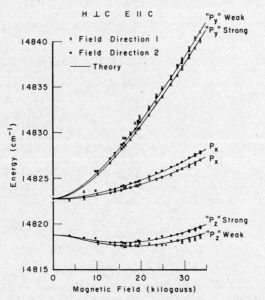

FIG. 7. The $n_1=2$ states are shown in the $E\|C$, $H\perp C$ configuration as compared with the theory. Photon momentum effects are observed in this configuration similar to those reported for CdS by Hopfield and Thomas.[11]

manipulations necessary to obtain the various parameters.

Having determined the electron masses, it is now a trivial calculation to obtain the series limit of the second series as well as the hole mass associated with the first Γ_7 valence band, from the observed energies of the $n_2=1$ and $n_2=2$ states. The series limit of the third series has been estimated by assuming a hole mass equal

TABLE III.

Series 1.		

From $H\|C$ diamagnetic shift, $\mu_x=0.100\pm0.005$.
   From zero field positions, $n_1=2$ states

$$\alpha=[1-\mu_x\epsilon_x/(\mu_z\epsilon_z)]=0.32\pm0.02,$$
   hence
$$\mu_x/\mu_z=0.75\pm0.04, \quad \mu_z=0.13\pm0.01.$$

From $H\perp C$ diamagnetic shift,

$$\mu_x/\mu_z=0.77\pm0.04, \quad \mu_z=0.13\pm0.01,$$
$R_y=106\pm5$ cm^{-1}, $a_0'=54$ A,
$E_{gap}=14\,850.5\pm2.0$ cm^{-1}.

State	Experimental	Calculated
$1S$	$14\,727 \pm1$ cm^{-1}	$14\,734 \pm6$ cm^{-1}
$2P_0$	$14\,818.6\pm0.3$ cm^{-1}	
$2P_{\pm1}$	$14\,822.5\pm0.2$ cm^{-1}	
$3P_{\pm1}$	$14\,839 \pm1$ cm^{-1}	$14\,838.0\pm1.5$ cm^{-1}

Mass parameters

$m_{ex}{}^*=0.13\pm0.01\ m$,	$m_{hz}{}^*=0.45\pm0.09\ m$,
$m_{ez}{}^*=0.13\pm0.03\ m$,	$m_{hz}{}^*\geq m$.

Series 2.	

State	Experimental
$1S$	$14\,931\pm3$ cm^{-1}
$2S-2P_{\pm1}$	$15\,032\pm2$ cm^{-1}
$2P_0$	$15\,022\pm3$ cm^{-1}

Calculated series limit	$15\,050\pm15$ cm^{-1}
$R_y=120\pm10$ cm^{-1}	
$\mu_x=0.11\pm0.01$	
$m_{ex}{}^*=0.13\pm0.01\ m$	$m_{hx}{}^*=0.9\pm0.2\ m$

Series 3.	

$1S$ $18\,218\pm10$ cm^{-1}	
Calculated series limit	$18\,340\pm20$ cm^{-1}
Crystal field splitting	200 ± 15 cm^{-1}
Spin-orbit splitting	3490 ± 20 cm^{-1}
Electron g values[a]	

$$|ge_x|=0.51\pm0.05 \quad |ge_z|=0.6\pm0.1$$

[a] There is some evidence that the electron g values are negative. The Zeeman splitting between the $2P_{\pm1}$, $\Gamma_1-\Gamma_2$ and Γ_5 states indicates a negative ge_z.

to m. A summary of all the band parameters and relevant energies is given in Table III.

In conclusion we note that other phenomena associated with intrinsic excitons in cadmium selenide have been observed. We have experimentally observed the longitudinal exciton[10] associated with the $n_1=1S$ and $n_1=2S$ and P_0 states. The effect of photon mo-

[10] J. J. Hopfield and D. G. Thomas, J. Phys. Chem. Solids 12, 276 (1960).

mentum is observed in the $n_1=2P$ states. Large spectral intensity changes occur upon $180°$ rotation of the magnetic field when perpendicular to the C axis.[11] Zeeman data on the impurity excitons has been obtained. A calculation has also been made relating the electron effective masses and g values to the spin-orbit splitting. However, due to the space limitations of the

conference, these as well as the details of the present work will be published elsewhere.

ACKNOWLEDGMENTS

The authors wish to thank Dr. Richard Bube of the RCA Laboratories for supplying the crystals used in this work, and to Dr. D. Berlincourt of the Clevite laboratories for the dielectric constant data before its publication.

[11] J. J. Hopfield and D. G. Thomas, Phys. Rev. Letters **4**, 357 (1960).

JOURNAL OF APPLIED PHYSICS   SUPPLEMENT TO VOL. 32, NO. 10   OCTOBER, 1961

Exciton States and Band Structure in CdS and CdSe

J. J. HOPFIELD

Bell Telephone Laboratories, Murray Hill, New Jersey

The theoretical effects of a finite slope condition band crossing on the direct exciton energy levels of wurtzite compounds is investigated. The lack of an experimental confirmation of these effects places an upper limit of about 10^{-10} ev-cm on the slope of the conduction band at $k=0$ in CdS, and a slightly larger limit in CdSe. Theoretical estimates of these slopes are also calculated. Another possible method of observing the magnetic effects due to the nonzero wave vector of the excitons is noted. The striking increase of the exciton oscillator strength in CdSe in a magnetic field is attributed to the magnetic compression of the excitons. A brief comparison of experimental energy band parameters for CdS and CdSe is given.

I. INTRODUCTION

THE energy band structure, effective mass values, and g values of electrons and holes in CdS have previously been investigated by a study of the exciton energy levels in CdS. Three p-like valence bands (split by spin-orbit coupling and the hexagonal crystal field) were located. The $1S$ and $2S$ states of excitons from these three valence bands were studied in reflection[1] and certain $n=1,2,3$, and 4 states were studied in transmission for the excitons formed from a hole in the top valence band[2] The excitons formed from this valence band (but not the other two) are easily investigated in crystals of reasonable thickness ($\sim 10\,\mu$) in light polarized parallel to the c axis because of the Γ_9 symmetry of this top valence band.

The present work concerns this same exciton series and its analog in CdSe. The effects of a finite slope band crossing in the conduction band on the exciton energy levels is calculated in Sec. II and the order of magnitude of this slope estimated. The lack of an experimental confirmation of these calculated effects places an upper limit of about 10^{-10} ev cm on the slope dE/dk of the conduction band at $k=0$ in CdS and CdSe. In Sec. III, another possible experimental effect of the wave vector of the exciton on the exciton energy levels in a magnetic field is noted. Section IV is devoted to the magnetic field dependence of the line strengths. Finally, a comparison is made between the band parameters of CdS and CdSe.

II. THE ENERGY BAND STRUCTURE VERY NEAR $k=0$

In crystals without spin-orbit coupling, all energy bands necessarily exhibit a two-fold spin degeneracy. This two-fold degeneracy is not lifted by spin-orbit coupling for crystals having inversion symmetry. Casella[3] and others have pointed out that this two-fold spin degeneracy will exist for all energy bands in crystals of the wurtzite structure along the line $k_x=k_y=0$ (i.e., for wave vectors along the hexagonal axis), but will not exist for a general point of the Brillouin zone. The group theoretic analysis shows that for bands of symmetry Γ_7 (including the conduction band and the two lower valence bands), the energy bands should have the form[3]

$$E=A(k_x^2+k_y^2)+Bk_z^2\pm C(k_x^2+k_y^2)^{\frac{1}{2}}.$$

The third term, linear in k, represents a splitting of the two spin states by an amount $2C(k_x^2+k_y^2)^{\frac{1}{2}}$.

Since the Γ_7 states transform under the symmetry operations of the crystal like the two spin states of an electron, it is useful to write the kinetic energy operator for a Γ_7 band in terms of these two pseudo-spin states, quantized along the c axis (z axis). One then obtains a 2×2 effective mass tensor

$$\begin{pmatrix} A(k_x^2+k_y^2)+Bk_z^2 & iC(k_x+ik_y) \\ -iC(k_x-ik_y) & A(k_x^2+k_y^2)+Bk_z^2 \end{pmatrix}. \quad (1)$$

This effective mass tensor has the same physical inter-

[1] D. G. Thomas and J. J. Hopfield, Phys. Rev. **116**, 573 (1959).
[2] J. J. Hopfield and D. G. Thomas, Phys. Rev. **122**, 35 (1961).

[3] R. C. Casella, Phys. Rev. Letters **5**, 371 (1960); Phys. Rev. **114**, 1514 (1959).

pretation as the 4×4 effective mass tensor for holes in germanium or silicon. In the representation in which the magnetic moment Hamiltonian in the presence of a magnetic field is

$$\tfrac{1}{2}[g_{\parallel}H_z\sigma_z+g_{\perp}H_x\sigma_x+g_{\perp}H_y\sigma_y],$$

the constant C must be real. (σ_x, σ_y, and σ_z are the Pauli spin matrices.) (Off-diagonal terms in the effective mass tensor quadratic in k are also possible, but these terms for small k will be small compared with the terms linear in k, which are themselves small compared with the diagonal terms for most of the k's involved in exciton states.)

The effective mass tensor can thus be written

$$T=[A(k_x^2+k_y^2)+Bk_z^2]I+C[k_x\sigma_y-k_y\sigma_x], \quad (2)$$

where I is the 2×2 identity matrix. The second term has a simple intuitive interpretation. A fast electron travelling in an electric field has, in addition to its kinetic and potential energy, a term in its Hamiltonian of the form $\tfrac{1}{2}\mu_0\boldsymbol{\sigma}\cdot(\mathbf{E}\times\mathbf{v})$ due to spin-orbit coupling. A wurtzite crystal may be regarded, from the point of view of symmetry, as having a unique direction to the c axis, equivalent to an electric field in the z direction. The extra term in T has exactly the form of $\boldsymbol{\sigma}\cdot[\mathbf{E}_z\times\mathbf{v}]$.

For a dielectric constant and electron and hole masses having axial symmetry around the c axis, a simple Hamiltonian may be written down to express the exciton energy states in terms of a distorted hydrogen atom. In terms of the wave vector q of the internal exciton motion and the wave vector K of the exciton center-of-mass motion, the off-diagonal terms in the electron mass tensor introduces the perturbation

$$\frac{m_{e\perp}}{m_{e\perp}+m_{h\perp}}C[K_x\sigma_y-K_y\sigma_x]+\frac{m_{h\perp}}{m_{e\perp}+m_{h\perp}}C[q_x\sigma_y-q_y\sigma_x] \quad (3)$$

into the exciton Hamiltonian. The second term couples excitons states of even and odd parity with electron spin flip. The first term polarizes the electron spin in the direction perpendicular to $\mathbf{K}$ and the c axis, and gives to the two different electron spin orientations different energies.

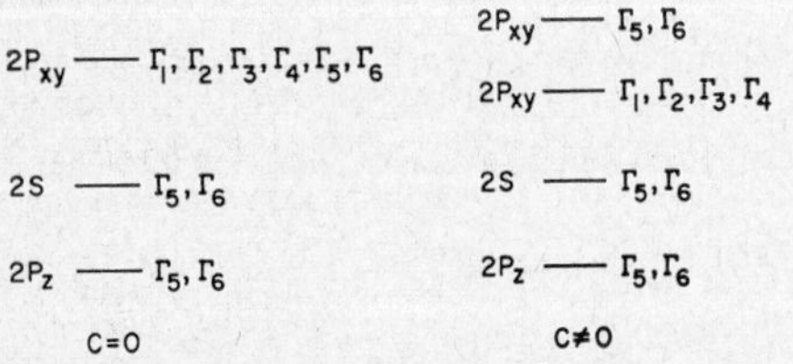

$$2P_{xy} \;\text{---}\; \Gamma_1,\Gamma_2,\Gamma_3,\Gamma_4,\Gamma_5,\Gamma_6$$

$$2S \;\text{---}\; \Gamma_5,\Gamma_6$$

$$2P_z \;\text{---}\; \Gamma_5,\Gamma_6$$

$$C=0$$

$$2P_{xy} \;\text{---}\; \Gamma_5,\Gamma_6$$
$$2P_{xy} \;\text{---}\; \Gamma_1,\Gamma_2,\Gamma_3,\Gamma_4$$

$$2S \;\text{---}\; \Gamma_5,\Gamma_6$$

$$2P_z \;\text{---}\; \Gamma_5,\Gamma_6$$

$$C\neq0$$

FIG. 1. The energy levels and exciton state symmetries of the $n=2$ exciton states from the top valence band in CdS or CdSe in the effective mass approximation. When C is nonzero, the coupling between the $2S$ and $2P_{xy}$ states mixes the $2S$ Γ_5, Γ_6 states with the $2P_{xy}\Gamma_{5,6}$ states and partially lifts the degeneracy of the $2P_{xy}$ states. (For simplicity, the longitudinal-transverse energy separation of the $2S$ states has been neglected.)

The effect of the second term can be most easily understood by considering the $n=2$ states of an exciton belonging to the top valence band in CdS. In the effective mass approximation with $C=0$, the energy levels, symmetries, and wave functions would be as described in Fig. 1. The extra term

$$\frac{m_{h\perp}}{m_{e\perp}+m_{h\perp}}C(q_x\sigma_y-q_y\sigma_x)$$

in the Hamiltonian will mix the $2S\Gamma_5$ (or Γ_6) and $2P_{xy}\Gamma_5$ (or Γ_6) states, thus shifting the energies of the $2S$ levels and breaking the eight-fold degenerate $2P_{xy}$ levels into two sets of four-fold degenerate levels. Because of the near-degeneracy of the $2S$ and $2P$ levels, this amount by which the $2P_{xy}$ Γ_5, Γ_6 states and the $2P_{xy}$ Γ_1, Γ_2, Γ_3, Γ_4 states would have their degeneracy lifted by the existence of a nonzero C can be approximately calculated by considering the $2S$ and $2P_{xy}$ states alone. The energy

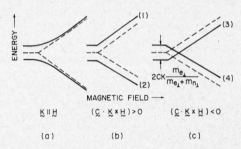

FIG. 2. The electron spin splitting of an exciton state of wave vector K in a magnetic field perpendicular to the c axis. K is also perpendicular to the c axis. (a) is for $H\|K$, while (b) and (c) represent the spin splitting for $H\perp K$. The dotted lines show the energy level behavior when $C=0$. The zero-field level splitting is $2CKm_{e\perp}/(m_{e\perp}+m_{h\perp})$. An arbitrary sign of C has been assumed. The states (1), (2), (3), and (4) have been marked for reference in the text.

splitting thus calculated is

$$\left[(E_{2S}-E_{2P_{xy}})^2+\frac{C^2}{32a_0^2}\right]^{\tfrac{1}{2}}-(E_{2S}-E_{2P_{xy}}), \quad (4)$$

where E_{2S} and $E_{2P_{xy}}$ are the energies of the $2S$ and $2P_{xy}$ states for $C=0$, and a_0 is the exciton Bohr radius.

The experiments of D. G. Thomas,[2] in which exciton states of symmetries appropriate to each component of the split $2P_{xy}$ state are observable, exhibit no obserbable splitting of the $2P_{xy}$ levels at zero magnetic field. A splitting of less than 3×10^{-4} ev would be undetectable. One concludes from this estimate that C is less than 5×10^{-10} ev cm.

The perturbation due to the finite slope conduction band and the finite wave vector of the light [the first term in (3)] is in practice easier to observe with precision than the effect previously described. For an exciton of wave vector K_y, this effect is equivalent to an applied magnetic field in the x direction proportional to C. For an exciton level split only by the interaction between the electron spin and an applied magnetic field perpendicular to the c axis, three types of Zeeman patterns would be observable for the three orientations of the magnetic field $H\|K$, $H\perp K$ $(C\times H\cdot K)>0$, and $H\perp K$, $(C\times H\cdot K)<0$. The splitting of the pair of levels as a function of magnetic field for these three possible field orientations is shown in Fig. 2.

Thomas[2] has studied the spin splitting of the $2P$ exciton states in a magnetic field in the three geometries described. Within an estimated possible experimental error of 3×10^{-5} ev, all three geometries yield the same spin splitting for the same magnitude of magnetic field. The wave vector of the excitons is, of course, the wave vector of the light in CdS at the exciton energy, about 3×10^{5} cm^{-1} in the present case. From mass parameters already published, $m_{e\perp}\approx0.20$, $m_{h\perp}\approx0.7$, one can calculate an upper limit of 1×10^{-10} ev cm for C. This corresponds to a displacement of the energy minimum below the $\mathbf{k}=0$ energy of less than 3×10^{-6} ev, and a removal of the position of the conduction band minimum from $\mathbf{k}=0$ by less than 10^{-4} times the distance to the zone boundary.

Thomas[4] has performed similar experiments in CdSe, where it was believed that the slope of the conduction band at $\mathbf{k}=0$ should be larger than in CdS. Some experimental results for the separations of levels (1) and (4) and for (1) and (2) are shown in Fig. 3. Unfortunately, the change of line intensity with magnetic field reversal tends to limit the obtainable precision. It is clear, however, that the difference between the separation of levels (1) and (4) and the separation of levels (1) and (3) is not more than 2.5×10^{-5} ev, placing an upper limit on C in CdSe of about 2×10^{-10} ev cm. The electron g value $|g_{e\perp}|$ is seen to be $0.70\pm10\%$.

An order of magnitude estimate of the magnitude of C can be simply obtained. Both the conduction band and the valence band states at $k=0$ can be thought of as having the following form

$$\alpha(S)+\beta(P_z)+\gamma(P_x+iP_y),$$
$$\alpha^*(S)+\beta^*(P_z)+\gamma^*(P_x-iP_y),$$

where α, β, and γ are constants, and S, P_z, etc., represent orbital wave functions of the atomic type in a unit cell. For the conduction band (dominantly S like) α is of the order of 1. The P_z state of the same spin is mixed with the S state by the directional crystal field, whereas the P_x and P_y states are mixed with the same state through the spin-orbit interaction. $k\cdot P$ perturbation theory then yields a band slope at $k=0$ of

$$(\hbar/m)|\alpha\gamma^*|\langle S|\prod_x|P_x\rangle,$$

where $\prod_x$ is the momentum operator. The matrix element $\langle S|\prod_x|P_x\rangle$ can be crudely estimated from the electron effective mass, and is about 1×10^{-19} for most II–VI compounds. The value of C is thus about $(6\times10^{-8})|\alpha\gamma^*|$ ev cm.

[4] D. G. Thomas (private communication).

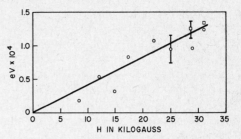

FIG. 3. The separation of a pair of spin-split exciton levels (1) and (4) (the points $\bigcirc$) and of (1) and (2) (the points $\square$) as a function of magnetic field in CdSe. The levels investigated here are $2P$ states. Estimated experimental errors are shown. The slope of the straight line represents a g value of 0.70.

For the conduction band, γ is of the order of $\delta\beta/3E_g$, where δ is the spin-orbit splitting of the valence band. For the Γ_7 valence bands, β and γ are of the order of 1. For the conduction band, β is difficult to estimate, but should at most be of the order of $(\Delta/E_g)^{\frac{1}{2}}$, where Δ is the crystal field splitting of the valence band. (This is an over-estimate of β, for it ascribes all the hexagonal crystal field splitting of the valence band to the mixing of the S and P_z states by the *odd* part of the crystal potential.) For the Γ_7 valence bands, α should be not most of the order of $(\Delta/E_g)^{\frac{1}{2}}$.

The following are estimates [based in β or $\alpha=(\Delta/E_g)^{\frac{1}{2}}$ and therefore certainly overestimates] of the possible Γ_7 conduction and valence band slopes in CdS and CdSe.

Substance	Band	$|C|$ (in ev cm)
CdS	Γ_7 conduction	6×10^{-11}
CdS	Γ_7 valence	6×10^{-9}
CdSe	Γ_7 conduction	1×10^{-9}
CdSe	Γ_7 valence	9×10^{-9}

The failure to observe a finite slope band crossing in the conduction band in CdS greater than 10^{-10} is thus to be expected. The failure to observe the finite conduction band slope at $\mathbf{k}=0$ in CdSe suggests that our estimate of β (α for the valence bands) is an order of magnitude too large, and that the best experimental estimates of all the band slopes of the above table are an order of magnitude less than those listed. The observation of any physical effect due to finite band-slope crossings at $\mathbf{k}=0$ in the conduction bands of the II–IV wurtzite structure compounds appears extremely doubtful with the possible exception of effects in the spin resonance of free carriers.

III. EFFECT OF EXCITON WAVE VECTOR ON g VALUES

The effective mass Hamiltonian for the exciton energy levels in the presence of a magnetic field parallel to the c axis has been given in reference 2. There are three magnetic perturbations, the ordinary Zeeman term, the diamagnetic perturbation, and the perturbation due to the exciton motion through the crystal, equivalent to the perturbation of an electric field $(\mathbf{v}/c)\times\mathbf{H}$ acting on the exciton, where v is the velocity of an exciton of wave vector **k**. (This electric field has been directly measured in CdS.[5])

For excitons of zero wave vector, the component of angular momentum in the direction of the c axis is a good quantum number. Consider two exciton states degenerate at $H=0$ which differ only in that one has an angular momentum $P_\varphi = n\hbar$ along the c axis, and the other has an angular momentum $P_\varphi = -n\hbar$ along this axis. For small magnetic fields, these lines will exhibit a linear Zeeman splitting and a quadratic diamagnetic shift. Because the angular component of momentum along the c axis is a good quantum number, these states will *always* be split in energy by $2\mu H$ (where μ is the magnetic moment of one of the states for zero magnetic field). This must continue to be true even when the magnetic field energy is comparable to the exciton binding energy, if it is possible to follow experimentally the original pair of lines.

Exciton motion and the consequent presence of an electric field destroy P_φ as a good quantum number. The exciton states $P_\varphi = n\hbar$ and $P_\varphi = -n\hbar$ will, in general, suffer different Stark effects for large magnetic fields (where the energy difference between $P_\varphi = n\hbar$ and $P_\varphi = -n\hbar$ is large), and the separation between these states will not be linear in the magnetic field. Since the radii of the excited exciton states in CdSe are large (the order of 170 A for the $n=2$ state) and the electric fields for 3×10^4 gauss are around 130 v/cm, deviations from linearity of about 10^{-4} ev should be readily detectable when one of the $P_\varphi = n\hbar$ lines is near to crossing a line with which it can be mixed via the electric field.

The measured "linear" Zeeman splittings of the $n=2$ states in CdSe[4] and the $n=3$ states in both CdS[2] and CdSe[4] exhibit small nonlinearities. In general, the measured splittings become sublinear for large magnetic fields. It is not yet clear whether the measured effect is real, or is associated with the failure to resolve nearly degenerate multiplets.

The only other obvious theoretical source of a failure of the measured splittings to be linear in magnetic field is associated with the nonquadratic shape of the valence band. The hole energy band has large quartic corrections in $E(k)$ due to the near proximity of the next lower valence band. The mean mass $\bar{m}$ for a bound hole in the top valence band (in the direction perpendicular to the c axis) is, in general, greater than the mass at $k=0$ by an amount of order $m^* T/\Delta$, where T is the kinetic energy of the hole, m^* the effective hole mass at $k=0$, and Δ is the separation between the top two valence bands.

In CdSe the changes in T due to the application of a magnetic field (and the resultant exciton compression) might cause an observable nonlinearity to the observed Zeeman splittings, since g will then depend on H. However, this will in general be a small effect, and would introduce a super-linearity to the observed Zeeman splittings.

The increase of the hole mass at large hole kinetic energies will introduce several other effects, which, unfortunately, are hard to isolate. Among these are a systematic increase in the binding energies of the lower exciton states and a systematic (but small) deviation between measured exciton g values for states of different zero-field binding energy.

IV. MAGNETIC FIELD DEPENDENCE OF TOTAL OSCILLATOR STRENGTH

The total oscillator strength of the $n=2$ lines in CdS for $E_{\parallel c}$ does not change appreciably in a magnetic field of 31 000 gauss. CdSe, on the other hand, exhibits a growth of a factor of 3 or 4 in the oscillator strength of the $n=2$ lines in a field of 31 000 gauss parallel to the hexagonal axis. The $3D_{\pm 2}$ lines in CdSe show an even larger effect of the magnetic field on their intensity, increasing from a strength invisibly small to a strength several times the strength of the $n=2$ lines at 31 000 gauss, an increase in strength of at least a factor of 50, and perhaps larger.[4]

This increase in line strength with magnetic field is due to the effective compression by the magnetic field. The ordinary diamagnetic term in the exciton Hamiltonian can easily become, for these weakly bound excitons, strong enough to modify appreciably the exciton mean radius. Since the oscillator strength of the $2P$ allowed or forbidden excitons is proportional to a^{-5} (where a is the state radius), and the oscillator strength of the $3D_{\pm 2}$ allowed exciton transitions to a^{-7}, even a small magnetic compression will produce relatively large effects in the line intensity.

The change in total oscillator strength of the $2P$ and $3D$ states as a function of magnetic field has been calculated approximately by using a variational wave function of the hydrogenic type with the Bohr radius as a variational parameter. The results for allowed or first forbidden transitions to S, P, or D exciton states are shown in Fig. 4. The dimensionless parameter ρ is the ratio (diamagnetic shift/H^2) for infinitesimal magnetic fields multiplied by the square of the actual magnetic field and divided by the zero-field binding energy.

For a magnetic field of 31 000 gauss parallel to the hexagonal axis, ρ is about 0.04 for $2P$ states in CdS and 0.4 for $2P$ states in CdSe. Figure 4 shows that the intensity increase to be expected at this magnetic field is only about 15% for the CdS $2P$ states, but is a factor of 2.7 for the CdSe $\cdot 2P$ states. For the $3D$ states, ρ is about 4 in CdSe, and the lines are expected to gain a factor of about 100 in strength at this magnetic field. The variational calculation performed is not very accurate, especially for large ρ. Nevertheless, it seems clear that the great increase in strength of the observed exciton lines in CdSe with magnetic field is simply due to the magnetic compression of the excitons.

[5] D. G. Thomas and J. J. Hopfield, Phys. Rev. Letters **5**, 505 (1960).

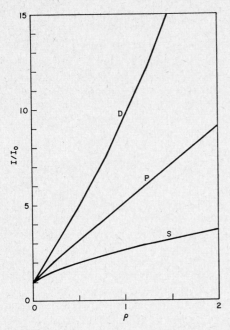

FIG. 4. The ratio I/I_0 of the sum of the intensities of the magnetic doublets for $H\|c$ divided by the line intensity at zero magnetic field for S, P, and D states. The dimensionless parameter ρ is the ratio of the diamagnetic shift which would be calculated if the state radius did not change divided by the zero-field binding energy of the state under consideration.

In a magnetic field parallel to the hexagonal axis, the $2P_{xy}$ state of CdSe splits into two lines. If only allowed optical transitions were observable, exactly this result would be expected. CdS, on the other hand, exhibited two allowed transitions and two forbidden transitions to the $2P_{xy}$ states in a magnetic field, and a forbidden transition to the $2P_z$ state as well.

In a magnetic field perpendicular to the c axis, and using (as always) light polarized parallel to the c axis, the observed intensities of the absorption lines depended strongly on the direction of the wave vector of the light with respect to the magnetic field. In particular, quite different absorption spectra were observed for H and $-H$ when H, c, and k were mutually perpendicular. In CdS, this effect could be quantitatively explained on the basis of the interference between allowed and forbidden optical transitions.[2] It was shown that similar effects would be expected in the exciton spectra of most II–VI compounds having the wurtzite structure. Wheeler and Dimmock[6] reported on the observation of this effect in CdSe. The effect, although less marked than in CdS, is qualitatively very similar. One is led then to inquire why the forbidden $2P_{xy}(\Gamma_5)$ exciton transitions, so clearly visible in CdS for $H\|c$, are not observable in this geometry for CdSe,[4] which shows the same interference effect between allowed and forbidden transitions. Two chief possibilities exist between which

it has not yet proved possible to choose. First, the observed interference effect is not as large as in CdS, and the relative strengths of the allowed and forbidden $2P_{xy}$ lines in $H\|c$ may be such that the weaker lines are simply not observable. Second, the energy separation between the allowed and forbidden $2P_{xy}$ lines in $H\|c$ may be too small to permit the resolution of the allowed and forbidden lines. Such a situation would exist if the electron g value parallel to the c axis were less than about 0.4. This state of affairs is far from impossible, for the electron g value $g_\perp$ is itself much less than 2.0.

V. SUMMARY

The band structure of CdS and CdSe are very similar. The conduction band minima and valence band maxima probably lie at $\mathbf{k}=0$. The essential features of the band structure may be regarded as obtained from a cubic crystal with a slight uniaxial crystal field.

The s-like conduction bands are virtually spherical, having effective masses 0.205 and 0.13[6] for CdS and CdSe, respectively. Although a finite slope band crossing at $\mathbf{k}=0$ is permitted for the conduction bands, this effect gives less than a 1 or 2×10^{-10} ev cm slope to the conduction bands at $\mathbf{k}=0$. The electron g value in CdS is about -1.8, and very nearly isotropic. In CdSe, $g_\perp$ is -0.7,[7] and $g_{e\|}$, although not definitely known, is probably similar.

The top valence bands in both materials has symmetry Γ_7, the exciton optical selection rules are identical, and the energy levels very nearly so.

The hole mass parallel to the axis is too heavy (greater than one) to be readily measured by examination of the exciton spectra. The hole masses perpendicular to the C axis are 0.7 (CdS) and 0.5 (CdSe). The hole g values are zero by symmetry considerations for $H\perp c$. For CdS, $g_{h\|}=-1.15$.

The behavior of the exciton states in a magnetic field are qualitatively similar. The great increase in strength of the $2P$ and $3D$ lines in CdSe in an applied magnetic field (an effect much smaller in CdS) is a simple consequence of the large radius and small binding energy and masses in CdSe. The exciton spectrum of CdSe shows the same peculiar intensity changes on reversal of the magnetic field as does CdS, as is of course expected. Unfortunately, the slope of the conduction bands at $\mathbf{k}=0$ of even CdSe is extremely small. All interesting effects due to a finite slope would seem to be unmeasurably small (at least in the compounds so far investigated) with the possible exception of measurements which might be made of electron spin-resonance.

ACKNOWLEDGMENT

The author wishes to thank D. G. Thomas for the information on his unpublished experiments in CdSe and many fruitful discussions.

[6] R. G. Wheeler and J. O. Dimmock, Bull. Am. Phys. Soc. 6, 148 (1961).

[7] The measurement of this g value seems unambiguous, but does not agree with the value reported by Wheeler and Dimmock.[6]

JOURNAL OF APPLIED PHYSICS   SUPPLEMENT TO VOL. 32, NO. 10   OCTOBER, 1961

Excitons and the Absorption Edge of ZnO

R. E. DIETZ, J. J. HOPFIELD, AND D. G. THOMAS

Bell Telephone Laboratories, Murray Hill, New Jersey

The absorption coefficient for polarized light at photon energies less than that of the lowest lying direct exciton was measured for single crystals of ZnO at temperatures ranging from 20° to 200°K. Analysis of the results shows that the absorption is in agreement with that calculated for a process involving the simultaneous creation of an exciton and absorption of a phonon, both particles having a small wave vector. This agreement provides evidence that the absolute minimum of the conduction band cannot lie lower than the lowest lying direct exciton level, and, therefore, that the absolute minimum probably occurs at the center of the Brillouin zone. Values for the hole and electron masses were estimated from the analysis.

INTRODUCTION

DESPITE considerable efforts to ascertain the energy band structure of ZnO during the past few years, the question of whether the absolute extrema of the energy bands occur at the center of the Brillouin zone is as yet unanswered. Transport measurements, particularly the measurement of a nonzero longitudinal magnetoresistance,[1] have tended to support a many-valley model for the conduction band (but see the following paper by Hutson), while optical studies of exciton lines by Thomas and Hopfield[2-4] have been interpreted successfully using direct processes at $k=0$.

Recently, Thomas, Hopfield, and Power[5] have shown that the form of the absorption edge in CdS for values of the absorption coefficient between 10 and 300 cm^{-1} can be ascribed to scattering of low lying excitons near $k=0$ by the longitudinal optical phonon. Since the exciton spectrum of ZnO is similar to that for CdS, one might expect that similar scattering processes would broaden the exciton lines in ZnO, and therefore account for the observed edge absorption.

Analysis of the exciton spectrum of ZnO has shown that near $k=0$ the P-like valence band of ZnO is split into three doubly degenerate bands by a combination of spin-orbit and crystal field interactions. The uppermost of these bands has the symmetry of Γ_7 and is split from the second band Γ_9 by about 0.007 ev. The third and lowest energy band, also of Γ_7, lies about 0.045 ev below the first Γ_7 band. The conduction band is considered S type and has the symmetry Γ_7; therefore direct transitions from the valence to the conduction band fall in the "allowed" classification. Excitons formed from holes in the upper Γ_7 band (exciton A in Thomas' notation[2]), and from the Γ_9 band (exciton B) are active chiefly in light with the electric vector perpendicular to the crystal c axis ($E \perp c$), while excitons formed from the lower Γ_7 band (exciton C) are active chiefly for the electric vector parallel to the c axis ($E \| c$).

If the difference in energy between the valence and conduction bands at $k=0$ is not an absolute minimum

for all points in the Brillouin zone, then the crystal should absorb light at energies below that of the direct band gap at $k=0$. In this paper we shall demonstrate that the absorption below the lowest lying exciton peak in ZnO can be accounted for by the exciton broadening processes discussed for CdS, and that any absolute minimum in the conduction band must lie not lower than the lowest lying direct exciton level.

EXPERIMENTAL

Transmission measurements were made by the sample in-sample out technique as described in reference 5. The monochromator was a Perkin-Elmer model 99 double pass-single beam instrument equipped with a NaCl prism. A slitwidth of 0.1 mm provided a spectral slit-width of about 0.002 ev, as determined from broadening of emission lines from a mercury spectral lamp. The instrumental wavelength calibration was obtained using mercury and thallium spectral lamps, and photon energies quoted in this paper are accurate to about 0.001 ev. The sample was indirectly cooled by soldering one end of the crystal to a copper-bronze plate which was then screwed to a copper block in thermal contact with refrigerant in the inner part of a double Dewar. The copper block was fitted with a heating coil through which an electric current could be passed to stabilize the temperature of the block above the temperature of the refrigerant (liquid N_2). For temperatures above 100°K, the sample was cooled by radiation to refrigerant in the outer Dewar only, again adjusting the temperature with the heating coil. The temperature of the crystal was measured directly by soldering a thermocouple to the free end of the crystal.

Single crystals of ZnO grown from the vapor were obtained in the form of flat plates or long flat needles with the hexagonal c axis along the long dimension of the crystals. Measurements were made on crystals as grown and on thin slabs which had been polished down from thicker crystals. In the latter case, the surfaces of the crystals were etched in concentrated H_3PO_4 until the reflectivity spectrum of the excitons equaled that of as-grown surfaces. Imperfections on the surfaces of all crystals studied resulted in transmission measurements at energies far below the threshold of the edge absorption which were smaller than the maximum transmission calculated for reflections from front and back plane

[1] A. R. Hutson, J. Phys. Chem. Solids **8**, 467 (1959).
[2] D. G. Thomas, J. Phys. Chem. Solids **15**, 86 (1960).
[3] J. J. Hopfield, J. Phys. Chem. Solids **15**, 97 (1960).
[4] J. J. Hopfield and D. G. Thomas, J. Phys. Chem. Solids **12**, 276 (1960).
[5] D. G. Thomas, J. J. Hopfield, and M. Power, Phys. Rev. **119**, 570 (1960).

parallel surfaces using the refractive index data of Mollwo.[6] Therefore, for the purpose of computing values of the absorption coefficient α, all transmission data were corrected for surface scattering losses by normalizing the transmission measurements to the theoretical values in a wavelength region where the transmission varied in proper relation with the refractive index, assuming no absorption losses. In calculating values for α, the variation of the refractive index with temperature and wavelength for each mode of polarization was taken into consideration.

RESULTS

The problem of background absorption encountered in the similar study of CdS is more severe for the case of ZnO. Examination of the edge absorption of many crystals of ZnO showed that the edge absorption could be characterized by two aspects: a relatively flat, temperature independent base absorption for low values of α, which varied widely from crystal to crystal; and a steeply rising edge, common to all crystals, which shifted to higher energies with increasing temperature. The low base absorption was considered to be an extrinsic effect, possibly arising from internal strain or impurities, while the steeply rising edge was considered to represent an intrinsic property of the crystal. Consequently, crystals were selected for further investigation using the criterion that the most perfect crystals were those which had the *least* absorption near the excitons. Figure 1 presents the absorption edge at 78°K for three crystals having thicknesses less than 0.5 mm. Crystal F1-79-8, which was 0.126 mm thick, was used to obtain most of the data presented in this paper. However, other crystals were found which equaled its performance.

In Fig. 2, the absorption edge for light polarized with $E \perp c$ of crystal F1-79-8 is plotted for temperatures ranging from 23° to 190°K. At 23°K, it is seen that the edge absorption rises steeply from the base very near to the exciton while at higher temperatures the edge progressively broadens to longer wavelengths and becomes more difficult to separate from the base absorption. Therefore, corrections were made for the base absorption by subtracting the base measured at 23°K from the absorption coefficients measured at higher temperatures. Since the base absorption rises exponentially near the edge (contrast the absorption edges of the three crystals in Fig. 1), it seems likely that the mechanism for the base absorption is coupled to the processes responsible for the edge absorption. Therefore, the base absorption at 23°K was shifted with the exciton peaks for corrections at higher temperatures. However, the details of the base correction are not really significant since the essential features of the edge absorption may be derived from the data whether the correction for the base is made or not. The arrows on the abscissa of Fig. 2 indicate the position of the lowest energy

[6] E. Mollwo, Z. angew. Phys. **6**, 257 (1954).

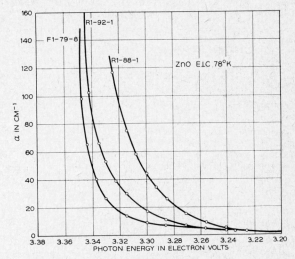

FIG. 1. The absorption edge of three single crystals of ZnO at 78°K for light polarized $E \perp c$. The crystal showing the least absorption, F1-79-8 was selected for further studies.

exciton, exciton A, active for $E \perp c$, from which 0.073 ev, the energy of the longitudinal optical phonon, as measured by Collins and Kleinman,[7] has been subtracted. The temperature dependence of this exciton was estimated from the shift with temperature of the position of the reflection peak of exciton B. At temperatures above 100°K, the reflection peaks of excitons A and B broaden to form a single peak, which more accurately indicates the position of exciton B because of its appreciably greater oscillator strength. The

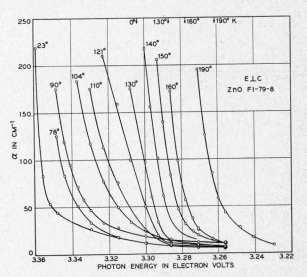

FIG. 2. The variation of the absorption edge with temperature for $E \perp c$. For temperatures above 100°K, the curves extrapolate to the abscissa at energies corresponding approximately to the positions of the lowest lying exciton (A) minus the energy of the longitudinal optical phonon. These positions, indicated by arrows at the top of the figure for four temperatures, were obtained from the temperature dependence of the positions of the reflectivity peaks of exciton (AB).

[7] R. J. Collins and D. A. Kleinman, J. Phys. Chem. Solids **11**, 190 (1959).

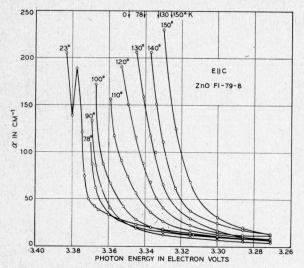

FIG. 3. The variation of the absorption edge with temperature for $E\|c$. This figure is analogous to Fig. 2 except that the curves are shifted to higher energies and do not appear to extrapolate as well to the arrows, which here indicate the temperature dependence of the position of exciton C diminished by the longitudinal optical phonon energy. The sharp spike in the edge at 23° is the absorption peak of exciton A, seen in the "wrong" mode of polarization.

uncertainty in the determination of the positions of the excitons at temperatures above 100°K is probably not more than 0.003 or 0.004 ev. The absorption curves of Fig. 2 bear a marked resemblance to the analogous data for CdS of reference 5. It is seen that the curves extrapolate to the approximate position, and with the proper temperature dependence of the position of the exciton less the optical phonon energy as observed for CdS. This suggests that the threshold of absorption for $E\perp c$ can be described by the scattering process proposed by Thomas, Hopfield, and Power for CdS, in which excitons are formed by the simultaneous absorption of a photon and a phonon having a small wave vector.

Analogous data are depicted for $E\|c$ in Fig. 3. Here again, a parallel may be drawn to the results obtained for CdS. The steeply rising portions of the curves are shifted to higher energies by about 0.035 ev. However, this shift is in better agreement with the position of exciton C less the energy of the longitudinal optical phonon, than the corresponding data for CdS. The lower portions of the curves tail off more slowly than for $E\perp c$, and finally merge with the background absorption. The sharp peak in the edge at 23°K is exciton A seen in the "wrong" mode of polarization. This occurs because excitons A and C are slightly mixed by spin-orbit coupling.

DISCUSSION

Theoretical Form of the Absorption Edge

Except for minor changes, the theory which will be applied to account for the edge absorption in ZnO is identical to that given in reference 5. A more complete

discussion of exciton broadening will be presented by Hopfield at the 1961 International Conference on Photoconductivity at Ithaca, New York.

The scattering processes which are considered to make the predominant contributions to the edge absorption below the energy of the lowest exciton are depicted in Fig. 4, in which the energy of low-lying exciton states is given as a function of the wave vector. The ground state of the crystal is represented by a point; the exciton states by bands. The transitions are represented by two steps involving an intermediate state for which energy is not conserved, but momentum is. The energy defect is then accounted for by the second step in which a phonon is absorbed. For the same reasons presented for the case of CdS, we shall assume that the lowest lying exciton states are the only important intermediate states. During the following discussion excitons A and B will be treated as a single combined band AB because of their near degeneracy and since thay have similar polarization properties. Thus for $E\perp c$, the absorption will be calculated assuming exciton AB is scattered in its own band, although near the threshold the absorption would be more accurately characterized as a scattering of exciton A in band A alone. Assuming spherical energy surfaces for parabolic electron and hole bands, one obtains from second-order perturbation theory for longitudinal optical mode scattering of the exciton in its own band,

$$\alpha = \alpha_0 \left(\frac{E_0}{E_0 - E}\right)^2 x^{\frac{1}{2}}$$

$$\times \left\{\frac{1}{x^{\frac{1}{2}}}\left[\frac{1}{(1+\gamma^2 x)^2} - \frac{1}{(1+\gamma'^2 x)^2}\right]\right\}^2 \frac{1}{e^{\hbar\omega/kT}-1}, \quad (1)$$

where

$$x = (E_0 - E + \hbar\omega)/E_0, \quad (2)$$

$$\gamma' = (m_e/m_h)\gamma, \quad (3)$$

$$\gamma = \frac{a_0}{2}\left[(m_e + m_h)2E_0/\hbar^2\right]^{\frac{1}{2}}\frac{m_h}{m_e + m_h}, \quad (4)$$

E_0 is the energy of the lowest exciton band at $k=0$, E is the photon energy, $\hbar\omega$ is the energy of the longitudinal optical phonon, m_e and m_h are the electron and hole masses, respectively, and a_0 is the exciton Bohr radius. [The formula given in reference 5 is an approximation to Eq. (1).] Equation 1 has been written as a product of five factors. The first of these α_0 is a collection of constants independent of E and T, which are associated with the factors and contains the square of the optical matrix element, a constant. The theoretical espression for α_0 is given in the Appendix. The denominator of the second factor is proportional to the square of the intermediate state transition probability. $x^{\frac{1}{2}}$ is the density of states function, while the squared fourth factor is the

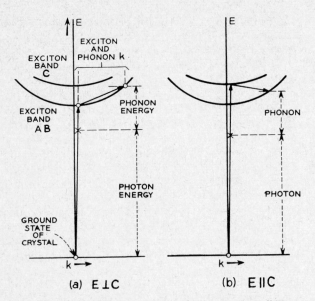

FIG. 4. $E(\mathbf{k})$ diagrams for electron-hole pairs. In these diagrams the ground state is represented as a point, and excitons as bands. Because of the near degeneracy and common polarization properties of excitons A and B, they are represented here, as in the text, as comprising a single band of states AB. Part (a) of this figure illustrates the process where the intermediate state exciton AB is scattered in its own band for excitation with $E \perp c$. Part (b) depicts another process, possible for $E \| c$ in which the intermediate state exciton C is scattered into band AB.

phonon scattering matrix element. The final factor is the Boltzman factor for phonons of energy $\hbar\omega$ ($\hbar\omega$ is approximately independent of $\mathbf{k}$ for small values of $\mathbf{k}$).

In the other mode of polarization $E\|c$ two final states may be reached through the $n=1$ state of exciton C. The intermediate state exciton C may be scattered in its own band, in which case the absorption coefficient will have the form given by Eq. (1). The other final state involves scattering the intermediate state exciton C into exciton band AB. The matrix element for this (interband) mode of scattering is, in principle, different from that given in Eq. (1); although it cannot be computed from ordinary optical mode scattering formulas, it is expected to be considerably smaller than that for the intraband scattering process. Since the oscillator strength of exciton AB is approximately equal to that of exciton C, one would expect that the absorption edge for $E\|c$ should appear similar to that for $E \perp c$ but shifted by the valence band splitting (about 0.045 ev) to higher energies, and have a tail extending down to the threshold for $E \perp c$ absorption.

Comparison with Experiment

A detailed comparison between theory and experiment can be made only for $E \perp c$, for which the scattering matrix element has been computed. However, it is apparent that the experimental data for $E\|c$ are in qualitative agreement with the theoretical predictions outlined in the previous section. Therefore, the remain-

der of the discussion will be devoted to analysis of the $E \perp c$ data.

Of the various parameters in Eq. (1), only m_e and m_h are arbitrary in the sense that they cannot be obtained from presently available experimental data. However, for the purpose of curve fitting, γ and γ' have been selected as the adjustable parameters since their ratio gives the electron-hole mass ratio directly, and does not involve the exciton Bohr radius. α_0 has been taken as an adjustable parameter, although a value for it is estimated in the Appendix. The temperature dependence of the absorption coefficient enters Eq. (1) chiefly through the Boltzmann factor, but a temperature dependence is also implicit insofar as the intermediate state exciton energy E_0 varies with temperature. Since the Boltzmann factor does not contain the adjustable parameters, it was divided into the experimental values of α taken from Fig. 2 after making the background correction as previously described. The data were then further corrected for the shift of the excitons with temperature by subtracting the shift from 0°K (as measured from the shift in the reflection peaks) from the photon energy appropriate to each curve. These corrected data are plotted in Fig. 5 on a semilog scale. The excellent agreement for the data above 100°K offers convincing evidence for the scattering role played by the longitudinal optical phonon. The solid curve represents a best fit to Eq. (1) using the parameters $\gamma=8$, $\gamma'=1.7$, and $\alpha_0=10.2$. It should be noted that on a semilog plot, γ and γ' alone determine the shape of the curve, while α_0 scales the ordinate uniformly. The value obtained for α_0 is in satisfactory agreement with the theoretical value computed in the Appendix, $\alpha_0=18.3$. Since Eq. (1) is symmetrical with respect to an interchange of γ and γ' it is impossible to distinguish between the two parameters. However, since the binding energies of the excitons in ZnO are almost identical, one concludes that the reduced exciton mass is dominated by an electron mass (common to excitons formed from holes in all three valence bands) which is considerably lighter than the hole mass. Taking, then, the ratio of γ'/γ to be the ratio of the electron to hole mass, as indicated in Eq. (3), and using the values of γ and γ' obtained from the experimental fit, one computes $m_e/m_h=0.21\pm0.03$. Finally, using the exciton reduced mass determined by Thomas from the exciton binding energy, $\mu=0.31$, m_e is calculated to be 0.38 ± 0.01, $m_h=1.8\pm0.2$ where the indicated errors represent the uncertainty in determining γ and γ'.

Because of the small exciton radius in ZnO, the exciton binding energy and hence the reduced mass may be larger than the hydrogenic values by a substantial central cell error; thus the true (average) inertial carrier masses might conceivably be some 20 or 30% smaller than the values stated. It should be noted that the electron mass is in reasonable agreement with the density of states mass determined by Hutson[1] from donor binding energies ($0.5m$), and with an electron

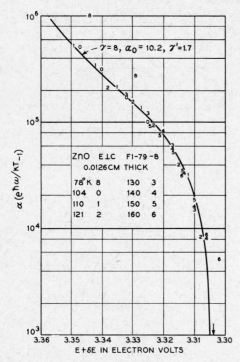

FIG. 5. A plot of the data of Fig. 2 from which the temperature dependence has been removed according to Eq. (1) in the text. The ordinates of points represent values of the absorption coefficient (corrected for the base absorption) divided by the appropriate Boltzman factor; the abscissas represent the photon energy to which the shift in the position of exciton A from 0°K has been added. For temperatures above 100°K, the data superpose very well on a single curve. For temperatures below 100°K, the agreement is not so good, but the scatter in these low-temperature data is considered to arise from experimental difficulties, chiefly temperature control. The solid line is a best fit of the theory using the parameters $\gamma = 8$, $\gamma' = 1.7$, and $\alpha_0 = 10.2$. The arrow at the bottom of the figure indicates the theoretical threshold for absorption.

mass determined from recent transport studies described in the following paper by Hutson. It is not consistent, however, with the value $(0.06m)$ determined by Collins and Kleinman[7] from the infrared reflectivity.

Because of the closeness of the fit attained using reasonable values of the adjustable parameters, it is considered that all of the absorption (except for the weak base absorption) can be accounted for by the optical mode scattering process outlined above for energies ranging from the threshold to some 0.020 ev below the energy of exciton A. Although the exact mechanism for the base absorption is unknown, indirect transitions can be excluded because of the weak temperature dependence of the base, and because fluorescence is frequently observed near the threshold for the intrinsic edge absorption (corresponding to the recombination of the direct excitons with emission of a longitudinal optical phonon). Thus, any indirect exciton associated

with a conduction band valley not at $\mathbf{k}=0$ must occur at an energy not lower than the energy of the lowest lying direct exciton.

CONCLUSIONS

The optical absorption observed in ZnO below the lowest energy direct exciton peaks has been shown to arise from exciton transitions near $\mathbf{k}=0$ in which the direct excitons are scattered by longitudinal optical phonons. These considerations allow one to specify the energy of the lowest lying direct exciton as a lower limit for the energy minima of any other conduction band valleys not at $\mathbf{k}=0$. Parameters obtained from the fitting of the theoretical expression for the form of the absorption edge to the experimental values for α also enabled an estimate to be made of the effective hole and electron masses.

ACKNOWLEDGMENTS

The authors are indebted to F. H. Doleiden for valuable technical assistance during the course of this work.

APPENDIX

The complete expression for α_0 is obtained under the same assumptions as Eq. (1):

$$\alpha_0 = \left(\frac{4\pi\beta}{\epsilon^{\frac{1}{2}}}\right)\left(\frac{\omega}{16c}\right)\left(\frac{1}{n_0^2} - \frac{1}{\epsilon'}\right)$$

$$\times \left(\frac{2M}{\hbar^2}\right)^{\frac{3}{2}} \frac{(m_e - m_h)^2}{(m_h^4 - m_e^4)} \hbar^4 E_0^{-\frac{1}{2}} e^2.$$

Definitions of the various parameters in this equation, and values pertinent to ZnO (taken from the exciton studies of Thomas and Hopfield, and the restrahl study of Collins and Kleinman) are as follows: $4\pi\beta =$ the sum of the zero frequency polarizabilities of excitons A and $B = 3.1 \times 10^{-2}$ for $E \perp c$; $\epsilon =$ contribution of all sources except the exciton under study to the energy dependent dielectric constant at $E = E_0$ ($\epsilon \approx 5.0$); ω is the angular frequency of the $k=0$ longitudinal optical phonon $= 591$ cm^{-1}; $c =$ velocity of light; $n_0^2 =$ high-frequency dielectric constant $= 4.0$; $\epsilon' =$ low-frequency dielectric constant $= 8.5$; $m_e =$ electron mass $\approx 0.38m$; $m_h =$ hole mass $\approx 1.8m$; $M =$ density of states mass for excitons $= m_e + m_h$; $\alpha_0 =$ Bohr radius of the $n=1$ exciton state $= 14$ A; and $E_0 =$ energy of the intermediate state exciton ≈ 3.38 ev. When substituted into the above equation these parameters yield a theoretical value for α_0 of 18.3.

JOURNAL OF APPLIED PHYSICS   SUPPLEMENT TO VOL. 32, NO. 10   OCTOBER, 1961

Piezoelectric Scattering and Phonon Drag in ZnO and CdS

A. R. HUTSON

Bell Telephone Laboratories, Incorporated, Murray Hill, New Jersey

Piezoelectric scattering of conduction electrons by acoustical phonons is discussed for ZnO and CdS, and approximate values of the mobilities determined by this mechanism alone are derived. The phonon drag contribution to the Seebeck effect in ZnO is assumed to arise from crystal-momentum exchange between electrons and acoustical phonons by way of the piezoelectric interaction alone. Comparison of the results of this assumption with the data has led to the discovery of strong piezoelectric phonon scattering from neutral donor states. These two piezoelectric scattering mechanisms and an effective electron mass of about $0.32m$, derived from other experiments, provide a model for phonon drag in ZnO which agrees with the temperature dependence and "impurity" dependence of the data and gives the correct magnitude of the effect to within the uncertainties of the approximations employed.

I. INTRODUCTION

A LARGE phonon-drag contribution to the Seebeck coefficient Q_p has been encountered in an investigation of the semiconducting properties of n-type zinc oxide.[1] The magnitude of the phonon-drag effect appears to be too large to be satisfactorily explained by electron-phonon coupling through a deformation potential of reasonable size. For this reason, the piezoelectric effect was measured in ZnO and was found to be very strong[2]—sufficiently strong to account for the electron-acoustical-phonon interaction required by the phonon-drag data. However, the temperature dependence of Q_p is *not* satisfactorily explained by the substitution of piezoelectric scattering for deformation potential scattering of the electrons (despite an incautious phrase to the contrary in reference 2). This difficulty with the temperature dependence suggested a re-examination of the phonon-scattering mechanism (which provides an essential part of the temperature dependence of Q_p). As a result, a new phonon-scattering mechanism peculiar to piezoelectric crystals with semiconductor-type impurity states has been discovered.

In the previous interpretation of the phonon-drag data[1] which did not take any piezoelectric effects into account, effective masses for the conduction band were inferred (including a light inertial mass $\approx 0.07\ m$), and a rather complicated conduction-band structure was postulated. Interpretation of the phonon-drag data on the basis of the piezoelectric effects no longer supports an extraordinarily light electron mass.

In Sec. II of this paper a brief account is given of the electric fields accompanying plane acoustic waves in a piezoelectric crystal with semiconductivity. Section III outlines the piezoelectric scattering of conduction electrons by thermal vibrations and gives some semiquantitative results for ZnO and CdS using the approximation that the relaxation time depends only on electron energy. Phonon-drag and phonon-impurity scattering are discussed in Sec. IV. A brief summary of the experimental data bearing upon the conduction band mass of ZnO is given in Sec. V, and the Appendix provides what we

believe are the best present estimates of the relevant piezoelectric and elastic constants and the spherical averages of the squares of electromechanical coupling constants employed in Secs. III and IV.

II. ELECTRIC FIELDS OF ACOUSTIC WAVES[3]

An acoustic wave propagating in a piezoelectric material is, in general, accompanied by some electrical disturbance. If the material is an insulator, and we are dealing with plane waves, a *longitudinal* electric field accompanies the wave

$$E_i = -e_{ijk}S_{jk}/\epsilon, \qquad (\text{II.1})$$

where i is the propagation direction, e_{ijk} are components of the piezoelectric e tensor, S_{jk} is the sinusoidally varying strain, and ϵ is the dielectric permittivity. (We shall assume both dielectric and elastic isotropy here for simplicity, and employ mks units.) The transverse electric fields accompanying the acoustic wave can be shown to be negligible by comparison with the longitudinal fields. If the material is a semiconductor, electric currents and space-charge may also accompany the wave, and if both drift and diffusion of the mobile charge carriers are taken into account, relation (II.1) becomes

$$E\left[1 + \frac{j(\omega_C/\omega)}{1 + j(\omega/\omega_D)}\right] = -\frac{eS}{\epsilon}. \qquad (\text{II.2})$$

Here $\omega_C = \sigma/\epsilon$ is a conductivity frequency (whose reciprocal is commonly called the dielectric relaxation time) and ω_D is that frequency above which the wavelength of sound is sufficiently short for diffusion to smooth out carrier density fluctuations having the periodicity of the acoustic wave.

$$\omega_D = \frac{q}{k_0 T\mu}\left(\frac{\omega}{k}\right)^2 = \frac{qv^2}{k_0 T\mu}.$$

(q is the electronic charge, k_0 is Boltzmann's constant, k the wave vector, and v the sound velocity.) A modifi-

---

[1] A. R. Hutson, J. Phys. Chem. Solids **8**, 467 (1959).
[2] A. R. Hutson, Phys. Rev. Letters **4**, 505 (1960).

[3] The relations briefly presented here are derived and discussed in an article, "Elastic wave propagation in piezoelectric semiconductors", by A. R. Hutson and Donald L. White, J. Appl. Phys. (to be published).

cation of (II.2) which is useful for the frequency range of interest in electron scattering is obtained upon recognizing that

$$\frac{\omega_C \omega_D}{\omega^2} = \frac{nq^2}{\epsilon k_0 T} \cdot \frac{1}{k^2} = \frac{1}{\lambda^2 k^2},$$

where λ is the Debye length, and noting that $(\omega/\omega_D) \gg 1$. Thus, for present purposes

$$E[1 + (1/\lambda^2 k^2)] = -eS/\epsilon \qquad (II.3)$$

includes the effects of screening.

III. PIEZOELECTRIC SCATTERING OF CONDUCTION ELECTRONS

The scattering of conduction electrons by the longitudinal electric fields which accompany plane-strain waves has been treated for crystals of the zinc-blende structure by Meijer and Polder[4] and by Harrison,[5] and the result of an adaptation of the former calculation to hexagonal ZnO and CdS has been given by Hutson.[2] We shall briefly sketch the calculation here, pointing up the differences between this scattering mechanism and the more familiar deformation-potential scattering of Bardeen and Shockley.[6]

The reciprocal of the relaxation time for an electron of wave vector k for one-phonon scattering processes is

$$\frac{1}{\tau} = \frac{m^* k}{2\pi^2 \hbar^3} \int_0^{2\pi} \int_0^{\pi} |M(\mathbf{k},\mathbf{k'})|^2 (1-\cos\theta)\sin\theta d\theta d\varphi, \quad (III.1)$$

where θ is the angle between $\mathbf{k}$ and $\mathbf{k'}$. For deformation-potential scattering, the potential energy of an electron contains a term $E_1 S$, where E_1 is the deformation potential (energy) and S is the local strain. In this case[6]

$$|M(\mathbf{k},\mathbf{k'})|^2 = E_1^2 k_0 T/2c,$$

where c is the appropriate elastic stiffness constant. In a piezoelectric material, the longitudinal electric field accompanying a plane wave contributes a term to the potential energy of an electron:

$$-q\varphi = -qeS/p\epsilon.$$

(Here p is the wave vector of the phonon, and we neglect screening.) The square of the matrix element for the piezoelectric case is then

$$|M(\mathbf{k},\mathbf{k'})|^2 = \frac{q^2 e^2}{p^2 \epsilon^2} \cdot \frac{k_0 T}{2c} = \left(\frac{e^2}{\epsilon c}\right) \cdot \frac{q^2}{p^2 \epsilon} \cdot \frac{k_0 T}{2}.$$

As usual, $\mathbf{p} = \mathbf{k'} - \mathbf{k}$ and $k \cong k'$.

[4] H. J. G. Meijer and D. Polder, Physica 19, 255 (1953).
[5] W. A. Harrison, Phys. Rev. 101, 903 (L) (1956); and thesis, Department of Physics, University of Illinois (1956).
[6] J. Bardeen and W. Shockley, Phys. Rev. 80, 72 (1950).

Since e, and to a lesser extent, ϵ and c depend upon the direction of $\mathbf{p}$, the relaxation time is really a function of $\mathbf{k}$ and the integrations over angles can become quite complicated. For the zinc-blende case, Meijer and Polder[4] took a weighted average of the magnitude of $(e^2/\epsilon c)$ in the [100], [110], and [111] directions and treated it as a constant in the integrations. Harrison[5] carried out the integrations in these same directions and found approximate isotropy for $\tau(\mathbf{k})$ and for the relative contributions of shear and logitudinal modes. The resulting τ (energy) was approximately the same for the two procedures, with the justification for the averaging coming from Harrison's integration.

It is worth noting that the dimensionless quantity $(e^2/\epsilon c)$ is just K^2, the square of the electromechanical coupling, since the elastic constant enters this expression via the (piezoelectrically stiffened) wave velocity and is, hence, that for constant D.

Once committed to an averaging procedure for the anisotropic quantities, we have an approximate relaxation time which is a function of energy only; thus

$$\frac{1}{\tau} = \frac{q^2 m^{*\frac{1}{2}}}{2^{\frac{3}{2}} \pi \hbar^2} \left[\frac{(K^2)_{av}}{\epsilon}\right] \cdot \frac{k_0 T}{\mathcal{E}^{\frac{1}{2}}}, \qquad (III.2)$$

where $\mathcal{E} = \hbar^2 k^2/2m^*$.

The drift mobility resulting from the operation of piezoelectric scattering alone is then obtained by summing contributions to τ^{-1} from the longitudinal and transverse modes and taking the proper average of the energy dependence of τ over the Maxwellian electron distribution.

$$\frac{\langle \mathcal{E}\tau \rangle}{\langle \mathcal{E} \rangle} \sim \frac{\langle \mathcal{E}^{\frac{3}{2}} \rangle}{\langle \mathcal{E} \rangle} = \frac{8(k_0 T)^{\frac{1}{2}}}{3(\pi)^{\frac{1}{2}}}$$

so that finally

$$\mu = \frac{16(2\pi)^{\frac{1}{2}}}{3} \cdot \frac{\hbar^2}{q m^{*\frac{3}{2}}(k_0 T)^{\frac{1}{2}}} \left\{ \sum_{\text{modes}} \left[\frac{(K^2)_{av}}{\epsilon}\right] \right\}^{-1}, \quad (III.3)$$

or in the laboratory units of cm²/v sec

$$\mu = 1.44(m/m^*)^{\frac{3}{2}}(300/T)^{\frac{1}{2}} \left\{ \sum_{\text{modes}} [(K^2)_{av}/\kappa] \right\}^{-1}. \quad (III.4)$$

If we use the spherical averages of K^2 given in the Appendix, we find that shear waves are responsible for the majority of the scattering for both ZnO and CdS. (The same result was obtained for cubic ZnS.[4,5]) The previous numerical values for piezoelectric scattering mobility[2] were apparently too low; we now find

$$\mu \cong 160(m/m^*)^{\frac{3}{2}}(300/T)^{\frac{1}{2}}, \quad \text{for} \quad \text{ZnO} \qquad (III.5)$$

and

$$\mu \cong 400(m/m^*)^{\frac{3}{2}}(300/T)^{\frac{1}{2}}, \quad \text{for} \quad \text{CdS}. \qquad (III.6)$$

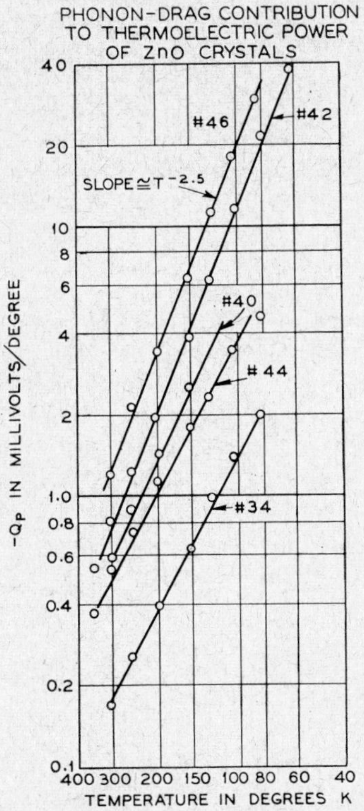

PHONON-DRAG CONTRIBUTION TO THERMOELECTRIC POWER OF ZnO CRYSTALS

FIG. 1. Q_p as a function of temperature for a number of "as-grown" ZnO crystals obtained by subtracting the electronic contribution to Q computed from carrier density and a density-of-states mass of $0.5\ m$.

solely to piezoelectric scattering. The Q_p data for ZnO are shown in Fig. 1. First, consider the $T^{-2.5}$ temperature dependence for the samples of highest purity. Substituting the $T^{-\frac{1}{2}}$ dependence of μ piezoelectric, in (IV.1) we find that $\bar{\tau}_p \sim T^{-2}$ is required to fit the data. The phonon scattering processes (other than boundary scattering) which randomize momentum in Herring's theory[7] are phonon-phonon scatterings due to the anharmonic lattice force constants. Herring has found the temperature and wave vector dependences for these processes to be

$$\tau_p \sim p^{-2-s} T^{-3+s+\gamma},$$

where

for small p $\begin{cases} s = 0, \text{ longitudinal} \\ s = -1, \text{ shear} \end{cases}$

and

for $T < \Theta$, $\gamma \to 0$ and $T > \Theta$ $\begin{cases} \gamma \to 2, \text{ longitudinal} \\ \gamma \to 3, \text{ shear.} \end{cases}$

Thus, for phonons whose wave vector scales with that of a thermal electron,

$$\tau_p \sim T^{-4+(s/2)+\gamma}$$

and a $\bar{\tau}_p \sim T^{-2}$ by this mechanism would require that γ be at its high-temperature asymptote for longitudinal waves or very close to it for shear waves for temperatures down to $60°K$. A different phonon scattering mechanism is required.

Rayleigh scattering, since it is proportional to p^4, would give the required $\bar{\tau}_p \sim T^{-2}$ for p scaled with the thermal electron wave vector. Purely mechanical scattering from point lattice imperfections or impurities is far too weak to limit τ_p in the wave vector range of interest. However, scattering of piezoelectric phonons by highly polarizable neutral donor states *can* be of sufficient magnitude. We may understand this process physically and derive a rough order of magnitude for the effect by analogy with the Rayleigh scattering of light.

For optical Rayleigh scattering, an incident wave produces a dipole of moment $\alpha E \exp(-j\omega t)$, where α is the polarizability. This dipole radiates spherical outgoing waves. The rate of energy radiation is[8] $v^{-3}\omega^4\alpha^2 E^2/12\pi\epsilon_0$. Since the energy density of the incoming wave is $(\frac{1}{2})\epsilon_0 E^2$, the proportional rate of loss of energy density is then

$$1/\tau = N\omega^4\alpha^2/12\pi\epsilon_0^2 v^3. \tag{IV.2}$$

In the piezoelectric case, an electric dipole radiates acoustic waves, and since this amounts to transducing from electrical to mechanical energy, the expression for the dipole radiation must be multiplied by some average

Screening, if important, can be taken into account very simply by reducing the appropriate values of K^2 by $\lambda^4 p^4 (1+\lambda^2 p^2)^{-2}$. Since p is roughly equal to the electron wave vector for phonons which contribute to the scattering, the screening affects the longitudinal and shear contributions to the scattering equally.

IV. PHONON DRAG

Herring[7] has shown that the phonon-drag contribution to thermoelectric power may be written

$$Q_p = -v^2 f \bar{\tau}_p / \mu T, \tag{IV.1}$$

where (f/μ) is the contribution to μ^{-1} from scattering mechanisms which transfer crystal momentum between long-wavelength acoustical phonons and electrons, v is the appropriate sound velocity, and $\bar{\tau}_p$ is a mean relaxation time for the long-wavelength phonons characterizing the rate at which they can get rid of crystal momentum.

The previous analysis of Q_p,[1] based upon deformation-potential electron scattering, required a very small electron effective mass $(0.07\ m)$ in order to explain the apparent decrease in Q_p with increasing carrier concentration by the "saturation effect." It simultaneously required an extraordinarily large deformation potential. We now wish to ascribe the phonon drag in ZnO

---

[7] C. Herring, Phys. Rev. **96**, 1163 (1954); and *Semiconductors and Phosphors* (Friedrich Vieweg & Sohn, Braunschweig, Germany, 1958).

[8] J. C. Slater and N. H. Frank, *Electromagnetism* (McGraw-Hill Book Company, Inc., New York, 1947), p. 159.

square of an electromechanical coupling constant for the outgoing wave. Note that v is the *outgoing* sound velocity. Similarly, since the oscillating dipole is driven only by the electrical part of the incoming wave, Eq. (IV.2) must also be multiplied by a K^2 characterizing the incoming wave. These complications with respect to incoming and outgoing waves come about because a dipole, in general, radiates both shear and longitudinal waves and may be driven by either. Also, because of the anisotropy of K, the scattering has a much more complicated directional dependence. In order to sidestep directional complications, we simply write the relaxation time to within geometrical factors as

$$\frac{1}{\tau} \cong \frac{N\omega^4 \alpha^2 (K_i^2)_{av}(K_o^2)_{av}}{v_0^3 \epsilon^2}, \quad (IV.3)$$

where i and o stand for "in" and "out."

To obtain the $\bar{\tau}_p$ applicable to phonon drag, the average wave vector of the incoming wave is set equal to that of an electron of energy $2.26 \, k_0 T$ (taking into account the energy dependence of the electron scattering), and the frequency can be written as

$$\omega^4 = v_i^4 p_i^4 = v_i^4 m^{*2} (2 \times 2.26)^2 (k_0 T)^2 \hbar^{-4}.$$

The polarizability of the donor centers may be written in terms of binding energy and effective mass as

$$\alpha = \hbar^2 m^{*-1} (E_D/q)^{-2}.$$

Making these substitutions in (IV.3),

$$\frac{1}{\tau} \cong \frac{N v_i^4 (2 \times 2.26)^2 (k_0 T)^2 (K_i^2)_{av} (K_o^2)_{av}}{v_0^3 (E_D/q)^4 \epsilon^2} \quad (IV.4)$$

is the phonon relaxation time. For this approximate treatment, the spherically averaged values from the Appendix may be used for K^2.

Since the longitudinal velocity is about 2.5 times the shear velocity for ZnO and CdS, and $(K^2)_{av}$ (shear) is considerably greater than $(K^2)_{av}$ (longitudinal), the dipole radiation is almost entirely shear. Furthermore, longitudinal waves have a relaxation time which is about 10% of that for the shear waves. Since Q_p may be divided into shear (s) and longitudinal (l) parts,

$$Q_{ps} = -\frac{v_s^2 \bar{\tau}_{ps}}{\mu_s T} \quad \text{and} \quad Q_{pl} = -\frac{v_l^2 \bar{\tau}_{pl}}{\mu_l T}, \quad (IV.5)$$

so that

$$Q_{ps}/Q_{pl} \cong (v_l/v_s)^2. \quad (IV.6)$$

A note of caution should be sounded with respect to (IV.6), and most of the other results as well, since they have all been derived by ruthless assumptions which have forced isotropy on an inherently anisotropic problem.

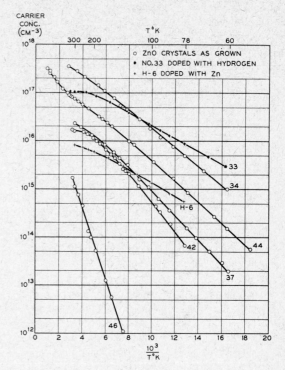

FIG. 2. Carrier concentration vs $10^3/T$ for some ZnO samples on which Seebeck measurements were made.

Returning to the data once more, we see that the T dependence of (IV.4) is in satisfactory agreement with the observed slope for the samples with large Q_p. A thorough comparison of theory with experiment would require a knowledge of neutral donor concentration and binding energy. This information can only be estimated for these samples, since the Seebeck data exist only for as-grown crystals. However, some information may be gleaned from the carrier concentration data obtained from Hall measurements and shown in Fig. 2. The trend of the data is unmistakable: samples with high Q_p have smaller donor concentrations and higher binding energies (though compensation may also be present in some cases, particularly for sample No. 46). It seems reasonable, therefore, to ascribe the variations in Q_p between samples to differences in the phonon relaxation time rather than the electron saturation effect.

The samples with the lowest Q_p, Nos. 44 and 34, have a distinctly smaller temperature dependence. This might come about through screening; however, a calculation based on $m^* = 0.32m$ (see Sec. V) indicates that screening is negligible throughout the entire temperature range. Furthermore, for these samples, the Debye length actually increases with increasing temperature for the higher temperatures. A more attractive explanation is that as the temperature increases, the electron (and hence phonon) wavelength starts to become comparable with the diameter of the neutral donors, so that the donors are less effective scatterers.

If samples Nos. 44 and 34 contain shallow donors ($E_D \cong 0.05$ ev), this hypothesis is born out by a calculation assuming $m^* \cong 0.32m$. Their log n vs T^{-1} slopes must then be the result of compensation.

Sample No. 42 is the only sample for which there is any real indication of the number of donors. It appears to have $N_D \cong 2 \times 10^{16}$ cm^{-3}, and assuming no compensation $E_D \cong 0.11$ ev. If these values are used together with $m^* = 0.32m$ in (IV.4) and (IV.5), order of magnitude agreement is obtained between the observed and calculated Q_p.

V. CONDUCTION BAND EFFECTIVE MASS IN ZnO

The history of the conduction band effective mass in ZnO was summarized in reference 1. At that time, there existed a density-of-states mass, for both spins, of $\approx 0.5m$ (assuming a simple 2-fold spin degeneracy for donor ground states); a mass of $0.27m$ from the binding energy of 0.51 ev of a number of *different* shallow donor states and the dielectric constant of 8.5 measured at 24 kMc; an apparent inertial mass of $0.07m$ from what was thought to be electron saturation of phonon-drag; and an inertial mass of $0.06m$ from a measurement of free-carrier reflectivity in the infrared by Collins and Kleinman.[9] The present theory of phonon drag in ZnO no longer supports the very small inertial mass. Furthermore, the binding energy of the exciton of 0.059 ev recently estimated by Thomas[10] from the $n=1$ and $n=2$ exciton states yields a reduced mass for the exciton of $0.29m$ using $\kappa^s = 8.2$,[9] and since the hole is presumed to be considerably heavier, this should not be too much smaller than the electron mass (which probably includes some of the polaron effects). The exciton work[11] and studies of the phonon-assisted exciton transitions which make up the long wavelength tail of the absorption edge[12,13] indicate rather conclusively that the conduction band minimum at $k=0$, which is observed in exciton formation, is the lowest conduction band minimum for both ZnO and CdS.

The mass obtained from the binding energy of shallow donors should be corrected (to some extent) for the piezoelectric contribution to the dielectric constant. It is tempting to apply this correction to the value of 8.5 found at 24 kMc, which may be larger than the infrared value[9] because of polarization associated with additional lattice modes between the acoustical and optical branches (resulting from the fact that ZnO possesses *two* zinc atoms and *two* oxygen atoms per unit cell). Thus, one arrives at a true κ (average) at zero *stress* of about 9.4 and, consequently, an effective mass of $0.33m$.

The mass of $0.38m$ deduced by Dietz et al.[13] from the phonon-assisted exciton absorption is in fair agreement with these considerations. We believe that the semiconducting properties of ZnO may be properly understood with a nearly isotropic conduction band mass of $\approx 0.32m$ (including polaron effects) for a minimum at $k=0$. The longitudinal magnetoresistance measured[1] along the hexagonal axis may be due to slight warping of the energy surface due to higher bands. The density-of-states mass may be somewhat increased by this warping, and may also be influenced by temperature dependence of the donor binding energy. We have at present no explanation for the infrared reflectivity mass.[9]

ACKNOWLEDGMENTS

The author is pleased to acknowledge a number of helpful discussions with C. Herring and J. J. Hopfield.

APPENDIX

For the approximate calculations of this paper, we use spherical averages of the square of the piezoelectric constant e, which determines the *longitudinal* electric field accompanying a plane strain wave, propagating in an arbitrary direction. These averages are obtained as follows:

For crystals of the Wurtzite structure, crystal class 6mm, there are three independent piezoelectric constants, e_{33}, e_{31}, and e_{15}. The three axis is the six-fold axis, and axes one and two are any pair of orthogonal axes lying in the plane perpendicular to the three axis. The nonzero components of the e_{ijk} tensor are then

$$e_{333} = e_{33}$$
$$e_{311} = e_{322} = e_{31}$$
$$e_{113} = e_{131} = e_{223} = e_{232} = e_{15}.$$

A spherical-polar coordinate system, r, θ, φ is now introduced where θ is the angle between the radius vector and the hexagonal crystal axis, and φ is the angle about that axis. For a longitudinal wave propagating in the r direction, e_{rrr} determines the longitudinal electric field. The longitudinal electric field accompanying a shear wave propagating along r is determined by $e_{r\theta r}$ and $e_{r\varphi r}$, where these constants are weighted by the cosines of the angles between the displacement direction of the shear wave and the θ and φ axes. Upon transforming the piezoelectric tensor to the new coordinate system, we find that

$$e_{rrr} = e_{33} \cos^3\theta + (e_{31} + 2e_{15})\cos\theta \sin^2\theta$$
$$-e_{r\theta r} = (e_{33} - e_{31} - e_{15})\cos^2\theta \sin\theta + e_{15} \sin^3\theta$$
$$e_{r\varphi r} = 0.$$

[9] R. J. Collins and D. A. Kleinman, J. Phys. Chem. Solids **11**, 190 (1959).

[10] D. G. Thomas, J. Phys. Chem. Solids **15**, 86 (1960).

[11] J. J. Hopfield and D. G. Thomas, Phys. Rev. **122**, 35 (1961).

[12] D. G. Thomas, J. J. Hopfield, and M. Power, Phys. Rev. **119**, 570 (1960).

[13] R. E. Dietz, J. J. Hopfield, and D. G. Thomas, J. Appl. Phys. **32**, 2282 (1961).

Table I. CdS "isotropic model".[a]

$v_l = 4.3 \times 10^3$ m/sec	$c_l = 8.8 \times 10^{10}$ Newtons/m²
$v_s = 1.8 \times 10^3$	$c_s = 1.54 \times 10^{10}$
$e_{33} = 0.49$ coul/m²	$\kappa^s = 9.7$
$e_{31} = -0.25$	$(K_s^2)_{av} = 0.031$
$e_{15} = -0.21$	$(K_l^2)_{av} = 0.0042$

[a] Elastic constants measured by H. J. McSkimin, T. B. Bateman, and A. R. Hutson, Abstract S-6, Sixty-first Meeting of Acoustical Society of America, May 10, 1961. Piezoelectric e constants computed from d constants measured by Hutson[2] and Jaffe et al.; κ^s from H. Jaffe, D. Berlincourt, H. Krueger, and L. Schiozawa, Proceedings of the 14th Annual Symposium on Frequency Control (U. S. Army Research and Development Laboratory, Fort Monmouth, New Jersey, May 31, 1960).

For longitudinal waves, the spherical average is then

$$(e_{rrr}^2)_{av} = \frac{1}{2} \int_0^\pi e_{rrr}^2 \sin\theta d\theta$$

$$= \frac{1}{7} e_{33}^2 + \frac{4}{35} e_{33}(e_{31} + 2e_{15}) + \frac{8}{105}(e_{31} + 2e_{15})^2.$$

For the shear waves, we first average over the angle between the displacement direction and the θ direction, taking into account the fact that two shear waves with orthogonal displacement vectors propagate in any direction. The spherical average for the two shear waves

Table II. ZnO "isotropic model".[a]

$v_l = 5 \times 10^3$ m/sec	$c_l = 14.1 \times 10^{10}$ Newtons/m²
$v_s = 2.1 \times 10^3$	$c_s = 2.47 \times 10^{10}$
$e_{33} = 1.1$ coul/m²	$\kappa^s = 8.2$
$e_{31} = -0.16$	$(K_s^2)_{av} = 0.062$
$e_{15} = -0.31$	$(K_l^2)_{av} = 0.012$

[a] Longitudinal sound velocity measured by Hutson[2] remaining elastic constants scaled from CdS. Piezoelectric d constants measured by Hutson[2] and e constants computed using scaled c's. κ^s from Collins and Kleinman.

taken together is then

$$(e_{r\theta r}^2)_{av} = \frac{1}{2} \int_0^\pi e_{r\theta r}^2 \sin\theta d\theta$$

$$= \frac{2}{35}(e_{33} - e_{31} - e_{15})^2$$
$$+ \frac{16}{105} e_{15}(e_{33} - e_{31} - e_{15}) + \frac{16}{35} e_{15}^2.$$

For our approximate model in which the crystal is assumed to be elastically and dielectrically isotropic, the spherically averaged squares of electromechanical coupling constants are

$$(K_l^2)_{av} = (e_{rrr}^2)_{av}/\epsilon c_l; \quad (K_s^2)_{av} = (e_{r\theta r}^2)_{av}/\epsilon c_s.$$

The data employed in this "isotropic model" for CdS and ZnO are given in Tables I and II.

JOURNAL OF APPLIED PHYSICS   SUPPLEMENT TO VOL. 32, NO. 10   OCTOBER, 1961

Double Phonon Processes in Cadmium Sulfide*

Minko Balkanski and Jean Michel Besson
Laboratoire de Physique de l'Ecole Normale Supérieure, Paris, France

The bands of infrared absorption in hexagonal cadmium sulfide have been studied over a wide range of temperatures and are assigned to double-phonon processes. An analysis of these absorption peaks on the basis of four frequencies is proposed. The temperature-dependence of the location and intensity of these bands give indications upon the anharmonic parameter and Grüneisen constant of the crystal. Anharmonic forces are shown to be responsible for the coupling between normal vibrational modes.

I. INTRODUCTION

THE properties of hexagonal cadmium sulfide have already been extensively investigated by spectroscopic methods, both in emission and in absorption. Those measurements were, as a rule, directed towards the study of the intrinsic absorption edge and its fine structure. Numerous data are also available on emission properties at different temperatures in pure or doped crystals.

Those studies did not extend far beyond the wavelength limit of visible light. Collins[1] was the first to give

information upon optical absorption on fundamental modes of vibration of the lattice. Afterwards, Francis and Carlson[2] studied the refractive index of cadmium sulfide up to 15 μ, where a zone of absorption takes place.

In this work, we have extended the study of the infrared absorption spectrum of hexagonal cadmium sulfide between 15 and 30 μ. The absorption peaks which we observed are assigned to coupled phonons in the crystalline lattice. This coupling is rendered possible by the presence of anharmonic terms in the expression of the crystalline field.

* This work was supported by the Office of Scientific Research, Air Research and Development Command, U. S. Air Force.
[1] R. J. Collins, J. Appl. Phys. 30, 1135 (1959).

[2] A. B. Francis and A. I. Carlson, J. Opt. Soc. Am. 50, 118 (1960).

In crystals which do not exhibit a first-order electric moment, Lax and Burstein[3] showed that phonon combination bands could occur through higher-order terms in the expression of the electric moment. This was verified experimentally by the analysis of the infrared spectrum of silicon by Johnson.[4]

On the other hand, radiation damage may materialize into a local destruction of the crystal symmetry and thus induce an over-all electric moment capable of interfering with electromagnetic field. Nazarewicz[5] has thus observed absorption on the simple modes of silicon, the combination of which had previously been calculated by Johnson.

Ionic crystals which possess an electric moment show an absorption on normal modes, which is observable in the infrared. Those normal modes may be coupled through anharmonic terms in the expression of the lattice potential. This possibility was first discussed by Born and Blackman,[6] then investigated in more detail by Barnes, Brattain, and Seitz.[7] Recently, the data of Kleinman and Spitzer[8] on gallium phosphide, and of Johnson[9] on indium antimonide confirmed the theory, at least as regards the addition of normal frequencies.

Cadmium sulfide is only partially ionic, and there is no way to discriminate between the contribution of anharmonic forces, and that of second-order electric moment. Nevertheless, one may notice that the absorption due to coupled modes is much weaker in homopolar crystals than in ionic crystals.[6] For this reason, we studied cadmium sulfide under the assumption that all absorption came from anharmonic forces, which is verified by the intensity of the observed peaks.

II. THEORETICAL BASIS

In a crystal the atoms of which are bound by harmonic forces, the vibrational states of the lattice may be represented by those of an assembly of independent oscillators through the normal transformation,[10–12]

$$\mathbf{r}_{sn} = (2N_0 m_n)^{-\frac{1}{2}} \sum_k \sum_{t=1}^{3n_0} (a_{kt} + ia_{kt}') \mathbf{B}^n_{kt} \exp[i(\mathbf{k}\cdot\mathbf{s})], \quad (1)$$

where $\mathbf{r}_{sn}$ is the displacement from equilibrium of the nth atom in cell s, N_0 is the number of unit cells, m_n is the mass of atom n, n_0 is the number of atoms per unit cell, t is the branch of the dispersion curve, $\mathbf{k}$ is the wave vector of the mode, $\mathbf{B}^n_{kt}$ is the complex polarization and phase vector of the plane wave, and $a_{kt} + ia_{kt}'$ is the complex amplitude of the harmonic oscillator.

[3] M. Lax and E. Burstein, Phys. Rev. 97, 39 (1955).
[4] F. Johnson, Proc. Phys. Soc. (London) 73, 265 (1959).
[5] M. Balkanski and W. Nazarewicz, J. Chem. Phys. Solids (to be published).
[6] M. Born and M. Blackman, Z. Physik 82, 551 (1933).
[7] B. B. Barnes, R. R. Brattain, and F. Seitz, Phys. Rev. 48, 582 (1935).
[8] W. G. Spitzer and D. A. Kleinman, Phys. Rev. 118, 110 (1960).
[9] S. J. Fray, F. A. Johnson, and R. H. Jones, Proc. Phys. Soc. (London) 76, 939 (1960).

When one of those oscillators, of energy $(n_{kt}+\frac{1}{2})\hbar\omega$, changes its quantum number by one unit of energy $\hbar\omega$, it is said to have absorbed or emitted one phonon. Operators of creation and annihilation of phonons may be defined as follows:

$$a_{kt} = \tfrac{1}{2}(\omega_{kt}/\hbar)^{\frac{1}{2}}(a_{kt} - ia_{kt}') + \tfrac{1}{2}i(\hbar\omega_{kt})^{-\frac{1}{2}}(p_{kt} - ip_{kt}') \quad (2)$$

$$a_{kt}^+ = \tfrac{1}{2}(\omega_{kt}/\hbar)^{\frac{1}{2}}(a_{kt} + ia_{kt}') - \tfrac{1}{2}i(\hbar\omega_{kt})^{-\frac{1}{2}}(p_{kt} + ip_{kt}'), \quad (3)$$

ω_{kt} is the pulsation and p_{kt} the conjugate momentum of a_{kt}. Using the inverse forms of (2) and (3), the harmonic Hamiltonian becomes:

$$H = V + \tfrac{1}{4} \sum_k \sum_{t=1}^{3n_0} \hbar\omega_{kt}(a_{kt}^+ a_{kt} + \tfrac{1}{2}), \quad (4)$$

V is the zero energy of the crystal and the sum on $\mathbf{k}$ is extended to the first Brillouin zone. Operators a^+ and a, which express a transition from a state n_k to a state n_k+1, have the following influence on functions $\Psi(n)$:

$$\begin{cases} a_k^+ | \Psi(n_k)\rangle = (n_k+1)^{\frac{1}{2}} | \Psi(n_k+1)\rangle \\ a_k | \Psi(n_k)\rangle = (n_k)^{\frac{1}{2}} | \Psi(n_k-1)\rangle \end{cases} \quad (5)$$

$$\begin{cases} \langle \Psi(n_k-1) | a_k | \Psi(n_k)\rangle = (n_k)^{\frac{1}{2}} \\ \langle \Psi(n_k-1) | a_k^+ | \Psi(n_k)\rangle = (n_k+1)^{\frac{1}{2}}. \end{cases} \quad (6)$$

The method of independent oscillators may be used in the case where a weak coupling exists between them, that is, if terms of the third degree and above are small in the expression of lattice potential. The general form for an anharmonic potential of the third degree is:

$$V' = \sum_k \sum_t \mathbf{B}_{kt}(a_{k_1,t_1} + a^+_{-k_1,t_1})(a_{k_2,t_2} + a^+_{-k_2,t_2}) \\ \times (a_{-k_3,t_3} + a_{k_3,t_3}^+) \times \Delta_{k_1+k_2+k_3}. \quad (7)$$

The quantity Δ is defined as:

$$\frac{1}{N} \sum_s \exp i(\mathbf{k}\cdot\mathbf{s}) = \Delta_k. \quad (8)$$

Here, it means that only those states which have the same wave vector as the initial state, may be generated.

In this work, we only considered the case of a crystal with two kinds of atoms where anharmonic forces are central and affect only the four nearest neighbors of opposite sign. A calculation which proves satisfactory, considering the data we could obtain, has been made by Kleinman[13] for gallium phosphide. An application of this calculation to the case of cadmium sulfide where the four nearest neighbors are not the same distance, is discussed further on.

[10] J. M. Ziman, Electrons and Phonons (Clarendon Press, Oxford, England), Chap. III.
[11] L. Van Hove, Tech. Rept. No. 11, Massachusetts Institute of Technology, Cambridge, Massachusetts, 1959.
[12] F. Seitz, Théorie moderne des solides (Masson et Cie, Paris, France, 1949).
[13] D. A. Kleinman, Phys. Rev. 118, 118 (1960).

III. EXPERIMENTAL PROCESS

Preparation of Cadmium Sulfide Samples

The samples we used for final measurements were not intentionally doped. They exhibited a high resistance at all temperatures and illuminations, lying between 2×10^8 and 10^5 ohms$\times$cm. In preliminary measurements, we checked the samples for free carriers. The absorption coefficient due to free carriers should follow[14] a law in $\lambda^{3.5}$ for a dispersion on impurity centers, and a law in $\lambda^{1.5}$ for a dispersion on lattice vibrations. Measurements up to three microns on intentionally doped crystals showed a variation of the form $\alpha = A\lambda^a$, with a lying between 2.5 and 3.2. At higher wavelengths, a lies between 3 and 3.5, which points to a dispersion on impurity centers, for the most part.[15] We found the absorption coefficient between 3 and 14 μ to be less than 0.5 cm^{-1}. Using the values given by Francis and Carlson,[2] and the above wavelength dependence, we took the absorption on free carriers, as negligible to within one cm^{-1} in the 20 to 25-μ region where we observed peaks of combination. We made measurements on different samples, and on various regions of a non-uniformly doped crystal. Those measurements failed to show a difference in intensity of the observed peaks.

The samples we used came from different sources. Final data were obtained on slabs cut from crystals grown by D. R. Reynolds. The crystals were cut either parallel or perpendicular to the c axis, to detect any anisotropy in absorption.

At high temperatures, the sample was fixed to a copper heating rod and the temperature was given by a thermocouple pressed against the sample itself. Low-temperature data were obtained with a double-envelope Dewar.

Experimental Technique

We used a 112-C, Perkin-Elmer, single-beam, double-pass, prism monochromator with a Globar source. Except for a few inches, all the optical path was flushed with dry nitrogen. The detection was made with a cesium bromide window thermocouple. The modulated signal is amplified by a Perkin-Elmer 13 cycles amplifier and fed into a Philips P.R. 4060 recorder.

The calibration of the prisms was made according to the method of Downie, Magoon, Purcell, and Crawford,[16] using the molecular absorption bands of different compounds.

In those experiments, slabs ranging from 1.5 to 0.2 mm were used. Although we ground slabs down to 0.05 mm, it seems difficult to obtain good quality thin samples because of the fragility of cadmium sulfide crystals.

At low temperatures, we compensated for thermal expansion displacement of the sample-holder by noting the values of apparent transmission at different temperatures in the wavelength regions where the sample is well transparent and where the crystal transmission is not affected by the temperature-dependence of the absorption coefficient. At high temperatures, we were limited by the emission of the sample itself, which mixes with transmitted radiation, since the light beam was modulated only after passing through the crystal.

Reflection and transmission data have shown that reflectivity does not vary with wavelength in the region of the combination bands. Therefore, the temperature-dependence of the intensity of absorption peaks, arises mostly from variations of the absorption coefficient of the crystal. Thus, we studied reflectivity at room temperature only. The absorption coefficient α, and reflectivity ρ, were calculated from transmission T and reflection R by the relations[17]

$$T = \frac{(1-\rho)^2 e^{-\alpha x}}{1 - \rho^2 e^{-2\alpha x}} \quad R = \rho\left[1 + \frac{(1-\rho)^2}{e^{-2\alpha x} - \rho^2}\right]. \quad (9) \quad \text{and} \quad (10)$$

Measurements on Evaporated Layers

The transmission of thin films has also been investigated in the region where, according to Collins,[1] one should find absorption on the T.O. mode.

Evaporated films of cadmium sulfide on high purity silicon layers, were annealed at high temperature in CdS powder. Such layers were shown by Chaves and Pastel[18] to be oriented with a c axis lying normal to the surface. On such layers, a strong absorption peaks must occur fairly close to the calculated T.O. frequency.

IV. EXPERIMENTAL RESULTS

Reflectivity

On the short wavelengths side, the reflectivity curve meets the values obtained from the refractive index

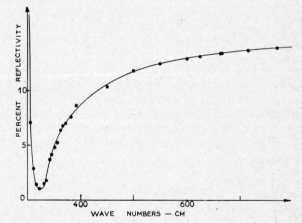

FIG. 1. Reflectivity of CdS at room temperature.

[14] H. Y. Fan, Repts. Progr. in Phys. **19**, 107 (1956).

[15] W. W. Piper and D. T. F. Marple, Bull. Am. Phys. Soc. **6**, 17 (1961).

[16] A. R. Downie, M. C. Magon, T. Purcell, and B. Crawford, Jr., J. Opt. Soc. Am. **43**, 941 (1953).

[17] F. Oswald and R. Schade, Z. Naturforsch. **9a**, 611 (1954).

[18] M. R. Chaves and C. A. J. Pastel (unpublished).

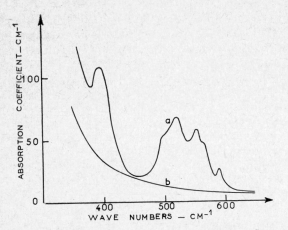

FIG. 2. Absorption coefficient at room temperature. Curve a represents the experimental data. Curve b is the absorption coefficient due to fundamental resonance, calculated from Eq. (12).

data of Francis and Carlson,[2] up to 700 cm^{-1} (Fig. 1). The 0.01 minimum at 320 cm^{-1} fits with Collins' result, of 305 cm^{-1} for the L.O. mode. Reflectivity data were used to compute the value of n, using relation (11) which holds for low extinction coefficients,

$$R = \left(\frac{n-1}{n+1}\right)^2. \qquad (11)$$

Correction for the Fundamental Absorption Tail

In order to evaluate the absorption coefficient due only to combination bands, the absorption due to fundamental resonance must be substracted. On the edges of the resonance, it is given by:

$$\alpha_f = F(1/n)[\lambda_0\lambda^2/(\lambda_0^2-\lambda^2)^2], \qquad (12)$$

where n is the refractive index, $\lambda_0 = 41.5\ \mu$ is the wavelength of the T.O. mode, and F is a constant depending of the width of resonance and dielectric constants.

Absorption Coefficient at Room Temperature

No difference was detected between slabs cut at various angles with the c axis. Therefore, the following results apply both to samples cut parallel, and perpendicular with respect to the axis.

We observed four well-defined peaks, flanked by weak shoulders (Fig. 2). Those peaks are located at wave numbers 590.5, 553.5, 520.5, and 398 cm^{-1}. It is difficult to analyze them into separate peaks since those maxima correspond to singularities in the dispersion curve and may not have a simple mathematical shape.

Temperature-Dependence of the Absorption Spectrum

As the temperature is lowered, the intensity of the peaks diminishes, and their maxima are shifted towards higher wave numbers. Moreover, the peaks become more separated (Fig. 3). We studied this shift in temperature for all peaks except the one at 398 cm^{-1}, for lack of energy in this region. The maxima of the bands which appear weakly have been determined graphically, assuming a symmetrical shape.

We could not study precisely the temperature-dependence of the intensity of all peaks since it is difficult to separate them, because of their strong overlap. Only for the most intense peak, at 520.5 cm^{-1}, is it possible to compare the temperature-dependence of the absorption coefficient (Fig. 4) with theory. For this peak, the agreement is good between experimental data and a theoretical curve for two phonons of equal energy.

Fundamental Absorption on Evaporated Layers

On layers ranging from 0.5 and 1.8 μ in thickness, we found a strong absorption peak at 240.5 cm^{-1}, at room temperature. The order of magnitude of this peak and its location, point to assigning it to the T.O. vibration. As it compares favorably with the 241 cm^{-1} value given by Collins,[1] we accepted this value for subsequent calculations.

V. ANALYSIS OF THE RESULTS

Assignment of Combination Bands

From our preliminary measurements, we concluded that it was not possible to assign the observed peaks to intraband transitions, or to transitions towards impurity states. Thus we were led to assigning them to lattice vibrations transitions which are made possible by coupling of normal modes through anharmonic terms in the expression of the lattice potential.

At liquid hydrogen temperature, the resolution of bands is good enough to allow analysis. In Table I, we propose a decomposition mechanism of the peaks based on four frequencies, and comparison with experiment.

The following remarks should be made, concerning

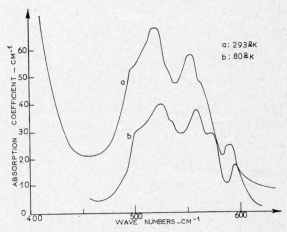

FIG. 3. Absorption coefficient at two different temperatures. This diagram does not include the last peak towards small wave numbers, which was not studied at low temperatures.

TABLE I. Proposed assignment of the peaks on the basis of four frequencies, at 21°K. Basic frequencies: $\alpha = 299.5 \pm 0.5$ cm^{-1}, $\beta = 281.5 \pm 0.5$ cm^{-1}, $\gamma = 262.5 \pm 1$ cm^{-1}, $\delta = 201.5 \pm 2$ cm^{-1}.

Frequency combinations	Calculated maxima cm^{-1}	Estimated error cm^{-1}	Measured maxima cm^{-1}	Experimental error cm^{-1}	Remarks
$\alpha+\alpha$	599	0.5	599	0.5	
$\alpha+\beta$	581	1	579	2	
$\alpha+\gamma$	562	1.5	562	1	$\alpha+\gamma$ and $\beta+\beta$ give the same frequency
$\alpha+\delta$	501	2.5	500	1	
$\beta+\beta$	563	1	562	1	
					$\alpha+\gamma$ and $\beta+\beta$ give the same frequency
$\beta+\gamma$	544	1.5	540	6	weak peak
$\beta+\delta$	483	2.5	...	...	not observed
$\gamma+\gamma$	525	1	524	1	
$\gamma+\delta$	464	3	...	...	not observed
$\delta+\delta$	403	4	403	4	extrapolated from room temperature data

this analysis: Peak $\delta+\delta$ could not be studied, at low temperatures, as pointed out before. The value 403 cm^{-1} was extrapolated, at 20°K, from its 398.5 cm^{-1} value at room temperature.

Out of ten predicted combinations, eight are observed, and careful measurements between 490 and 450 cm^{-1} failed to exhibit the other two combinations. Those two combinations imply frequency δ, which only gives a weak band with frequency α.

If we compare frequencies α, β, and γ to the values of optical frequencies for zero wave vector, we find that they fall in the probable optical mode region. The first frequency $\alpha = 295$ cm^{-1} is taken as the L.O. frequency.

On the other hand, δ is well separated from the other frequencies and seems to combine with difficulty to the other modes. This led us, from lack of other information, to consider it as an L.A. mode.

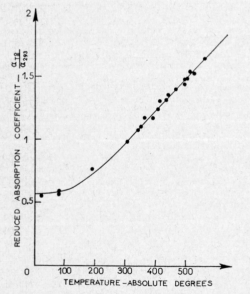

FIG. 4. Temperature dependence of the absorption coefficient. Peak $\gamma+\gamma$ at 520 cm^{-1}. ——: theoretical curve from Eq. (14). Dark circles: experimental data.

At room temperature, we propose the values of Table II for the frequencies of modes on the edges of the first Brillouin zone.

Temperature-Dependence of the Absorption Coefficient

The bands that we studied come from a combination of two phonons of opposite and equal wave vectors. The matrix element for creation of one pair is from Eqs. (5) and (6), proportional to $(n_{+k}+1)^{\frac{1}{2}}(n_{-k}+1)^{\frac{1}{2}}$ for absorption and to $(n_{+k})^{\frac{1}{2}}(n_{-k})^{\frac{1}{2}}$ for emission, with:

$$n_k = [\exp(h\nu_k/KT) - 1]^{-1}. \qquad (13)$$

The absorption coefficient will be proportional to the difference of the squares of matrix elements, which gives, for two phonons of frequencies ν_1 and ν_2:

$$\alpha(T) \propto 1 + [\exp(h\nu_1/KT) - 1]^{-1}$$
$$+ [\exp(h\nu_2/KT) - 1]^{-1}. \qquad (14)$$

The variation of $\alpha(T)$ is given in Fig. 4, compared to the theoretical curve for $\nu_1 = \nu_2 = 260$ cm^{-1}.

Evaluation of the Anharmonic Parameter

Although cadmium sulfide has the wurtzite structure, our data do not show a dependence on orientation of the samples. Since the anisotropy appears to be weak, it is possible, for verifications of orders of magnitude, to make calculations assuming the nearest four neighbors to be at the same distance, and in the same way as for a cubic crystal. Kleinman[13] calculated the relation between the anharmonic parameter, and the integrated absorption on optical modes, assuming the different optical-optical combinations produce equal absorption.

Using his results and taking $\nu_k = 1.15 \nu_f$, we have the relation:

$$I = \int \alpha d\lambda = \frac{8\pi^3 B^2 N \hbar \hat{e}^2}{m^4 \omega_f^7 \epsilon_0^{\frac{1}{2}}}(1+2n), \qquad (15)$$

ν_k = mean value of optical modes at the edge of the Brillouin zone, found experimentally equal to 277 cm^{-1};

ν_f = fundamental mode wave number, 241 cm^{-1};

N = concentration of ion pairs, 1.99×10^{22} cm^{-3};

$\hat{e}$ = effective charge, $\hat{e} = [(\epsilon_\infty - \epsilon_0)m/(2\pi N)]^{\frac{1}{2}}\omega_f = 2.27e$;

m = reduced mass, 41.4×10^{-24} g;

ϵ_0 = high-frequency dielectric constant, 5.8 (reference 2);

ω_f = fundamental pulsation, 4.54×10^{13} sec^{-1};

n = number of phonons [Eq. (13)].

$\int \alpha d\lambda$ is the integrated absorption. We derived it from the absorption curve at room temperature, discarding the peaks at 398 and 503 cm^{-1}, which arise from acoustic modes. We get:

$$\int \alpha d\lambda = 16\times 10^{-3}.$$

This value is much stronger than for silicon[4] and double than for gallium phosphide,[8] which fits with the strongly ionic bonding of cadmium sulfide. It compares with that of silicon carbide,[19] which is 20×10^{-3}. With the values given above, we obtain, for the anharmonic parameter:

$$B=2\times10^{12}\,\text{g}\times\text{cm}^{-1}\times\text{sec}^{-2}.$$

This order of magnitude seems acceptable.

Temperature Shift of Absorption Peaks

In a linear model of atoms bonded by anharmonic forces, it is possible to evaluate the modification of the anharmonic constant caused by thermal expansion.[13] For a potential of the form Ax^2+Bx^3, an increase ϵ in the crystal parameter will lead to an increase $\Delta\omega$ in frequency:

$$\Delta\omega/\omega = -3(B\epsilon/A). \qquad (16)$$

From (16), we may take, for a linear chain[13]:

$$(1/\omega)(\Delta\omega/\Delta T) = -12(Ba\alpha/M\omega_{LA}^2), \qquad (17)$$

where a is the crystal parameter, α the thermal expansion coefficient, M the mass of atoms of opposite signs, ω_{LA} the frequency of the L.A. mode on the edges of the Brillouin zone. We took $\alpha_m = \frac{1}{3}(\alpha_{11c}+2\alpha_{\perp c})$. Dilatometric measurements gave us: $\alpha_{11c}=3.4\times10^{-6}/°\text{K}$, which compares with the values of Francis and Carlson[1] which are: $\alpha_{11c}=3.5\times10^{-6}/°\text{K}$ and $\alpha_{\perp c}=5\times10^{-6}/°\text{K}$.

We took $\alpha_m=4.5\times10^{-6}/°\text{K}$, $a=5.82\times10^{-8}$ cm, $M=2.4\times10^{-22}$ g, $\omega_{LA}=3.81\times10^{13}$ sec^{-1}, $B=2\times10^{12}$ g$\times$cm$^{-1}\times$sec^{-2}. The value for thermal shift is then

$$(1/\omega)(\Delta\omega/\Delta T) = -2\times10^{-5}/°\text{K}.$$

Experimental values (Fig. 5) range from 1.5 to $5.5\times10^{-5}/°\text{K}$.

From those data, an approximate value for the Grüneisen constant may be derived, in an isotropic approximation

$$\gamma = -(d\log\omega/d\log V) \approx -3(1/\omega)(d\omega/dT)(1/\alpha), \qquad (18)$$

TABLE II. Frequencies of vibration at the edges of the first Brillouin zone at room temperature, on the basis of the analysis of Table I.

Frequency	Wave number	Assignment
α	295 cm^{-1}	L.O.
β	276 cm^{-1}	} Optical
γ	260 cm^{-1}	} branches
δ	199 cm^{-1}	L.A.

[19] W. G. Spitzer, D. A. Kleinman, and D. Walsh, Phys. Rev. **113**, 127 (1959).

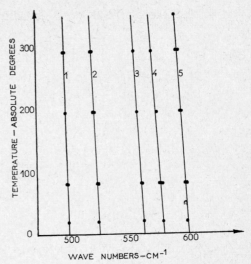

FIG. 5. Temperature shift of peaks.

Peaks Nos.	1	2	3	4	5
$\dfrac{1}{\omega}\dfrac{\Delta\omega}{\Delta T}=$	1.5×10^{-5}	(3×10^{-5})	(4×10^{-5})	5×10^{-5}	(5.5×10^{-5})

The safest values are in parentheses.

with $(1/\omega)(d\omega/dT)=4.2\times10^{-5}/°\text{K}$, average value of the best three results (Fig. 5), we get for the Grüneisen parameter:

$$\gamma \approx 3\pm1.$$

VI. CONCLUSIONS

Temperature-dependence of the observed bands tends to assign the absorption bands of cadmium sulfide in the 15 to 30-μ region, to double-phonon transitions. It is possible to analyze them on the basis of four frequencies. Thus, one obtains reasonable values for the anharmonic parameter and the temperature shift of the bands. Moreover, the magnitude of integrated absorption on optical modes shows the following progression:

$$\text{Si}\ll\text{GaP}<\text{CdS}<\text{SiC}.$$

These four materials are, in this way, written in order of increasing ionicity, which suggest the role played in this absorption by anharmonic forces.

The spectroscopic method does not allow discrimination in different directions of **k** space, for dispersion. Nevertheless, it might be interesting to look for substraction peaks in the far infrared, since such peaks, which are predicted by theory have seldom if ever been observed. Also, a more careful study in polarized light may show anisotropy effects in the double-phonon spectrum and reststrahlung curves.

ACKNOWLEDGMENT

It is a pleasure to acknowledge the help of Dr. D. R. Reynolds for providing CdS single crystals and encouraging this work.

JOURNAL OF APPLIED PHYSICS   SUPPLEMENT TO VOL. 32, NO. 10   OCTOBER, 1961

Excitons and Band Splitting Produced by Uniaxial Stress in CdTe

D. G. Thomas

Bell Telephone Laboratories, Incorporated, Murray Hill, New Jersey

A single exciton peak may be seen in the reflection spectrum of CdTe, a semiconductor with the cubic zinc-blende structure. Some properties of this exciton and of the associated fluorescent phenomena are described. Under uniaxial compressive stress, the single exciton peak splits into two peaks corresponding to the splitting of the $J=\frac{3}{2}$ valence band. The movement of the peaks as a function of stress has been determined for four directions of stress in the (110) plane. Within experimental error, the splittings are identical for a given stress applied in any direction. Thus, although the material is elastically anisotropic, the splitting may be described by one rather than two deformation potentials. These conclusions are consistent with the polarization properties of the exciton transitions. The polarization properties also show that under compressive stress the $M_J=\pm\frac{1}{2}$ band moves "up" and the $M_J=\pm\frac{3}{2}$ band moves "down."

INTRODUCTION

IN the study of the direct exciton absorption spectrum of germanium, it was found that strain introduced by gluing a thin slab of crystal to a glass substrate caused the exciton line to split into two lines. The mean energy of these lines was also shifted from the position of the single line of an unstrained crystal. Kleiner and Roth[1] pointed out that the strain in the crystal could split the degenerate valence band at $k=0$ (the center of the Brillouin zone) into two bands, each of which could give rise to separate exciton states. Suitable measurements enable three deformation potentials to be obtained. Although the sharp exciton lines are well suited to the measurement of band gap energies, gluing a slab to a substrate has certain disadvantages as a method of introducing strain. It is not easy to vary the stress, and since an isotropic strain in the plane of the slab is equivalent to the presence of a uniaxial stress perpendicular to the slab, it is only possible to use light which is polarized perpendicular to the stress direction. Frequently, direct exciton transitions are associated with high absorption coefficients so that thin samples must be used and it is difficult to introduce compressional strain into such samples except by the method described. An exception to this are the weak lines of the yellow exciton series of Cu_2O which can be split and shifted by the application of uniaxial stress to bulk samples.[2] If, however, the exciton lines are sufficiently strong and narrow, they can appear as anomalies in the reflection spectrum of the material; since the thickness of the crystal is no longer of importance for the optical measurements, crystals suitable for the application of stress can now be prepared and can be studied with light polarized parallel and perpendicular to the stress direction. This paper describes the application of this method to determine some of the band parameters of CdTe which has the cubic zinc-blende structure. Some of the exciton characteristics in CdTe are also discussed.

EXPERIMENTAL PROCEDURE

Crystals of CdTe can be grown from the melt[3,4] and from the vapor.[4] Ingots obtained by lowering the material through a temperature gradient are usually polycrystalline, but sizable single crystals can usually be cleaved from the ingot; frequently, such crystals contain twinned regions which can be recognized in light reflected from sandblasted surfaces. Vapor grown crystals usually occur as untwinned distorted dodecahedra, growth taking place by addition to (110) faces; under certain circumstances, vapor growth can take place on (111) faces leading to microtwinning and crystals which superficially appear to have hexagonal symmetry.

The amplitude of the reflection anomalies caused by the excitons depends critically on the crystal perfection and particularly on the state of the surface at which reflection occurs. A polished surface gives no indication of reflection peaks. The natural surfaces of vapor grown crystals give good results as do cleaved surfaces from most melt grown crystals. Etched surfaces may also be used[5] but for the bulk of this work cleaved surfaces were employed. The crystals cleave to give (110) faces which is fortunate for the present purposes, since these faces contain the [100], [111], and [110] directions which are the preferred directions for the application of stress. The cleavage properties may be used to orient the crystals; if two cleaved surfaces meet with an interior angle of 120°, the line of intersection is a [111] direction, and if the angle is 90° the line of intersection is a [100] direction. The [110] direction lies at right angles to the [100] direction. Crystals were oriented in this manner and the results were checked with x-ray analysis. Sometimes, cleaved crystals contained twinned regions which had a common cleaved (110) face with the remainder of the crystal. Under these circumstances, twinning can only occur on the (111) planes which lie at right angles to the [111] directions which are contained in a particular (110) face. Twinning takes place by a 60° rotation of the lattice about one of these directions

[1] W. H. Kleiner and L. M. Roth, Phys. Rev. Letters **2**, 334 (1959).

[2] A. F. Gross and A. A. Kapylanskii, Fiz. Tverdogo Tela **2**, 1676 (1960) [translation Soviet Physics—Solid State **2**, 1518 (1961)]; **2**, 2968 (1960) [translation, Soviet Physics—Solid State **2**, 2637 (1961)].

[3] D. de Nobel, Phillips Research Repts. **14**, 361, 430 (1959).

[4] R. T. Lynch (to be published).

[5] T. Ichimiya, T. Niimi, K. Mizuma, O. Mikami, Y. Kamuja, and K. Ono, *Solid State Physics in Electronics and Telecommunications* (Academic Press Inc., New York, 1960), Vol. 2, Part 2, p. 845.

giving a (110) face in the twin parallel to the original (110) face. The twinning planes therefore lie at right angles to the cleaved (110) faces.

The spectra were obtained by imaging a tungsten filament onto the crystal which was immersed in the refrigerant (usually liquid hydrogen). The reflected beam was then focussed onto the entrance slit of a Bausch and Lomb spectrograph equipped with a 15 000 line/inch grating blazed at 1 μ giving a linear dispersion of about 8 A/mm. The spectra were recorded photographically on Kodak 1N plates and microphotometer traces of the plates were made. Short wavelengths were removed with a filter, and polaroid sheet was used as an analyzer. Fluorescence was excited using filtered light from a high-pressure mercury arc.

The crystals were strained by applying a compressive stress in the [100], [111], and [110] directions. For

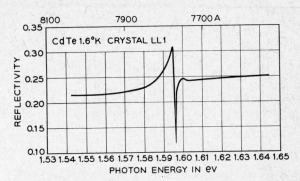

FIG. 2. The reflectivity of an unstrained vapor grown crystal at 1.6°K as determined from a photographic plate. Similar curves are obtained from melt grown crystals. At 20°K the reflection peak is only slightly broadened.

this purpose, a large crystal was selected with a cleaved surface of good quality. It was oriented in the fashion just described. This surface was protected by waxing it to a piece of glass; the opposite surface was ground and polished to give a slab about 0.043 cm thick. A rectangular ultrasonic cutter was used to cut oriented pieces from the slab 0.109 by 0.158 cm. Stress was subsequently applied along the longer direction of the rectangles parallel to the cleaved face. The crystal faces normal to the applied stress were polished flat and parallel by waxing the crystal into a slot cut in a piece of steel the thickness of which was slightly less than the length of the crystal. The ends of the crystal projected slightly from both metal surfaces and these were polished parallel to the metal surfaces reducing the length to 0.125 cm. In these operations, care was taken not to damage the cleaved surface.

Stress was applied to the crystals in the apparatus illustrated in Fig. 1. At each end of the crystal there was a piece of paper, then a piece of hardened steel with a rounded end which was in contact with another rectangular piece of steel. These precautions were taken in an attempt to produce uniform stress throughout the crystal. The plunger and rod transmitted the force to the crystal from weights applied outside the Dewar. The stress experiments were performed near 20°K. A stream of purified helium was passed down the supporting tube and entered the liquid hydrogen just above the crystal. This prevented the occurrence of gas bubbles near the crystal which could obscure the light reflected from the rather small crystal surface.

RESULTS

I. Unstrained Crystals

Reflection

Figure 2 shows the reflectivity from a vapor grown crystal at 1.6°K. Similar curves could be obtained from melt grown crystals. The values of the reflectivities are not very accurate since the data is taken from a calibrated but highly nonlinear photographic plate. The

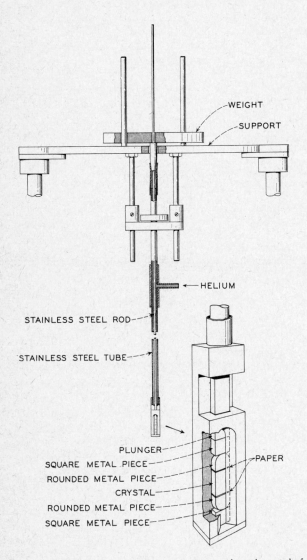

FIG. 1. Apparatus used to apply stress to an oriented crystal of CdTe. The lower section of the crystal holder is immersed in liquid hydrogen. Light is reflected from the exposed crystal surface.

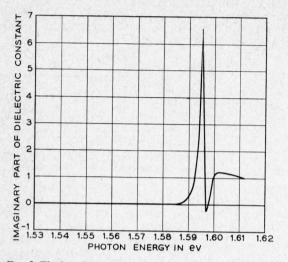

FIG. 3. The imaginary part of the dielectric constant of crystal LL1 at 1.6°K as derived from a Kramers-Kronig analysis of the data of Fig. 2. The absorption coefficient has a peak value of about 1.8×10^5 cm^{-1}.

absolute reflectivities were estimated by extrapolating the value of the infrared refractive index (2.67) as given by Fisher and Fan[6] to 2.74 at 1.41 ev using an approximate formula. From this point, the refractive index could be determined from an analysis of interference fringes seen in reflection from a thin flake cleaved from a larger crystal. The absolute reflectivity could thus be obtained in a region near the exciton peak where it was known that the absorption coefficient was high enough to prevent light reflected from the back surface making a contribution, and yet not high enough to influence the reflectivity by means of the extinction coefficient becoming comparable to unity. A Kramers-Kronig analysis[7] of the data ignoring other structure in the reflec-

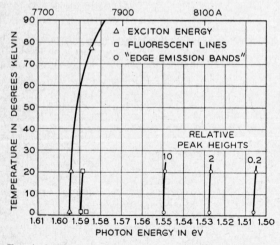

FIG. 4. A summary of the excitation energies and fluorescent peaks as a function of temperature for crystal LL1. The full width at half-height of the "edge emission bands" is about 6.5×10^{-3} ev.

[6] P. Fisher and H. Y. Fan, Bull. Am. Phys. Soc. 4, 409 (1959).
[7] D. G. Thomas and J. J. Hopfield, Phys. Rev. 116, 573 (1959).

tion spectrum[6,8] yielded the imaginary part of the dielectric constant ϵ_i shown in Fig. 3, and an oscillator strength for the transition of 1.3×10^{-3}. No further structure could be seen in the reflectivity curves which could be interpreted as corresponding to excited exciton states.

Rather similar reflection curves were obtained at 20°K, but at 77°K there was considerable thermal broadening although the transitions were still recognizable. The estimated resonant energies of the exciton transitions are plotted as a function of temperature in Fig. 4. The exciton energy of 1.5945 ev at 20°K is in good agreement with the value of 1.594 ± 0.001 ev already reported by Halsted et al.[9] who used the method of ellipsometry to derive the optical constants.

The increase in ϵ_1 seen in Fig. 3 a few millielectron volts above the exciton peak is probably to be ascribed to exciton formation with the emission of phonons, because it commences at a lower energy than the expected energy of the excited exciton states.

Fluorescence

The fluorescence shown by crystal LL1 at 1.6°K is also summarized in Fig. 4. Three emission peaks were seen between 1.55 and 1.50 ev with separations of 0.022 ev, which value is in good agreement with the longitudinal optical phonon energy of 0.0221 ev as calculated from the data of Fisher and Fan.[6] The full widths at half-height of the peaks are about 0.0065 ev and do not vary much in going from 1.6° to 20°K for the crystals studied; the positions of the peaks shifted with temperature in the same way that the exciton peaks shifted. This emission is closely analogous to the so-called green "edge emission" of CdS. For one crystal the relative intensities of the peaks are shown in Fig. 4. This edge emission has been investigated in greater detail by Halsted et al.[9] These workers found two series of peaks separated by about 0.008 ev, the lowest energy series becoming more intense at low temperatures, there being again a close correspondence with CdS. It is likely that the series reported here is the high-energy series, the other lines remaining weak for some unknown reason.

Sharp emission lines have been detected at photon energies slightly less than the exciton energies at 1.6°K; these probably correspond to "impurity" or "bound" exciton states associated with some crystal imperfection, again in analogy with CdS and other materials.

II. Stressed Crystals

Under compressive uniaxial stress in the (110) plane the single exciton reflection peak splits into two peaks. Figure 5 shows microphotometer traces of these peaks as the stress is increased; the resonant energies have

[8] S. Yamada, J. Phys. Soc. Japan 15, 1940 (1960).
[9] R. E. Halsted, D. T. F. Marple, M. Lorenz, and B. Segall, Bull. Am. Phys. Soc. 6, 148 (1961).

TABLE I. Splittings and shifts of center of gravities of exciton lines in CdTe in millielectron volts at a stress of 1000 kg/cm².

Stress direction Crystal	$\theta=0$ [100] Splitting	Shift	Crystal[a]	$\theta=54°42'$ [111] Splitting	Shift	Crystal[a]	$\theta=90°$ [110] Splitting	Shift	Crystal[a]	$\theta=70°$ Splitting	Shift
2761F'A	12.6	2.0	2761F'C	15.7	3.8	32161CA*	14.9	3.9	32161AB*	16.5	3.8
2761F'B	13.15	1.7	2761HA	11.4	3.1	32161CB*	8.65	2.0	32161AD*	14.6	3.8
123060B	13.0	1.9	31061A	13.6	3.9	32161CB*	15.5	4.2			
32161GC	16.9	2.8	31061CB	14.7	3.3	32161CD*	19.7	4.3			
			32161DA	14.4	3.6	32161EA	16.9	3.9			
			32161DB*	14.2	3.9						
			2761HA	11.9	2.8						
Average	13.9	2.1		13.7	3.5		15.1	3.7		15.5	3.8

[a] Crystals marked with an asterisk contain small twinned regions.

been taken to lie on the rapidly falling part of the reflectivity curves near the reflection maximum. The use of an analyzer showed that the lowest energy peak A was active in light polarized with its electric vector perpendicular and parallel to the direction of stress z, whereas peak B was active only for light with $E\perp z$. For all stress directions used in this work no trace of peak B could be detected for $E\|z$, indicating that it was probably at least 10 times stronger for $E\perp z$ than for $E\|z$. Figure 5 shows that peak A for $E\|z$ is the strongest transition; peak B for $E\perp z$ is somewhat weaker and peak A for $E\perp z$ is weakest of all.

Figure 6 shows the variation of the resonant energies of the various peaks as a function of the applied stress in a particular run. It is to be noticed that within experimental error the points fall on straight lines which extrapolate to the single peak at zero stress. Several runs were made in a number of crystal directions, principally the [111], [100], and [110] directions. There was considerable scatter in the results, which are presented in Table I, in which the energy difference between the peaks, and the shift of the center of gravity of the peaks is shown at a stress of 1000 kg/cm² determined from plots similar to Fig. 6. It is probable that the scatter in these results arises from uneven stress distribution in the crystals. The photographic plates

often showed that the splitting varied somewhat from one end of the crystal to the other, but this variation could not account for the largest variations of Table I. It is likely that some crystals were bowed, producing uneven strain from one side of the crystal to the other; as the reflection peaks are controlled by a thin layer of crystal at the surface, it is clear that reflection peaks are particularly sensitive to this source of trouble. The situation might be improved by changing the aspect ratio of the crystal and by having closer tolerances between the components inside the crystal holder, shown in Fig. 1, and the holder itself. The bowing would be eliminated if the crystals were in tension rather than compression.

As indicated in Table I, some of the crystals contained small regions of a twin. It is not considered likely that this leads to serious difficulties since, as shown in the experimental section, the twinning planes are normal to the crystal surface. No systematic effects produced on the photographs could be traced to the small twinned regions, and this is not surprising, as Table I shows that the effects produced by a fixed stress are very similar for all directions of stress.

On some stressed crystals, weak additional structure

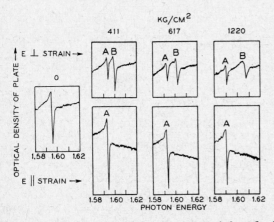

FIG. 5. Some typical microphotometer traces of the reflection spectra of unstressed and stressed CdTe crystals at 20°K. The background spectra changes somewhat as the polarization is changed. The stress direction is $\theta=70°$.

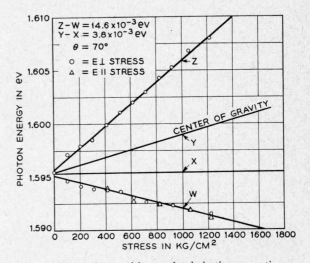

FIG. 6. The exciton positions and polarization properties as a function of applied stress. The splitting and shift of the center of gravity at 1000 kg/cm² are as indicated.

TABLE II. Expressions for the total splittings Δ of a $P\frac{3}{2}$ multiplet valence band produced by a uniaxial stress T applied in the (110) plane in the direction indicated. C_{11}, C_{12}, and C_{44} are the usual cubic elastic stiffness constants. Du and Du' are the deformation potentials defined by Kleiner and Roth. If compressive stress along the [100] or [111] directions moves the $M_J = \pm\frac{3}{2}$ band "up" (i.e., towards the conduction band), and the $M_J = \pm\frac{1}{2}$ band "down," then Du and Du' are negative.

Angle between stress direction and 100 axis	Valence band splitting
0° [100]	$\Delta = -\dfrac{4}{3} Du \dfrac{T}{C_{11}-C_{12}}$
54° 42′ [111]	$\Delta = -\dfrac{4}{3} Du' \dfrac{T}{2C_{44}}$
90° [111]	$\Delta = -\dfrac{4}{3}\left\{ \dfrac{Du^2}{4(C_{11}-C_{12})^2} + \dfrac{3}{16}\dfrac{Du'^2}{C_{44}^2} \right\}^{\frac{1}{2}} T$
θ	$\Delta = -\dfrac{4}{3}\left\{ \dfrac{Du^2(\cos^2\theta - \frac{1}{2}\sin^2\theta)^2}{(C_{11}-C_{12})^2} + \dfrac{3}{16}\dfrac{Du'^2 \sin^2\theta(1+3\cos^2\theta)}{C_{44}^2} \right\}^{\frac{1}{2}} T$

has been seen near the reflection peaks. This has not been investigated in detail.

Using the Hamiltonian of Kleiner and Roth,[1] J. C. Hensel has derived the expressions given in Table II for the total splitting, Δ, of a $P_{\frac{3}{2}}$ multiplet valence band produced by a uniaxial stress T in the (110) plane.[10] These relations are only true when the $J=\frac{1}{2}$ doublet is far removed by spin orbit interaction from any of the $J=\frac{3}{2}$ multiplets. This condition is certainly satisfied for all stresses attained in CdTe. For any direction of uniaxial stress, the quantity $\bar{E} - E_0$, where $\bar{E}$ is the mean gap between the valence and conduction bands and E_0 is the gap at zero stress, is given by

$$\bar{E} - E_0 = \frac{T}{c_{11} + 2c_{12}}(D_d{}^c - D_d{}^v), \qquad (1)$$

where $D_d{}^c$ and $D_d{}^v$ are the shifts per unit dilatation of the conduction and valence bands, respectively. If the band gap increases under compression $(D_d{}^c - D_d{}^v)$ is negative. If T is replaced by a hydrostatic pressure P, the expression becomes

$$\bar{E} - E_0 = \frac{3P}{c_{11} + 2c_{12}}(D_d{}^c - D_d{}^v). \qquad (2)$$

The zero field elastic stiffness constants for CdTe at room temperature have been determined by McSkimin and Thomas[11] and are

$$c_{11} = 5.351 \times 10^{11} \text{ d/cm}^2$$
$$c_{12} = 3.681 \times 10^{11} \text{ d/cm}^2$$
$$c_{44} = 1.994 \times 10^{11} \text{ d/cm}^2.$$

These constants together with the splittings in the [100] and [111] directions enable Du and Du' to be deter-

mined. The splittings in the other directions of stress may then be calculated. The results are shown in Table III(A). Within the probable experimental uncertainty there is agreement between the predicted and observed splittings for $\theta = 90°$ and $\theta = 70°$, where θ is the angle between the stress direction in the (110) plane and the [001] direction in the same plane.

If the condition

$$Du/(c_{11} - c_{12}) = Du'/2c_{44} \qquad (3)$$

is fulfilled, then the splittings are identical for a fixed stress for any value of θ. Values of Du and Du' have been chosen which obey Eq. (3) and these values and the corresponding splitting for 1000 kg/cm² are shown in Table III(B). These values probably represent as good a fit to the data as those in Table III(A). It may be noted that CdTe is not isotropic; the anisotropy factor is

$$2c_{44}/(c_{11} - c_{12}) = 2 \cdot 39. \qquad (4)$$

Table III(C) presents the values of $(D_d{}^c - D_d{}^v)$ as calculated from the shifts of the bands for the different directions of applied stress. These values are expected to be the same. The fact that the value for [100] stress appears to be less than the values for the other stress directions may arise from experimental error.

It was observed that reflection peak B became more diffuse as it moved further away from peak A (see Fig. 5), indicating that its half-width was increasing. Peak A was not similarly affected. No clear threshold for the onset of the broadening was convincingly demonstrated. Since the position of peak B is more sensitive to pressure than that of peak A, it is expected that uneven stress distribution would affect the appearance of B more than A. However, it did not seem likely that this factor alone could have accounted for the broadening of B.

DISCUSSION OF RESULTS

I. Band Structure

The reflection peaks seen in CdTe are very similar to those seen in CdS and ZnO and other materials which have been identified as transitions to $n=1$ exciton states,[12] and there can be no doubt that this is also the case for CdTe. We may assume that at $k=0$ the valence band is p-like and the conduction band s-like. In unstrained CdTe the $J=\frac{3}{2}$ branch of the valence band will be completely degenerate at $k=0$, and a hole from this band gives rise in combination with an electron from the conduction band to a single exciton reflection peak. The $J=\frac{1}{2}$ branch of the valence band is split off by the spin-orbit perturbation; a crude extrapolation of this splitting in ZnO, CdS, and CdSe to CdTe yields a splitting of the order of 1 ev for CdTe. An exciton arising from this band would be broadened by autoionization and

[10] J. C. Hensel and G. Feher, Phys. Rev. Letters 5, 307 (1960); J. C. Hensel (private communication).
[11] H. J. McSkimin and D. G. Thomas (to be published).

[12] D. G. Thomas and J. J. Hopfield, Phys. Rev. 116, 573 (1959); D. G. Thomas, J. Phys. Chem. Solids 15, 86 (1960).

TABLE III. Deformation potentials and band splitting in CdTe.

	Deformation potentials ev		Splittings in millielectron volts at a stress of 1000 kg/cm² Stress direction			
	Du	Du'	[100]	[111]	[110]	$\theta = 70°$
Average observed splittings			13.9	13.7	15.1	15.5
A Calculated deformation potentials and predicted splittings	+1.77	+4.18	...	...	13.8	13.8
B Deformation potentials chosen to satisfy Eq. (3)	+1.86	+4.45	14.6	14.6	14.6	14.6
C ($Dd^c - Dd^v$) values obtained from various stress directions	[100] −2.7	[111] $(Dd^c - Dd^v)$ −4.5	[110] −4.8	$\theta = 70°$ −4.9		

largely obscured by the background absorption, and it was not sought. By analogy with other 2–6 semiconductors, the $k=0$ conduction band minimum is probably an absolute minimum. No detailed optical absorption work near the exciton is reported here; however, Davis and Shilliday[13] have measured absorption coefficients between about 10 and 200 cm^{-1} very close to the exciton at 78°K. These authors interpreted their results in terms of direct and indirect band-to-band transitions and did not consider the formation of direct excitons with phonon cooperation, which must occur.[14] It would be of interest to reconsider their data in the light of the present knowledge of excitons in this material.

The band gap at $k=0$ is the exciton energy plus the exciton ionization energy. In the crystals examined so far, no anomaly in the reflection spectrum which could be ascribed to the formation of an $n=2$ state has been seen. Presumably, this transition is weaker than that of the $n=1$ state and is probably more susceptible to broadening because of the large size of the $n=2$ state. Hence, no direct estimate of the exciton binding energy can be made. An indirect estimate can be made by plotting the reduced exciton mass of ZnO,[15] CdS,[16] and CdSe[17] against the respective band gaps, extrapolating to the band gap of CdTe and so obtaining the value of 0.011 ev for the exciton binding energy in CdTe. The direct band gap is thus estimated to be 1.606 ev at 1.6°K.

II. Strain Effects

From the shifts and splittings of the exciton peaks as a function of applied stress, and from the elastic stiffness constants, three deformation potentials have been determined for the CdTe band structure. The uniaxial stress splits the $J = \frac{3}{2}$ multiplet valence band into two

bands, one with $M_J = \pm \frac{3}{2}$, the other with $M_J = \pm \frac{1}{2}$. The $J = \frac{1}{2}$ doublet valence band is far removed by spin orbit coupling. In the analysis, it has been assumed that the exciton binding energies are independent of stress and the same for the excitons arising from each valence band. In particular, complications might be expected at low stresses where the valence band splitting is comparable with or less than the kinetic energy of the hole in the exciton; since the hole is probably heavier than the electron, this energy will be some small fraction of the binding energy which has been estimated to be 0.011 ev. Figure 6 shows that the exciton energies vary linearly with stress at separations well above the expected hole kinetic energy and extrapolate to a common point, the observed exciton energy, at zero stress. From this we may conclude that the exciton binding energies are approximately the same for both bands, as follows logically from the assumption of a light electron. In fact, it is expected from $k.P$ perturbation theory that the masses of the holes derived from both valence bands should obey the relation

$$\frac{2}{(m_{h\perp})_A} + \frac{1}{(m_{h\parallel})_A} = \frac{2}{(m_{h\perp})_B} + \frac{1}{(m_{h\parallel})_B},$$

where A and B are the two valence bands. As a consequence of this relation, the differences in the reduced exciton masses[16] would not be expected to be large even if the electron were heavy. Thus, under any circumstances the exciton binding energies will be approximately the same for both bands. In addition, no departures from a linear splitting at low stresses could be detected in the experiments described here.

The signs of the deformation potentials, Du and Du', are determined from the polarization properties of the excitons. For stress along the [100] or [111] directions, the band Hamiltonians are diagonal, that is, the elements of the strain Hamiltonian matrix which connect different bands vanish in the angular momentum representation quantized along the stress direction.[10]

[13] P. W. Davis and T. S. Shilliday, Phys. Rev. 118, 1020 (1960).
[14] D. G. Thomas, J. J. Hopfield, and M. Power, Phys. Rev. 119, 570 (1960).
[15] D. G. Thomas, J. Phys. Chem. Solids 15, 86 (1960).
[16] J. J. Hopfield and D. G. Thomas, Phys. Rev. 122, 35 (1961).
[17] J. O. Dimmock and R. G. Wheeler, Bull. Am. Phys. Soc. 6, 148 (1961).

Thus we may speak of $M_J = \pm\frac{1}{2}$ and $M_J = \pm\frac{3}{2}$ bands. Optical transitions are thus made from an $M_J = \pm\frac{1}{2}$ or an $M_J = \pm\frac{3}{2}$ valence band to an $M_J = \pm\frac{1}{2}$ state of the $J = \frac{1}{2}$ conduction band. Transitions from the $M_J = \pm\frac{1}{2}$ valence band to the conduction band can satisfy the selection rules $\Delta M = 0$ and $\Delta M = \pm 1$, and so can occur for light polarized both parallel and perpendicular to the stress direction. Transitions from the $M_J = \pm\frac{3}{2}$ valence band can only satisfy the selection rule $\Delta M = \pm 1$ and so can only occur for light polarized perpendicular to the stress. Since it is observed that for compression along the [100] and [111] axes the exciton which moves to lower energies is active in both modes of polarization, while that which moves to higher energies is active only for perpendicular light, it is concluded that Du and Du' are both positive. Thus for compression the $M_J = \pm\frac{1}{2}$ band moves "up" (towards the conduction band) and the $M_J = \pm\frac{3}{2}$ band moves "down." Such simple polarization properties are not expected for stress in an arbitrary direction, for which band mixing occurs. However, Hensel[10] has shown that if a given stress along the [100] or [111] directions produces splittings of the same magnitude and sign [that is, if Eq. (3) is obeyed], then the strain Hamiltonian becomes diagonal for stress in the [110] direction so that the simple polarization properties are preserved. It was found that for the stress in the [110] direction (and also for $\theta = 70°$), the polarization properties of the excitons are identical to those for stresses in the two principal directions. Thus, the polarization effects indicate that Eq. (3) is obeyed, and this is also the conclusion reached from the magnitudes of the observed splittings. Within the experimental accuracy there is, therefore, internal consistency in the results.

The conclusion is therefore reached that despite the fact that CdTe is elastically anisotropic, the splitting of the valence band can be at least approximately described by one, not two, deformation potentials. It is also of interest to notice that the deformation potentials and elastic constants are such that in CdTe the splitting of the valence band produced by a uniaxial stress is greater than the shift of the center of gravity of the bands. This is also true for Cu_2O[2] but is apparently not the case for germanium.[1]

Hopfield's quasi-cubic model[18] for the band structure of crystals which are nearly but not quite cubic, may be used to estimate the ratio of the strengths of the various exciton transitions. For stress directions which produce no band mixing (i.e., for all directions used in the case of CdTe) the relative strengths expected for the transition are as follows:

Exciton	A	B	C
$E \perp$ Strain	$\frac{1}{6}$	$\frac{1}{2}$	$\frac{1}{3}$
$E \parallel$ Strain	$\frac{2}{3}$	...	$\frac{1}{3}$

[18] J. J. Hopfield, J. Phys. Chem. Solids **15**, 97 (1960).

Exciton C refers to the exciton derived from the $J = \frac{1}{2}$ band which is not observed in the present work. The strength of the single exciton peak in the unstrained cubic crystal is expected to be $\frac{2}{3}$ in the same units. The results of Fig. 5 show that these relations are at least qualitatively obeyed. A quantitative comparison could be made by performing a Kramers-Kronig analysis of the reflectivity data of Fig. 5.

It was mentioned that, as the excitons were split further apart by stress, the high-energy exciton transition became broadened. A broadening of this state is to be expected as it crosses into the continuum of the low-energy exciton since autoionization can then occur. A threshold of the broadening is therefore expected which should give information concerning the exciton binding energy. Although splittings were produced which exceeded the estimated exciton binding energy, and broadening certainly did occur, no clear-cut threshold for the broadening could be detected in the reflection spectrum. Uneven stress distribution could partially obscure such a threshold, and so the threshold might be more easily detected for crystals in tension rather than compression.

SUMMARY

The major part of this work has consisted in the determination of certain deformation potentials of the band structure of CdTe. This was done by studying the shifts and splittings of exciton lines seen in reflection produced by the application of uniaxial compressive stress. Despite some scatter in the experimental results, probably arising from bowing of the crystals, it was apparent that a given stress in any direction in the [110] plane at least, produced approximately identical splittings of the valence band. This conclusion is consistent with the polarization properties of the excitons in the stressed crystals, and these properties also indicate that the $M_J = \pm\frac{1}{2}$ valence band moves "up" while the $M_J = \pm\frac{3}{2}$ band moves "down". The fact that the valence band splitting can in effect be described by one rather than two deformation potentials may be of interest in connection with the microscopic theory of elasticity. One might expect that other materials with similar valence band structures, such as silicon and germanium, would show the same relation between the deformation potentials as does CdTe.

ACKNOWLEDGMENTS

The author wishes to thank J. C. Hensel and J. J. Hopfield for many helpful discussions. Thanks are also due to E. A. Sadowski for much technical assistance, to Mrs. M. H. Read for providing the x-ray results, and to R. T. Lynch who grew the crystals.